Radiochemical Methods
in Analysis

Radiochemical Methods in Analysis

Edited by

D. I. COOMBER

Department of Industry
Laboratory of the Government Chemist
London, England

PLENUM PRESS • NEW YORK AND LONDON

Library of Congress Cataloging in Publication Data

Main entry under title:

Radiochemical methods in analysis.

Includes bibliographical references and index.
1. Radiochemical analysis. I. Coomber, D. I.
QD605.R3   545'.82   72-95069

ISBN-13: 978-1-4613-4403-2   e-ISBN-13: 978-1-4613-4401-8
DOI: 10.1007/978-1-4613-4401-8

©1975 Plenum Press, New York
Softcover reprint of the hardcover 1st edition 1975
A Division of Plenum Publishing Corporation
227 West 17th Street, New York, N.Y. 10011

United Kingdom edition published by Plenum Press, London
A Division of Plenum Publishing Company, Ltd.
Davis House (4th Floor), 8 Scrubs Lane, Harlesden, London, NW10 6SE, England

Contributors

M. S. Baxter · Department of Chemistry, The University, Glasgow G12 8QQ, Scotland

H. J. M. Bowen · Department of Chemistry, The University, Whiteknights Park, Reading, England

D. I. Coomber · Department of Industry, Laboratory of the Government Chemist, Cornwall House, Stamford Street, London S.E.1, England

D. F. Covell · U.S. Naval Ordnance Laboratory, White Oak, Silver Spring, Maryland 20910, U.S.A.

J. G. Cuninghame · Chemistry Division, Atomic Energy Research Establishment, Harwell, Didcot, Oxfordshire, England

W. G. Duncombe · The Wellcome Research Laboratories, Langley Court, Beckenham, Kent, England

R. P. Ekins · Institute of Nuclear Medicine, The Middlesex Hospital Medical School, Nassau Street, London W1N 7RL, England

P. Johnson · The Wellcome Research Laboratories, Langley Court, Beckenham, Kent, England

P. Martinelli · Commissariat à l'Energie Atomique, C.E.N. Saclay, Boite Postale No. 2 Gif-Sur-Yvette, France

J. W. McMillan · Applied Chemistry Division, A.E.R.E., Harwell, Didcot, Oxfordshire, England

C. T. Peng · Department of Pharmaceutical Chemistry, School of Pharmacy, University of California, San Francisco, California 94122, U.S.A.

R. K. Webster · Applied Chemistry Division, A.E.R.E., Harwell, Didcot, Oxfordshire, England

Preface

The aim of this book is to give an account of the principal radiochemical methods used in chemical analysis. It is assumed that the reader already has some background knowledge of radioactivity, available from several general textbooks. For this reason some subjects, e.g. the fundamentals of radioactivity, the properties of radiation, statistics of counting procedures, the precautions needed in working with radioactive materials, which could have occupied half the text, are not considered in detail. The different aspects of radiochemical analysis have been covered by specialized books and reviews, e.g. on activation analysis, gamma spectrometry, radiometric titrations. A good deal of information is in the form of reports of meetings and symposia and liquid scintillation counting, for instance, has been mainly covered in this way. There are also a large number of journals. It is therefore hoped that this book will help fill the gap between the introductory texts and the specialized sources, many of which are referred to in the chapter references.

The first three chapters in the present volume deal with the methods of measurement of radioactive nuclides. Chapter 1 gives a general account of detection and measurement techniques. The next two chapters are devoted to two specialized techniques: gamma-ray spectrometry and liquid scintillation counting. Chapter 4 describes the use of computers which are needed to make full use of the developments in these techniques – the complexity of the data in the former and the amount of it in the latter. Chapter 5 outlines separation methods for inorganic species while Chapter 6 gives a full account of radiochromatography and radioelectrophoresis, considered mainly from the point of view of their use in organic and biochemical analysis.

At one time radiochemical analysis and activation analysis were almost synonymous. After losing ground to rival techniques it has become more important again due to the increasing interest in determining trace elements in the environment. Chapter 7 is on activation analysis. Chapter 8 is on radiotracers in inorganic analysis, considered both from the point of view of their use in checking analytical procedures, even though the ultimate method of analysis may not be a radiochemical one, and for their use in analysis *per se*. Chapter 9, on the use of tracers in organic and biochemical analysis, includes a detailed description of radioimmunoassay, immunoradiometric methods and related techniques. Radioimmunoassay is becoming of increasing importance as it becomes possible to apply it to smaller molecules not initially immunogenic. The penultimate chapter is on the determination of radioactivity in the environment and the applications of these determinations in dating, as tracers in the earth sciences and from the point of view of environmental health. Finally there is a chapter on the use of sealed sources in those analytical methods which depend on the interaction of radiation with matter; absorption and backscattering methods, X-ray fluorescence analysis, Mössbauer spectroscopy and ionization methods.

The Editor would like to acknowledge the advice and encouragement given him by Dr. J. G. Cuninghame and to thank Dr. J. E. Johnstone of A.E.R.E., Harwell, and Dr. S. J. Lyle of the University of Kent at Canterbury, for advice given at the planning stage of the book. Finally he would like to thank the publishers for their help with the many problems which arise in dealing with a multi-author book.

D. I. Coomber

Contents

Contents

Chapter 1

Methods of Detection and Measurement of Radioactive Nuclides

J. G. Cuninghame

Chemistry Division, Atomic Energy Research Establishment
Harwell, Didcot, Oxfordshire

1.1 INTRODUCTION

1.1.1 The Different Kinds of Radiation

If an atom can achieve a lower energy level by re-arrangement or alteration of its nucleus, it may well do so, and it is then said to be 'radioactive'. The change in the nucleus may take the form of emitting either a particle or electromagnetic radiation, and by making suitable measurements of this radioactivity the analyst may be able to identify the radioactive isotope qualitatively or determine it quantitatively.

If we consider a nuclear reaction, for example where an isotope is bombarded by neutrons, emission of particles or rays from the radioactive nuclei produced may occur over a time scale ranging from 10^{-18} sec up to many years. When the time scale is less than about 0.1 sec we usually talk about 'prompt' emission, while for times greater than 0.1 sec we use the term 'delayed'.

Until fairly recently, analysts were concerned almost exclusively with delayed emissions, and these are still by far the most important to them. The majority of this chapter therefore, will deal with the measurement of this type of radioactivity, i.e. with α- and β-particles and with γ- and X-rays.

With the increasing use of particle accelerators and neutron generators however, it is becoming more necessary to be able to measure prompt particles and rays, both in actual prompt analytical methods and for monitoring beams used in the bombardment of targets to be analysed. This may involve the analyst in the measurement of neutrons, protons, deuterons and heavier ions in addition to those mentioned above, and these measurements are dealt with also, but only briefly.

1.1.2   The Layout of This Chapter

The chapter discusses in detail the detectors themselves and, less fully, the associated electronics, mounting of sources for counting, low level counting, and calculations. The discussions of the detectors deal first with the various types in a general way, the emphasis being on how they operate. After this there is a section in which the detectors are grouped according to the kind of radiation to be detected, the object being to enable the analyst to assess readily the most convenient means of detecting the radiation in which he is interested. The methods of γ-counting are only outlined since these are treated in Chapter 2, and similarly with liquid scintillators and Čerenkov counting which will be found in Chapter 3 and auto-radiography which is in Chapter 6. Brief mention is also made of multi-channel pulse

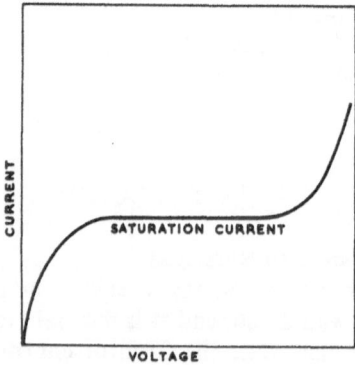

Fig. 1.1.   Ionisation chamber current as a function of applied voltage.

height analysers used in spectroscopy of radioactive sources, and of the current trend to replace them by small real-time computers. However, a detailed account of computer use will be found in Chapter 4.

The Appendix consists of a summary of the majority of detector types in current use, grouped according to the radiation to be detected. A selection of references is given which is not intended to be exhaustive; as far as possible, books and review articles have been quoted. Only a small proportion of the detectors mentioned in the tables are described in detail in the ensuing pages. References given in the table have not been repeated in the text.

1.2   TYPES OF DETECTORS

1.2.1   General Principles of Detection

The general principle behind all detection methods is that the particle or ray to be detected gives up some or all of its energy to the medium of the detector, either by ionising it directly, or by causing it to emit a charged particle, which in its turn produces ionisation in the medium.

The primary ionisation is then detected by one of several methods. The charge may be collected directly by electrical means, as it is in gas ionisation counters and semiconductor counters; it may cause the emission of light quanta, as in scintillation counters: it may leave a trail of nuclei in the medium along which bubbles may condense or chemical change occur, as in cloud and bubble chambers and photographic emulsions; and finally it may cause permanent damage to the medium which can then be made visible, as in track detectors. The first two of these are by far the most important to analysts, and detectors based on them are the only ones extensively discussed in this book.

1.2.2   Gas Ionisation Counters

In a gas ionisation counter the counting gas fills a space containing two electrodes, between which there is an electric field. When radiation causes ionisation in the gas, the electrons produced are drawn rapidly to the anode while the heavy positive ions move more slowly to the cathode. The result is that current flows through the circuit and may be used as a measure of the amount of radiation which is causing the ionisation.

Ionisation Chambers

If the current flowing in a circuit containing an ionisation counter is plotted against the applied voltage, a curve similar to that shown in Fig. 1.1 will result.

The initial increase in current occurs because the field is at first too weak to prevent the ion pairs from recombining, but as the voltage increases a point is reached where there is virtually no recombination and all the charge is collected. This is the start of the plateau in Fig. 1.1: the plateau current is called the 'saturation current'. As the voltage is increased further, eventually the current will start to rise again due to the onset of secondary ionisation.

If a counter is operated in the plateau region it is called an 'ion-chamber' — it is, of course, simply a capacitor. Ion chambers are of two main kinds, d.c. and pulsed. The basic difference between them is that in a d.c. chamber the charge generated on the electrodes dies away slowly compared with the rate at which ionisation is occurring, and so a steady current proportional to the intensity of the source producing the ionisation is developed. The rate at which the charge is dissipated depends on the time constant of the circuit in which the ion-chamber is incorporated and this is a function of the capacity and resistance to earth of the circuit. A d.c. ion-chamber has a long time constant, while in a pulse ion-chamber it is short.

D.c. ion-chambers are used mainly for health physics purposes. The simplest possible type is the electroscope in which a gold covered quartz fibre or a piece of gold leaf is connected to one electrode of the ion-chamber, which is then charged up. Electrostatic force makes the fibre move against the restoring force of its own elasticity to a zero position on a scale. Radiation causes ionisation which gradually discharges the chamber and the fibre moves across the scale. Examples of other types of health physics d.c. ion-chambers are those used for monitoring γ and neutron levels. These are usually parallel plate chambers, filled with boron tri-fluoride for neutron counting or with argon for γ-counting. Fig. 1.2 is a diagrammatic view of a standard chamber used for neutron monitoring; ionisation in it is caused by the α-particles from the nuclear reaction $^{10}B(n, \alpha)^7Li$.

For analytical purposes the pulse ion-chamber has far more relevance than the d.c. chamber. Its time constant is much shorter and so, given the proper conditions, a pulse can be produced each time an ionising event occurs in the chamber. In a typical chamber the electrons from one ionising event will be collected in a time of the order of 10^{-6} sec., but the positive ions, being much heavier, will take about 10^{-3} sec. The result of this is that the time between pulses accepted, the 'dead-time', must be made rather long. In addition to this, the positive ions left behind after the collection of the electrons affect the size of the pulse produced by the latter and the magnitude of this effect depends on their distance from the anode and hence on the position of the ionisation track in the chamber, thus destroy-

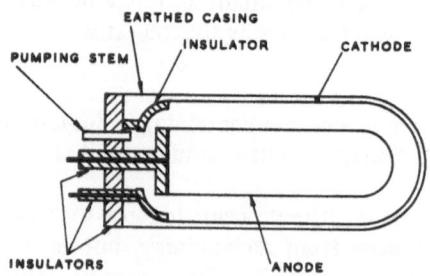

Fig. 1.2.   D.c. ion chamber used for neutron monitoring.

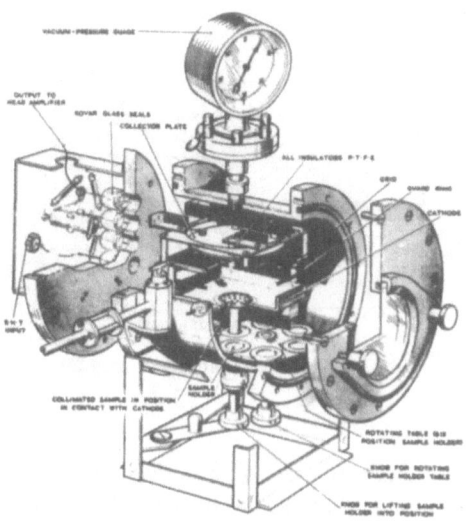

Fig. 1.3.   Gridded ion chamber used at A.E.R.E. Harwell, for α-pulse height analysis.

ing the proportionality of pulse height and energy. These difficulties are surmounted by screening the anode by means of a grid having a negative charge which allows the electrons to pass but repels the positive ions. Gridded ion chambers are used extensively in α-pulse height analysis. Fig. 1.3 shows a design which has been satisfactorily employed at A.E.R.E. Harwell for many years, both for absolute counting against a known standard and for α-spectroscopy (p. 14).

The counting gas must not give rise to negative ions and the electrons must have as high a drift velocity as possible so that the charge can be collected quickly. Argon containing 10% of either methane or carbon dioxide is commonly used.

Proportional Counters

If the pulse height developed by a gas ionisation counter is plotted against increasing applied voltage, it will be found to change as shown diagrammatically in Fig. 1.4.

Region A is the ion-chamber region discussed above In region B, called the proportional region, the pulse height at any particular applied voltage is proportional to the energy of the radiation, provided that the counter is large enough to contain the whole of the ionisation track from the event. What is happening is that after the ion-chamber region, in which only primary ionisation occurs, the acceleration given to the primary electrons by the field in the proportional region is sufficient to cause 'avalanches' of

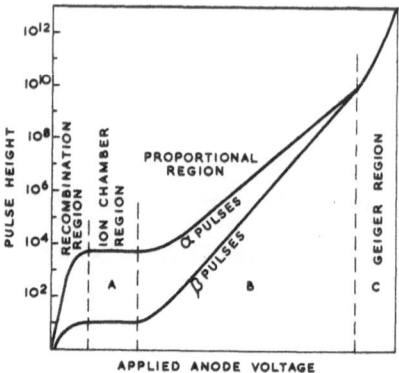

Fig. 1.4.   Variation of pulse height in gas ionisation counters as a function of applied high voltage.

secondary ionisation along the tracks of individual primary electrons due to collisions between them and molecules of the counting gas. The pulse height therefore increases because more charge is collected. This process is called 'gas multiplication'; the value of the gas multiplication factor attained depends on the dimensions of the counter, the nature and pressure of the gas, and on the voltage. Values of between 10 and 1000 are commonly employed.

The anode of a proportional counter is usually a loop or straight piece of thin (~0.025 mm diameter) wire, since the high field strengths required for proportional operation cannot conveniently be achieved with large plate electrodes. The greater the distance from the anode, the weaker is the field and so the avalanches tend to take place very close to the wire. So long as they are independent of each other, the proportionality of pulse height and energy of the particle is maintained, whatever the position of the original particle track in the counter.

The electrical pulse in a proportional counter rises to half height in about 10^{-6} sec but is very much slower in achieving its maximum because of the long time taken by positive ions to reach the cathode. High counting rates are therefore attained by using a time constant of about $1\mu s$ in the associated amplifier circuits (p. 23) thus giving a counter dead-time of the same order.

Proportional counters may be operated either with the counting gas (commonly argon or argon-methane, but other gases are used for particular purposes) sealed in at high or low pressures, or with the gas flowing slowly through at atmospheric pressure. It is important that impurities with which electrons can form negative ions are not present in the counting gas and hence electro-negative gases such as oxygen and the halogens must be removed by careful purification if they are present. It is also essential to exclude traces of corrosive impurities which may attack the anode, such as sulphur compounds sometimes present in commercial methane, and dust which may form spikes on the anode and cause an electrical discharge.

A simple method of attaining the required purity of the gas used in flow proportional counters is described below (p. 17).

While proportional counters are still used for spectroscopy, particularly in special applications such as the xenon filled proportional counters sometimes employed in X-ray spectroscopy, their importance for analysts lies mainly in the normal assay of radioactive sources. Counters used for this purpose are not necessarily of a size to contain the whole of the track of the primary particle since it is only necessary that the particle to be counted produces a pulse large enough to be separated from the background noise by means of a discriminator. Designs of three proportional counters used at A.E.R.E. Harwell, are shown in Figs. 1.5–1.7. All these counters use pure methane as counting gas, although they will also operate satisfactorily with a mixture of 90% argon + 10% methane.

Fig. 1.5 shows a low-geometry counter used for absolute counting of α-particles [59]. The counter must be evacuated after insertion of the source and then filled with the counting gas to a known pressure – see also Section 1.3.1 (p. 13).

Fig. 1.6 illustrates a flow proportional counter [55], designed for routine assay of both α- and β- or mixed αβ-sources. The sample is placed externally to the window and the counter is thus easy and convenient to use (p. 16). The counter has almost 2π geometry.

Fig. 1.7 shows a flow proportional 4π counter used in absolute counting [60]. The source is placed inside the counter, but since it operates at atmospheric pressure, counting can begin only two minutes after it has been closed up (p. 15).

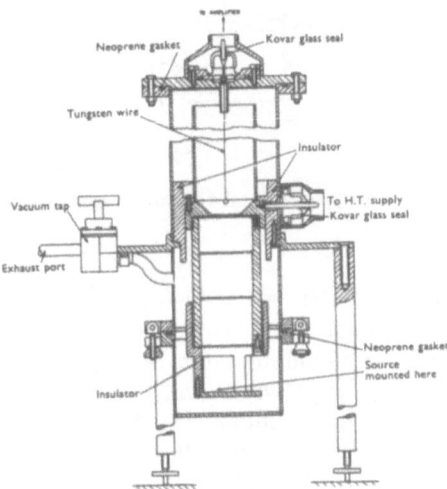

Fig. 1.5. Low geometry counter used at A.E.R.E. Harwell for absolute α-counting.

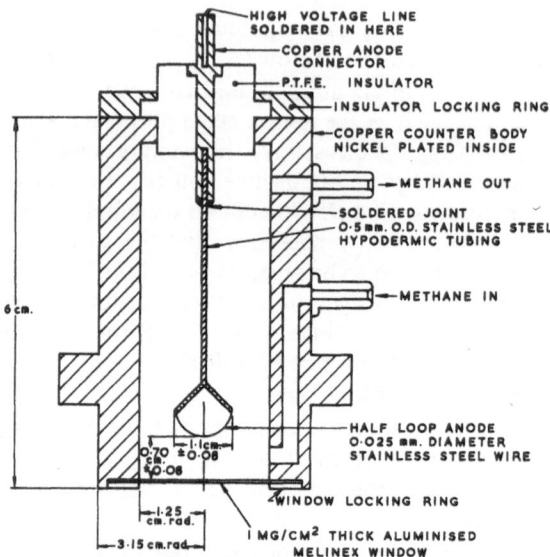

Fig. 1.6.   End window flow proportional counter designed for routine assay of both α and β particles.

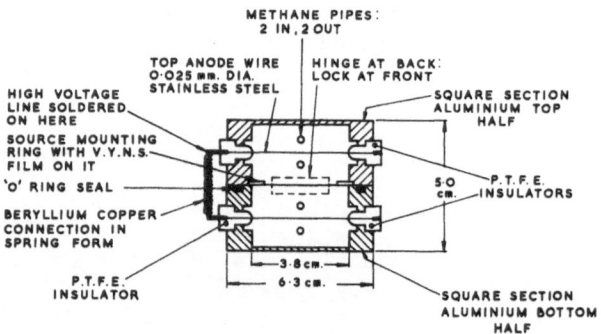

Fig. 1.7.   Flow proportional 4π counter used for β-assay.

When proportional counters are used for assay purposes it is usual to make a plot of count-rate for a particular source against applied high voltage. It will then be found that the count-rate will be substantially constant over a voltage range which will depend on the energy of the particles being counted, the counting gas, the design of the counter and other factors. It is customary to operate the counter at the voltage corresponding to the middle of this 'plateau'. A detailed account of the flow counter shown in Fig. 1.6 with particular reference to plateau measurements has been published [55].

Geiger-Müller Counters

Referring back to Fig. 1.4 (p. 6) if the voltage is increased to that of region C, the counter will become a Geiger-Müller counter. In the G.M. region the secondary avalanches expand along the anode under the influence of the emission of photoelectrons and a large pulse, independent of the energy of the original ionising particle is produced. When the positive ions reach the cathode they cause further ionisation, the electrons from which are accelerated towards the anode. This will result in the production of a second pulse and the process now starts all over again. In order to prevent the counter oscillating in this fashion, a 'quenching agent', usually alcohol or halogen, which absorbs the positive ions before they can reach the cathode, is added to the counting gas. Alternatively, the quenching can be performed externally by means of a circuit which, when the pulse occurs, reduces the counter voltage below that of the G.M. region for a time sufficiently long for the positive ions to have reached the cathode.

G.M. counters produce large pulses and so require little amplification. They are thus simple and cheap to use, but they have disadvantages which render them unsuitable for accurate work. They have rather short plateaux with slopes at least a factor of ten worse than a similar proportional counter; the simple permanently-filled types have rather thick windows; the necessity for quenching results in dead-times of the order of 200 μs to 1ms so that high intensity sources cannot be counted; and finally, their life is usually fairly short. Fig. 1.8 is a diagrammatic view of a typical G.M. counter.

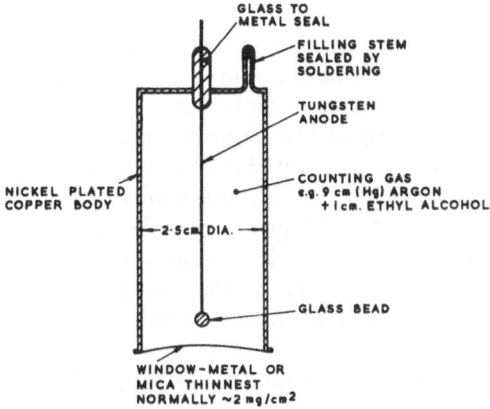

Fig. 1.8.   A typical G.M. counter.

1.2.3 Semiconductor Detectors

Semiconductor detectors are merely another form of ionisation chamber, but in recent years their convenience and the rapid improvement in their performance has had the result that they have virtually replaced the old types of detectors for some purposes and this trend is continuing.

In crystalline material the sharp energy levels of the component atoms have, as a result of their proximity and the consequence of the Pauli principle, become broad energy bands. Electrons fill up these bands from the lowest energy up to the top of the 'valence' band. The band above the valence band is the 'conduction' band and conduction can occur because of the mobility of any electrons which may have been excited into this band. The electron population of the conduction band depends on the temperature and on the difference in energy between the valence band and the conduction band: this energy difference is called the 'energy gap'.

Semiconductors used as detectors have a large energy gap; in the most frequently used materials, silicon and germanium, it is 1.1 eV and 0.66 eV respectively. Because of this there are normally only a few electrons in their conduction bands, the number being an equilibrium value depending on the temperature, but the much smaller gap for germanium means that detectors made of it have to be operated at low temperatures in order to keep this number down and so reduce 'noise'. In spite of this disadvantage, germanium is the preferred material for γ-ray detectors because its higher atomic number makes it much more sensitive than silicon and because for practical manufacturing reasons, much thicker detectors can be made, at least up to the present time.

When ionisation occurs in a semiconductor, electrons are excited into the conduction band and the charge may be collected at the electrodes by application of an electric field. If all the ionising events in a detector may be regarded as being independent of each other then resolution depends on the square root of the number of ion pairs, and so it will be about three times as good in a semi conductor as in a gas. However, Fano [63] showed that events are not really independent and introduced the 'Fano factor'. F, to correct for this effect. The true intrinsic resolution is thus $F\sqrt{n}$. F is about 0.1 for Ge and Si and somewhat better for gases.

A pure crystal of the type just described is called an 'intrinsic' semiconductor. In it there is an equal number of electrons (n) and of holes (p) in the conduction and valence bands respectively. If impurity atoms are present in the crystal lattice the picture is different because the impurity band level usually comes within the energy gap. Some impurities called n-type or donors, donate electrons to the conduction band, while others, p-type or acceptors, accept them from the valence band. In either case the crystal is no longer intrinsic since the number of holes and electrons in the valence and conduction bands is now not equal. A detector composed of either pure p-type or pure n-type material is called a 'homogeneous conduction counter', while one which has a combination of the two types is a 'junction counter'.

Junction counters are the most commonly used semiconductor counters for analytical purposes, the main types being the surface barrier, the lithium drifted and the diffused junction. The general principle behind all of them is that if n- and p-type materials are in contact and a bias voltage is applied so that it is positive to the n-type material, the free charge carriers (electrons for the n-type material, holes for the p-type) are drawn away from both sides of the junction, leaving a layer called the 'depletion layer' which is an excellent intrinsic semiconductor: Fig. 1.9. The thickness of the depletion layer is proportional to $\sqrt{V\rho}$ where V is the applied voltage and ρ the resistivity in ohm-cm.

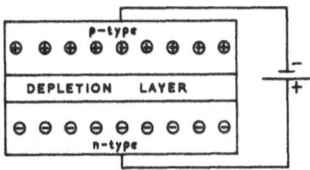

Fig. 1.9. Diagrammatic illustration of the operation of a junction counter.

When ionisation occurs in the depletion layer the resulting electrons and holes are rapidly swept out by the applied field and the charge collected.

Diffused junction detectors may be made by starting with either n- or p-type material. For example, p-type material may be prepared by diffusing boron into the silicon. The n-layer is then made by painting on phosphorus and diffusing it into the material at high temperature simultaneously with diffusion of aluminium from the other face to make an electrical contact. Surface barrier detectors are made from n-type material whose surface is allowed to oxidise, and on to which is evaporated a thin layer of gold which acts as a contact and influences the oxide film in such a way as to bring about the formation of a p-layer over a period of about a week. Depletion layer thicknesses are limited to about 2 mm for both these types of junction detector.

Lithium drifted detectors are prepared from p-type material into which lithium is diffused at high temperature with bias applied. Depletion depths of several centimetres can be achieved and hence lithium drifted germanium detectors are rapidly becoming the preferred means for conducting γ-ray spectroscopy. For charged particle spectroscopy, however, silicon detectors are normally used because great thicknesses are not usually needed and because of the convenience of being able to operate at room temperature. For such work surface barrier and diffused junction detectors are better because their resolution is higher, but lithium drifted silicon detectors are often used for spectroscopy of high energy charged particles.

1.2.4   Scintillation and Čerenkov Counters

Scintillation counters in one form or another have been used for the detection or spectroscopy of practically every form of radiation. They are extensively treated in Chapters 2 and 3 of this book and consequently only a very brief mention is given here. As in all other types of counters, the basic principle is that ionisation is caused in the medium of the counter by the slowing down of the particle, but whereas in gas and semiconductor detectors the ionisation is measured electrically, in a scintillation counter it first causes the emission of light which is then viewed optically, as in the original spinthariscope of Crookes, or is converted to an electrical pulse by means of a photo-multiplier as in modern practice.

In general, scintillation counters are cheap and efficient, simple to operate, and have such a tremendous variety of forms that it is always advisable to consider their use in any counting problem before finally deciding how to solve it.

1.2.5   Other Types of Counters

Cloud Chambers, Bubble Chambers and Spark Chambers
In these chambers the ionisation affects the medium of the counter in such a way that the track of the particle becomes visible and so may be photographed. In a cloud chamber the medium is a supersaturated gas and droplets of condensation form on the ions; a bubble chamber is filled with a superheated liquid and vapour bubbles form along the path of the particle; while in a spark chamber a series of parallel plates with a gas between them has a pulsed high voltage applied to alternate plates. A spark passes when ionisation is produced in the gas. All these devices are cumbersome and expensive and their use is mainly confined to high energy physics experiments.

Magnetic and Electrostatic Spectrometers
In one form or another these instruments permit the most accurate spectroscopy of radiations. They operate by magnetic or electrostatic analysis of the charged particles being examined or, in the case of γ-ray spectroscopy, of the electrons produced by interaction of the rays with a reflector. The bent crystal spectrometer is somewhat different, however, in that it makes use of the diffraction of γ-rays by a crystal. The energy resolution of the spectrometers is usually better than that of any other type of counter but they are very expensive and awkward to use and so are not in general employed for normal analytical purposes.

Photographic Emulsion
If radiation passes through a photographic emulsion the ionisation causes chemical changes in the material which can then be made visible by development, as in a normal photograph. Generally speaking, the examination of particle tracks in this way is mainly of interest to particle physicists, but integration of the effect and measurement of the blackening of emulsion is used in the film badges worn by those working with radioactivity, and the effect has also found application to some methods of neutron spectroscopy (p. 21).

Track Detectors
When an ionising particle passes through certain types of material it causes radiation damage which may then be made visible by chemical or physical processes. Many substances, including glass, mica, and polycarbonate film, can be used and it is often possible to select the material in such a way that

unwanted radiations leave no trace in it. For example, fission fragments produce tracks in polycarbonate film whereas lighter charged particles, neutrons, or γ-rays do not. After exposure, the film is etched by immersing it in caustic soda and the tracks of the fission fragments can then be seen and counted by using an optical microscope. The method is extremely simple but the counting of the tracks is tedious and time-consuming, although various mechanical aids to alleviate this do exist. Analysts may find track detectors useful in neutron monitoring by placing them close to a fissile target, or in examination of fission fragments in nuclear reactions.

1.3   METHODS USED IN DETECTION AND SPECTROSCOPY OF VARIOUS KINDS OF RADIATION

1.3.1   Alpha Counting
When an α-particle passes through a medium it produces dense ionisation along its path. The energy loss per collision is small and therefore there are a large number of collisions per track, with the result that α-particle tracks are straight and, for particles of the same energy, substantially equal in length. Their range is, however, quite small: it is, for a 5 MeV α-particle, for example, only 3.5 cm in air at 15° and 760 mm pressure, or 0.03 mm in silicon. From the point of view of counting α-particles this has the great practical advantage that gas counters used in spectroscopy need not be very large and that silicon semiconductor detectors can be used rather than germanium, but the disadvantage is that the sources used must be extremely thin to prevent serious deterioration of the resolution. The mounting of sources is described later in this chapter (p. 28).

Absolute α-Counting
The necessity for using thin sources makes 4π counting, used extensively in β-assay (p. 15), rather difficult and so it is common practice to arrive at the absolute emission rate of the source by using a low geometry counter. In such a device a carefully designed collimator reduces the angle of acceptance of the α-particles from the source to about 10^{-3} of 4π. Great care is taken over the design and manufacture and the geometry can be calculated to ~ 0.1%. The actual detection of the particle can be accomplished in many ways, the more usual being to use a scintillator or semiconductor detector, of area considerably greater than that of the collimation system, mounted in a proportional counter (Fig. 1.5, p. 7).

A low geometry counter is the most accurate instrument for absolute α-counting but it is fairly expensive to make and the low efficiency makes it unsuitable for sources having low activity or a short half-life. The other commonly used technique is to employ a counter with much higher geometry and to calibrate it by means of a source of known activity. Gridded ion chambers (p. 5) and semiconductor detectors (p. 9) are the most frequently used for this purpose. The reference source, if possible of the same nuclide as the unknown to be counted, is usually standardised by counting in a low geometry counter, but it may also be prepared by weighing a quantity of source material, dissolving, diluting to a suitable strength, and then mounting a weighed amount of the diluted solution. This procedure does not result in very thin sources however. The specific activity of the source material must be known, and this will have been measured, either by using a low geometry counter, or by calorimetry.

α-Spectroscopy

The first essential for good spectroscopy of any type of radiation is to have a thin source (p. 28). Having achieved this, the best instruments for routine α-spectroscopy are the gridded ion chamber using methane as counting gas, resolution $\sim 0.4\%$ (Fig. 1.3, p. 5) and the silicon surface barrier detector, resolution $\sim 0.2\%$ (p. 10). Magnetic spectrometers (p. 12) give somewhat better resolution, about 0.1%, but unless this slight improvement is essential, are not worth the extra time and expense involved in using them.

For spectroscopy of beams of α-particles (^{4}He ions) surface barrier detectors are the most convenient, the thickest available being usable up to about 50 MeV: above this it is necessary to use lithium drifted detectors.

Simple Counting of α-Emitters

If one only requires to know the count-rate of an α-source, as distinct from the absolute disintegration rate, there are many suitable types of counters, as listed in the Appendix. Gas flow proportional counters, running on argon/methane mixture, are commonly employed in analytical work; a design in use for many years at the A.E.R.E. Harwell [49] is shown in Fig. 1.10.

The counter requires the source to be inserted into it; an end window type, also used for β-counting, is shown in Fig. 1.6 (p. 8). Another convenient assay method is to employ a semiconductor detector, usually a surface barrier. Their only disadvantage compared with the flow counter is that they are somewhat more fragile and have a shorter life.

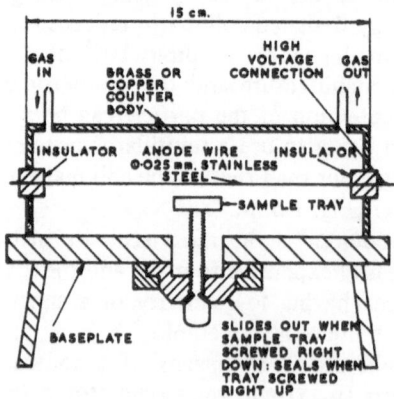

Fig. 1.10. A simple windowless flow proportional counter for routine α-assay.

Liquid scintillators can also be used and they have the advantage that very low specific activity sources can be counted provided they are soluble in the scintillating liquid. This method has been used in some very elegant experiments on the weakly α-emitting rare earth nuclides[50].

ZnS/Ag scintillators are commonly used in health physics α-monitoring instruments. The fact that ZnS emits a flash of light when struck by an α-particle was employed by Crookes and other early experimenters with radioactivity. In a modern probe the ZnS, with a little Ag activator, is spread on a thin perspex sheet mounted on a small photo-multiplier. Such an arrangement is cheap and simple and can easily be incorporated in a portable instrument.

1.3.2   Beta Counting

A β-particle has a very much higher velocity than an α-particle of the same energy because it is so much lighter. Consequently it spends less time in the vicinity of atoms of the counting medium and so it produces far fewer ion pairs per unit length of track. However, since the ionisation is produced by a collision between two particles of equal mass, the energy transfer in a single collision can be as much as half the original particle energy. The net result of these facts is that β-particles have a much longer range than α-particles, a 5 MeV β-particle has a range of about 10 mm in silicon for example, and their tracks are very erratic.

Absolute β-Counting

To measure the absolute disintegration rate of a β-source it is usual to employ a 4π counter. Many types, Geiger-Müller and proportional, gas flow and filled, have been used. The commonest is the gas flow proportional counter, an example of which is illustrated in Fig. 1.7 (p. 8). The source must be mounted on a very thin film so that particles emitted in any direction are counted. This is difficult and source mounting, described below (p. 28) is the most serious cause of error. A good 4π counter has a long (200–400 volts) plateau which is absolutely flat and has no background counts due to electronic noise whatsoever. With such a counter operated in the middle of the plateau and after subtraction of the true background, there is no way in which a count-rate higher than the disintegration rate can be obtained and so the effect of changes in source-mounting procedures can easily be checked — the method giving the highest count-rate being the best.

For β-particles having a maximum β-energy over about 1 MeV it is easy to get accurate disintegration rates, but below this energy the results become more and more erratic due to the difficulty of producing thin even sources, until for energies below 100 keV only rough answers — perhaps to an accuracy of about ± 10% — are possible. One way of overcoming these difficulties is to use the method of 4π βγ coincidence counting. The principle here is that if a source emits both β-particles and γ-rays in coincidence and both can be counted, the errors caused by source thickness are eliminated. It is necessary to know the decay scheme of the nuclide and this, together with the fact that many nuclides do not emit γ-rays, means that the method cannot always be used.

The calculation is as follows: Let R_β, R_γ and $R_{\beta\gamma}$ be the observed count rates for β-particles, γ-rays and βγ coincidences; let E_β and E_γ be the efficiencies of the β- and γ-counters respectively; let D be the required disintegration rate. Then

$$\begin{aligned} R_\beta &= DE_\beta \\ R_\gamma &= DE_\gamma \\ R_{\beta\gamma} &= DE_\beta E_\gamma \end{aligned} \qquad (1.1)$$

$$\therefore D = \frac{R_\beta R_\gamma}{R_{\beta\gamma}} \qquad (1.2)$$

This equation is correct only for the simple case in which the nuclide decays by emission of a β-particle to a single level of the daughter nuclide which then decays to the ground state by means of γ-ray in coincidence with the β-particle. Provided the efficiency E_β is near to 100%, i.e. it is a 4π counter, corrections need to be made only for chance coincidences, background, dead-time, and internal conversion. A full discussion of the procedure, the corrections, application to nuclides with complex decay schemes, and illustrations of counters will be found in refs. [51] and [52]. The counter shown as Fig. 1.7 (p. 8) can readily be adapted for 4π βγ-coincidence counting by standing it on top of a 7.5 cm × 7.5 cm sodium iodide crystal.

β-Spectroscopy

Unless one is prepared to go to the expense of a magnetic spectrometer, the only reasonably simple and accurate method of measuring the spectra of beta particles and conversion electrons with good resolution is to use a semiconductor detector. A 5 mm depletion depth lithium drifted silicon detector is thick enough to stop beta particles having energies up to about 2.5 MeV. All that is necessary is that the detector and thin source are mounted in a vacuum chamber with a suitable collimator in front of the detector to prevent the particles from entering it at the edges. Resolutions of 30 keV f.w.h.m. (full width half maximum) are easily attained and this can be improved by cooling the detector or, for lower energies, by using thinner detectors.

Simple β-Counting

The cheapest device for simple counting of beta emitters is the G.M. counter, but an end window flow proportional counter such as that shown in Fig. 1.6 (p. 8) is not very much more complicated and has many advantages, with the result that it is the most important instrument for this purpose today. The main advantage of the G.M. counter is that it requires the minimum of electronics and does not need a supply of counting gas. However, the proportional counter has a much longer, flatter β-plateau (~ 0.1% per 100 volts is normal, compared to 2.5% for a typical G.M.

counter) has a thinner window, and so is suitable for α-counting, is far more stable, and has a much longer life. It is advisable to purify the counting gas but this is easily done by passing it through copper turnings at about 550°C followed by silica gel and then magnesium perchlorate.

The usual procedure for assay with either of these two counters is to carry out any necessary purification of the activity after the addition of a known amount of about 10 mg of inactive carrier of the same element and then to precipitate and filter the carrier and activity on to a small filter paper through a filter stick which defines the shape of the precipitate exactly. The latter is then dried, usually at 110°C and weighed before counting. The chemical yield can thus be calculated and corrected for. The sample is then counted, if possible, at intervals over a period of about five half-lives, and the decay curve extrapolated back to some suitable zero time. Ref. [53] gives details of such procedure as well as for purification of many commonly used β-emitters.

If the absolute disintegration rate of the sample is required, it may be obtained from a previously prepared self-absorption and self-scattering curve with an absolute accuracy of about ± 5% for a β-emitter having a maximum energy greater than 0.5 MeV. This is prepared by first obtaining a quantity of the activity concerned in carrier-free form and standardising it as described above (p. 15). A number of portions containing a known amount of this activity with differing weights of inactive carrier are then mounted as described above. The overall efficiency of the counter, that is the ratio of the counting rate under the standard conditions to the true disintegration rate for each source weight, can then be calculated and plotted to give the self-absorption curve. It is also helpful to prepare a general efficiency curve for the particular counter by the method of Bayhurst and Prestwood [54] to use in cases where a self-absorption curve is not available. A large number of curves for the counter shown in Fig. 1.6 as well as a considerable amount of information about the counter will be found in ref. [55].

Scintillation counters of various types are also used for β-counting, but the bulk of the photo-multiplier makes them rather less convenient than the gas ionisation counters. However, for very low energy β-emitters such as ^{12}C, ^3H, etc. the only really satisfactory method is liquid scintillation counting, described in Chapter 3.

1.3.3 Gamma-Ray and X-Ray Counting
This subject is extensively treated in Chapter 2 and only a brief account is given here. Photons lose energy mainly by three processes, the photoelectric effect, the Compton effect, and pair production. In all three the bulk of the ionisation is secondary and is caused by the primary electrons, which are the direct result of the photon interactions.

For analytical purposes the lithium drifted germanium detector is by far the best spectroscopic tool. It is comparatively cheap and easy to use and gives a resolution of ~ 0.3%. If, however, a resolution of ~ 6% is acceptable, the sodium iodide scintillation crystal is a somewhat cheaper substitute. For the best resolution, the magnetic spectrometer, giving about 0.2%, or the bent crystal spectrometer, about 0.05%, may be used, but these are very expensive instruments and inconvenient to operate. If the photon energy is very low, such as it is in X-rays, the xenon filled proportional counter is also capable of good resolution.

Scintillation methods of one kind or another are useful for simple photon counting but true absolute counting is hard to achieve because it requires a knowledge of the decay scheme of the nuclide concerned. It is best to use a relative method by comparing the counting rates in the same full energy peak of the unknown and of a source of the same nuclide which has been standardised by absolute α- or β-counting. It is also possible to prepare an efficiency curve for the counter by counting a variety of standardised sources having different photopeak energies under identical conditions as the unknown.

1.3.4   Counting of Protons, Deuterons, Tritons, etc.

Apart from α- and β-particles and fission fragments (p. 18) the interest of the analyst in the counting of charged particles stems exclusively from their use as projectiles in accelerator experiments, i.e. in beam monitoring (p. 33), or from the fact that they are produced in nuclear reactions employed in prompt analytical methods. The behaviour of the lighter charged particles is not very different from that of α-particles and essentially similar counting techniques can be used. For spectroscopy semiconductor detectors are easily the most convenient, while for simple counting scintillators will also be found useful.

For particles of the same energy, provided this is above about 5 MeV, the heavier the particle the shorter is its range. This means that energy losses in counter windows and other absorbing materials are more serious for heavier ions and account must always be taken of them in making measurements. The ranges of the lighter positively charged particles in silicon, taken from ref. [56], are shown in Fig. 1.11. For ranges of heavier particles the reader is advised to refer to the work of Northcliffe [57, 58].

1.3.5   Fission Fragment Counting

Fission fragments are simply highly charged heavy ions, having an average mass of perhaps 100 a.m.u. but they are treated separately here for two reasons. In the first place, their spectroscopy and counting has played a large part in studies of the fission process whose importance in the field of nuclear energy is very great; in the second, fission fragment sources differ from those of most other heavy ions in that fragments cover a whole spectrum of both mass and energy and themselves emit neutrons, γ-rays and other particles during the first $\sim 10^{-11}$ sec of their existence. Another difference is that radioactive nuclides which emit fission fragments spontaneously (i.e. without being excited by nuclear bombardment) exist, and so fission experiments can be performed on them without having to use a reactor or accelerator.

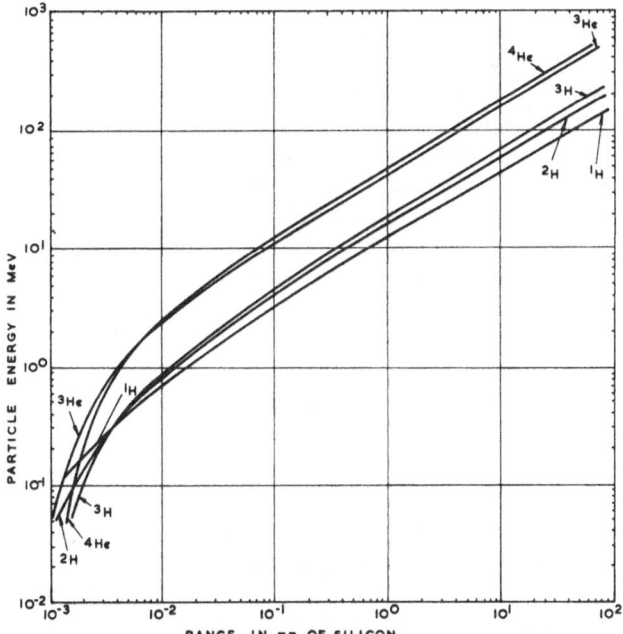

Fig. 1.11. The range of charged particles in silicon (courtesy Ortec instruction manual for surface barrier detectors).

Spectroscopy of Fission Fragments

The kinetic energy spectrum of fission fragments has been measured by three main methods. The earliest experimenters used gas ionisation counters, normally a pair of them mounted back to back with the source in the middle. Since there are two fragments from any one fission event, and since they are expelled from the source substantially at 180° to each other, this means that the fragment pair from a single event may be counted in coincidence. If this is done, it is possible to obtain the mass spectrum as well as the kinetic energy from the results. It is not necessary to use the double counter if only the kinetic energy spectrum is required.

Gas counters used in this way have energy resolutions of ~3%: this was improved to ~1% with the advent of time of flight methods in which the time taken by the fragments to travel along an evacuated tube is used as a measure of their velocity and hence their energy. Equally good resolution, with much less complication of apparatus, is obtained by using semiconductor detectors. These are usually surface barrier counters specially made with a very thin (~0.04 mm) front electrode so as not to degrade the fragment energy.

Simple Fission Fragment Counting

The fission fragment emission rate of a source may be needed either in research on fission itself or as a method for measuring neutron flux (p. 22).

Both gas ionisation and semiconductor detectors may be used for the
purpose, or one can employ photographic emulsion or track detectors. In
both these methods a permanent change is induced in the detector material
by the ionisation caused by the fragments. The damage in the materials is
made visible by development and the tracks produced by individual fragments may then be counted by optical means. The methods have the advantage of extreme simplicity, but the disadvantage that track counting is
tedious and slow. Of the two methods, the track detector is probably the
better and it has the advantage that very large pieces of the material can be
used in circumstances where the intensity of the fragment source is very
low.

1.3.6   Neutron Counting

Neutrons present a counting problem totally different from that of any
other type of radiation. The reason is that, being uncharged, they interact
with atoms through nuclear reactions and do not lose their energy by
collision with atomic electrons. Detection methods for them are normally
based on the principle that some nuclear reaction is chosen which has a high
cross-section and which has a charged particle as a reaction product. This
charged particle then causes ionisation in the counter medium exactly as
already described in the sections on detection of such particles.

Counters in which neutrons collide with protons in hydrogenous material
and eject them from it form one large group of neutron detectors: they are
known as hydrogen recoil counters. In another group which includes BF_3
and fission counters, a nuclear reaction takes place and the reaction
products are detected.

The counters described so far are prompt detectors, requiring that the
measurements are made 'on-line' but reactions may be such that the
reaction products are delayed and the measurements are made after removal
of the detector from the neutron source. Examples of these are (1) the
activation methods, in which the nuclear reaction gives rise to α-, β- or
γ-rays which are counted, sometimes after chemical processing of the
counter material, and (2) the placing of fission track detector material next
to a thin film of a fissile nuclide, exposing it to the neutrons and then
developing and counting the fission tracks (p. 12).

Spectroscopy of Neutrons

There are two ways of measuring the energy of neutrons which do not
involve nuclear reactions: the time of flight method is exactly the same in
principle as already described for fission fragments (p. 19). Pulsed neutron
sources are normally used because it is then easy to time the start of the
flight, but this time may also be obtained by causing a nuclear reaction to
occur at the beginning of the flight tube and observing some particle
emitted in this reaction. The second direct method, suitable only for
neutrons in the eV energy range, is crystal diffraction in which the angle of
diffraction of the neutrons depends on their energy.

Both of the direct methods are cumbersome and quite unsuitable for neutron spectroscopy under most experimental conditions. Counters employing nuclear reactions, while not giving such good energy resolution are far more practical for most purposes. In one technique a proportional counter is filled with hydrogen and the energy of the protons recoiling after collision with the neutrons is measured. The range covered is 20 keV to 1 MeV, but measurements must be taken at several different filling pressures and the whole procedure is quite slow. Only a low neutron flux, about 10^3 n/cm^2/sec, is required.

In a proton recoil telescope the protons come from a thin foil of plastic placed in the neutron beam and the proton energies are measured by means of semiconductor detectors in a direction defined by a coincidence system. The range covered is 1–10 MeV, but the sensitivity is quite low.

Proton energies are also measured in the photographic emulsion method by examination of their tracks. The range covered is 1–10 MeV, but the time taken for scanning the plates, say 14 man-days for 15% accuracy, restricts the use of this procedure.

On the whole, the best high resolution method for general use is probably the tritium spectrometer, or lithium sandwich counter as it is sometimes called. Two surface barrier semi conductor detectors face each other, with a thin film of ^{6}LiF between them. α-particles and tritons are emitted in accordance with the reaction

$$^6\text{Li} + \text{n} \rightarrow {}^3\text{H} + {}^4\text{He}$$

The energy spectra of the two reaction products are taken and one obtains the sum of the energies and also the energy of the tritons. From the sum energy it is possible to calculate the neutron spectrum in the range 500 keV to 5 MeV reasonably easily. The range can be extended down to 200 keV if the triton energy is also included, but this involves a great deal of calculation. Backgrounds can be as high as 30% and so it is necessary to measure them separately with no lithium present. An energy resolution of 5% at 1 MeV is attainable. A very similar method involves making use of the reaction

$$^3\text{He} + \text{n} \rightarrow {}^3\text{H} + {}^1\text{H}$$

Finally, if only a crude spectrum is required, threshold detectors provide a very simple method of achieving it. The principle is that a series of detectors is employed, each one having a different threshold for some neutron reaction. Count rates are measured and the spectrum may be calculated from them and a knowledge of the cross-sections. The range covered is 0.5–5 MeV.

Neutron Flux Measurement

Simple counting of neutrons is complicated by the fact that the cross section for the nuclear reaction employed by the counter will be energy dependent. It is therefore necessary to know its value for the neutron spectrum being measured, or to calibrate the counter in a similar spectrum whose intensity is already known.

The most convenient counters for general use are probably the BF_3 counter and the fission counter. The first uses the reaction

$$^{10}B + n \rightarrow {}^7Li + {}^4He$$

and is usually in the form of a proportional counter whose counting gas is BF_3, normally enriched in ^{10}B. The ^{10}B cross section is very high for thermal neutrons, but decreases rapidly at higher energies, and so a modified version of it, called a long counter, is more suitable for fast neutrons. In it the BF_3 counter is surrounded by a shield of paraffin wax which thermalises a high proportion of the fast neutrons. The shield is carefully shaped so that the counter response is reasonably flat for neutrons in the range of $0.1-10$ MeV.

Fission counters are proportional counters or ion chambers with a thin film of a fissile material, usually ^{235}U, ^{238}U, or ^{239}Pu on the cathode. The fission fragments produced by the neutron interactions are counted in the usual way. ^{235}U and ^{239}Pu are fissionable by neutrons of all energies, while ^{238}U has a fission threshold of about 1 MeV and one can use such threshold differences to measure different parts of the neutron spectrum. It is also possible to shield the counter with thermal neutron absorbers such as cadmium or ^{10}B.

Other neutron counting methods employ scintillators, both solid and liquid, the latter sometimes loaded with ^{10}B to improve the sensitivity. Activation methods in which a radioactive daughter is counted are also used. They are simple but require counting off-line. One such reaction frequently used is

$$^{34}S + n \rightarrow {}^{35}S$$

Finally, the method of sandwiching fission track material with a fissile foil (p. 12) is convenient and easy, but it does have the disadvantage that the counting of the tracks is a slow process.

1.4   ANCILLARY EQUIPMENT

1.4.1   The Essential Electronic Equipment for Counting
For simple counting of pulses from a detector, all that is required is that one count should be recorded for every particle entering the counter and no counts should be recorded from any other cause. It is not difficult to arrange for the first requirement, but very difficult indeed to achieve the second when counting most kinds of radiation. What can be done is to reduce the 'background' to a level as low as possible, measure it in a separate experiment and then subtract it from the total count rate. It is usually quite easy to eliminate almost all the background due to most of the potential causes, such as that from the electronic equipment itself, but one eventually arrives at a level which is mainly caused by the effect of cosmic rays on the

counter and by intrinsic factors in the detector. For most purposes one usually puts the counter inside a reasonably thick shield, say 2 in. of lead, and accepts whatever residual background is left. Attempts at further reduction beyond this bring in diminishing returns, but very low backgrounds may be necessary for some purposes and this topic is discussed in Section 1.5.3 (p. 32).

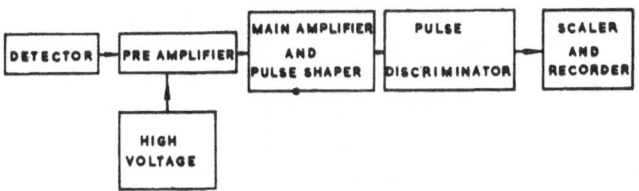

Fig. 1.12. Block diagram of a typical electronic system for simple counting.

The essential parts of any simple electronic counting system are shown in the block diagram, Fig. 1.12. The electrical pulses produced by the detector may range in size from about 1 mv for a gas ionisation chamber to 50 v for some G.M. counters and so it is obvious that the amplification system will not be the same in all cases. Indeed, all that is needed for G.M. counters is a small unit which provides the external quenching if wanted, fixes the dead-time so that it is known for calculation purposes, shapes the pulses to suit the discriminator, and routes the high voltage to the detector anode. For most other types of counter a more sophisticated amplifier system is required and it is common practice to split it into two parts, a head and a main amplifier. A detector is a high impedance device and the head amplifier, which is placed as close to it as possible to avoid pulse distortion, may do some amplifying and pulse shaping but its chief purpose is to match the pulses into low impedance cables so that the main amplifier and other electronics can be more conveniently situated, for example, outside a radiation field in which the counter may have to be placed. It also normally serves as the point of entry for the high voltage supply to the counter. This may range from the few volts needed for the bias supply for some semi-conductor detectors to several kilovolts used by proportional counters (and, indeed, some semiconductors) under some circumstances. The supply must be carefully smoothed so as to be free from ripple and may have to be extremely stable, e.g. a voltage stability of at least ~0.1% is required for most scintillation counters.

The main amplifier amplifies and shapes the pulses to the required degree, both gain and time constants usually being variable. The necessity for pulse shaping arises because the pulses from the detector usually have a long tail which makes them quite unsuitable for transmission through the subsequent stages of the equipment. They may 'pile up', that is, a new pulse may arrive while the system is still responding to the old one and this may cause the amplifier to overload and will, in any case, result in a drastic reduction in the capability of the system for dealing with high counting rates. In addition to this, in order to deal with the long pulses the amplifier must have such a good low frequency response that it becomes very susceptible to hum and microphony, etc.

As we have seen, in detector design the whole emphasis is on collecting the useful information about a counting event as quickly as possible. Thus the important parts of a detector pulse are the leading edge and the top, the long tail being mainly caused by slow effects such as, for example, the late arrival of the positive ions at the cathode. The object of the pulse shaping then, is to preserve the useful information but to reduce the duration of the pulse as much as possible, a process often known as 'clipping'. The commonest way of clipping pulses is by means of a resistance-capacity (RC) circuit. The pulse passes through networks such as those shown in the Fig. 1.13(a) and (b), often called differentiating and integrating circuits.

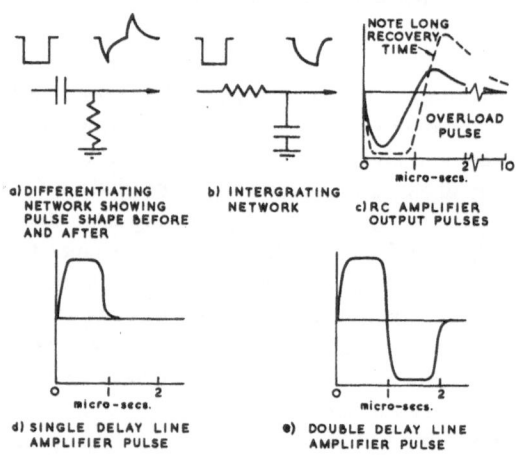

Fig. 1.13. Pulse shaping in R.C. and delay line amplifiers.

Differentiation affects mainly the top of the pulse and causes an overshoot, while integration alters its leading and trailing edges. The total effect of a practical network of these two types is shown in Fig. 1.13(c). In this case conditions of equal differentiation and integration time constants of 1 μs have been used. To shorten the tail to only a few μs as in this case, it is necessary to differentiate twice – 'double differentiation' as it is called. Clipping can also be done by utilisation of delay lines in the amplifier in which the pulse is made to travel along a length of delay cable and back again. Single delay lines produce pulses as shown in Fig. 1.13(d) and double delay lines as shown in Fig. 1.13(e). The advantages of such systems over RC shaping amplifiers arise because the rectangular pulses are very suitable for further processing and their height is readily measured.

The other main control on an amplifier is the gain switch. It is not necessary that the gain is linear, i.e. output pulse height is directly proportional to input pulse height, for simply recording numbers of counts, but since linearity is essential for pulse height analysis (Section 1.4.3, p. 27) it is usual to employ linear amplifiers for most purposes. The gain switch is set so that the largest pulses do not overload the amplifier: such overloading is easily detected in RC shaping amplifiers by the fact that the pulse becomes flattened as illustrated in Fig. 1.13(c). Delay line amplifiers show a widening of the rectangular pulse, but overloading does not affect them so seriously. Overloading destroys the information contained in the height of the pulse and tends to paralyse or 'block' the amplifier and so cause loss of pulses, but this blocking is of shorter duration in the delay line amplifiers.

The amplifier output will have one pulse for each particle which triggered the detector, but will also include a large number of very small pulses due to electronic noise or even, as in the case of a proportional counter detecting α-particles in the presence of $\beta's$, due to particles which produce a smaller amount of ionisation in the counter medium than the required ones. These unwanted pulses are removed by inserting a discriminator before the scaling equipment. This is merely a device which generates an output pulse when triggered by an input pulse which is higher than some pre-set level.

From the discriminator the pulses pass into a scaler which registers one count for every pulse reaching it. The counts may simply be indicated on an internal register of some kind or may pass to more sophisticated recording equipment (p. 27).

1.4.2   Coincidence and Anti-coincidence Units
While simple electronic arrangements such as have just been described are all that is needed for many experiments, more complicated systems involving coincidences between pulses and careful timing of them may well be required for some purposes. Examples of the use of such techniques are the anti-coincidence counters employed in the reduction of cosmic ray backgrounds in β-counting (p. 32) and the elaborate fast coincidence methods used in on-line double energy measurements of fission fragment pairs [46, 47].

It is important to make a clear distinction in one's mind between analogue and digital pulses. The pulses from the detector and amplifier are analogue pulses whose shape carries information about the radiation which caused the counter to produce them, and it is therefore vital that any manipulations of such pulses do not distort this shape. The shape of digital pulses carries no information and what matters about them is their position in time relative to other pulses and the fact of their existence: they are, therefore, merely switching pulses. In a coincidence arrangement two pulses, which in most analytical applications are analogue pulses, arrive simultaneously at the inputs of the unit and trigger a pulse generator which causes a digital pulse to be emitted at the output; if the pulses do not arrive simultaneously there is no output pulse. If the system is an anticoincidence an output pulse is only generated when a pulse arriving at the first input is not accompanied by one at the other. Many other possible coincidence and anti-coincidence arrangements can be made, for example, a triple coincidence of three input pulses or a coincidence between any two of three may be demanded.

The most important parameter to be considered in coincidence work is the 'coincidence (or anticoincidence) resolving time'. In a simple two input coincidence unit a pulse arriving at one input opens the 'gate' for a period of time τ. If a pulse arrives at the other input within this time an output pulse is generated. A glance at Fig. 1.14 shows immediately that a coincidence occurs if the two pulses are within a time 2τ, which is the coincidence resolving time of the unit.

There will always be a statistically determined number of chance coincidences between uncorrelated pulses, and the number of these is a function of the resolving time. If the counting rates in the two channels are R_1 and R_2, the chance coincidence rate will be $2\tau R_1 R_2$. It is thus vital to reduce the resolving time as much as possible, but there is a limit to how short it can be, which is set by the fact that variations in time of arrival of the two correlated events at the detectors, and of the response of the detector and that of the subsequent analogue pulse handling circuits, result in an uncertainty in the arrival time of pairs of correlated pulses at the coincidence unit. The resolving time must not be shorter than this uncertainty or coincidence counts will be lost. These considerations also lead to the conclusion that detectors which have slow pulse rise-times or ill-defined pulse shapes are not suitable for coincidence work. Thus semiconductors and scintillation counters are especially suitable while G.M. counters are really only satisfactory where low counting rates are involved. Resolving times ranging from a few nanoseconds to a few microseconds are commonly used.

The easiest way to set up a coincidence system initially, is to view the two pulses simultaneously on a double beam oscilloscope and to trigger the oscilloscope by some related pulse which comes earlier in time, or by one of the pulses themselves. By adjusting the time delays of the pulses they can then easily be brought into visual coincidence. Having done this, one carries out two series of measurements. In the first experiment one counts the output pulses while varying the time delay of one of the input pulses, keeping the delay of the other constant. A curve as in Fig. 1.15(a) whose

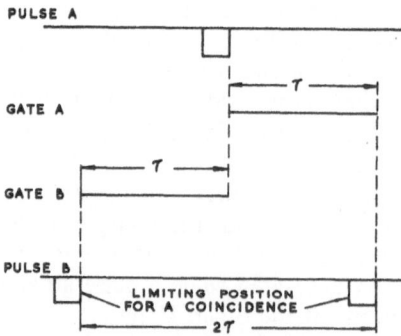

Fig. 1.14. Diagram to show what is meant by coincidence resolving time.

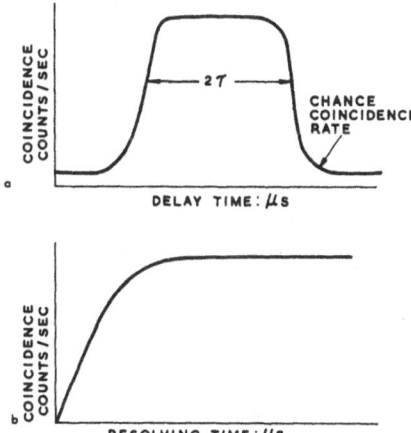

Fig. 1.15. The two series of measurements which must be made when setting up a coincidence system.

width is equal to the resolving time 2τ, is produced. The correct operating point is thus the centre of the flat top and the experiment also measures the chance coincidence rate.

Having found the correct settings for the time delay controls, one then carries out a series of counts of coincidences at various resolving times. The minimum resolving time which should be used is then that setting at which the count rate reaches the maximum as shown in Fig. 1.15(b).

Finally, by making a measurement of how many events which should give rise to a coincidence count are present in the channel having the smallest number of such events and comparing this with the observed number of coincidences, one obtains the coincidence efficiency of the system. What this should be must depend to some extent on the experimental arrangement and the detectors, but if it is below 90% it is worth considering whether or not there is a fundamental error in the design of the experiment.

1.4.3 Pulse Height Analysers and Real Time Computers

If one wishes to measure the energy spectrum of the particles emitted by a source a suitable counter such as has been described earlier in this chapter must be employed. Provided this is done, the height of the pulses produced will be proportional to their energies and so, if they can be sorted according to height the result is the required spectrum: this is exactly what a pulse height analyser does.

The pulse, perhaps after passing through coincidence gates and over threshold levels, enters the analogue to digital converter (ADC) which gives out a digital signal proportional to the pulse height. This signal is used to add one unit into the memory word or 'channel' corresponding to the size of the digital signal. The memory may have a number of channels, 4096 ($= 2^{12}$) being a common size in use today, and it can usually be split up into a number of smaller channel groups. The remainder of the analyser consists of units performing such functions as automatic timing, calculations and output of results.

There has been a trend in recent years to use an ADC coupled to a small computer instead of an analyser. Such computers are called 'real-time' computers since they must handle the data as it appears and as quickly as possible. This practice has many advantages such as extreme flexibility, the possibility of performing extensive calculations on the data, the ability to change the configuration at will, etc., but the disadvantage that the initial programming of the computer may take a considerable time. This subject is discussed extensively in Chapter 4, and the detailed use of pulse height analysers is dealt with in Chapter 2.

1.4.4   Data Handling Equipment

One result of the improvement in counting techniques and the increased sophistication of experiments over the years has been that the amount of data, whether in the form of complete pulse height spectra or of simple counts, which one experimenter produces, has increased enormously.

In simple counting experiments one usually makes a series of measurements of the total number of counts, the period over which this total was taken, and the real time at which this was done. The duration of the count can conveniently be recorded by counting clock pulses of say 10^{-2} min. into a scaler which is started and stopped automatically with the scaler which is registering the counts. The time of day can be recorded in the same way by using a scaler which is never reset, perhaps employing units of 10^{-4} days. Such a system can be made fully automatic by attaching it to a suitable control unit which causes the data to be recorded in some way, such as by means of a card-punch or magnetic tape deck, and then resets and re-starts the count and duration scalers.

Pulse height analysers and real-time computers normally record data on magnetic tape or punched paper tape, as well as printing it out. Computations can therefore be performed internally, or the data can easily be fed into another computer if desired.

1.5   TECHNIQUES

1.5.1   Source Mounting

The production of suitable sources is frequently one of the most important and difficult operations in simple counting of charged particles and it is even more important in their spectroscopy. The reason for this lies in the fact that the particles lose energy so readily by collision with atoms that their spectra are rapidly degraded and some of the particles may be lost altogether. The heavier the charged particle, the more important the thickness of the source becomes, as a glance at Fig. 1.11 will show. β-particles (both positrons and electrons) behave somewhat differently because they are very light, but even they have a comparatively short range in matter, as shown in Fig. 1.16 and hence quite small amounts of absorbing material

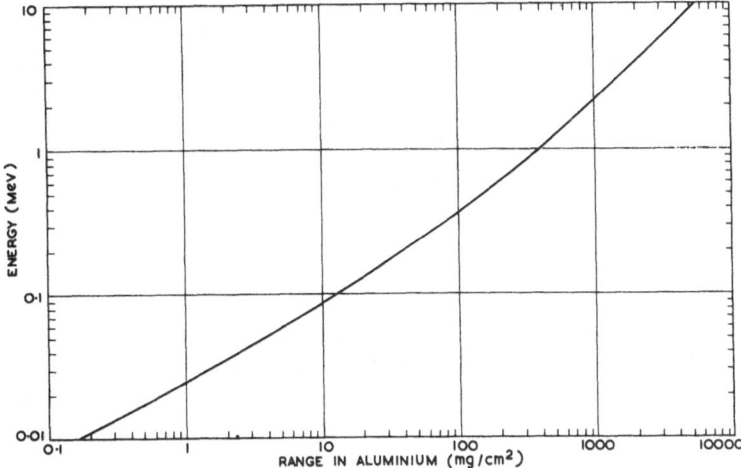

Fig. 1.16. Range-energy curve for β-particles in aluminium. (Reprinted by permission from L. E. Glendenin, Nucleonics, Jan. 1948.)

have a serious effect on their counting characteristics. Electromagnetic radiation presents much less of a problem; indeed this is one of the advantages of γ-counting over counting of charged particles. This subject is discussed in Chapter 2.

In mounting sources for counting and for use as targets for on-line accelerator measurements, the usual requirement is that the source itself should be very thin relative to the range of the particle. In addition, one often wishes to measure particles which have passed through the source backing as in 4π β-counting or double energy fission fragment experiments, and in this case the backing itself must also be very thin. The whole subject was reviewed by Yaffe [61] in 1962 and so only a few of the most important points will be mentioned here.

One of the most useful thin backing materials is VYNS resin, self-supporting films of which can be made as thin as 2 μg/cm² and then rendered conducting by vacuum evaporation of 2–4 μg/cm² of gold on the surface: such films are almost always used for 4π β-counting. Unfortunately, the resin does not stand up to bombardment by charged particles and so VYNS films are not suitable for use in accelerator target backings. For this purpose, one of the most convenient materials is undoubtedly nickel foil, which is available commercially in thicknesses down to 50 μg/cm². The thinner foils have a backing of copper which is removed by dissolution after deposition of the target material and so allows the use of more violent methods in the preparation of the target than is the case with VYNS. Another very useful material is carbon, commercially available in foils down to about 4 μg/cm². Such foils are not so strong as VYNS, but stand up well to accelerator beams.

Mounting of sources on VYNS for 4π counting normally involves weighing a solution of the activity on to the film and then drying it slowly under an infra-red lamp. The difficulty is that the source does not crystallise evenly over the film and so its thickness is very variable, and this gives erratic counting results. Some improvement ensues if a protein such as insulin is deposited on the film before the activity is put on. This causes the crystals to spread better, but if the maximum energy is below 100 keV only rough answers can be expected. It is much better to use 4π $\beta\gamma$-coincidence counting if this is possible (p. 15). Other methods of deposition, such as by vacuum evaporation have also been tried, but it is difficult to do these quantitatively.

For absolute α-counting (p. 13) thick backings of stainless steel or platinum are normally used and a small correction is applied for back-scattering. The source can be weighed on to the mount, dried, and then, provided it is not volatile, fired at high temperature. This procedure gives much thinner sources than drying alone does. Thick backings are also used in α-spectroscopy, but since the weight of the material in the source is not required, techniques such as vacuum evaporation or electro-spraying can now be used for depositing the activity.

For preparation of β-sources for routine counting it is usual to mount the activity in a completely different way. The source is deliberately made quite thick (at least 5 mg/cm^2) and correction curves are used to relate the observed count rate to the disintegration rate. The reason for going from very thin sources to very thick ones is that the correction curves tend to smooth out and duplicate samples to agree much better for the thick sources than they do for the intermediate ones. The procedure, which is discussed in Section 1.3.2 (p. 15) gives results accurate to perhaps ± 5% for a β-emitter having a maximum β energy > 0.5 MeV.

1.5.2   Calculations

Apart from those associated with a particular counting method such as 4π $\beta\gamma$-coincidence counting (p. 15) or neutron spectroscopy (p. 20) for details of which the reader is advised to consult the quoted references, most counting calculations are concerned with the growth and decay of radio-activity.

The basis of all such calculations is that radioactive decay is a purely random process, and so can be treated statistically. Thus the rate of decay of a large number of atoms, N, is proportional to N

$$-\frac{dN}{dt} = \lambda N \tag{1.3}$$

Where λ is the disintegration constant of the radioactive species concerned. It can easily be shown that

$$\lambda = \frac{\ln 2}{t_{1/2}} \tag{1.4}$$

where $t_{1/2}$ is the 'half-life' of the species, that is, the time taken for the number of atoms to be reduced to half its original value. Integration of equation (1.3) gives

$$N = N_0 e^{-\lambda t} \tag{1.5}$$

where N is the number of atoms remaining at time t and N_0 is the initial number. Activities can be substituted for numbers of atoms, giving

$$A = A_0 e^{-\lambda t} \tag{1.6}$$

which states the well-known fact that a radioactive source decays exponentially.

If the daughter product of radioactive decay is also radioactive, equation (1.5) becomes

$$N_2 = \frac{\lambda_1}{\lambda_2 - \lambda_1} N_1^{\,0} \left(e^{-\lambda_1 t} - e^{-\lambda_2 t} \right) + N_2^{\,0} e^{-\lambda_2 t} \tag{1.7}$$

where $N_1^{\,0}$ and $N_2^{\,0}$ are the amounts of parent and daughter originally present at $t = 0$ and λ_1 and λ_2 are their disintegration constants. The general solution for a chain of decays was given by Bateman in 1910[62].

$$N_n = C_1 e^{-\lambda_1 t} + C_2 e^{-\lambda_2 t} + \ldots + C_n e^{-\lambda_n t} \tag{1.8}$$

where

$$C_1 = \frac{\lambda_1 \lambda_2 \ldots \lambda_{(n-1)}}{(\lambda_2 - \lambda_1)(\lambda_3 - \lambda_1) \ldots (\lambda_n - \lambda_1)} N_1^{\,0}$$

$$C_2 = \frac{\lambda_1 \lambda_2 \ldots \lambda_{(n-1)}}{(\lambda_1 - \lambda_2)(\lambda_3 - \lambda_2) \ldots (\lambda_n - \lambda_2)} N_1^{\,0} \text{ etc.}$$

When the half-life of the daughter is less than that of the parent the condition called 'transient equilibrium' occurs. Equation (1.7) then reduces to

$$N_2 = \frac{\lambda_1}{\lambda_2 - \lambda_1} N_1^{\,0} \tag{1.9}$$

or, in terms of activities

$$\frac{A_2}{A_1} = \frac{\lambda_2}{\lambda_2 - \lambda_1} \tag{1.10}$$

while, if the parent half-life is so long that its decay rate is negligible, equation (1.7) becomes

$$\frac{A_2}{A_1} = 1 \tag{1.11}$$

that is, the activities of parent and daughter are equal, a condition found in undisturbed chains of natural radioactivity and known as 'secular equilibrium'.

Finally, when a radioactive species is produced by irradiation in a constant flux of particles, we have

$$N_2 = \frac{nv\,\sigma}{\lambda_2} N_1{}^{\circ} \left(1 - e^{-\lambda_2 T}\right) \tag{1.12}$$

Here N_2 is the number of atoms of product present at the end of an irradiation of duration T in a flux nv, while σ is the cross-section for production of N_2 from N_1.

1.5.3 Low Level Counting

The main problem in low level counting is how to reduce the background of the equipment to, at worst, no greater than the count-rate of the source. A secondary problem is to ensure that the equipment is stable enough to allow counting for what may be very long periods of time.

The three main causes of background are electronic, counter and shield materials, and cosmic rays. It is usually easy enough to reduce the first of these virtually to zero. One starts by employing carefully selected equipment in good condition, extra care being taken to ensure cleanliness of high voltage components and that good electrical contacts are made on the high voltage lines, particularly in the final connection to the counter anode. The equipment is then connected to a very clean mains supply which may have to be specially filtered to remove mains-borne noise. The minimum number of earth contacts are then made, every care being taken to avoid 'earth-loops', that is, duplicate paths to earth. The effect of extra earths may then be tried, for example, it is sometimes advantageous to earth the cathode of a counter to its shield and sometimes it is not. Finally, the equipment may be surrounded by an earthed Faraday cage which sometimes reduces noise due to radiation from such things as relays or electric motors. It is far better, however, to remove these to a safe distance or at least attempt to suppress their interference at source.

Counter and shield materials should be free from radioactive contamination, a requirement which is becoming increasingly difficult to realise as time goes on. Electrolytic copper is a good counter material because it can be obtained very pure, while magnesium has also been found to be suitable. Lead is not a good shield material because of its natural contamination. Pre-war steel such as that from scrapped naval vessels or old railway lines or axles is usually excellent but awkward to handle and extremely expensive. The author has found that the cheapest way to build a shield is to make it from bright steel bar, which is about one tenth the price of pre-war steel. Suppliers can usually be persuaded to segregate a sufficient quantity all from one batch while a sample is tested for contamination by the user: if it is unsuitable another batch can be tried. An easy way to test the steel is to build it into a shield of a low background counter which is already operating and note the effect.

Cosmic rays are practically always eliminated by some sort of anti-coincidence sheath round the counter. There have been innumerable counters designed in this way, the principle being that the source counter is surrounded by a sheath of other counters which are as sensitive as possible to cosmic rays. The two counters are connected in anti-coincidence (p. 25) so that if a count is caused in the source counter by a cosmic ray, it will be cancelled because it will be in coincidence with a count caused in the guard counter by the same ray. An example of an anti-coincidence counter will be found in ref. [55].

1.5.4 Beam Monitoring

The simplest form of beam monitoring involves measurement of the charge collected in a Faraday cup, placed directly in the beam. When the target being irradiated is thin, the cup may be situated behind the target as the beam stopper and monitoring can then be continuous, while for thick targets periodic measurements must be made by pushing the cup into the beam in front of the target.

Exact flux measurements are harder to carry out and must usually be made as a separate experiment. Detectors such as those discussed in section 1.3 are used, semiconductors being particularly suitable for most purposes. However, if a relative measurement of beam intensity is all that is required, there will often be some scattered product of the primary nuclear reactions which can be utilised by placing a detector to one side of the beam and counting this product.

REFERENCES

1. W. FRANZEN and L. W. COCHRAN, Nuclear Instruments and Their Uses. Vol. 1 (A. H. Snell, ed.), Wiley, New York (1962).
2. D. H. WILKINSON, Ionisation Chambers and Counters. Cambridge University Press, Cambridge (1950).
3. J. SHARPE, Nuclear Radiation Detectors. Methuen, London (1964).
4. H. W. FULBRIGHT, Encyclopaedia of Physics. (S. Flügge, ed.), Vol. 45/2, p. 1 Springer-Verlag, Berlin (1958).
5. S. A. KORFF. Encyclopaedia of Physics. (S. Flügge, ed.), Vol. 45/2, p. 52 Springer-Verlag, Berlin (1958).
6. W. E. MOTT and R. B. SUTTON, Encyclopaedia of Physics. (S. Flügge, ed.), Vol. 45/2, p. 86 Springer-Verlag, Berlin (1958).
7. S. C. CURRAN, Encyclopaedia of Physics. (S. Flügge, ed.), Vol. 45/2, p. 174 Springer-Verlag, Berlin (1958).
8. C. M. YORK, Encyclopaedia of Physics (S. Flügge, ed.), Vol. 45/2, p. 260 Springer-Verlag, Berlin (1958).
9. M. M. SHAPIRO, Encyclopaedia of Physics. (S. Flügge, ed.), Vol. 45/2, p. 342 Springer-Verlag, Berlin (1958).
10. H. H. BARSCHALL, Encyclopaedia of Physics. (S. Flügge, ed.), Vol. 45/2, p. 437 Springer-Verlag, Berlin (1958).
11. R. T. SPIEGEL, Encyclopaedia of Physics. (S. Flügge, ed.), Vol. 45/2, p. 487 Springer-Verlag, Berlin (1958).
12. K. SIEGBAHN (ed.), Alpha, Beta and Gamma Spectroscopy. Vol. I, North Holland, Amsterdam (1965).
13. G. DEARNALEY and D. C. NORTHROP, Semi-conductor Counters for Nuclear Radiations. Spon, London (1964).

14. Instruction Manual for Surface Barrier Detectors. Ortec Inc. Oak Ridge (1969).
15. Nuclear Enterprises Ltd. Catalogue 1970, Nuclear Enterprises, Edinburgh (1970).
16. G. BERTOLINI and A. COCHE (eds.), Semiconductor Detectors, North Holland, Amsterdam (1968).
17. L. YUAN and C. S. WU, (eds.), Methods of Experimental Physics. Vol. 5A, Academic Press, New York (1961).
18. M. WIDGOFF, Techniques of High Energy Physics. (M. Ritson, ed.), Interscience, New York (1961).
19. R. SCHLITER, Techniques of High Energy Physics. (M. Ritson, ed.), Interscience, New York (1961).
20. W. BARKAS, Nuclear Research Emulsions. Academic Press, New York (1963).
21. C. G. BELL and F. N. HAYNES, (eds.), Liquid Scintillation Counting. Pergamon, London (1958).
22. J. B. BIRKS, The Theory and Practice of Scintillation Counting. Pergamon, Oxford (1964).
23. J. M. DU MOND, *Ann. Rev. Nucl. Sci.*, 8, 163 (1958).
24. D. WEST and E. F. BRADLEY, *Phil. Mag.*, 2, 957 (1957).
25. J. A. HARVEY, (ed.), Experimental Neutron Resonance Spectroscopy. Academic Press, New York (1970).
26. G. F. J. LEGGE and P. VAN DER MERWE, *Nucl. Instr. and Methods*, 64, 157 (1958).
27. M. G. SILK, A.E.R.E. Report R-5183 (1966).
28. M. G. SILK, A.E.R.E. Report R-5438 (1967).
29. M. G. SILK, A.E.R.E. Report R-2009 (1968).
30. P. W. BENJAMIN, C. D. KEMSHALL and A. BRICKSTOCK, A.W.R.E. Report 09/68 (1968).
31. P. W. BENJAMIN, C. D. KEMSHALL and J. REDFEARN, A.W.R.E. Report NR 1/64 (1964).
32. P. W. BENJAMIN and G. S. NICHOLLS, A.W.R.E. Report NR5/63 (1963).
33. D. HATTON, Ph.D. dissertation. The Application of Activation Techniques to the Measurement of Epi-thermal and Fast Neutron Spectra. Birmingham University (1970).
34. K. BODDY, J. A. DENNIS and R. C. LAWSON, *Phys. Med. Biol.*, 14, 471 (1969).
35. A. O. HANSON and J. L. MCKIBBEN, *Phys. Rev.*, 72, 673 (1947).
36. R. L. FLEISCHER, P. B. PRICE and R. M. WALKER, *Ann. Rev. Nucl. Sci.*, 15, 1 (1965).
37. L. M. BOLLINGER, Proc. 1st Intern. Conf. Peaceful Uses At. Energy, United Nations, New York (1956).
38. W. HORNYAK, *Rev. Sci. Instr.*, 23, 264 (1952).
39. W. S. EMMERICK, *Rev. Sci. Instr.*, 25, 69 (1954).
40. H. O. ZETTERSTROM, S. SCHWARZ and L. G. STRÖMBERG, *Nucl. Instr. and Methods*, 42, 277 (1966).
41. J. CAMERON, L. M. HARRISON and J. B. PARKER, *Nucl. Instr. and Methods*, 64, 157 (1958).
42. W. E. STEIN, *Phys. Rev.*, 108, 94 (1957).
43. W. E. STEIN and S. L. WHETSTONE, *Phys. Rev.*, 110, 476 (1958).

44. J. C. D. MILTON and J. S. FRASER, *Phys. Rev.*, 111, 877 (1958).
45. G. ANDRITSOPOULOS, T. CORNELL and A. L. RODGERS, Proc. 1st Intern. Symp. on the Physics and Chemistry of Fission, Vol. 1 p. 481 I.A.E.A., Vienna (1965).
46. H. W. SCHMITT, W. M. GIBSON, J. H. NEILER, F. J. WALTER and T. D. THOMAS, Proc. 1st Intern. Symp. on the Physics and Chemistry of Fission, Vol. 1 p. 531 I.A.E.A., Vienna (1965).
47. H. W. SCHMITT and F. PLEASONTON, *Nucl. Instr. and Methods*, 40, 204 (1966).
48. D. C. BRUNTON and G. C. HANNA, *Canad. J. Res.*, 28A, 190 (1950).
49. A.E.R.E. Unpublished Manual for Equipment Type 1077B (1949).
50. P. M. WRIGHT, E. P. STEINBERG and L. E. GLENDENIN, *Phys. Rev.*, 123, 205 (1961).
51. P. J. CAMPION, *Int. J. Appl. Rad. and Isotopes*, 4, 232 (1959).
52. R. GUNNINK, L. J. COLBY and J. W. COBBLE, *Anal. Chem.*, 31, 796 (1959).
53. J. G. CUNINGHAME, MRS. J. A. B. GOODALL, G. P. KITT, C. B. WEBSTER and H. H. WILLIS, A.E.R.E. Report R-5587 (1967).
54. B. P. BAYHURST and R. J. PRESTWOOD, *Nucleonics*, 17, (3) 82 (1959).
55. J. G. CUNINGHAME, MISS M. P. EDWARDS, MRS. J. A. B. GOODALL, G. P. KITT, C. B. WEBSTER and H. H. WILLIS, A.E.R.E. Report R-4727 (1964).
56. C. F. WILLIAMSON, J. P. BOUJOT and J. PICARD, Fr. A.E.C. Report CEA-R-3042 (1966).
57. L. C. NORTHCLIFFE, *Phys. Rev.*, 120, 1744 (1960); *Phys. Rev.*, 120, 1758 (1960).
58. L. C. NORTHCLIFFE and R. F. SCHILLING, *Nucl. Data Tab. 'A'*, 7, 233 (1970).
59. R. HURST, G. R. HALL and MRS. K. M. GLOVER, A.E.R.E. Report C/R 647 (1951).
60. R. C. HAWKINS, Unpublished Work at Atomic Energy of Canada Ltd. (1956).
61. L. YAFFE, *Ann. Rev. Nucl. Sci.*, 12, 153 (1962).
62. H. BATEMAN, *Proc. Cambridge Phil. Soc.*, 15, 423 (1910).
63. U. FANO, *Phys. Rev.*, 72, 26 (1947).
64. J. ADAMS and B. W. MANLEY, IEEE Trans. on Nucl. Sci., NS-13 No. 3 88 (1966).
65. L. A. J. VENVERLOO, Practical Measuring Techniques for β-Radiation, Macmillan, London (1971).

APPENDIX

Summary of Detectors Which May Be Used for Different Types of Radiation

Particle or ray	Detector	Remarks	Ref.
α	*Gridded ion chamber	Commonly used; resolution $\sim 0.4\%$	1, 2, 3, 4, 12
	*Proportional counter	More awkward to use; resolution $\sim 1\%$	1, 3, 7, 12, 17
	*Semiconductor detector (surface barrier)	Very convenient; resolution $\sim 0.2\%$	3, 12, 13, 14, 15, 16
	*Magnetic spectrometer	Very bulky and expensive but the best resolution $\sim 0.1\%$	12, 17
	*Photographic emulsion	Simple and provides permanent record but interpretation very time-consuming	9, 17, 18, 20
	*Cloud chamber	Bulky and complicated	8, 17, 19
	Liquid scintillator	Scintillation methods generally not suitable for spectroscopy, but very simple for ordinary counting; sample can be dissolved in liquid, giving high geometry	3, 12, 15, 17, 21, 22
	Gel scintillator		3, 12, 15, 17, 21
	Plastic scintillator		3, 6, 12, 17, 22.
	CsI scintillator		3, 6, 12, 15, 17, 22
	Anthracene		3, 6, 15, 17, 22
	ZnS/Ag phosphor		3, 6, 12, 15, 17, 22
β	*Proportional counter	Low energy spectroscopy; resolution $\sim 3\%$	1, 3, 7, 12, 65
	*Semiconductor detector (Li/Si)	Very convenient; resolution $\sim 1\%$	3, 12, 13, 14, 15, 16, 65
	*Magnetic spectrometer	Very bulky and expensive but the best resolution $\sim 0.1\%$	12, 17

*Commonly used in spectroscopy of the particle or ray and may also be used for simple counting

APPENDIX   continued

Particle or ray	Detector	Remarks	Ref.
β (*cont.*)	*Photographic emulsion	Simple and provides permanent record but interpretation very time-consuming	9, 17, 18, 20
	*Cloud chamber	Bulky and complicated	8, 17, 19
	*Čerenkov counter	For high energy particles	3, 6, 12, 15,
	Liquid scintillator	Scintillation methods generally not suitable for spectroscopy, but very simple; sample can be dissolved in liquid; glass inert to liquids	17, 21, 22, 65
	Plastic scintillator		3, 6, 12, 15, 17, 22, 65
	CaF$_2$ scintillator		3, 6, 17, 22, 65
	Anthracene		3, 6, 17, 22, 65
	Glass scintillator		15, 17, 22, 65
	G. M. counter	The simplest and cheapest counter but less stable and accurate than the proportional counter	2, 3, 5, 65
γ	*Semiconductor detector (Ge/Li)	Cheap and convenient; fairly high efficiency; resolution $\sim 0.3\%$	12, 13, 14, 15, 16
	*Magnetic spectrometer	Bulky and expensive; low efficiency; resolution $\sim 0.2\%$	12, 17
	*NaI (Tl) scintillator	Cheap and convenient; high efficiency; resolution $\sim 6\%$	3, 6, 15, 17, 22
	*Bent crystal spectrometer	Complicated and very low efficiency but the best resolution $\sim .05\%$	12, 17, 23
	Ion chamber	Used mainly for health physics measurements	2, 3, 4
	G.M. counter	Not very efficient	2, 3, 5

APPENDIX   continued

Particle or ray	Detector	Remarks	Ref.
γ (*cont.*)	Liquid scintillator Plastic scintillator Glass scintillator	Scintillators simple to use and convenient; plastic scintillator gives very fast pulses; glass inert to liquids	3, 12, 15, 17, 21, 22 3, 6, 15, 17, 22 15, 17, 22
	Photographic emulsion	Used mainly in medical, industrial and health physics work	9
X-rays	*Semiconductor detector (Ge/Li, Si/Li)	Cheap and convenient, high efficiency; resolution $\sim 3\%$	12, 13, 14, 15, 16
	*Magnetic spectrometer	Bulky and expensive, low efficiency; resolution $\sim 1\%$	12, 17
	*NaI (Tl) scintillator (and other inorganic crystals)	Cheap and convenient, high efficiency; resolution $\sim 20\%$	3, 6, 15, 17, 22
	Bent crystal spectrometer	Complicated, very low efficiency but the best resolution $\sim 0.2\%$	12, 17, 23
	*Xe filled proportional counter counter	Fairly simple, good efficiency; resolution $\sim 3\%$	12, 17, 23
	Liquid scintillator Plastic scintillator Glass scintillator	Scintillators simple to use and convenient	3, 15, 17, 22 3, 6, 12, 15, 17, 22 15, 17, 22
	Photographic emulsion	Used mainly in medical industrial and health physics work	9

APPENDIX   continued

Particle or ray	Detector	Remarks	Ref.
X-rays (*cont.*)	Channel electron multiplier	Useful for very soft X-rays; can also be used for charged particle detection	64
p, d, T, etc.	*Semiconductor detector (Si/Li, Ge/Li or surface barrier)	Cheap and convenient except that Ge/Li detectors need cooling in liquid N; resolution $\sim 0.1\%$	14, 15, 16
	*Photographic emulsion	Simple and gives permanent record but interpretation difficult and time-consuming	9, 17, 18, 20
	Proportional counter	Used in counter telescopes	1, 3, 7, 12
	NaI (Tl) or other inorganic scintillators	Scintillators easy and convenient to use	3, 6, 15, 17, 22
	Plastic scintillator		3, 6, 12, 15, 17, 22
	Glass scintillator		15, 17, 22
n	*Time of flight using pulsed source or chopper	Complicated and expensive; covers range thermal − 5 MeV; typical resolution at 1 MeV $\sim 1\%$.	11, 25
	*Proton recoil + neutron time of flight	Proton recoil method, complicated electronics and calculations; range 300 keV−7 MeV resolution at 1 MeV $\sim 20\%$	26
	*^{6}Li spectrometer	Uses reaction ^{6}Li (n,α) ^{3}T; range 20 keV−1 MeV; resolution at 1 MeV $\sim 5\%$	4, 2, 4, 3, 44

APPENDIX   continued

Particle or ray	Detector	Remarks	Ref.
n (*cont.*)	*Proportional counter	Proton recoil method: range 20 keV−1 MeV; resolution 1 at MeV ∼ 3%	10, 11, 30, 31
	*Proton recoil telescope	Efficiency very low; range 1−10 MeV	11
	*Photographic emulsion	Proton recoil method; simple but time-consuming to interpret; range 1−10 MeV	10, 11, 17, 18, 32
	*Resonance detectors	Suitable for low energy neutrons in the resonance region; simple but gives only a crude spectrum	33
	*Threshold detectors	Very simple but gives only in a crude spectrum; range 0.1−5 MeV	26
	BF_3 counter	For thermal neutrons	3, 10
	Long counter	BF_3 counter surrounded by specially shaped hydrogenous shield; range 0.1−10 MeV	3, 11, 35
	Fission track detector + coating of fissile material	Very simple, but scanning detector very time-consuming	36
	Liquid scintillator	To improve efficiency may be loaded with ^{10}B, etc.	10, 15, 37
	Solid scintillators	Glass, plastic and various combinations of scintillators and particle production coatings	10, 15, 38, 39, 40, 41
	Activation methods	Easy to use but require chemical processing and/or counting off-line	10, 11

APPENDIX   continued

Particle or ray	Detector	Remarks	Ref.
Fission fragments	*Time of flight	Complicated: resolution ~ 1%	42, 43, 44, 45
	*Semiconductor detector (surface barrier)	The easiest method of fission fragment spectroscopy Resolution ~ 1%	3, 13, 14, 15, 16, 46, 57
	*Gas ionisation counters	More trouble to use than semiconductor detectors; resolution ~ 3%	3, 48
	Track detectors	Very simple but scanning time-consuming	36
	Photographic emulsion	Very simple but scanning time-consuming	17, 18, 20

Chapter 2

Gamma-Ray Spectrometry

D. F. Covell

U.S. Naval Ordnance Laboratory
White Oak, Silver Spring, Maryland 20910, U.S.A.

2.1 INTRODUCTION

Gamma-ray spectrometry by the method of pulse-height analysis makes possible the direct determination of individual radionuclides in a gamma-emitting sample. Such determinations are possible because the method provides a basis for the identification of specific nuclear transitions, and these, in turn, are characteristic of specific radionuclides. The method has proven to be easy to use, highly sensitive and fast, and has been applied routinely, with good success, to analytical problems in radiochemistry. Thus, samples containing complex mixtures of radionuclides, with activities ranging from the nanocurie to the picocurie level, are readily measured, non-destructively, in periods of time ranging from a few seconds to several hours.

For the successful utilization of the method, several instrumental qualities must be optimized. In order to effect such optimization, know-

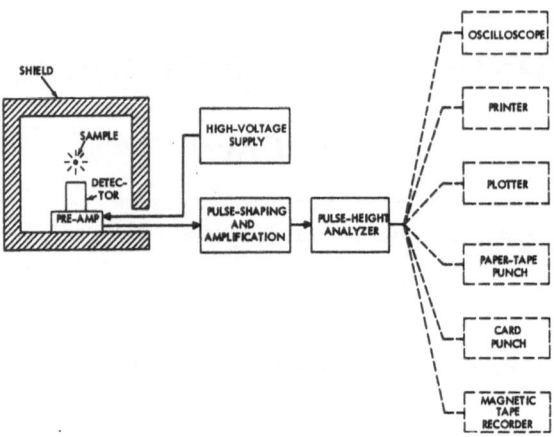

Fig. 2.1. Block diagram showing the elements of a typical gamma-ray pulse-height spectrometer. Note the various devices available for displaying or recording the spectral data.

ledge and understanding on the part of the user is essential in regard to processes of gamma-ray interaction, processes of detection, the form and statistical significance of the spectral data obtained and techniques for the interpretation of such data. It is not the purpose of this chapter to treat each of these factors in depth, but rather to provide a basic orientation for a potential user so that he has an awareness of the important facets of this technique. Excellent in-depth studies have been made of each of the several factors involved and references to representative studies will be provided.

Fig. 2.1 is a simplified block diagram of the instrumental system most commonly used in this application. When a sample is measured, gamma-rays interact with the detector, and electrical impulses are obtained whose amplitudes are approximately linearly proportional to the energy expended by the interacting gamma-rays. The pulses are shaped, amplified, and then electronically sorted according to their amplitudes so that a pulse-height distribution (or pulse-height spectrum) is obtained. The pulse-height spectrum may be displayed or recorded on various devices, as shown in Fig. 2.1, and various techniques can be used to make a determination either of the gamma-ray energy spectrum or of the radionuclides which are involved.

2.2   GAMMA-RAY INTERACTIONS AND THE PULSE-HEIGHT SPECTRUM

Interactions within the detector do not always result in a total expenditure of the gamma-ray energy, so the resultant pulse-height spectrum does not correspond exactly to the energy spectrum of the incident gamma-rays. Three types of interaction are possible: (1) photoelectric; (2) Compton; and (3) pair production. The following is a brief description of each of these types and of the effect they each have on the shape of the pulse-height spectrum. A more complete description of these interaction processes is given by Davisson [1].

2.2.1 Photoelectric Interaction

This is an interaction between the gamma-ray and an atom such that the entire energy of the incident gamma-ray photon, $E_\gamma = h\nu_0$, is absorbed and an atomic electron (usually K or L) is ejected from the atom with kinetic energy T_p such that

$$T_p = E_\gamma - B_e$$

where B_e is the binding energy of the ejected electron. The probability, or cross-section for photoelectric interaction varies approximately as Z^5, where Z is the atomic number of the interacting material.

2.2.2 Compton Interaction

Here, as in the photoelectric interaction, an atomic electron is ejected. In this case, however, only part of the energy of the incident photon is imparted to the electron; the remainder resides with a scattered photon. The electron is referred to as a Compton electron and the scattered photon as a Compton scattered photon. The geometrical relationships are shown in Fig. 2.2. In this interaction, energy and momentum are conserved. The respective energies of the scattered electron and photon are dependent upon the energy of the incident photon and on the scattering angle, θ. The energy of the scattered photon, $h\nu_s$, is given by the following relationship:

$$h\nu_s = \frac{m_0 c^2}{1 - \cos\theta + m_0 c^2/E_\gamma}$$

where $m_0 c^2$ is the electronic rest energy (0.51 MeV). From this relationship, it is seen that the energy of the scattered photon approaches a minimum value of $m_0 c^2/2 = 0.25$ MeV at $\theta = 180°$ for incident photon energies such that $E_\gamma \gg m_0 c^2$. The kinetic energy of the Compton electron is $T_c = E_\gamma - h\nu_s$. The cross section for Compton interaction varies directly as the Z of the interacting material.

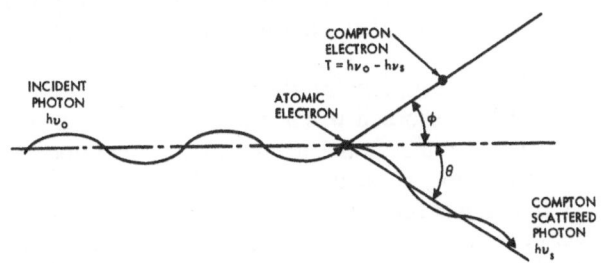

Fig. 2.2. The geometrical and energy relationships of the Compton interaction.

2.2.3 Pair Production

In this interaction, the gamma-ray photon is completely absorbed and in its place appears a positron-negatron pair. The energy disposition is as follows:

$$E_\gamma = [T(-) + m_o c^2] + [T(+) + m_o c^2]$$

where $T(-)$ and $T(+)$ are the kinetic energies of the negatron and positron, respectively. From this relationship, it is seen that pair production is not a competing interaction mode at incident photon energies below $2m_o c^2 = 1.02$ MeV. The cross section for pair production varies as the square of the Z of the interacting material.

2.2.4 The Pulse-Height Spectrum

The pulse-height spectrum is usually represented by a plot on which events per unit time per pulse-height increment are plotted as a function of pulse-height increment. Such a plot is shown in Fig. 2.3. It is characterized by a series of approximately gaussian-shaped peaks superposed on a nearly characterless continuum. These features (the peaks as well as the continuum) are a direct manifestation of the various interaction processes which have taken place within the detector/shield assembly.

If all of the energy of a gamma-ray photon, representative of a specific nuclear transition within the sample, were deposited within the detector, a 'full energy pulse' would be observed. These full energy pulses would appear in the pulse-height spectrum distributed about a 'full energy peak', the center of which would be displaced from zero on the pulse-height scale a distance proportional to the photon energy. Pulses which would result from interactions in which only part of the energy of the gamma-ray photon was deposited within the detector would appear in the pulse-height spectrum proportionately closer to zero. These 'partial energy pulses' may produce 'partial energy peaks', the positions of which would not correspond directly to the energy of the primary gamma-ray photons.

In the case of the photoelectric interaction, nearly all of the energy of the incident photon is imparted to an atomic electron which is ejected from its atomic orbit in the process. The vacancy left by the ejected electron results in a rearrangement of the remaining atomic electrons, and in the process, X-rays or Auger electrons are emitted. The energy of the ejected electron as well as of the X-rays and Auger electrons is generally quickly absorbed in secondary processes so that for practical purposes, it is sufficiently accurate to consider that the total energy of the incident photon, E_γ, is absorbed within the detector.

In the case of the Compton interaction, the energy of the Compton electron is also quickly absorbed within the detector. For incident photon energies such that $E_\gamma \gg m_o c^2$, the energy of the Compton electron may extend from zero to a value of $T_c = E_\gamma - \frac{1}{2} m_o c^2$. This energy continuum, as it is represented in the pulse-height spectrum, is generally discernable. Its upper energy trace is terminated more or less abruptly at a point equivalent to $T_c = E_\gamma - \frac{1}{2} m_o c^2$. This feature of the pulse-height spectrum (i.e., the upper energy termination of the Compton electron spectrum) is referred to as the 'Compton edge'. The Compton-scattered photon may escape from the detector without further interaction, or it may undergo one or more secondary interactions, depositing an additional increment of energy in the detector with each interaction. These sequential interactions take place very quickly and their effects are summed within the detector so that they appear as a single interaction. Thus, the total amount of energy deposited in a Compton interaction may extend from zero to the full energy of the incident photon, E_γ.

In the case of pair production, the kinetic energy of both the positron and the negatron is quickly absorbed within the detector. In addition, after the positron has yielded its kinetic energy, it is annihilated by combination with a negatron, and two photons are emitted, each having an energy of 0.51 MeV. Either or both of these photons may escape from the detector without further interaction or they may interact, either by the photoelectric process or by the Compton process, as described above. Thus, for pair production in the detector, the amount of energy absorbed by the detector is variable and extends from a value of $(E_\gamma - 2m_0c^2)$ to (E_γ).

In addition to the energy absorbed by the detector directly from the sample, some additional photons will interact with the detector as a result of scattering from the shield or other structures or materials surrounding the detector. Thus, a gamma-ray photon may be emitted by the sample, interact with the shield, undergo a Compton interaction and the resultant Compton-scattered photon may impinge on the detector where it may undergo a secondary photoelectric or Compton interaction. Similarly, such a gamma-ray photon may result in pair production with subsequent positron annihilation within the substance of the shield, and part of the annihilation radiation may impinge on the detector and undergo secondary interactions.

Partial energy peaks may be formed as a result either of (a) the escape of certain degraded gamma-ray photons from the detector, (b) scattering within the shield, or (c) summation within the detector. As a result of the escape of one or both photons of the annihilation radiation accompanying pair production, peaks may be observed at positions corresponding to gamma-ray energies of $(E_\gamma - m_0c^2)$ and $(E_\gamma - 2m_0c^2)$, respectively. As a result of the escape of the X-ray emitted in the secondary process accompanying a photoelectric interaction, a peak may be observed (provided the detector resolution is good enough) at a position corresponding to a gamma-ray energy of $(E_\gamma - E_X)$, where E_X is the energy of the X-ray.

A broad peak may be observed at a position corresponding to a gamma-ray energy of 0.25 MeV as a result of $180°$ Compton scattering angle (backscatter) from the shield. A peak may also be observed at 0.51 MeV as a result of pair production in the mass of the shield, the subsequent escape of one photon of the annihilation radiation from the shield, and the interaction of this photon with the detector.

Summation may take place as the result of rate effects, i.e., interactions occurring within the detector at such a rapid rate that the instrument is unable to recognize a time difference between them (pulse pile-up). Such summation may also occur when there are coincident nuclear transitions,

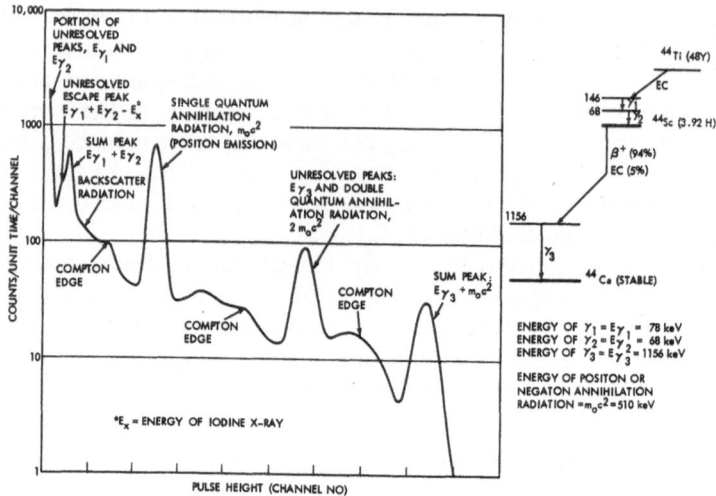

Fig. 2.3. A scintillation gamma-ray spectrum of ⁴⁴Ti, showing the various features that are typical in gamma-ray pulse-height spectra. Note the several features (viz., backscatter peak, escape peaks, sum peaks and annihilation radiation peaks) which do not bear a simple proportional relationship to the gamma-ray energies involved (see decay scheme). These non-proportional features complicate the interpretation of such spectra and may completely obscure the desired spectral information.

i.e., when the lifetime of an intermediate energy level is short compared to the time resolution capability of the detector. In the case of coincident summing, a peak will generally be observed for each of the two transitions involved, and in addition, a 'sum peak' will be observed on the pulse-height scale at a position corresponding to the sum of the energies of the two nuclear transitions.

The various spectral characteristics described in the foregoing are well illustrated in the several spectra contained in the Gamma-Ray Spectrum Catalogue compiled by Heath [2].

2.3   SPECTROMETER DETECTORS

Detectors suitable for gamma-ray spectrometry, as applied to problems in radiochemical analysis, must possess qualities of resolution, efficiency, linearity, uniqueness of response and speed. From the foregoing discussion on gamma-ray interaction processes, it is noted that interaction probability (and hence detection efficiency) is enhanced by choosing a detector of high Z (high density) material. Such a preference is also desirable in respect to uniqueness of response, since the photoelectric interaction, which provides the most unique response of any of the interaction processes, also has the greatest correlation with Z.

Scintillation detectors have generally been used in the past and continue to be most useful for a variety of gamma-ray spectral measurements, but with the rapid progress which has been made in recent years in the development of the semiconductor detector, an increasing interest and acceptance is being accorded this relatively new device. In the respective pulse-height distributions from these detectors, several qualities of interest are significantly different, and the advantages and disadvantages of each type must be considered in a proposed application.

In the following are brief descriptions of these detectors and of the important factors which affect their performance as gamma-ray spectrometry detectors.

2.3.1 The Scintillation Detector

The scintillation detector makes use of the processes of luminescence, photoelectric conversion, and secondary electron emission. The complete detector consists of a scintillator and one or more multiplier phototubes. The scintillator, through the mechanism of luminescence, converts interacting gamma-ray photons into a series of low energy photons whose wavelengths are in the visible region. The phototube converts these low energy photons into photoelectrons which are multiplied by secondary emission. The secondary emission electrons are collected and are observed as an electrical signal.

The Scintillation Process

Many materials demonstrate the property of luminescence and the subject is treated in depth in a number of excellent books and review articles [3–11]. Types of materials which demonstrate this property include certain inorganic and organic crystals, organic liquids and plastics. The mechanisms whereby luminescence takes place in each of these types of material, however, is somewhat different. The following semiquantitative description of the process is sufficiently non-specific so that it might apply, with minor corrections, to any type of luminescent material. Luminescent materials suitable for gamma-ray spectrometry have been mostly in the class of inorganic crystals, however, and the description is most appropriate to this class of materials.

In general, a gamma-ray having energy, E_γ, incident upon any material will dissipate a fraction, f, of its energy as a result of one or more interactions within the material. One or more highly energetic electrons will be produced and the energy which has been imparted to these electrons will, in turn, be dissipated in the ionization, electronic excitation, dissociation, or thermal vibration, rotation, or translation of the molecules of the interacting material. In the case of non-luminescent materials, all of the energy, with the exception of the energy absorbed in molecular dissociation, is ultimately completely transformed into thermal vibrational, rotational, or translational energy of the molecules (phonons), and de-excitation takes place by radiationless transitions. In the case of the scintillator, however, a small part, ϵ_p, of this molecular energy will be converted into photons and re-emitted. Each of the photons thus created will have an average energy, E_p, derived from direct energy transitions from excited electronic levels to the ground state. The number of photons, p, thus emitted is

$$p = \frac{E_\gamma f \epsilon_p}{E_p}$$

Thallium activated sodium iodide NaI(Tl) is frequently used as a phosphor in the scintillation detector. For this phosphor, $E_p \approx 3$ eV, and $\epsilon_p \approx 0.08$, so that for an interaction involving a 1 MeV gamma-ray such that all of the energy is deposited in the phosphor (i.e., f = 1), $p \approx 2.5 \times 10^4$.

The energy transitions which give rise to the luminescent photons do not all take place simultaneously. The transition rate and hence the photon emission rate, varies exponentially with time with a time constant, T, which is characteristic of the scintillant being used. With an initial emission rate, R_o, the emission rate, R, after a time, t, is

$$R = R_o e^{-t/T}$$

and the number of photons emitted in the time interval, t, is

$$p_t = p(1 - e^{-t/T})$$

Photoelectric Conversion and Secondary Electron Emission
The multiplier phototube consists of a photocathode, a series of electrodes called dynodes, and a collection electrode called the anode. An electrostatic field is established between the photocathode and the first dynode and between each succeeding dynode, and finally between the last dynode and the anode. Light, entering through the glass envelope of the tube and falling on the photocathode, ejects photoelectrons which are intercepted at the first dynode. Photoelectrons impinging on the dynode surface produce secondary electrons, which are accelerated towards the next dynode by the electrostatic field. The process is repeated for each successive stage and the electrons emitted from the last dynode are collected on the anode. The number of secondary electrons produced, and hence the overall gain of the tube is determined by the inter-dynode voltages. If the phototube has m dynodes, each with an electron multiplication factor of A, the overall gain of the tube is

$$G = A^m$$

The multiplier phototube is an integral part of the scintillation detector and a portion, a, of the photons originating within the scintillator are transmitted to and absorbed by the photocathode where they are converted, with an efficiency, ϵ_{pe}, into photoelectrons. These photoelectrons traverse the dynode structure and are ultimately collected at the anode. The total charge collected at the anode is

$$Q = pa\epsilon_{pe}\xi G = \frac{E_\gamma f\epsilon_p a\epsilon_{pe}\xi A^m}{E_p} \text{ coulombs}$$

where ξ is the electronic charge.

Thus, the output information from the scintillation detector is a quantity of charge, Q, which is proportional to the energy lost by the interacting gamma-ray within the sensitive volume of the scintillant. The collected charge produces a voltage signal of magnitude $V = Q/C$, where C is the total circuit capacitance at the anode of the phototube.

Thallium Activated Sodium Iodide (NaI(Tl))

Of all luminescent materials, NaI(Tl) possesses a combination of properties which have made it the most popular scintillant for use in gamma-ray spectrometry. Hofstadter [12] first demonstrated the potentialities of NaI(Tl) in this application. Subsequent work by Van Sciver [5, 6], Zerby et al. [7], Meyer et al. [8], and Murray et al. [9] provided an early basis for the understanding of the scintillation phenomenon in NaI(Tl). Engelkemeir [13] reported that the intensity of the light emitted by NaI(Tl) per unit gamma-ray energy expended within the crystal was not a constant. This finding has since been substantiated by Nemilov et al. [14], Managan [15], and Heath [2]. Deviations from constancy are most prominent for gamma-ray energies below 100 keV, but are not of such magnitude as to preclude the use of NaI(Tl) for spectral measurements.

In the intervening years since Hofstadter's early work, a variety of NaI(Tl) detection systems have been assembled in which crystals of widely varying sizes and shapes have been used, with consistently good resolution [2, 10, 16–19]. The popularity which NaI(Tl) has enjoyed through the years has been due to: (a) its ready availability at low cost in single crystals in a wide range of crystal sizes which can be formed into a variety of shapes; (b) good luminescent properties (i.e., high conversion efficiency (ϵ_p), high transparency to its own emissions and conveniently fast fluorescent decay time (T = 250 ns)); (c) fair resolution over a wide range of gamma-ray energies [2]; (d) high density (3.67 g/cm^3); and (e) fair linearity in response as a function of incident gamma-ray energy [13].

Calculated efficiencies have been given for NaI(Tl) detectors for various crystal sizes, different source configurations and different source/detector geometrical arrangements [18, 20–25]. These calculated efficiencies are useful in the quantitative analysis of gamma-ray spectral data, and in the estimation of detection limits for proposed measurements.

Multiplier phototubes are designed and manufactured for various applications. They have been built having as many as 16 dynodes, and gains as high as 10^6. Photocathodes have been built as large as 24 in. in diameter, and are available in standard sizes of 3/4, 2, 3, and 5 in. in diameter. Phototubes generally used for gamma-ray spectrometry have 10 or 11 dynode stages and gains in the order of 10^5 or 10^6. Variations in performance from tube to tube have frequently been significant in the past. Although production control tests have served to reduce such variations, it is still necessary in certain applications to select tubes on the basis of individual testing. Examples of production control testing techniques which have been used are described by Engstrom *et al.* [26]. More recently established standards for testing have been defined [27].

Various performance characteristics are of interest in selecting a phototube for a particular application. Photocathodes can be constructed to provide a variety of spectral responses [28], and selection should be such that the phototube is most responsive to the emissions from the scintillant. The photocathode spectral response is included in the catalogue specifications provided by the various manufacturers.

Photocathode sensitivity is related to conversion efficiency (ϵ_{pe}) and is usually specified in microamperes per lumen. A photocathode sensitivity of 1 microampere per lumen corresponds approximately to a conversion efficiency of 0.0025 electron per photon.

Noise, or dark current can limit the sensitivity of the phototube. Noise is related to the operating inter-dynode voltages and to the temperature at which the tube is operated. It comes primarily from two sources: (1) ohmic noise, due to electrical leakage caused by traces of conductive materials on the internal insulators or in the base, and (2) thermionic noise due to thermionic emission from the photocathode or from the several dynodes. Ohmic noise varies directly as the applied voltage and can be minimized by care in the manufacturing processes, special treatment of the base or total removal thereof, and by operating the tube at as low a voltage as possible consistent with required gain. Thermionic noise can be reduced by several orders of magnitude by cooling the phototube. The gain of the phototube remains relatively unaffected by the reduced temperature, so that a major improvement in sensitivity can be achieved by this technique. More detailed discussions of noise in phototubes are given by Sharpe [29] and Eberhardt [30].

There is a transit time involved for the electrons to be accelerated from the photocathode, to traverse the dynode structure, and ultimately to be collected at the anode. Some control of transit time can be achieved by careful design of the phototube and by operating the tube with high inter-electrode voltages (i.e., high gain). Variation in the transit time (i.e., 'transit time spread') will determine the lower limit of time resolution where the phototube is to be used for making precise time measurements (e.g., coincidence measurements). The overall r.m.s. deviation in transit time of the tube is usually specified in the manufacturer's catalogue information. Factors which have an effect on transit time spread are discussed and methods for its measurement are described by Roth [31] and by Kerns [32]. Values as low as a few nanoseconds can be obtained with corresponding time resolutions less than 10^{-9} sec [33].

In the past, some phototubes have shown gain changes which have correlated with sample counting rate [34, 35]. Although currently available tubes show less of this effect, high counting rates should be avoided where possible, particularly at high phototube gains, not only to minimize such gain changes but also to minimize pulse summing effects (pile-up). In general, for routine radiochemical applications, in which NaI(Tl) is used as the scintillant, rates as high as 15,000—20,000 counts per second can be tolerated provided phototube gains are kept relatively low. With increasing counting rates beyond this level, rate-induced distortions in the pulse-height spectrum become increasingly noticeable, and where phototube gain changes are involved, it may take hours for the tube to return to its normal operating gain after such an exposure.

For routine measurement, it is essential to stabilize the voltage supply and the phototube temperature in order to achieve a stable phototube gain. The value of G varies approximately as V^n, where V is the total voltage applied to the dynode structure, and n, for a typical phototube, has a value in the order of 6 or 7. Variations of G in the order of 0.5—1.0% per degree Celsius at normal room temperatures (21°C) are typical. Studies of temperature effects on phototube gain have been made by Webb *et al.* [36], Ball *et al.* [37], and Kinard [38]. A study of temperature effects on the composite scintillant and phototube was made by Rohde [39] who observed that in some circumstances it was possible to specify an optimum ambient temperature such that the scintillant/phototube showed a minimum sensitivity to changes in the ambient temperature.

Some phototubes show a strong sensitivity to magnetic fields. The effect is most noticeable in tubes which have a large space between the photocathode and the first dynode, but it is noticeable to some extent on all tubes. The effect is observed as a substantial loss of gain, depending upon the magnitude of the field and the orientation of the tube within the field. The susceptibility to magnetic fields can usually be reduced to insignificant proportions by the use of magnetic shielding (e.g., mu-metal shields) around the tubes [10, 26].

Integral assemblies of NaI(Tl) scintillants and phototubes are commercially available in a variety of scintillant sizes, with phototubes selected to satisfy particular usage requirements. Phototube selection can be made on the basis of resolution, gain, noise, background, speed, stability, photocathode size and spectral response, or matched performance between two or more tubes. Where useable, these integral units, which are assembled under well controlled conditions, eliminate the need for selection, testing and assembly of crystals and phototubes in the laboratory.

Resolution

Resolution, as the term is applied to the gamma-ray pulse-height spectrum, is a measure of the minimum energy difference that must exist between two gamma-rays for their corresponding full-energy peaks to be distinguishable. Peak-width (or line-width) is related to resolution capability but is not, in itself, a direct measure of resolution. Nonetheless, usage has made the terms 'resolution' and 'line-width' interchangeable. For the scintillation detector, the convention has been to define resolution as the full width of the full-energy peak at half the maximum value of the peak (FWHM), expressed as a percentage of the peak location on the pulse-height scale. In order to compare detectors, the 0.662 MeV gamma-ray of ^{137}Cs has been adopted as a standard, and resolution figures are usually quoted with reference to this gamma-ray. Figures ranging from 7–9% are typical for NaI(Tl) assemblies.

The pulses observed at the phototube anode are the product of a series of processes which take place within the scintillant and the phototube. These include (a) the generation of photons in the detector, (b) the collection of photons on the photocathode, (c) the photoelectric conversion at the photocathode, (d) the collection of photoelectrons on the first dynode, and (e) electron multiplication on each of the m dynodes. The statistical variances associated with each of these processes make a large contribution to the observed line-width in the pulse-height spectrum. Additional factors which also contribute to line-width, but for which a statistical variance is not as readily definable, include (f) varying transfer efficiency, (g) voltage instabilities, (h) inhomogeneous luminescence yield throughout the crystal, (i) interaction, edge, and scattering effects, and (j) non-proportional scintillation response.

Breitenberger [40] gives an excellent discussion of and statistical analysis for each of these factors in respect to their contribution to line broadening. This and subsequent work concerned with line broadening is reviewed in an article by Managan [15]. In more recent work, Takhar [41] measured the sensitivity profiles for the photocathodes of various phototubes and observed significant nonuniformities. These nonuniformities were found to have an adverse effect on resolution. Prescott et al. [42] observed that a large contribution to line broadening could result from variations in the light collection efficiency from different regions of the scintillator.

Recent developments in phototube technology are described in papers by Simon et al. [43], Morton et al. [44] and Krall et al. [45]. Excellent discussions on scintillation counting methodology are presented by Birks [4], Murray [10] and Neiler et al. [18].

2.3.2 The Semiconductor Detector

Compared to the scintillation detector, the semiconductor detector has qualities which are particularly attractive for application in gamma-ray spectrometry. Such qualities include superior energy resolution, superior linearity, and greater tolerance for magnetic fields. It also has qualities which detract from its suitability for this application. These deterrent qualities include small sensitive volume, low atomic number, low detection efficiency, small output signal, and a requirement for operating at reduced temperatures (liquid nitrogen).

As in the scintillation detector, the interaction of gamma-ray photons results in the ionization, excitation and possible dissociation of the molecules of the detector material. However, the complex mechanisms of the scintillation detector (viz., multiple conversion, transmission, electron multiplication and charge collection) are reduced to a simple and direct process of charge collection. Incident gamma-ray photons create pairs of charge carriers which are caused to drift in an electrical field established between electrodes attached to the detector. The charge carriers are eventually collected on the field-forming electrodes where they are observed as electrical signals.

The semiconductor detector functions as a solid-state ionization chamber and is analogous to a gas ionization chamber. Compared to the gas ionization chamber, the solid-state ionization chamber has characteristics which make possible an improved detection efficiency for gamma-rays and a superior resolution capability. These characteristics include higher density, larger numbers of charge carriers, greater carrier mobility, faster charge collection, and smaller physical dimensions.

The potential advantages of a solid-state ionization chamber were early recognized. Such a detector, in concept, would consist of a rectangular or cylindrical piece of homogeneous material placed between two electrodes. Dearnaley *et al.* [46] list several material qualities which would be important in the design of such a detector. These qualities include: (1) low carrier density in order to minimize current noise; (2) freedom from carrier traps in order to minimize loss of pulse rise time and possible loss of signal; (3) high carrier mobilities in order to achieve short pulse rise times; (4) a low value for the mean energy required to create an ion pair in order to enhance energy resolution; (5) high atomic number for good stopping power; and (6) long carrier lifetimes so that efficient charge collection could be easily achieved. Some of these requirements, it may be noted, are in conflict with each other.

Dearnaley *et al.* [46] also list several candidate materials which would be suitable for the construction of a solid state ionization chamber, and the important relevant properties of these materials. From such a list, various materials have been selected and attempts have been made using them in building homogeneous solid state detectors. Imperfections, however, either in the form of chemical impurities or of crystalline lattice structural defects, caused large variations in charge collection efficiency and the detectors were unusable. Silicon and germanium emerged as particularly appealing candidates: both materials were readily available, highly pure, and with a high degree of crystalline perfection; they were well understood materials whose technology was highly developed, and they were known to have long carrier lifetimes and large carrier mobilities. When an electric potential was applied across detectors made of these semiconducting materials, however, leakage currents were excessive and these detectors were also unusable.

Although the quest for a simple, homogeneous, solid-state ionization chamber met with extremely limited success, it was found that a reverse-biased solid-state diode structure did permit a high electric potential to be applied, that under these conditions noise and leakage currents were acceptably low, and that a sufficiently large sensitive volume could be defined so that detection of ionizing radiation could take place.

Thus, theoretical and practical concepts which are important in the design and manufacture of the semiconductor diode have become equally important in the design and manufacture of the semiconductor detector, and some understanding of these concepts is desirable for effective utilization of such a detector. The following is a brief description of the principles of operation of the semiconductor detector and a discussion of some of the factors which affect its suitability for gamma-ray spectrometry. Excellent review articles describing the principles of operation and the early work in the development and application of the semiconductor detector are presented by Bromley [47], Miller et al. [48], Ewan et al. [49], and Dearnaley et al. [50]. In addition, a book by Dearnaley and Northrop [46] provides an excellent pre—1966 summary of semiconductor detector technology and describes the application of these detectors to measurement problems in nuclear physics.

Principle of Operation

The atoms of a perfect crystal are bound into an orderly array referred to as the crystal lattice. In an isolated atom (i.e., one not bound into such a lattice structure), the atomic electrons occupy discrete and easily identifiable energy levels. The crystal lattice structure, however, is itself a quantized system and thus is bound by the rules governing such systems. In accord with the Pauli exclusion principle which forbids any two electrons in a quantized system to occupy identical energy levels, slight changes take place in the energy levels of the several atomic electrons in the lattice structure. The result is that instead of sharply defined energy levels, the atomic electrons in the lattice structure occupy bands of permissible energy levels. The highest band of permissible energies is that occupied by the outer-shell (valence) electrons of the bound atoms, and in this valence band there exists an energy level for each of the valence electrons in the lattice structure.

If the excitation level of a valence electron in a crystal lattice is sufficiently high, it will move up to the next higher band of allowed energy levels. This band is known as the conduction band, and electrons in this band are no longer bound into the lattice structure. These unbound electrons are free to move, and will do so in a direction parallel to the gradient of an applied electric field, or in random directions in the absence of such a field. The removal of an electron from the valence band would leave a vacancy (or 'hole') in the band. Holes are also free to move and will do so in a manner identical to that of unbound electrons, except that the direction of motion of a hole in an electric field will be opposite to that of an unbound electron. The energy differential between the highest energy level in the valence band and the lowest energy level in the conduction band is known as the band-gap energy. Its value is characteristic of the atoms (or molecules) comprising a particular crystalline material.

It may be possible, as a result of impurities or imperfections in the crystal lattice, for electrons to occupy energy levels within the band-gap. Such impurities or imperfections are usually introduced, either accidentally or intentionally, as the crystal is grown. If intentionally introduced, the process is known as doping. If, for example, pentavalent elements of Group Va of the periodic table (e.g., phosphorus, arsenic, or antimony) are introduced during the growth of silicon or germanium crystals, some atoms of these elements will substitute for some of the silicon or germanium atoms and will be bound into the lattice structure. Such atoms possess one more valence electron than the four required for binding into the lattice structure of either silicon or germanium. The valence electrons of these extraneous atoms occupy an energy band below the normal conduction band but well above the normal valence band, and a relatively low activation energy is sufficient to cause one of them to be excited to (or 'donated' to) the conduction band. Thus, these impurity atoms are termed 'donors'.

If trivalent elements of Group IIIa of the periodic table (e.g., boron or gallium) were used as dopants during the growth of silicon or germanium crystals, these atoms would also be bound into the crystal lattice. Their valence electrons would occupy an energy band above the normal valence band of the crystal, but well below the conduction band. In addition, such atoms would possess one valence electron less than the four required for binding into the lattice structure. In this case, a relatively low activation energy is sufficient to cause one of the regular valence electrons (i.e., an electron from an atom of silicon or germanium) to be excited to the energy band of the extraneous atom where it would be 'accepted' and would satisfy the local need for an extra electron in the crystal lattice. Thus, these impurity atoms are termed 'acceptors'.

Semiconducting material is termed p-type material if it contains acceptor impurities and n-type material if it contains donor impurities. A p-n junction can be formed within a crystal. Such a junction occurs at the boundary between regions of p-type and n-type impurities. At the boundary interface, the respective impurity atoms diffuse across the junction and the acceptors in the n-type region and the donors in the p-type form a charged double layer. The resultant electric field is such that free charges are repulsed and prevented from crossing the boundary. If a voltage is applied such that the n-type material is biased positively with respect to the p-type material (reverse biased) the resultant electric field enhances the existing field at the junction and any free charges which may be produced in the junction interface are swept clear and collected on the field-forming electrodes.

The charge-free region in the junction interface is referred to as the charge depletion layer, and the thickness of this layer determines the suitability of the junction for use as a detector of ionizing radiation. Dearnaley *et al.* [46] show derivations of equations for estimating the approximate thickness of the depletion layer in both n- and p-type materials. It becomes apparent from an examination of these equations that layers greater than 1 or 2 millimeters in thickness would be difficult to achieve. Such thicknesses would be suitable for the detection of alpha- and beta-particles, but very poorly suited to the detection of gamma-rays.

For gamma-ray detection, the p-i-n device has been developed. Instead of a p-n interface, the p- and n-type materials are separated by an electrically neutral region called an 'intrinsic' or 'compensated' region through which charge carriers can move freely. The device is made by diffusing lithium into a piece of p-type material. Lithium acts as a donor impurity in silicon and germanium. The face of the semiconductor material where the lithium ion enters becomes n-type material because of the high concentration of donor impurities, and a p-n junction is formed where the donors and acceptors are present in equal numbers. Lithium is a very light ion with a high diffusion coefficient, and if a bias is applied to the semiconductor material and the material is heated, the lithium ions will continue to drift through the material. As the lithium drifts past the junction, it will tend to pair with the p-type impurities and an electrically neutral material is formed. In germanium, the lithium diffuses very rapidly unless the temperature is reduced, so that as soon as the p-i-n structure has been formed, it is necessary, in order to stabilize the device, to cool it to, and maintain it at, the temperature of liquid nitrogen.

The depth of the intrinsic material can be controlled in manufacture, and sensitive volumes of $30-50$ cm^3 can be readily achieved. Because of the higher Z, lithium-drifted germanium (Ge(Li)) is more suitable as a gamma-ray detector than is lithium-drifted silicon (Si(Li)), although the latter is useful as a detector in X-ray spectrometry. A comparison of Ge(Li) and Si(Li) as gamma-ray detectors is made by Camp [51], and a demonstration of the excellent resolution attainable with Si(Li) detectors for X-rays is given by Bowman et al. [52]. The analytical sensitivity which can be achieved for certain elements with an X-ray fluorescence spectrometer in which a Si(Li) detector is used is demonstrated by Yamamoto [53].

Resolution
The superior resolution of the semiconductor detector is its most outstanding feature. The convention for this detector, in contrast to that of the scintillation detector, has been to define resolution as the FWHM of the full-energy peak expressed as an equivalent amount of gamma-ray energy. The 1.33 MeV gamma-ray of ^{60}Co has frequently been used as a standard for comparison purposes, and resolution figures are usually quoted with reference to this gamma-ray. Figures ranging from 2 to 4 keV are typical for currently available Ge(Li) detectors.

The gamma-ray energy resolution attainable with a semiconductor detector is affected by three principal factors: (1) the statistical variance associated with the number of electron-hole pairs generated by a gamma-ray interaction in the detector; (2) electrical noise originating either in the detector or in the associated amplifier; and (3) carrier trapping and recombination effects which can result in various types of pulse distortion.

The fractional variance in the amount of charge collected in the semiconductor detector is directly related to the number of electron-hole pairs generated by a gamma-ray interaction in the detector. The number of such carriers depends directly upon the amount of energy deposited within the sensitive volume of the detector and inversely upon the amount of energy required to create an electron-hole pair. The energy required to create an electron-hole pair is related to the band-gap energy and for silicon or germanium detectors, approximately 2–4 eV are required per electron released. This value is to be compared to a value of approximately 300 eV which must be deposited in NaI(Tl) for each photoelectron released from the photocathode of the multiplier tube. Compared to the scintillation detector, the greater number of charges generated by the semiconductor detector results in a substantial improvement in pulse-height (energy) resolution. In addition to this fundamental contribution to an improved resolution, the semiconductor detector, with its simple and direct collection of charge, generates a signal which is completely free of those other sources of variance which are introduced directly into the scintillation signal as a result of multiple conversion and collection processes.

Electrical noise in a system may be defined as any variation in the electrical output which originates independently of, and which tends to obscure, the desired signal. Basically, noise in the semiconductor arises because the detector is not a perfect dielectric. Electrons and holes, in addition to those generated by incident radiation, exist within the volume of the detector. Such extraneous carriers appear, in part, as a result of simple thermal excitation which may raise the excitation level of valence electrons sufficiently so that they cross the band-gap and enter the conduction band. The velocity distribution of these carriers is such that a fluctuating and non-uniform distribution of charge exists within the volume of the detector which results in a fluctuating voltage across the detector. This source of noise can be reduced by cooling. In the case of germanium, thermal generation of charge carriers is particularly likely because germanium has a band-gap energy of only 0.66 eV. Thus, cooling of the germanium detector is essential, not only for reduction of the lithium mobility as described earlier, but also for the reduction of thermal noise.

Noise is also generated when an electric field is applied to the semiconductor detector. This noise is identified as 'current noise', and is itself made up of three separate types of noise, as follows: (1) shot noise; (2) generation-recombination noise; and (3) excess noise. Each of these types is the result of charge displacement within the detector. Shot noise is the result of discrete movements of electrons and holes which may be introduced and withdrawn at electrodes or generated thermally within the detector. It is proportional to

$$\left(\frac{\tau}{\tau_c}\right)^{1/2}$$

where τ is the circuit integration time constant and τ_c is the transit time for the carrier to traverse the detector.

Generation-recombination noise results from the generation of pairs of carriers within the detector, the partial traversal of the detector by these carriers under the influence of the applied field, and the recombination with other carriers before the traversal is complete. The resultant current pulses from such partial traversals are smaller than those due to electrons and holes which make a complete traversal of the detector.

Excess noise is that part of current noise not attributable to either shot noise or generation-recombination noise, and although its contribution to current noise is frequently the dominant one, very little is known of its physical origin. It is predominantly a low-frequency effect and therefore can be selectively reduced by reducing the low-frequency response of the associated amplifiers. Variations in certain manufacturing processes have been observed to have a significant (albeit somewhat unpredictable) effect on the amount of excess noise present in a detector.

Trapping and recombination of charge carriers may take place at any of a number of localized centers which exist throughout the crystal lattice. Charge carriers have a finite life-time within the semiconductor and recombination while trapped in such centers is the principal mechanism whereby carriers are neutralized. These localized centers are the result of impurities or lattice imperfections which, as described earlier, introduce localized energy levels, usually within the band-gap. Either a hole or an electron may become localized (trapped) at such a center for a period, after which it may be returned to the appropriate band where it again will contribute to the conductivity. If such a center captures first an electron and then a hole (or vice versa), recombination will take place, and neither the electron nor the hole will reappear to contribute to the conductivity.

From the foregoing, it is seen that trapping produces a variation in the time of collection of charge carriers, whereas recombination results in a loss of charge. Either effect can be detrimental to pulse-height resolution. Variation in the time of charge collection results in a signal pulse with a corresponding variation in rise time. In this circumstance, if either the circuit integration or differentiation time-constants are too short, a variation in pulse-height will occur which will correspond to the variation in rise time. Loss of charge, on the other hand, will have a direct effect on pulse-height, and hence on resolution, since pulse-height is directly proportional to the amount of charge collected.

Various other factors will have an effect on semiconductor detector resolution. Reduced temperatures will generally increase carrier mobility and will reduce the generation rate of carriers. A high bias voltage will also increase carrier mobility, but will also increase current noise. The presence of magnetic fields increases the probability of recombination and thus can effect resolution. Light of quantum energy greater than the band gap energy is strongly absorbed in semiconductors and results in the generation of electron-hole pairs.

In order to optimize resolution, it is necessary to minimize noise and trapping and recombination effects, but the techniques for accomplishing such optimization are not always mutually compatible. Noise can be reduced by cooling the detector, operating the detector with a low bias voltage, using short integration and differentiation time-constants, and operating the detector in near darkness. Trapping and recombination effects can be minimized by techniques which will tend to decrease charge collection time, e.g., cooling the detector and using a high bias voltage. Trapping effects can also be minimized by using circuit time-constants which are long with respect to trapping times. The presence of magnetic fields is usually not a serious problem with semiconductor detectors. According to criteria established by Shockley [54], germanium should show very little sensitivity to fields up to a few hundred oersteds, and silicon to fields up to several thousand oersteds.

Efficiency

A major deficiency of the semiconductor detector, particularly in regard to its application to analytical problems in radiochemistry, has been the small sensitive volumes that have been possible. Various techniques have been employed to overcome this inadequacy. The two most successful techniques have been to drift coaxially and to assemble arrays of small volume detectors to form a single large-volume detector.

The so-called planar detector is made from a right cylindrical crystal of germanium or silicon, and lithium is drifted from one face so that one face is n-type material and the other is p-type. It is difficult to obtain quality crystals or to make quality detectors with diameters greater than 4.5–5.0 cm, and drift depths greater than 1 cm, so that sensitive volumes greater than $15-18$ cm^3 are rare for the planar detector. Phelps [55] describes a planar Ge(Li) detector which has an 18 cm^2 area and a 0.95 cm drift depth. A resolution of 2.64 keV was obtained with this detector and it has been used routinely to count biological samples as large as 8.5 cm in diameter. Dearnaley *et al.* [56] review recent progress which has been made in the development of large diameter germanium crystals. They describe planar diodes which have an area of 40 cm^2 and a depletion thickness of 0.8 cm. A resolution of 3.7 keV has been attained with such detectors.

The coaxially drifted detector is made either from an ingot of germanium or a right cylindrical crystal. In either configuration, lithium is drifted from the outer surface toward the center. Thus, the outer surface becomes n-type material and a small central core remains p-type. With this technique, sensitive volumes as large as $45-50$ cm^3 have been achieved. Malm *et al.* [57] describe the early techniques of coaxial drifting and present results which were obtained with these detectors. They describe results obtained with more recently drifted coaxial detectors [58]. In their more recent work, a 44 cm^3 detector was constructed with a resolution of 2.8 keV.

Saunders *et al.* [59] describe a detector made by using four planar Ge(Li) detectors in parallel to achieve a total active volume of 28 cm^3 with a resolution of 2.90 keV. Other arrays are described by Lalovic [60], in which coaxial detectors were used. A dual element array was constructed with an active volume of 40 cm^3 and a four element array with an active volume of 180 cm^3. A resolution for the four element array of 15.5 keV was observed. The poor resolution was believed to be due to poor matching of the individual detector elements (i.e., they were made of different samples of crystals, and the depletion depths were not uniform). For the 1.33 MeV full energy peak, the four element array showed an efficiency which was approximately 25% of that shown by a 3 in. $\times$ 3 in. NaI(Tl) scintillation detector.

Curves of measured efficiencies of Ge(Li) detectors of various sizes and types for gamma-ray energies from 50 keV to 10 MeV are presented by Cline [61] and Ewan *et al.* [62]. Calculated values of the intrinsic efficiency (i.e., the probability that an incident photon will interact and deposit at least part of its energy in the detector), photoelectric efficiency, photofraction, and single and double escape probabilities are presented for various detector sizes and shapes for various energies by Wainio *et al.* [63], and by de Castro Faria *et al.* [64]. Heath [65] and Walters [66] discuss difficulties in establishing a consistent set of definitions that would allow meaningful cross-comparisons of various detectors.

Timing with the Semiconductor Detector
Fast coincidence measurements are more difficult with the semiconductor detector than with many scintillation detectors. This is so because the shape of the semiconductor pulse is not constant. Variations in pulse rise time occur as a result of variations in charge collection time which, in turn, are the result of influences which have previously been discussed herein. Graham *et al.* [67] discuss measurement techniques and give some quantitative information on timing difficulties which can be experienced with large coaxial Ge(Li) detectors. Quaranta *et al.* [68] review the problem involved in obtaining a precise timing signal from semiconductor detector signals, and Miehe *et al.* [69] report that half-lives of delayed transitions of 5 ns and shorter can be measured with these detectors.

2.4   APPLICATION TECHNIQUES

Gamma-ray spectral measurements can be made without the imposition of any special conditions other than the specification of an energy threshold and an energy range. Various other restraints can be imposed, however, to improve either the selectivity or the sensitivity of the method. The imposition of such special restraints is accomplished through the use of auxiliary detector elements and discriminators which, together, generate 'gating' pulses for the spectrometer. The spectrometer is placed under the control of these gating pulses so that only those gamma-rays are measured which have been emitted in accordance with the specially selected conditions. The choice of technique (i.e., gated or ungated) and the types of restraints which are imposed, will depend on the nature of the sample to be measured and the expected complexity of the spectrum.

2.4.1   Ungated Spectrometry
This is accomplished by using an instrumental assembly essentially as shown in Fig. 2.1. A sample is placed adjacent to the detector and all detectable gamma-rays within the specified energy limits are accepted for analysis. The only controls of selectivity and sensitivity are in the choice of detector and in the optimization of efficiency and resolution over the specified energy range.

Ungated spectra are simple to obtain and easy to interpret if the spectra are not too complex (i.e., do not have too many components). If, however,

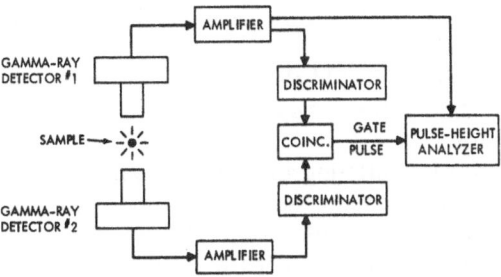

Fig. 2.4. The basic elements of a simple coincidence-gated gamma-ray pulse-height spectrometer.

the spectrum contains many different peaks, it may be difficult to interpret, especially if the peaks are poorly resolved, or if a nuclide of interest is a minor constituent with its peaks possibly obscured by other spectral features.

2.4.2 Gated Spectrometry
In the following are brief descriptions of some gating techniques which have been used. These are presented only as examples; restraints other than those indicated can be imposed, as well as combinations of such restraints.

Coincidence Gating
The gating technique which is probably most frequently used is that of 'coincidence gating'. Here a gate pulse is generated only when gamma-rays are emitted in cascade. The instrumental arrangement of a typical coincidence-gated spectrometer is shown in Fig. 2.4. Since two (or more) gamma-rays must be observed to satisfy the condition of coincidence, and since these gamma-rays generally will occur within a time which is short compared to the detector decay time constant, two (or more) detectors are required for such measurements. In addition to the restraint of cascade gamma-rays, discriminators can be used so that gamma-ray energy range restraints are also imposed as a condition for generation of a gate pulse.

The advantages of coincidence gating are the improvement in selectivity and in sensitivity that can be achieved for those nuclides which emit gamma-rays in cascade. The restraints imposed also serve to reduce the normal 'cosmic' and noise background. In addition to the residual cosmic and noise background, a unique 'coincidence' background will be observed which results from 'accidental' coincidences, escape radiation, and scattering. The coincidence background is generated primarily by the sample itself, and is considered to be a background only in the sense that it is an undesired response which detracts from the ability to observe the desired response.

The observed accidental coincidence rate, N_a, is related to sample counting rates as follows:

$$N_a = 2N_1 N_2 \tau$$

where N_1 and N_2 are the average rates at which pulses are presented to the coincidence detection circuit from the respective detectors, and 2τ is the coincidence resolving time.

Unwanted coincidences could also arise as a result of the escape of X-rays or annihilation radiation from one detector and subsequent interaction with the second detector.

Coincidences due to scattering arise when a gamma-ray interacts with one detector but part of its energy is scattered out of this first detector and interacts with the second detector.

Each of these contributions to the coincidence background are dependent upon sample counting rate. Accidental coincidences can be reduced by reducing the coincidence resolving time, 2τ, and where possible, by keeping sample activities as low as possible, consistent with counting times of convenient length. Collimators of various types have been employed to reduce unwanted coincidences from escape radiation and from scattering. The design of the collimator must be such that an optimum sample-detector geometry is maintained, however, and only the escape radiation and scattering geometries are reduced. This is important since coincidence counting efficiency is related to the product of the counting efficiencies of the two detectors, and therefore, if maximum sensitivity is desired, it is essential that the counting efficiency of each detector be as high as possible.

An extension of the concept of the coincidence-gated spectrometer is the 'two-parameter' spectrometer shown in Fig. 2.5. With this instrument, correlations between the several values of any two parameters can be measured. In gamma-ray coincidence spectrometry, correlations between the corresponding pulse-height values from each of two detectors is desired. Whereas the simple coincidence gated spectrometer required many measurements and much laborious analysis to make a determination of all of the energy relationships, their relative abundances, and ultimately the nuclides involved, the two-parameter spectrometer makes possible such determinations in a single measurement. In addition, an apparent three-dimensional display is provided (pulse-height (X), pulse-height (Y), and counts per unit time per XY increment (Z)) which greatly simplifies interpretation of the data.

Euler [70] describes a versatile pulse-height analyzer capable of two-parameter measurement, and shows examples of two-parameter spectra, such as might be encountered in analytical measurements. The work of

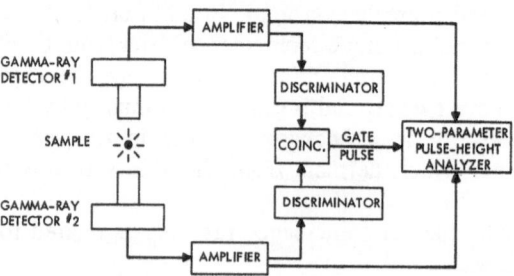

Fig. 2.5. The basic elements of a gated two-parameter gamma-ray pulse-height spectrometer.

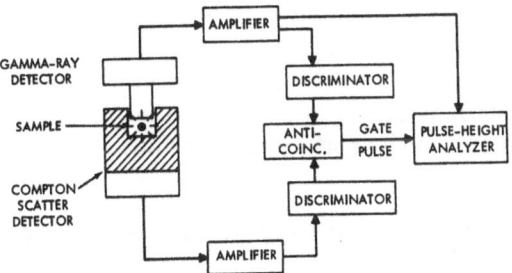

Fig. 2.6. The basic elements of a Compton-suppressed gamma-ray pulse-height spectrometer.

Delucchi *et al.* [71] demonstrates the excellent utility of the two-parameter analyzer in making observations on particular isotopes of a given nuclide.

Anti-Compton Gating

By adding suitable detector elements, it is possible to reduce the contribution of the Compton-scattered gamma-rays to the pulse-height spectrum. The usual technique is to surround a central detector, either with NaI(Tl) or plastic scintillant, and place the outputs of the two detector elements in anti-coincidence (i.e., a gate pulse is obtained only when there is an output from one of the two detector elements). In Fig. 2.6 is shown a detector schematic and block diagram of the instrumental arrangement for such a system. Here a gate pulse is generated only when a pulse is obtained from the central detector; if such a pulse is accompanied by a pulse from the outer detector, a gate pulse is not generated. In addition to a large reduction of the Compton contribution to the spectrum, this system also provides a large reduction in background.

A number of anti-Compton spectrometers have been assembled. NaI(Tl) systems are described, along with data indicative of performance, by Wogman *et al.* [17] and Euler *et al.* [19]. Ge(Li) systems are described by Camp [72], Sayres *et al.* [73], Palms *et al.* [74], and Cooper *et al.* [75]. A disadvantage of this technique for Compton suppression lies in the fact that for nuclides which decay by coincident gamma-ray emission, the sensitivity may be substantially reduced, since one gamma-ray may interact in the central detector and the other may interact in the anti-Compton detector, with the result that neither gamma-ray will be recorded. This effect would be particularly serious in the detector arrangement shown in Fig. 2.6, but would not be a troublesome factor in the two-crystal summing anti-Compton detectors described by Wogman *et al.* [17] and Euler *et al.* [19].

Beta Gating

By placing the sample between a beta-detector and the gamma-detector, and generating a gate pulse only when pulses are observed in both detectors, it is possible to achieve a substantial reduction in background. This system is shown in Fig. 2.7. The beta detector can be in the form of a plastic disk cut from a thin sheet of plastic scintillant. The scintillant disk can be mounted directly on a small phototube and if the sample to be measured is small, it can be mounted directly on the scintillant. The effectiveness of such a detection system is described by Euler *et al.* [19] and by Brauer and Connally [76]. Brauer and Connally also describe the use of the gating pulse so derived to route information to an appropriate pulse-height analyzer, so that several low-level activity samples can be counted simultaneously on the same gamma-ray detector.

2.5   ELECTRONIC INSTRUMENT REQUIREMENTS

The electronic instruments required for gamma-ray spectrometry include the following: (1) high voltage supplies; (2) pulse pre-amplifiers; (3) pulse amplifiers; (4) pulse-height analyzers; (5) stabilizers; and (6) output devices (e.g., printer, plotter, paper-tape punch, card punch, magnetic-tape recorder). If gating is required, fast discriminators and time-determining circuits are required. Auxiliary instruments, such as precision pulse generators and fast oscilloscopes are also required as an aid in the set-up of the spectrometer, and to monitor performance during actual measurements. All of this apparatus is readily available commercially, most of it in convenient modular form.

Crouch *et al.* [77] describe several measurement and testing procedures which are suitable for evaluating the performance of the gamma-ray spectrometer. Heath *et al.* [78] define performance requirements for high-precision spectrometry and describe the results of extensive tests that have been made on various components. These studies are of interest in that they define testing procedures and indicate the types of test equipment required. They also clearly show the types of instrumental insufficiencies that may be encountered.

Fairstein *et al.* [79—83] have described the general problems of amplification for nuclear pulse measurements. Although the concepts defined are of general applicability, the advent of the high-resolution detector (semiconductor) posed additional problems. Fairstein [84] describes the special design requirements for amplifiers to be used with high-resolution detectors. Four broad problem areas are identified: amplifier noise, detector

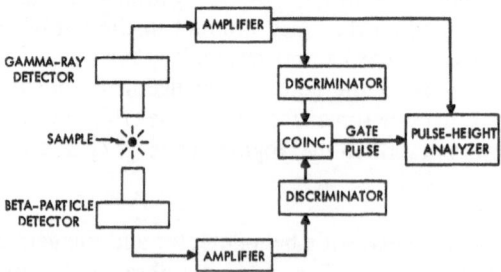

Fig. 2.7.   The basic elements of a 'beta-gated' gamma-ray pulse-height spectrometer. Beta-gating is a particularly effective method of background reduction, and thus makes possible the measurement of samples which have very low levels of activity.

rise time variations, count-rate capability, and gain stability. Further discussion of the special problems encountered in high-resolution spectrometry and a description of specific circuits to relieve these problems is presented by Williams [85]. Goulding *et al.* [86] describe the design features of an amplifier system specifically designed for high-resolution and high counting-rate spectrometry and show excellent performance data for the system.

For scintillation spectrometry, a pulse-height analyzer with 400–1000 channels is generally adequate, whereas the excellent resolution of the semiconductor detector cannot be fully realized over a very wide range of energies unless 4000–8000 channels are available. Stabilization of the coefficient of energy equivalence per channel (gain), and of the zero channel energy intercept (baseline) of the spectrometer may be desirable if high precision measurements are to be made on a routine basis, or if counting times are expected to be long (i.e., samples to be counted have low activity levels). Stabilization is particularly desirable in high-resolution spectrometers. A review of methods of pulse-height spectrometer stabilizer systems and a comparison of their relative merits is presented by Konrad [87]. These systems all work on the principle of negative feedback, and their success depends upon the ability to include all of the potentially unstable elements of the system within the feedback loop. Care must be exercised in the use of stabilization systems and in the selection of reference pulse sources so that the stabilization system does not itself become a source of noise and thereby contribute to the degradation of resolution.

For two-parameter spectrometry, a special pulse-height analyzer is required. Since pulses are digitized in both the X and the Y dimensions, the total number of storage channels required is $(N_X \cdot N_Y)$, where N_X is the number of channels in the X dimension and N_Y is the number of channels in the Y dimension. Thus, for an instrument with 128 channels in X and 128 channels in Y, it would be necessary to provide a storage capacity of 16,384 channels.

Various alternatives to large storage for two-parameter analyzers have been examined in attempting to circumvent the high cost. Most of these alternatives have been more appropriate to the requirements of nuclear physics research than to the requirements of radiochemical analysis. One technique has been referred to as delayed-time-totalizing. Here, a small 'buffer' memory is used which stores the address information of pairs of pulses from the two analog-to-digital converters (ADC's). The buffer memory can store a limited number of such address pairs and when it is full, it 'dumps' this digital information onto magnetic tape and then continues to store additional address pairs. The disadvantage of this method is that the pulse-height information is not placed on the tape as a frequency distribution (i.e., the data are uncompiled), and as a result, display and analysis of the data are more difficult.

Normally, the data accumulated by the two-parameter analyser are displayed on a rectilinear grid with pulses originating in the 'X' detector displayed along the X-axis and pulses originating in the 'Y' detector displayed along the Y-axis. With this arrangement, if a gamma-ray cascade occurred and gave rise to two gamma-rays of energies E_{γ_1} and E_{γ_2}, the position on the display would depend on whether γ_1 had been detected by the X detector and γ_2 had been detected by the Y detector, or vice versa. Thus if it was unimportant to know which detector had given rise to a particular pulse, identical spectral detail would appear in two places on the display, and would, in fact, require two regions of the memory for storage. Connally et al. [88], have modified the memory of such an instrument so as to eliminate this redundant use of core storage. As an example of the benefits which can be achieved, an instrument is described in which a single 4096-channel memory is used to store the data from four independent two-parameter detector systems, with no loss of channel resolution.

Another technique has been to use a small general-purpose computer in conjunction with separate ADC's to provide a multi-parameter analysis system. Such a system is described by Frederick et al. [89]. The advantage of this technique is that a more versatile instrument results and a substantial computing capability is provided in addition to the large storage capability. The technique of integrating a computer directly into the data acquisition system has been used in various applications in addition to multi-parameter analysis. Several such applications are reviewed by Lidofsky [90], and others are described by Gemmell [91], Robinson [92], and Broberg [93]. 'Dedicated' computers are required in these applications, however, and the computing capability is not available for general problems. In addition, the programming ('software') problems associated with these systems are not insignificant. Wyckoff [94] describes one approach to programming for such a system.

2.6   IMPORTANT STATISTICAL CONSIDERATIONS

Various methods, appropriate to the requirements for radiochemical analysis, have been devised for the analysis of gamma-ray pulse-height spectral data [95–99]. To be most effective, a method should be based on known statistical properties of the data, and should use correct statistical procedures. Certain reasonable assumptions can be made in respect to the statistical properties of data collected in the manner described herein. In actual practice, however, extraneous influences may be introduced into such data and these will tend to destroy the validity of the basic assumptions. The result may be a deterioration of accuracy and of precision of the determinations made.

The three most important assumptions regarding the data derived from the pulse-height spectrometer are the following: (1) the data in each of the several pulse-height channels are normally distributed; (2) the data in the several pulse-height channels are uncorrelated; and (3) the observed composite pulse-height spectrum is simply a linear summation of several constituent spectra. Influences which may make one or more of these assumptions invalid, and hence impair the accuracy and precision obtainable, include the following: (1) instrumental instabilities; (2) instrumental changes which effect the appropriateness of the established calibration data;

and (3) improper application of the data analysis method. In the following are brief discussions of each of these factors.

Instrumental Instabilities

The most noticeable instrumental instabilities consist of periodic drifts in baseline and in gain. During the period of a measurement, the instrument may be stable but offset from its normal calibration, or it may drift. Drifts which occur during measurement degrade the resolution, but in the case of an offset, the quality of the data is not impaired. Various techniques have been used either to minimize drifts by stabilization of the instrument, [100, 101] or, through correction procedures, to remove from the data the effects of offset [102, 104]. No satisfactory method exists for the correction of drifts which have occurred during measurement, and the accuracy with which such data can be analyzed is consequently poorer. Gain changes sometimes occur, particularly in the scintillation detector, which correlate with sample counting rate [35], and these changes will also tend to degrade resolution. Pulse pile-up and baseline variations which may result from high counting rate may not only degrade resolution, but may also introduce extraneous features into the spectrum. Minor deteriorations in the performance of the various elements of the spectrometer (viz., detector, amplifier, high-voltage supply, pulse-height analyzer, etc.) may produce more subtle effects. Such effects may be observed as small changes in the resolution, linearity, or background. These constitute minor effects and are not as likely to be as troublesome as are drifts in gain and baseline. If they do occur, however, they may introduce substantial error.

Changes in Instrument Calibration

The instrument response may be inadvertantly altered by changes in the detector/shield geometry or in the sample/detector geometry. These changes may in turn alter the photoelectric-to-Compton interaction ratio and thus render inappropriate calibration spectra which had been obtained before the changes. Calibration spectra could also be invalidated by changing elements of the spectrometer, such as the detector or the ADC, which might alter the resolution or linearity characteristics of the instrument.

Improper Data Analysis

Certain data analysis procedures may cause inaccuracies in the results. Transformation of the raw data to correct for shifts in gain and baseline or to smooth such data tends to introduce correlations among the several channels. Such correlations are also introduced by subtracting a common background from each member of a set of spectral data where such a set is to be used to effect a single analysis. In some data analysis methods, variances of the data in the several channels are used as weighting factors, and if the data have been time-normalized before the analysis is made (as is sometimes done in order to subtract a background), an incorrect estimate of the variance may be used. Error may also be introduced by assuming that the counts in the several channels are normally distributed when, in fact, they may not be (e.g., in those cases where the number of counts are very small or are zero).

2.7   CALIBRATION AND QUALITY CONTROL

In radiochemical analysis, the gamma-ray pulse-height spectrometer may be used to determine the energies and abundances of gamma-emitting nuclear transitions, to monitor separation procedures in order to determine product quality, or to determine constituent nuclides in a mixture. Because of the non-unique character of the features in the pulse-height spectrum, each of the above applications requires that the instrument be calibrated. If high precision is not required, or if measurements are made infrequently, the calibration may be performed as part of the measurement procedure. If, however, the instrument is to be used for making high-precision measurements routinely, a 'permanent' calibration is essential. The permanent calibration is a compilation of known pulse-height spectra. It is laborious to prepare, but once established, is considered to be a fixed (i.e., long-term) calibration of the instrument. In order to assure that the established calibration does, in fact, continue to represent instrumental response characteristics accurately, the instrument is brought under a program of quality control.

2.7.1   Calibration

Where energy and abundance estimates of gamma-emitting transitions are required, a series of gamma-ray standards are used to calibrate the instrument. The principal requirements for nuclides which are to be used as standards for this type of calibration are that they have simple but well-known decay-schemes, that they can be easily standardized, and that they have conveniently long half-lives. A pulse-height spectrum is obtained for each such standard, and relationships between gamma-ray energy and the position of the full-energy peak, and between abundance and peak-height or peak-area are determined. Heath [2] describes this procedure in greater detail, as applied to the NaI(Tl) detector, and suggests a set of nuclides which would be useful for calibration. A more extensive list of nuclides is presented by Marion [105], which covers the energy range from 25 keV to 11 MeV. Although of general utility, the list by Marion would be particularly appropriate for calibration of the Ge(Li) detector.

Where the instrument is to be used for monitoring separation procedures, usually only a rough estimate is sought of the presence of contaminating nuclides. Accuracy and precision are not dominant factors. Judgments may be made by comparing spectral shapes observed on an oscilloscopic display with reference spectra of individual nuclides. Such reference spectra may be obtained from an established compilation, or 'library' of known spectra, or they may be obtained as needed from a set of known sources.

To calibrate when high-precision determinations of radionuclide abundances are required, it is necessary to obtain a normalized pulse-height spectrum for each nuclide of interest. Such a set of spectra constitutes a spectral library, and for general applicability, a large library is required.

Candidate nuclides must be available with a high degree of purity and source preparations must be accurately standardized. In addition, it is important that all measurement procedures (e.g., sample/detector geometry, instrument gain and baseline, etc.) be highly standardized. The form of the standards (i.e., whether liquid or solid, planchet or test-tube, point-source or dispersed source, etc.), and the measurement procedures adopted should be consistent with the form of the samples to be measured and with the procedures which may be dictated by such samples. It is necessary, in order to specify optimum activity levels and counting times to have a knowledge of various performance qualities of the instrument. For example, in order to minimize statistical errors and the effects of background in the standard spectra, it is preferable to use source preparations with activity levels as high as possible, but these must be consistent with the counting-rate tolerance of the instrument. It is also essential to collect as many counts as possible, but this requirement must be consistent with practical counting times. The choice of counting times, in turn, may be restricted because of the drift characteristics of the instrument (i.e., rate and magnitude of drift). Computer techniques for spectral interpretation are generally used for high-precision determinations, and so the library spectra are usually catalogued in digital form. Plots of such spectra are also useful, however, for reference purposes and for making qualitative or semi-quantitative interpretations of sample spectra.

2.7.2 Quality Control

In order to assure long-term stabilization of instrument response, a sensitive method for indicating the presence of various types of aberrant response is necessary. The technique of statistical quality control has been found to be effective, both in helping to achieve long-term stability and in providing an indication of aberration. Methods for statistical quality control for the gamma-ray pulse-height spectrometer are described by Covell [106, 107]. The methods described consist of the regular counting of a standard and background, the calculation of important parameters in the resultant spectra, and the plotting of the values of the parameters on a control chart. From the patterns which develop on the chart, various types of instrument malfunction are promptly detected and corrected. These methods have been found to be effective, not only in maintaining long-term stability, but also in identifying causes of instability so that drift tendencies are gradually reduced. Approximately 15–20 minutes per instrument per day are required.

2.8 METHODS OF SPECTRAL ANALYSIS

For problems in radiochemical analysis, it is usually not as important to measure accurately gamma-ray energies as it is to determine constituent radionuclides. A variety of techniques have been proposed for 'unfolding' pulse-height spectra to accomplish this objective. The problem is complicated because of the non-unique nature of the gamma-ray interaction.

One of the earliest methods of spectral unfolding was referred to as spectrum stripping. In this method, the radionuclide responsible for the most energetic gamma-ray was identified (by inspection of the spectrum). Since the full energy peak associated with this gamma-ray was on or near the baseline, the abundance could be fairly well estimated from peak-height or peak-area measurements. Next, the spectrum of the identified nuclide, appropriately weighted according to the abundance estimate, was subtracted from the sample spectrum. This would make possible the identification and estimation of another 'most energetic' gamma-ray, and the process of identification, estimation, and subtraction would continue until an accounting had been made of all contributions to the sample spectrum. The stripping technique was laborious, and since determinations were made serially, it was time-consuming. It also resulted in the propagation of variance into the residual data after each subtraction, so that radionuclides with low energy gamma-rays were determined with less precision than those with high energy gamma-rays.

Covell [96] described a digital method which made possible determinations of abundance directly from the full-energy peaks. The method could be applied to any well-resolved peak, regardless of its position in the pulse-height spectrum, and thus did not propagate errors as did the stripping method. It also made possible the determination of any identifiable nuclide without having to determine the nuclides responsible for all higher-energy peaks as a prerequisite.

More elegant methods of unfolding were devised. Perhaps the most popular has been the multiple linear regression method (MLR), referred to more commonly as the 'least squares' method. Applications of this method to problems in radiochemical analysis have been described by Trombka [108], Schonfeld et al. [109], Parr et al. [103], Nicholson et al. [98], Yamamoto et al. [110], and Salmon [97]. Discussions of the theoretical bases of the method, and justifications for its use in this application are presented by Schonfeld et al. [109], Nicholson et al. [98], and Covell et al. [111].

The MLR method requires a library of known spectra, and proceeds best when the candidate set of known spectra is complete and exclusive (i.e., no real component of the unknown spectrum is omitted, and no presumed component is included which is not actually in the unknown spectrum). Hogan et al. [99] have proposed and evaluated a method for automatic candidate selection to be used in conjunction with MLR. In this method selection is based on statistical indicators derived directly from the unknown spectrum.

Data from Ge(Li) detectors are typically contained in 1000–4000 channels (or more) and spectra are complex. Gunnink et al. [112] have developed a computer code for unfolding such spectra. The code makes possible precise determinations of peak locations and peak areas. Then, from a library tape containing extensive calibration data and decay-scheme information, precise determinations are made of peak energies and intensities and, finally, nuclides are identified.

REFERENCES

1. C. M. DAVISSON, in Alpha, Beta and Gamma-Ray Spectroscopy. (K. Siegbahn, ed.) North-Holland Publ. Co., Amsterdam p. 37 (1965).
2. R. L. HEATH. Scintillation Spectrometry Gamma-Ray Spectrum Catalogue, 2nd ed., USAEC Report IDO-16880–1 (1964).
3. S. C. CURRAN, Luminescence and the Scintillation Counter, Academic Press, Inc., New York (1953).
4. J. B. BIRKS, The Theory and Practice of Scintillation Counting, The Macmillan Co., New York (1964).
5. W. J. VAN SCIVER, *IRE Trans. on Nucl. Sci.*, NS-3, No. 4, 39, 1956.
6. W. J. VAN SCIVER, *Phys. Rev.*, **120**, 1193 (1960).
7. C. D. ZERBY, A. MEYER and R. B. MURRAY, *Nucl. Inst. and Methods*, 12, 115 (1961).
8. A. MEYER and R. B. MURRAY, *IRE Trans. on Nucl. Sci.*, NS-7, No. 2–3, 22 (1960).
9. R. B. MURRAY and A. MEYER, *Phys. Rev.*, **122**, 815 (1961).
10. R. B. MURRAY, in Nuclear Instruments and Their Uses. (A. H. Snell, ed.) p. 82. John Wiley & Sons, Inc., New York (1962).
11. R. G. KAUFMAN, W. B. HADLEY and H. N. HERSH, *IEEE Trans on Nucl. Sci.*, NS-17, No. 3, 82 (1970).
12. R. HOFSTADTER, *Phys. Rev.*, **74**, 100 (1948).
13. D. ENGELKEMEIR, *Rev. Sci. Inst.*, **27**, 589 (1956).
14. J. A. NEMILOV, J. J. LOMONOSOV, A. N. PESAREVESKI, L. V. SOSHIN and E. D. TETERIN, *Izvestia Academia Nauk*, **SSSR**, **24**, No. 2, 257 (1959).
15. W. W. MANAGAN, *IRE Trans. on Nucl Sci.*, NS-9, No. 3, 1 (1962).
16. R. W. PERKINS, *Nucl. Inst. and Methods*, 33, 71 (1965).
17. N. A. WOGMAN, D. E. ROBERTSON and R. W. PERKINS, *Nucl. Instr. and Methods*, 50, 1, (1967).
18. J. H. NEILER and P. R. BELL, in Alpha, Beta, and Gamma-Ray Spectroscopy. (K. Siegbahn, ed.) p. 245. North Holland Publ. Co., Amsterdam (1965).
19. B. A. EULER, D. F. COVELL and S. YAMAMOTO, *Nucl. Inst. and Methods*, 72, 143 (1969).
20. E. A. WOLICKI, R. JASTROW and F. BROOKS, Calculated Efficiencies of NaI Crystals, NRL Report 4833 (1956).
21. S. H. VEGORS, L. L. MARSDEN and R. L. HEATH, Calculated Efficiencies of Cylindrical Radiation Detectors, USAEC Report IDO-16370 (1958).
22. C. M. DAVISSON and L. A. BEACH, A study of Photons in Sodium Iodide Scintillation Crystals, NRL Report 5408 (1959).
23. R. GUNNINK and A. W. STONER, *Anal. Chem.*, 33, 1311 (1961).
24. W. F. MILLER and W. J. SNOW, *Nucleonics*, 19, No. 11, 174 (1961).
25. M. L. VERHEIJKE, *Int. J. of App. Rad. and Isotopes*, 13, 583 (1962).
26. R. W. ENGSTROM, R. G. STOUDENHEIMER and A. M. GLOVER, *Nucleonics*, 10, No. 4, 58 (1952).
27. IRE Publication No. 62, IRE Standards on Electron Tubes, Methods of Testing, 1962, IRE 7. Sl (1962), Pt. 5, Sect 7.

28. JEDEC Publication No. 50, Relative Spectral Response Data for Photosensitive Devices, Electronic Industries Association, Washington, D.C. (1964).
29. J. SHARPE, *IEEE Trans. on Nucl. Sci.*, NS-9, No. 3, 54 (1962).
30. E. H. EBERHARDT, *IEEE Trans. on Nucl. Sci.*, NS-14, No. 2, 7 (1967).
31. S. J. ROTH, *IRE Trans. on Nucl. Sci.*, NS-7, No. 2–3, 57 (1960).
32. C. R. KERNS, *IEEE Trans. on Nucl. Sci.*, NS-14, No. 1, 449 (1967).
33. A. Z. SCHWARZSCHILD and E. K. WARBURTON, *Ann. Rev. Nucl. Sci.*, p. 265. Annual Reviews Inc., Palo Alto, Calif. (1968).
34. L. CATHEY, *IRE Trans. on Nucl. Sci., NS-5, No. 3, 109 (1958).*
35. D. F. COVELL and B. A. EULER, in Proceedings of the 1961 International Conference: 'Modern Trends in Activation Analysis', pp. 12–15 (1961).
36. L. A. WEBB and R. J. JOHNSON, *Phys. Rev.* 98, 234-A (1955).
37. W. P. BALL, R. BOOTH, and M. H. MACGREGOR, *Bull. Am. Phys. Soc.*, 31, 183 (1956).
38. F. E. KINARD, *Nucleonics*, 15, No. 4, 92 (1957).
39. R. E. ROHDE, *IEEE Trans. on Nucl. Sci.*, NS-12, No. 1, 16 (1965).
40. E. BREITENBERGER, *Prog. in Nucl. Phys.*, 4, 56 (1955).
41. P. S. TAKHAR, *IEEE Trans. on Nucl. Sci.*, NS-14, No. 1, 438 (1967).
42. J. R. PRESCOTT and P. S. TAKHAR, *IRE Trans. on Nucl. Sci.*, NS-9, No. 3, 36 (1962).
43. R. E. SIMON and B. F. WILLIAMS, *IEEE Trans. on Nucl. Sci.*, NS-15, No. 3, 167 (1968).
44. G. A. MORTON, H. M. SMITH, JR. and H. R. KRALL, *IEEE Trans. on Nucl. Sci.*, NS-16, No. 1, 92 (1969).
45. H. R. KRALL, F. A. HELVY and D. E. PERSYK, *IEEE Trans. on Nucl. Sci.*, NS-17, No. 3, 71 (1970).
46. G. DEARNALEY and D. C. NORTHROP, Semiconductor Counters for Nuclear Radiations, 2nd ed., E. & F. N. Spon Limited, London (1966).
47. D. A. BROMLEY, *IEEE Trans. on Nucl. Sci.*, NS-9, No. 3, 135 (1962).
48. G. L. MILLER, W. M. GIBSON and P. F. DONOVAN, *Ann. Rev. Nucl. Sci.*, 12, pp. 189–220. Annual Reviews Inc., Palo Alto, Calif. (1962).
49. G. T. EWAN and A. J. TAVENDALE, *Can. J. of Physics*, 42, 2286 (1964).
50. G. DEARNALEY, A. T. G. FERGUSON and G. C. MORRISON, *IRE Trans. on Nucl. Sci.*, NS-9, No. 3, 174 (1962).
51. D. C. CAMP, Applications and Optimization of the Lithium-Drifted Germanium Detector System, Lawrence Radiation Laboratory Report, UCRL-50156 (1967).
52. H. R. BOWMAN, E. K. HYDE, S. G. THOMPSON and R. C. JARED, Applications of High-Resolution Semiconductors in X-Ray Emission Spectrography, Lawrence Radiation Laboratory Report UCRL-16485 (1965).
53. S. YAMAMOTO, *Anal. Chem.*, 41, 337 (1969).
54. W. SHOCKLEY, Electrons and Holes in Semiconductors, p. 214, Van Nostrand, Princeton, N.J. (1950).
55. P. L. PHELPS, *IEEE Trans. on Nucl. Sci.*, NS-15, No. 1, 376 (1968).

56. G. DEARNALEY, P. E. GIBBONS, and R. ELLIS, *IEEE Trans. on Nucl. Sci.*, NS-17, No. 3, 282 (1970).

57. H. L. MALM and I. L. FOWLER, *IEEE Trans. on Nucl. Sci.*, NS-13, No. 1, 62 (1966).

58. H. L. MALM and I. L. FOWLER, Semiconductor Nuclear-Particle Detectors and Circuits. (W. L. Brown, W. A. Higinbotham, G. L. Miller, and R. L. Chase, eds.) p. 237, Publication 1593, National Academy of Sciences, Wash., D.C. (1969).

59. E. W. SAUNDERS and C. J. MAXWELL, *IEEE Trans. on Nucl. Sci.*, NS-15, No. 1, 423 (1968).

60. B. LALOVIC, in Semiconductor Nuclear-Particle Detectors and Circuits. (W. L. Brown, W. A. Higinbotham, G. L. Miller and R. L. Chase, eds.) p. 230. Publication 1593, National Academy of Sciences, Wash., D.C. (1969).

61. J. E. CLINE, in Semiconductor Nuclear-Particle Detectors and Circuits. (W. L. Brown, W. A. Higinbotham, G. L. Miller, and R. L. Chase, eds.) p. 241. Publication 1593, National Academy of Sciences, Wash., D.C. (1969).

62. G. T. EWAN and A. J. TAVENDALE, *Can. J. of Physics*, 42, 2286 (1964).

63. K. M. WAINIO and G. F. KNOLL, *Nuc. Inst. and Methods*, 44, 213 (1966).

64. N. U. DE CASTRO FARIA and R. J. LEVESQUE, *Nuc. Inst. and Methods*, 46, 325 (1967).

65. R. L. HEATH, in Semiconductor Nuclear-Particle Detectors and Circuits. (W. L. Brown, W. A. Higinbotham, G. L. Miller, and R. L. Chase, eds.) p. 247. Publication 1593, National Academy of Sciences, Wash., D.C. (1969).

66. F. J. WALTER, in Semiconductor Nuclear-Particle Detectors and Circuits. (W. L. Brown, W. A. Higinbotham, G. L. Miller, and R. L. Chase, eds.) p. 214. Publication 1593, National Academy of Sciences, Wash., D.C. (1969).

67. R. L. GRAHAM, I. K. MACKENZIE, and G. T. EWAN, *IEEE Trans. on Nucl. Sci., NS-13, No. 1, 72 (1966).*

68. A. ALBERIGI QUARANTA, M. MARTINI, and G. OTTAVIANI, *IEEE Trans. on Nucl. Sci.*, NS-16, No. 2, 35 (1969).

69. J. A. MIEHE and P. SIFFERT, *IEEE Trans. on Nucl. Sci.*, NS-17, No. 5, 8 (1970).

70. B. EULER, *Nucl. Inst. and Methods*, 61, 211 (1968).

71. A. A. DELUCCHI and A. E. GREENDALE, *Phys. Rev.* C, 1, 1491 (1970).

72. D. C. CAMP, in Semiconductor Nuclear-Particle Detectors and Circuits. (W. L. Brown, W. A. Higinbotham, G. L. Miller, and R. L. Chase, eds.) p. 693. Publication 1593, National Academy of Sciences, Wash., D.C. (1969).

73. A. R. SAYRES and J. A. BAICKER, *IEEE Trans. on Nucl. Sci.*, NS-15, No. 3, 393 (1968).

74. J. M. PALMS, R. E. WOOD, and O. H. PUCKETT, *IEEE Trans. on Nucl. Sci.*, NS-15, No. 3, 397 (1968).

75. J. A. COOPER, N. A. WOGMAN and R. W. PERKINS, *IEEE Trans. on Nucl. Sci.*, NS-15, No. 3, 407 (1968).

76. F. P. BRAUER and R. E. CONNALLY, *Trans. Am. Nucl. Soc.*, 6, 174 (1963).
77. D. F. CROUCH and R. L. HEATH, USAEC Rep., IDO-16923 (1963).
78. R. L. HEATH, W. W. BLACK, and J. E. CLINE, *IEEE Trans. on Nucl. Sci.*, *NS-13, No. 3*, 445 (1966).
79. E. FAIRSTEIN and J. HAHN, *Nucleonics*, 23, No. 7, 56 (1965).
80. E. FAIRSTEIN and J. HAHN, *Nucleonics*, 23, No. 9, 81 (1965).
81. E. FAIRSTEIN and J. HAHN, *Nucleonics*, 23, No. 11, 50 (1965).
82. E. FAIRSTEIN and J. HAHN, *Nucleonics*, 24, No. 1, 54 (1966).
83. E. FAIRSTEIN and J. HAHN, *Nucleonics*, 24, No. 3, 68 (1966).
84. E. FAIRSTEIN, in Semiconductor Nuclear-Particle Detectors and Circuits. (W. L. Brown, W. A. Higinbotham, G. L. Miller, and R. L. Chase, eds.) p. 411. Publication 1593, National Academy of Sciences, Wash., D.C. (1969).
85. C. W. WILLIAMS, *IEEE Trans. on Nucl. Sci.*, NS-15, No. 1, 297 (1968).
86. F. S. GOULDING, D. A. LANDIS, and R. H. PEHL, in Semiconductor Nuclear-Particle Detectors and Circuits. (W. L. Brown, W. A. Higinbotham, G. L. Miller and R. L. Chase, eds.) p. 455. Publication 1593, National Academy of Sciences, Wash., D.C. (1969).
87. M. KONRAD, in Semiconductor Nuclear-Particle Detectors and Circuits. (W. L. Brown, W. A. Higinbotham, G. L. Miller, and R. L. Chase, eds.) p. 731. Publication 1593, National Academy of Sciences, Wash., D.C. (1969).
88. R. E. CONNALLY, W. A. MITZLAFF and F. P. BRAUER, *IEEE Trans. on Nucl. Sci.*, NS-17, No. 3, 440 (1970).
89. D. E. FREDERICK and T. MARSHALL, *IEEE Trans. on Nucl. Sci.*, NS-13, No. 1, 144 (1966).
90. L. J. LIDOFSKY, *IEEE Trans. on Nucl. Sci.*, NS-15, No. 1, 93 (1968).
91. D. S. GEMMELL, *IEEE Trans. on Nucl Sci.*, NS-13, No. 1, 158 (1966).
92. L. ROBINSON, *IEEE Trans. on Nucl. Sci.*, *NS-13, No. 1, 161* (1966).
93. J. B. BROBERG, *IEEE Trans. on Nucl. Sci.*, NS-13, No. 1, 192 (1966).
94. J. M. WYCKOFF, *IEEE Trans. on Nucl. Sci.*, *NS-13, No. 1, 199* (1966).
95. W. LEE, *Anal. Chem.*, 31, 800 (1959).
96. D. F. COVELL, *Anal. Chem.*, 31, 1785 (1959).
97. L. SALMON, in Application of Computers to Nuclear and Radiochemistry. (G. D. O'Kelley, ed.) p. 165. Publication 3107, National Academy of Sciences, Wash., D.C. (1963).
98. W. L. NICHOLSON, J. E. SCHLOSSER and F. P. BRAUER, *Nucl. Inst. and Methods*, 25, 45 (1963).
99. M. A. HOGAN, S. YAMAMOTO and D. F. COVELL, *Nucl. Inst. and Methods*, 80, 61 (1970).
100. H. DE WAARD, *Nucleonics*, 13, No. 7, 36 (1955).
101. R. A. DUDLEY and R. SCARPATETTI, *Nucl. Inst. and Methods*, 25, 297 (1964).
102. R. L. HEATH, *Nucleonics*, 20, No. 5, 67 (1962).
103. R. M. PARR and H. F. LUCAS, Jr., *IEEE Trans. on Nucl. Sci.*, NS-11, No. 3, 349 (1964).

104. D. F. COVELL, *Nucl. Inst. and Methods,* **36**, 229 (1965).
105. J. B. MARION, *Nucl. Data, Sect.. A,* **4**, 301 (1968).
106. D. F. COVELL, *Nucl. Inst. and Methods,* **22**, 101 (1963).
107. D. F. COVELL, *Nucl. Inst. and Methods,* **47**, 125 (1967).
108. J. I. TROMBKA, Jet-Propulsion Laboratory Technical Report, JPL-TR-32-373 (1962).
109. E. SCHONFELD, A. H. KIBBEY and W. DAVIS, Jr., *Nucl. Inst. and Methods,* **45**, 1 (1966).
110. S. YAMAMOTO and M. BROWN, *Int. J. of Appl. Rad. and Isotopes,* **20**, 209 (1969).
111. D. F. COVELL, M. BROWN and S. YAMAMOTO, *Nucl. Inst. and Methods,* **80**, 55 (1970).
112. R. GUNNINK, R. A. MEYER, J. B. NIDAY and R. P. ANDERSON, *Nucl. Inst. and Methods,* **65**, 26 (1968).

Chapter 3

Liquid Scintillation
and Čerenkov Counting

C. T. Peng

Department of Pharmaceutical Chemistry
School of Pharmacy, University of California
San Francisco, California 94122, U.S.A.

3.1 LIQUID SCINTILLATION COUNTING: INTRODUCTION

Liquid scintillation counting is of relatively recent origin and has been widely used in various fields of research. The flexibility, versatility, and sensitivity of liquid scintillators allow their use in large volume as well as in small volume detectors. This type of counting technique is exploited not only in the field of nuclear physics, archaeology, biology, chemistry, engineering, and other branches of natural science, but also in the domain of medicine and medical research where its application has widened the scope of studies by rendering the counting of bulky samples possible and the use of weak β-emitting radionuclides as tracers more practical. In fact, liquid scintillation spectrometers have superseded Geiger-Mueller counters in a vast number of laboratories where measurements of radionuclides are routinely conducted.

The development of liquid scintillation from a laboratory curiosity into a dependable counting technique is a result of contributions from many laboratories. The measurement of fluorescence in solutions containing p-terphenyl activated with γ-rays by Ageno et al. [1] in 1949 marks the beginning of the era of liquid scintillation counting. In 1950, Reynolds et al [2] in a report under the title 'Liquid Scintillation Counters', studied the scintillation properties of various liquids including solutions of anthracene, naphthalene, and p-terphenyl in benzene or in m-xylene irradiated with ^{60}Co γ-rays, using a coincidence counter with two 1P21 multiplier phototubes as detectors. Kallmann and Furst [3—5] made extensive studies on the scintillation characteristics of many liquids and fluorescent organics and on the mechanisms involved in energy transfer processes. Hayes et al. [6—9] in subsequent years reported on the synthesis of substituted polyaryls as potential scintillators.

The first application of liquid scintillation counting was for the indirect detection of free neutrinos by Cowan et al. [10] using a large volume detector containing a solution of p-terphenyl and 1,4-bis-2-(5-phenyloxazolyl)benzene (POPOP) in triethylbenzene. The successful application of the technique of liquid scintillation counting in detecting this strange particle has laid the neutrino hypothesis on a solid foundation.

Large-volume liquid scintillation detectors have been subsequently used as γ-ray detecting devices to monitor environmental radioactive contamination in foodstuffs and in man [11—14], to study the potassium and radium contents in the human body [14], and to detect the disposition of radiotracers in animals in research [15]. The energy discrimination of liquid scintillators is inferior to that of a thallium-activated sodium iodide crystal, limiting their application in the analysis of γ-ray-emitting radionuclides.

Scintillation properties of solvents and solutes intended for liquid scintillation systems were studied by Hayes et al. [8, 9, 16—19] leading to the adoption of 2,5-diphenyloxazole (PPO) as a standard scintillator. Internal sample counting was introduced as an analytical tool in 1953 [20] and was immediately accepted as the preferred method for assaying β-emitters, for it afforded increased detection efficiency and ease in sample preparation. The incorporation of specimens to be assayed into the liquid scintillator frequently resulted in a loss of counting efficiency by quenching, which was corrected by the use of an internal standard.

The development of liquid scintillation counting into a reliable and sensitive analytical technique should be credited to the instrument manufacturers who supply highly dependable and sophisticated spectrometers. The first commercial instrument, which appeared in 1953, was a 2-

multiplier phototube coincidence spectrometer patterned after the original design from the Los Alamos Scientific Laboratory [21]. As demand for the instrument increased, other manufacturers entered the market. Spectrometers with a fixed mode of operation for quench correction, with computational capability, with large sample capacity and other refinements were introduced. Spectrometers currently marketed are completely transistorized, and are equipped with integrated circuitry and logic, improved readout devices, computation capabilities and more sensitive, low noise phototubes. Price [22] and Rapkin [23] have reviewed the current status and the development of modern liquid scintillation counters.

A number of international conferences and symposia have been held and the proceedings of these meetings have been published [24–29]. In addition to these proceedings, hundreds of reports appeared in the literature. The manifestation of interest and activity in this area attests the wide acceptance and usage of this nuclear counting technique.

3.2 LIQUID SCINTILLATION SYSTEMS

Many substances fluoresce in solution but only a limited number of compounds with extended π-electron conjugation such as substituted polyaryls are suitable scintillators. A liquid solution scintillator is sensitive to a-, β-, and γ-radiations. When used for detecting γ-rays, the liquid scintillator can be isolated from the radiation source. In contrast, the systems employed for measuring α- and β-radiations require direct incorporation of the source into the system. Since the photophysical processes accompanying the occurrence of a nuclear event are complex, the efficiency of a liquid solution scintillation system is dependent upon the outcome of a series of competing, consecutive and parallel processes among the components of the system. In the following we shall discuss the various stages involved.

3.2.1 Scintillation Processes
Accompanying the decay of a radionuclide nuclear emissions interact with atoms and molecules causing excitation and ionization. In this manner, the kinetic energy of the nuclear particles is converted into molecular excitation energy. When this excitation is localized in molecules with fluorescent properties light photons or scintillations are emitted upon deactivation. These sequentially related events are known as the scintillation processes [30, 31] and are discussed below.

Excitation Energy Transfer
Ionizing radiation interacts with matter giving rise along its path to isolated spurs, short tracks, and blobs [32] which are regions of various ionization density.

Ion pairs are formed:

$$M \rightsquigarrow M^+ + e^- \tag{3.1}$$

In addition, excited and superexcited molecules in their first and higher excited singlet states are produced:

$$M \to M^* ({}^1M, {}^2M \ldots {}^nM) \tag{3.2}$$

The geminate ion pair (M^+, e^-) can recombine within the spur to yield an excited molecule either in the singlet or in the triplet state:

$$M^+ + e^- \to {}^1M^* \tag{3.3}$$

$$M^+ + e^- \to {}^3M^* \tag{3.4}$$

The ratio of abundance of the triplet state to that of the singlet state produced by recombination is 3 to 1 [33].

The triplet state may also be formed from the ground state of M by impact with a subexcitation electron, i.e. an electron with energy below the threshold for excitation [34]:

$$M(\uparrow\downarrow) + e_{sb}(\uparrow) \to {}^3M(\uparrow\uparrow) + e(\downarrow) \tag{3.5}$$

Depending upon the ambient environment and intrinsic properties, the excited molecule M^* and ion-molecule M^+ may enter into chemical reactions through ionic and/or free radical mechanism [32, 35] or become deactivated by intermolecular energy transfer or by intramolecular relaxation [36].

A liquid solution scintillator, in its simplest form, consists of solvent and solute with the concentration of the latter at about $10^{-2}M$. The energy dissipated in the scintillator by the incident radiation will be practically completely absorbed by solvent molecules since they are present in an overwhelming number. The solvent molecules are promoted in the above-mentioned manner to the first and higher excited singlet states. From these states, energy migration among solvent molecules can occur according to the Birks-Conte model [37–9] or the Voltz model [40, 41]. The former requires the excited solvent molecules to exist in a dynamic equilibrium with excited dimers ('excimers') formed by the reaction of an excited singlet monomer with a second molecule in the ground state so that the excitation migration can be effected by successive formation and dissociation of the excimers thus:

$$
\begin{aligned}
{}^1M_A^* + {}^1M_B &\rightleftharpoons {}^1D_{AB}^* \rightleftharpoons {}^1M_A + {}^1M_B^* \\
{}^1M_B^* + {}^1M_C &\rightleftharpoons {}^1D_{BC}^* \rightleftharpoons {}^1M_B + {}^1M_C^*, \text{ etc.}
\end{aligned}
\tag{3.6}
$$

The condition for the excimer formation is that the two adjacent molecules should be oriented parallel. For molecules not so positioned towards their near neighbors, energy migration occurs by the mechanism proposed by Voltz involving Coulombic (octupole-octupole) and electron-exchange interactions, similar to exciton migration in crystals:

$$
\begin{aligned}
{}^1M_A^* + {}^1M_B &\rightleftharpoons {}^1M_A + {}^1M_B^* \\
{}^1M_B^* + {}^1M_C &\rightleftharpoons {}^1M_B + {}^1M_C^*
\end{aligned}
\tag{3.7}
$$

In liquid solution scintillators both mechanisms are operative in the migration of solvent excitation.

In addition to the first excited singlet state S_1, transfer of solvent excitation energy can occur via the third excited singlet state S_3 [41, 42].

The population of the first three excited states, S_1, S_2 and S_3, of benzene was reported to be formed to the extent of 0.3, 12.6 and 87.1 per cent upon irradiation with ionizing radiation [43]. Other scintillation solvents such as toluene, p-xylene and mesitylene, presumably yield a similar distribution, indicating that the S_3 state is predominant [42]. Laustriat et al. [41] gave evidence for the transfer of excitation energy from the solvent S_1 and S_n states to the corresponding F_1 and F_n states of the solute. The relative importance of either S_1 or S_3 as the transferring state depends upon the solute concentration. At low solute concentration ($\sim 10^{-3}$M) energy migration is solely via the solvent S_1 state; at intermediate concentration ($\sim 10^{-2}$M), commonly used in liquid scintillators, both the S_1 and the S_n states are equally important in energy migration; and at high concentration ($\sim 10^{-1}$M), energy migration is predominantly via the S_n state [44].

Excitation transfer via the higher excited states also affects the quenching property of liquid scintillation systems. For instance, chloroform and carbon tetrachloride cause severe quenching in the system excited by ionizing radiation and no quenching in that excited by uv light because these compounds do not accept excitation energy from the lower excited states of the aromatic solvent molecules [41, 45]. Other similar examples are also noted [46, 47].

The excited solvent molecules in the S_1 state may also undergo deactivation by emission of fluorescence. Since the fluorescence quantum yield of solvent is relatively low compared to that of solute, the solvent fluorescence makes negligible contribution towards the overall efficiency of a liquid scintillation system.

Excitation energy migration continues between solvent molecules until the energy becomes trapped in the solute molecule which possesses a high fluorescence quantum yield ($\phi_f > 0.8$) and can convert the excitation energy into light photons efficiently. The transfer of solvent excitation to scintillation solute occurs by collision and by the Förster resonance mechanism [48]. The latter allows excitation transfer to take place over a distance of several molecular diameters and requires a dipole-dipole interaction between the donor and the acceptor molecules and also an overlap of the emission spectrum of the donor with the absorption spectrum of the acceptor. The rate of transfer by the resonance mechanism is greater than that by diffusion. The long range transfer is implicated in the excitation migration between PPO and diphenylhexatriene in toluene and is postulated to be the main form of energy flow from solvent to solute molecules in liquid scintillation systems [30, 49].

The excited solute molecules deactivate by radiative transition from the first excited singlet state, F_1, to the ground singlet state, F_o, with emission of fluorescence; it is the reverse of the electronic transition accompanying

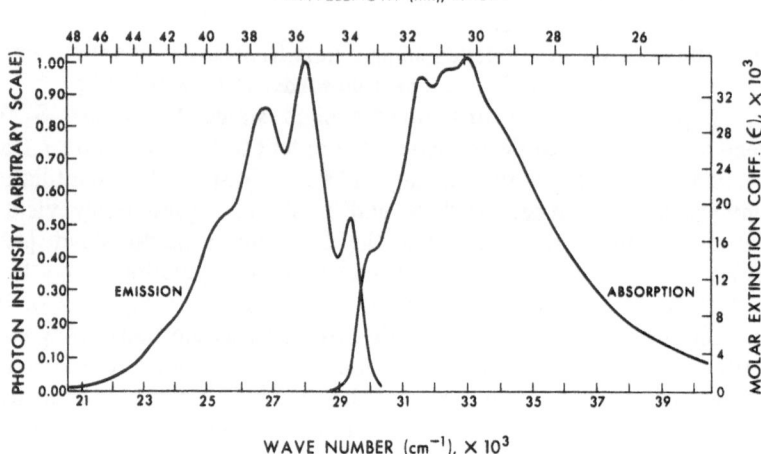

Fig. 3.1. Emission and absorption spectra of PPO (after Berlman [50]).

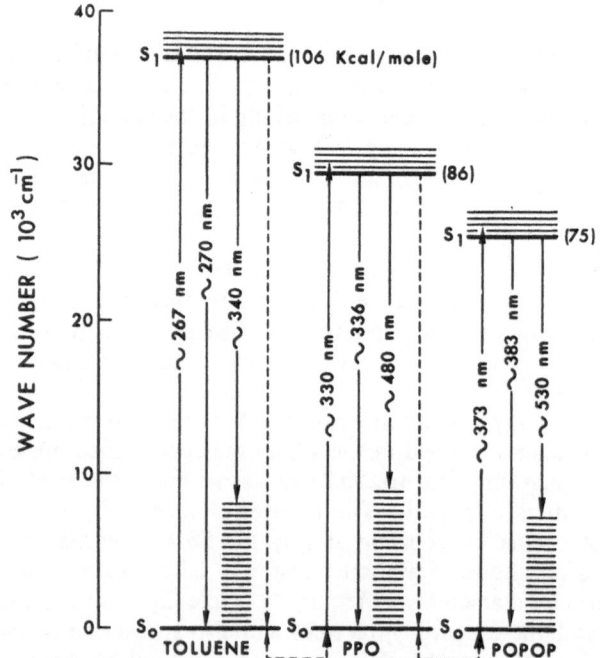

Fig. 3.2. Coupled energy transfer and energy level diagram for the components in a
ternary scintillation system.

energy absorption ($F_1 \leftarrow F_0$). The absorption and emission spectra of the
scintillator PPO are given in Fig. 3.1. The spectra show an approximate
mirror symmetry with a mirror point at 338 nm from which the energy gap
between the ground and the first excited state is calculated to be about
85 kcal/mole. The intermolecular energy transfer between the singlet states
requires the energy of the donor to be greater than that of the acceptor. In

Fig. 3.2, energy transfer between singlet energy levels of toluene, PPO and POPOP, which are components of a standard liquid scintillator, are shown. The energy values are from their fluorescence emission spectra reported by Berlman [50].

Fluorescence emitted by an organic scintillator undergoing prompt transition $F_o \leftarrow F_1$ is recognized as the fast scintillation component with a lifetime of the order of 10^{-9} sec [51].

A slow scintillation component due to delayed fluorescence is formed in the following manner. The excited singlet state of the solute can cross over to a corresponding lower lying triplet state by intersystem crossing. The radiative transition from the first triplet state to the ground singlet state is spin-forbidden and leads to the emission of phosphorescence with a radiative lifetime of 10^{-5} to 1 sec. The triplet state may yield delayed fluorescence by repopulation of the first excited singlet state by thermal activation or by triplet-triplet annihilation as shown below:

$$^3M \xrightarrow[\text{activation}]{\text{thermal}} {}^1M^* \tag{3.8}$$

$$^3M + {}^3M \longrightarrow {}^1M^* + {}^1M$$

$$^1M^* \longrightarrow {}^1M + h\nu_F \tag{3.9}$$

Delayed fluorescence arising from the mechanism described by (3.8) belongs to the E-type and that from (3.9) to the P-type [33]. The slow scintillation component is the delayed fluorescence of the P-type [51].

Energy transfer to non-fluorescent molecules present as impurities in liquid scintillator systems will be discussed in Section 3.2.3.

The photophysical processes involved in the liquid scintillator system may be summarily represented in Scheme 3.1.

Scintillation Efficiency

The power of a scintillation system to convert incident radiation energy into light photons is termed the scintillation efficiency. In a liquid solution scintillation system, this efficiency is dependent upon three distinct photophysical processes: (i) the number of excited solvent molecules formed per 100 eV of energy absorbed, conventionally expressed as the G-value; (ii) the efficiency of energy transfer from the excited solvent molecules to the solute molecules; and (iii) the fluorescence quantum yield of the excited solute molecules. The scintillation efficiency, S_x, can be defined as the ratio of the total photon energy released as photons to the exciting radiation absorbed [52]. Thus

$$S_x = \frac{h \int N(\nu) \, d\nu}{E_{ex}} = \frac{N_{ph} \, E_{ph}}{E_{ex}} \tag{3.10}$$

where h is Planck's constant, $N(\nu)$ the number of photons of frequency ν in the spectral range of ν_1 and ν_2, and E_{ex} the exciting energy absorbed. The total photon energy, represented by the integral, can be approximated by

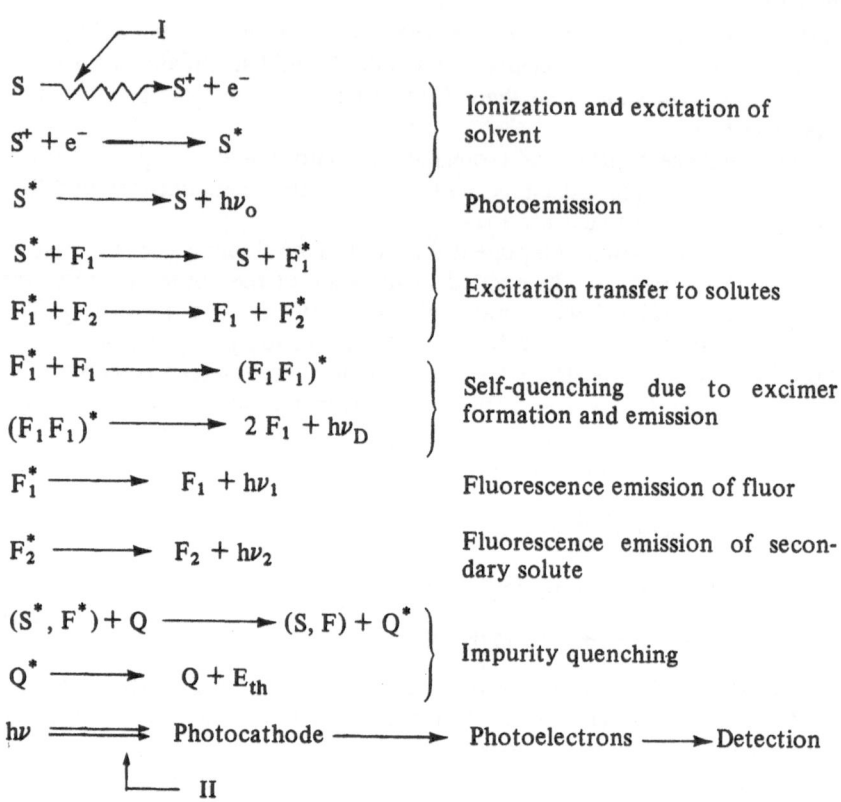

Scheme 3.1. S, solvent molecule; F_1 primary solute (fluor); F_2 secondary solute (wavelength shifter); Q, quenching agent; E_{th}, thermal energy; $\nu_1 > \nu_D$; $\nu_1 > \nu_2$. Interference at I results in 'photon' quenching and at II color quenching.

the product of the number of photons, N_{ph}, and the average photon energy, E_{ph}, obtained from the mean wavelength of the fluorescence spectrum [50].

The scintillation efficiency of a binary system containing benzene and p-terphenyl has been determined for unit energy transfer efficiency by Skarstad et al. [53] to be 0.042 (equivalent to 0.0116 ± 0.004 photon/eV). The efficiency of a deaerated solution of PPO and POPOP in toluene containing ^{14}C-hexadecane was reported by Hastings and Weber [54] to be 0.052 (0.018 photon/eV), and a similar system containing ^{3}H-hexadecane yielded a scintillation efficiency of 0.044 (0.014 photon/eV). The average wavelength of fluorescence emission of POPOP is 415 nm corresponding to an energy of 3.0 eV (E_{ph}). Insertion of these values into equation (3.10) leads to a relation between the number of photons generated and the exciting radiation; thus

$$N_{ph} = \frac{S_x E_{ex}}{E_{ph}} \qquad (3.11)$$

$$= 17.3 \, E_{ex} \text{ for } {}^{14}\text{C}\dagger$$

$$= 14.7 \, E_{ex} \text{ for } {}^{3}\text{H}$$

where E_{ex} is expressed in keV. The exciting energy varies from 0 to E_{max} of the β-radiation. In coincidence detection a sufficient number of light photons must impinge upon each of the two multiplier photocathodes to generate a total of two photoelectrons for a signal pulse. Consequently, the detection efficiency for one-electron pulses is zero and the probability of counting R-electron pulses is given by Swank [55] as $1-2^{(1-R)}$ from which Horrocks [52] computed the efficiency of coincidence counting of various numbers of photoelectrons.

The term photon yield is frequently used for scintillation efficiency. The photon yield of a scintillator solution is difficult to measure but the relative photon yield can be obtained by comparing the light yield of a given scintillator to that of a standard; thus

$$\text{Relative photon yield} = \frac{\alpha}{\alpha_o} \qquad (3.12)$$

where α and α_o are the photon yields per unit energy for the given and standard scintillators, respectively.

The photon yield is directly related to the pulse height measured by pulse height analysis. To achieve a maximum output from the phototube, the spectral distribution of the emitted photons must match the spectral response of the multiplier photocathode. The pulse height, therefore, becomes a function of both photon yield and spectral response matching factor [56].

Relative pulse height referred to a standard scintillator is a useful index for the evaluation of liquid scintillator systems. Hayes *et al.* [8, 9] reported the relative pulse heights of 102 primary solutes and the list has been extended by Kowalski *et al.* [56] to include those of newer fluors.

3.2.2 Scintillator Composition

Solvents
Scintillation solvents play an important role in determining the efficiency of a scintillator. The solvent forms the bulk of the system and interacts almost exclusively with the incident radiation, thereby converting the kinetic energy of the β-particle into electronic excitation energy resulting in the formation of the first and higher excited states. From these states excitation energy is eventually transferred to the solute molecule. An efficient solvent

---

†A lower value of 7 photons/keV was suggested by Swank [55] in 1958 as the basis for predicting the performance of liquid scintillation spectrometers. This value has been widely quoted.

should possess molecular properties that facilitate both energy conversion and energy transmission.

Furst and Kallmann [57] rated the solvents as effective, moderate, and poor, based upon the relative fluorescence intensity of a solution containing the solvent and a suitable solute under γ-ray irradiation. Hayes *et al.* [16] measured the relative pulse heights of 44 solvents containing PPO at a concentration of 3 g/l, activated by ^{137}Cs γ-rays. These results indicate that the best scintillation solvents are aromatics; the order of increasing efficiency being benzene < toluene < *p*-xylene. Their high efficiency is attributed to the fact that the aromatic π electrons, mobile within the molecular frame, are readily excited; the relative G values are higher for the more efficient solvents (1.0, 1.4 and 1.5 for benzene, toluene, *p*-xylene, respectively) [53] and they form excimers which act as intermediaries in excitation migration according to the Birks-Conte model. The fraction of excimer formation in these solvents and their excitation migration coefficients have been studied [31]. Recently, benzonitrile [58] has been found to be a very efficient scintillation solvent in quenched samples. The fluorescence quantum yield, the decay time and the relative pulse height of some aromatic solvents are listed in Table 3.1.

TABLE 3.1. *Scintillation Characteristics of Solvents*

Solvent	Relative pulse height[a,c]	Quantum yield[b]	Decay time (ns)[b]	Wavelength (nm) at which mean free path is 0.5 m[c]
p-Xylene	1.12	0.40	30	359
m-Xylene	1.09	0.17	30.8	356
Phenyl cyclohexane	1.02	0.15	26.4	360
Toluene	1.00	0.17	34	346
o-Xylene	0.98	0.19	32.2	364
Ethyl benzene	0.96	0.18	31	359
Triethyl benzene	0.96	0.12	24	373
Benzene	0.85	0.07	29	346
Anisole	0.83	0.29	8.3	383
Mesitylene	0.82	0.17	36.5	343
Isopropyl benzene	0.80	0.12	22	367
Fluorobenzene	0.67	0.13	8	340
p-Dioxane	0.65	–	–	346

[a] Measured at a fixed concentration of 3 g PPO/l.
[b] Data from Berlman [50].
[c] Data from Hayes *et al.* [16].

The aliphatic solvents such as cyclohexane, p-dioxane, mineral oil, etc., are inefficient scintillation solvents. At high concentrations of the solute, they yield approximately half the scintillation efficiency obtainable with aromatic solvents [57]. Among the aliphatic solvents, p-dioxane is 70 per cent as efficient as toluene [59] but is widely used because its water miscibility allows aqueous specimens to be incorporated into the scintillation system.

Primary Solutes (Fluors)

Organic scintillators are fluorescent substituted polyaryls which may be used in the liquid solution, solid, or plastic form for the detection of ionizing radiation. The scintillators employed in liquid solution scintillation systems as primary solutes are known as fluors.

An efficient fluor must possess the desirable properties of a high fluorescence quantum efficiency (or high photon yield), a maximal spectral matching to the multiplier photocathode, short fluorescence decay time, a large Stokes shift, sufficient solubility and a low sensitivity to quenching agents.

Synthetic efforts and search for efficient organic scintillators have been numerous. Many compounds have been synthesized and their fluorescence characteristics studied by Hayes and coworkers [6–9, 16–19, 60, 61] in search for efficient fluors. Out of these, p-terphenyl (TP), 2,5-diphenyloxazole (PPO) and 2-phenyl-5-(4-biphenylyl)-1,3,4-oxadiazole (PBD) have been selected for scintillation use. Wirth and coworkers [62, 63] have prepared substituted p-oligophenylenes and studied their scintillation properties. These fluors have large relative pulse height values and will be of considerable interest and use if commercially available at a reasonable cost. Among others, Heller [64, 65] has explored the properties of substituted styrylbenzenes; Kowalski et al. [56, 66] have studied the performance of some benzoxazoles, benzoxazolyl-thiophenes and oxadiazoles, all as potential new scintillators; and Leggate and Owen [67] have investigated the fluorescence and scintillation properties of new oxazoles, oxadiazoles and pyrazolines. Some of these are found to be very efficient fluors among which butyl-PBD has been extensively evaluated for liquid scintillation counting [68].

Fluors have considerably higher photon yields than scintillation solvents. In most cases, quantum yields (ϕ_f) of the fluors range from 0.85 to 1.0 relative to 9,10-diphenylanthracene as 1.0. The latter is selected as the standard because of the absence of self-quenching [50]. The absolute fluorescence quantum efficiency of 9,10-diphenylanthracene as determined in ethanol solution or in benzene against quinine bisulfate in dilute sulfuric acid, is 0.81 [69].

The photon yield of a fluor is dependent upon its concentration [56]. At optimum concentration (C_o), the light output is maximum, beyond which self- or concentration-quenching occurs, owing to formation of excimers. The probability of excimer formation is greater in fluors with planar molecular configuration than those offering steric hindrance with bulky substituent groups [70]. A salient example is PPO, exhibiting a reduced ϕ_f value of high concentrations due to excimer formation [71].

Fluors with short fluorescence decay times are preferred to those with long decay times, for the chances of quenching or of intermolecular energy transfer from the excited state will be less. A large Stokes shift ensures a minimal overlap of the absorption and fluorescence spectra of the fluor which is essential for maintaining the transparency of the solution scintillator.

As light output from a liquid scintillator depends upon the concentration of the fluor, those with a solubility below the optimum concentration (C_o) will be less desirable for liquid scintillation counting. The solubilities of substituted polyaryls and p-oligophenylenes can be substantially increased by introducing alkyl and alkoxy groups [72]. The alkyl groups tend to enhance the solubility in toluene, while the alkoxy groups favor an increased solubility in more polar solvents. The presence of *tert*-butyl groups in butyl-PBD increases its solubility over that of PBD with only a slight bathochromic shift of the fluorescence spectrum and a small decrease in photon yield. The introduction of methyl groups to the oxazolyl ring in POPOP yields dimethyl-POPOP with increased solubility and essentially the same fluorescence characteristics [73]. When the methyl groups are placed in the *para* position of the two terminal phenyl rings of POPOP, giving rise to TOPOT, the efficiency of the latter is slightly inferior to POPOP as a secondary solute [74].

Characteristics of the commonly used fluors are given in Table 3.2.

The relative ability of the fluors to resist quenching will be discussed in Section 3.2.3.

The photochemical stability of a number of scintillation solutes has been investigated. BBOT, PBBO, PBD and butyl-PBD are more stable than PPO, POPOP, and dimethyl-POPOP [56]. The photo-oxidation of PPO can be sensitized by methylene blue [75]. In general, the oxazoles undergo rapid photodecomposition, while the oxadiazole and benzoxazole derivatives are considerably more stable.

Secondary Solutes (Wavelength Shifters)

Scintillation solutes that are added to a liquid solution scintillation system in addition to the primary solute are called secondary solutes. Their function is to shift the emission spectrum of the system to longer wavelengths in order to match the spectral response of the multiplier photocathode; for this reason, they are known as wavelength shifters.

They have essentially the same scintillation characteristics as the primary solutes but with fluorescence maxima in the region of 415–430 nm [9]. The usual concentration of a wavelength shifter in liquid scintillator is about 10^{-4}M. The recent introduction of multiplier phototubes with bialkali photocathodes having peak spectral response at about 400 nm has vitiated the use of wavelength shifters but there is evidence that their presence in the liquid scintillator may serve to decrease the sensitivity of the latter to the action of a quencher [76] (see Section 3.2.3).

The characteristics of some of the secondary solutes are given in Table 3.3.

TABLE 3.2. *Scintillation Characteristics of Primary Solutes (Fluors)*[a]

Solute	Abbreviation	C_o(g/l)[b]	C_s(g/l)[c]	RPY[d]	Quantum yield[e]	Decay time (ns)	Fluorescence wavelength (ave) (nm)	L_o/L[f]
2,5-Diphenyl-O[g]	PPO	4	414	1.01	1.00	1.4	370.3	1.10
2,5-Diphenyl-D[h]	PPD		79		0.89	1.35	346.6	1.06
2-Phenyl-5-(4-bi-phenylyl)-D	PBD	10	13	1.31	0.83	1.0	366.9	1.05
2,5-Di(4-biphenylyl)-D	BBD		2.5		0.85	$(0.92)^{Bi}$	$(386.8)^{Bi}$	
2-(4-Biphenylyl)-5-(p-tert-butylphenyl)-D	Butyl-PBD	7	105	1.25	0.85	1.2	$(\lambda_{max} : 366.0)$	
2,5-bis[5'-*tert*-Butyl-benzoxazolyl(2')-thiophene	BBOT	6	58.2	1.05	0.74	1.1	439.5	1.02
p-Terphenyl	TP	8^{sj}	8	1.33	0.93	0.95	341.8	1.07
Naphthalene			265		0.23	96	334.4	6.4

[a] Data from References 30, 50 and 56.
[b] C_o = optimum concentration.
[c] C_s = solubility in toluene at 20°C.
[d] RPY = relative photon yield.
[e] Quantum yield of a dilute solution of the solute in cyclohexane.
[f] L_o/L = the ratio of the fluorescence intensity of a nitrogen-bubbled solution to the fluorescence yield of an aerated solution.

[g] O = oxazole.
[h] D = 1,3,4-oxadiazole.
[i] Measured in benzene solvent.
[j] s = at the limit of solubility.

TABLE 3.3. *Scintillation Characteristics of Secondary Solutes (Wavelength Shifters)*[a]

Solute	Abbreviation	C_s (g/l)[b]	RPY[c]	Quantum yield[d]	Decay time (ns)	Fluorescence wavelength (ave) (nm)	L_o/L_e[e]
2-(1-Naphthyl)-5-phenyl-O[f]	α-NPO	108	1.05	0.94	2.06	404.0	1.11
2-(1-Naphthyl)-5-phenyl-D[f]	α-NPD	49		0.70	1.7	377.3	1.07
2,5-Di(4-biphenylyl)-O	BBO	3.1	1.19	0.75	(1.16)[Bg]	(417.5)[B]	
2-(4'-Biphenylyl)-6-phenyl-benz-O	PBBO	4.2	1.25			(λ_{max} : 396)	
1,4-bis-2-(5-Phenyloxazolyl)benzene	POPOP	2.2	1.17	0.93	(1.26)[B]	(423.7)[B]	1.04
1,4-bis-2-(4-Methyl-5-phenyloxazolyl) benzene	DM-POPOP	3.9	1.17	0.93	1.5	427.3	1.06
1,6-Diphenylhexatriene		(~9)		0.80	12.4	455.5	1.57
p-bis(o-Methylstyryl) benzene	bis-MSB	4[h]	(75)[i]	0.94	1.35	421.9	1.08
bis(Isopropylstyryl) benzene	BPSB	2.5[h]	(75)[i]	0.94	1.1	418.4	1.07
2,5-bis[5'-tert-Butylbenzo-oxazolyl(2')]-thiophene	BBOT	58.2	1.05	0.74	1.1	439.5	1.02

a Data from References 30, 50 and 56.
b C_s = solubility in toluene at 20°C.
c RPY = relative photon yield.
d Quantum yield of a dilute solution of the solute in cyclohexane.
e L_o/L = the ratio of the fluorescence intensity of a nitrogen-bubbled solution to the fluorescence yield of an aerated solution.
f O = oxazole. D = 1,3,4-oxadiazole.
g Measured in benzene solvent.
h Solubility at 30°C, data from Reference 65.
i Relative pulse height based on anthracene crystal = 100 (Ref. 65).

Adjuncts

Adjuncts are a group of dissimilar substances added to the liquid scintillator to improve its efficiency or to enhance its solubility for sample specimens.

In scintillators with *p*-dioxane as the solvent, the efficiency can be improved by adding large quantities of naphthalene. Naphthalene in high concentration acts like an intermediate and a secondary solvent, for it can transfer efficiently the excitation energy from the less effective solvents such as *p*-dioxane to the fluor [77]. Evidence shows that the excited naphthalene molecules are less prone to impurity quenching than the excited solvent molecules and can thus result in an increased light output of the scintillator. Caution must be exercised in the use of *p*-dioxane and naphthalene for the combination leads to chemiluminescence [78].

Ethylene glycol, methyl glycol methanol, ethoxyethanol, 1,2-dimethoxyethane, etc., are the other adjuncts used to modify the non-polar character of the liquid scintillator. Their main function is to increase its solubility for aqueous sample specimens. These compounds are moderately effective solvents according to the classification of Furst and Kallmann. Among them, ethylene glycol has been found to enhance the fluorescence of sodium naphthionate by decreasing the formation of its non-fluorescent aggregate [79].

A number of surfactants have been added to toluene or *p*-xylene-based scintillators to enhance water incorporation (see Section 3.9.2). Currently, Triton X-100 and Triton N-101† (isooctyl- and nonylphenoxypolyethoxyethanol) are widely used. Triton X-100 can be purified by successive passage through an activated charcoal [80] and an activated silicic acid column [81] to remove a luminescent impurity; treatment with moist Dowex-1 (OH) resin has been found to be equally effective [82].

Cab-O-Sil (M-5)‡ consisting of submicroscopic particles of colloidal silica in chainlike formation and with large surface area (200 ± 25 m²/g), is frequently used to gel dioxane- or toluene-based scintillators for counting insoluble samples by suspension [83]. It has superb thixotropic properties and forms an almost transparent, rigid gel at 3—5 per cent concentration. Other thixotropic agents such as Thixcin,§ a castor oil derivative [84], and aluminium 2-ethyl hexanoate [85] have also been used for the same purpose.

3.2.3 Types of Quenching

The incorporation of a specimen into a liquid solution scintillator for internal counting frequently causes a decrease in its photon yield. This may be attributed to a disruption of the scintillation processes, to a decrease in the optical transparency of the medium or to an alteration of the source geometry due to nonhomogeneity in the sample. The consequence is a diminution of the light output of the system for the radiation energy absorbed and is generally known as quenching. For the causes given above,

---

†Trade name of Rohm & Haas Co., Philadelphia, Pa.
‡Trade name of Cabot Corporation, Boston, Mass.
§Trade name of Baker Caster Oil Co., New York, N.Y.

the types of quenching can be respectively categorized as impurity, color and 'photon' quenching. Molecules whose presence in the system causes a decrease in the photon yield are known as quenchers.

Impurity Quenching

The term 'chemical quenching' is used by many to indicate a reduction of the photon yield of a scintillant by the presence of an added specimen. Birks [31] has pointed out the erroneous usage of this term and recommends that the phenomenon be named 'impurity quenching' because the added impurity affects many rate parameters of the scintillation processes in which no chemical changes are involved.

From the point of view of energy transfer the liquid scintillation system is a donor-acceptor system; any process interfering with the donor-acceptor relationship leads to quenching. Impurity quenching can be divided into two categories:

(i)   Collisional quenching. In *dynamic* collisional quenching the excited solvent or solute molecules approach impurity or quencher molecules to a distance where chemical or molecular interaction may occur, and encounter complexes are formed. These include charge transfer complexes, exciplexes (excited complexes) and molecular complexes; the latter may rapidly dissociate into free radicals to enter chemical reactions; the excitation energy is deactivated via these complexes leading to a reduction of the fluorescence quantum yield of the scintillation system according to the following mechanisms [86]:

$$F^* + Q \rightarrow [F^+.Q^-] \rightarrow F + Q$$
$$F^* + Q \rightarrow [^3F.Q]^* \rightarrow {}^3F + Q \tag{3.13}$$
$$F^* + Q \rightarrow \text{chemical reactions}$$

where F^*, ^3F, and F represent the excited state, the triplet state and the ground state of the fluor, respectively, and Q is the quencher.

Tanielian [86] gave evidence that the above mechanism operates in the quenching by CCl_4 of scintillation solutes, TP, PPO, α-NPO and 1-methylnaphthalene in various solvents. Kallmann-Oster [87] showed that the quenching of benzene, naphthalene anthracene and 9,10-diphenylanthracene in cyclohexane by CCl_4 is by the formation of free radicals. Under similar conditions, PBD is unquenched by CCl_4. Oxygen, an ubiquitous quencher, reduces the fluorescence quantum yield by complex formation; its quenching action is attributed to the π-electron levels and the short lifetime of the solvent-oxygen complex in deactivation [88, 89]. The quantitative quenching effect of oxygen in a liquid scintillator has been studied [90].

In *static* collisional quenching, the encounter complexes consist of unexcited fluor and quencher molecules, and the effect is a decrease of the concentration of fluor molecules in the system. The encounter complexes may also cause dynamic quenching of the fluorescence of the solute monomers and excimers [91].

(ii)   Energy transfer quenching. In this type of quenching excitation energy can jump from an excited molecule or a donor over several molecular diameters (~50 Å) to a quencher or an acceptor molecule by a dipole-dipole interaction (Förster mechanism) [48]. The rate of energy transfer by the resonance mechanism is greater than by diffusion. The impurity, being non-fluorescent in nature, can undergo radiationless transition from the excited state to the ground state by internal conversion and convert the excitation energy into molecular translational and vibrational energy which eventually appears as heat. The requirement for excitation transfer by this mechanism is that the energy level of the acceptor molecule be lower than that of the donor molecule and that there be an overlap of the respective fluorescence and absorption spectra of the donor and acceptor. Many chemical compounds such as aldehydes, ketones, etc., may cause quenching by this mechanism.

In relation to energy transfer, impurity quenching can be reduced by increasing the solute concentration or by using a secondary solute. Kowalski *et al.* [56] have shown that at high solute concentrations and in the presence of a wavelength shifter, the half-value concentration $C_{1/2}$, defined as the concentration of a quencher necessary to reduce the sample count rate to half its original value, is increased. Their results on different scintillators are given in Table 3.4.

Color Quenching

Maximal optical transparency of a liquid scintillator or scintillant to the emitted light photons is essential for achieving a high counting efficiency. The photons from the liquid scintillant have spectral maxima in the range of 380–390 nm or of 415–435 nm if a wavelength shifter is employed. Compounds with absorption spectra in this region will decrease the transmittance of the medium and result in a reduction of the number of light photons reaching the multiplier photocathode. Since organic compounds with absorption maxima in this region of the spectrum are colored, the decrease in scintillation efficiency due to attenuation of the emitted photons is known as color quenching.

In general, color quenching prevails when a scintillation medium is tinted. Red and yellow coloration causes more quenching than blue coloration [92]. The extent of quenching caused by impurity and by color in a sample are difficult to measure separately, but a spectrophotometric method [93, 94] and an isolated internal standard method [95] have been recommended for color quench correction. Since the transmittance of a solution obeys Beer-Lambert's law, attenuation of the light intensity depends upon absorbance, degree of coloration, and distance traversed by the light photons. These parameters affect the pulse height spectra

TABLE 3.4. *Measured Half-Value Concentrations $C_{1/2}$ for Binary and Ternary Liquid Solution Scintillators. Samples Containing ^{14}C Were Quenched with CCl_4* [a]

Primary solute	Concentration (g/l)	Secondary solute[b]	$C_{1/2}$ (g/l)[c]
PPO	4	–	1.3
PPO	4	POPOP	1.3
PPO	10	–	1.8
PPO	10	POPOP	2.3
PBD	9	–	3.5
PBD	9	POPOP	3.6
BBOT	4	–	0.8
BBOT	4	POPOP	1.0
BBOT	9	–	1.8
BBOT	9	POPOP	2.2
Butyl-PBD	7	–	3.0
Butyl-PBD	7	POPOP	3.0
Dibutyl-PBD	10	–	2.2
Dibutyl-PBD	10	POPOP	2.5
TP	2.5	–	0.8
TP	2.5	POPOP	1.0

[a] Data from Kowalski *et al*. [56].
[b] The concentration of POPOP was 0.5 g/l throughout the experiment.
[c] Standard deviation from all the $C_{1/2}$ values was ± 0.2 g/l.

quenched by impurity and by color, in different ways. In color quenched spectra, there is an even pulse height distribution of all energy levels, whereas, in impurity quenched spectra, a preponderance of low energy pulses is observed [96]. This phenomenon is more distinct with samples containing ^{14}C than with those containing ^{3}H.

'Photon' Quenching

In liquid scintillation systems in which the added radioactive specimen is soluble and forms a homogeneous solution, the emitted β-radiation can interact with the fluor without experiencing self-absorption. When the added radioactive sample is either insoluble or partially soluble, precipitated, absorbed on the counting vial surface or dispersed at interphase boundaries, the geometry of the excitation source will be less than 4π, and there will be a reduction in light output. The decrease in photon yield per unit radiation energy absorbed as a result of adverse geometry of the source is termed 'photon' quenching [97].

This type of quenching occurs mainly in heterogeneous counting, e.g. of a suspension or on a solid support. Since the insoluble material is dispersed, its fineness of dispersion determines the degree of self-absorption, especially when β-radiation from ^{3}H is concerned. As the degree of dispersion cannot be controlled from sample to sample, the decrease in photon yield is unpredictable and not subject to correction. Bush [98] proposed the use of a double ratio method for ascertaining the occurrence of photon quenching where phase separation at microscopic levels occurs.

3.2.4 Interfering Processes

Chemiluminescence
When the alkaline solubilizer Hyamine 10-X hydroxide, NCS, Solulene (see Section 3.9.1) or 2N potassium hydroxide is added to a liquid scintillator consisting of PPO, POPOP and naphthalene in *p*-dioxane, light pulses are observed in the absence of a radioactive source [78, 99, 100]. These light photons are attributed to chemiluminescence resulting from the radiative decay of an excited electronic state produced by the energy from a chemical reaction. Diverse chemical reactions give rise to chemiluminescence. The chemiluminescent reaction associated with liquid scintillators appears to require the presence of oxygen or peroxide; the latter is readily formed in *p*-dioxane upon standing.

Kalbhen [78, 99] studied the chemiluminescence produced by mixing alkaline solubilizers with scintillation solvents or liquid scintillators. His results indicate that chemiluminescence is produced from dioxane-Hyamine and from dioxane-NCS. Addition of naphthalene and scintillation solutes causes an increase in the light produced, due to increased energy transfer and improved transparency of the medium. The decay of chemiluminescence is initially fast but may persist for many hours or even days to yield a count rate appreciably above the background. The decay is temperature dependent and proceeds faster at elevated temperatures. The spectrum of chemiluminescence has been studied by Kalbhen [101] who found that at high light intensities the spectrum appears in the ^{14}C channel and as the intensity decreases it 'walks down' to the tritium channel. For this reason chemiluminescence cannot be eliminated by increasing the baseline discriminator level or corrected by channels-ratio method. Although chemiluminescence gives rise to one-photon events, their vast number allow them to trigger the coincidence gate in the spectrometer as accidental coincident events and to be detected.

Chemiluminescence may be eliminated by acidifying solubilized tissue digests to a pH below 7.0 with acetic acid or hydrochloric acid before addition of liquid scintillator [99]. Other acids may increase impurity quenching and may not always succeed in eliminating the chemiluminescence; in fact, tissue samples digested with oxidizing acids ($HClO_4$, HNO_3), have been shown to luminesce [102]. Chemiluminescence can be avoided by using an acid solubilizing agent or combustion for sample preparation, by using the enzyme catalase to remove residual peroxide in bleached tissue digests, or by adding minute amounts of ascorbic acid or stannous chloride as oxygen scavenger to the scintillator [103]. Di-*t*-butyl-4-hydroxytoluene (BHT) has also been recommended [104] and may be used in combination with hydrochloric acid; it is not completely effective when used alone

[102]. Storing the samples at elevated temperatures to allow chemiluminescence to decay completely prior to counting is another method of avoiding chemiluminescence.

Phosphorescence

When proteinous specimens such as rat liver, horse serum, gelatin, egg albumin, trypsin, etc., are dissolved in Hyamine and diluted with toluene or other solvents, the solution exhibits a high count rate in the absence of radioactivity and scintillation solutes [105]. The apparent activity is due to phosphorescence; it is affected by the purity of the solubilizer itself, decays hyperbolically rather than exponentially with a short lifetime and can be photo-reactivated by exposure to fluorescent or sunlight. The cause of this phosphorescence in scintillation systems is not well understood. It can be eliminated by dark-adaptation or acidification of the sample before counting.

Proteinous materials do phosphoresce in the solid state; this is attributed to the presence of the amino acids tyrosine and tryptophane and of condensed amide linkages [106]. However, a solution of trypsin in 1.0 N KOH does not phosphoresce.

Toluene shows an increase in luminescence when saturated with oxygen. Toluene of high purity luminesces less than toluene of ordinary grade, and 5 per cent water quenches the luminescence [107].

Empty glass and plastic vials when photo-activated yield counts. The former phosphoresces because of traces of rare earths and the latter due to residual plasticizer. In the presence of scintillator molecule, the phosphorescence of the bottle, the solvent and the additive increases; this is attributable to a facilitation of the energy transfer to the fluor [108].

Phosphorescence yields one-photon events and can be suppressed by coincidence gating. Its pulse height spectrum resembles closely that of ^{3}H, causing interference in the measurement of samples containing ^{3}H. For samples containing ^{14}C, phosphorescence can be eliminated by raising the discriminator bias level.

3.3   INSTRUMENTATION

The success of liquid scintillation counting as an analytical technique largely depends on the sensitivity and reliability of the detection instruments. The use of two multiplier phototubes in coincidence circuitry to detect the low light yield from scintillation samples has been the standard practice for eliminating thermionic noise from phototubes and noise of other origin. The conversion of light photons from the sample into electronic pulses of different amplitudes and the sorting and counting of these pulses constitutes the core function of a liquid scintillation spectrometer. In the following, some of the salient features of the instrument will be discussed; it is necessary to understand the function of the parts in order to derive the maximum performance of the whole system.

3.3.1 Light Detection

The average energy of β-particles from ^{3}H is 5.6 keV. This, according to equation (3.11) will produce approximately 80 photons in a deaerated solution scintillator containing PPO and POPOP in toluene. An ordinary liquid scintillator is oxygen-equilibrated with a concentration of 41.4 ppm oxygen in toluene [90]. Quenching by oxygen will reduce this photon yield by 10 to 15 per cent and the detection of an even smaller number of photons in the presence of impurity or color quenching is indeed an exacting task.

Light photons from the sample are emitted isotropically and are reflected from a highly polished aluminium surface in the counting chamber onto the photocathodes of multiplier phototubes (see Chapter 2).

The multiplier phototube is the core in light detection. Quantum efficiencies of photocathodes may vary with the cathode and window materials from 20 to 30 per cent, i.e., 2 to 3 electrons are produced for every 10 photons reaching the photocathode. Phototubes with bialkali photocathodes are superior in having spectral response maxima at 380–400 nm, high quantum efficiency and low dark current. Their response is not improved by the use of wavelength shifters. Phototubes with quartz windows remain transparent at short wavelengths and also yield a lower background due to the absence of ^{40}K.

Dark current or thermionic noise from random events arising from thermally excited electrons released from the photocathode or dynode surfaces, can be minimized by cooling or eliminated by coincidence circuitry in two-phototube instruments. Phototubes with bialkali photocathodes having negligible dark current are preferably employed in spectrometers operating at ambient temperature.

Pair matching of multiplier phototubes for maximum performance, based upon their spectral response, sensitivity and efficiency, is a necessity for two-phototube spectrometers. Some of the photocathode characteristics are given in Table 3.5.

The efficiency of light detection may be increased by the use of a light guide. Horrocks and Studier [109] reported that a Lucite guide enhances the counting efficiency for ^{14}C by 20 per cent and doubles the efficiency for ^{3}H. Bensen and Maute [110] noted an approximate threefold increase in ^{3}H efficiency using a short Lucite guide with aluminium foil coupled to its wall as reflector. Light guides are not used, however, in modern commercial spectrometers because the block of Lucite provides a copious source of Čerenkov radiation which elevates the background and diminishes the figure of merit (see Section 3.3.3).

3.3.2 Pulse Height Analysis

Pulse height analysis is a technique by which pulses can be electronically sorted according to their amplitudes.

The pulse height depends on the scintillation efficiency of the scintillator, β-particle energy, and instrument parameters, namely, the gain of the amplifier and the discriminator setting. Kowalski *et al.* [56] have related pulse height to counting efficiency on the basis of these factors.

TABLE 3.5. *Characteristics of Photocathodes*[a]

Spectral response	Photo-cathode material	Window material	λ_{max} (nm)	Sensi-tivity (μA/lumen)	Quan-tum effic-iency % at λ_{max}	Dark emission at 20°C (10^{-15} A/cm^2)
107 (S-11)	Cs-Sb	glass	450	70	15.7	3
110 (S-20)	Na-K-Cs-Sb (tri-alkali)	glass	420	150	18.8	0.4
113	Na-K-Cs-Sb	pyrex	–	200	23.9	0.4
115	K-Cs-Sb (bialkali)	glass	405	67	24.5	0.02
116	K-Cs-Sb	pyrex	380	85	31.2	0.02
117	K-Cs-Sb	UV-glass	400	67	24.4	0.02
133	K-Cs-Sb	fused silica	400	60.	22	0.02

[a] From RCA PIT-701B Chart.

Amplifier Gain

The amplifier is an analog device whose function is to amplify the input pulses for the analog-to-digital converter in the form of a pulse height analyzer and to conserve the energy information during the amplification process. The amplifiers used in liquid scintillation spectrometers may have linear or logarithmic characteristics depending upon the relationship desired between the output and the input pulses. The logarithmic amplifier has the advantage of handling a very wide range of signal levels while the linear one has finer discrimination properties. A good amplifier should possess temperature independence, long-term gain stability, optimum signal-to-noise ratio, non-overloading characteristics and minimum sum effects resulting from pile-up of pulses.

Amplifiers may also be classed as voltage and current amplifiers. The latter are frequently used for low input impedances. Since the advent of transistors, integrated circuits in miniature have replaced bulky vacuum tube circuits. Transistors have low power dissipation, long lifetime, and fast response; they offer unsurpassed miniaturization in integrated circuitry.

The gain of an amplifier has a direct bearing on sample counting efficiency. Quenching diminishes the light output from a scintillation sample and decreases the pulse height of the input signal to the amplifier. To offset the effect of quenching, the counting may be performed with a higher voltage supply to the multiplier phototube or at an increased amplifier gain [111]. Quenching decreases the pulse amplitude but so long as it remains above the detection threshold, the pulse will be counted, thereby making the effect of quenching inconsequential.

Discriminators

The discriminator is a form of analog-to-digital converter; it sorts input pulses according to amplitude, processing those above a preselected threshold value for registration by a digital device or scaler and suppressing the rest.

In liquid scintillation spectrometers, two discriminators are provided for each counting channel; one discriminator is used to set the lower threshold or the channel position or the discriminator height and the other the upper threshold. The difference between the two discriminator levels is known as the acceptance slit or the channel width or simply the 'window'. When the two discriminators are connected in anti-coincidence to form a gating system accepting only pulses with amplitudes appearing within the 'window', they are known as differential discriminators. When the discriminator limiting the upper threshold is switched off, the lower one becomes the baseline discriminator and is functioning as an integral discriminator, in which case all the pulses with amplitudes above the baseline are counted.

The position of the baseline discriminator or the discriminator height determines the sample counting efficiency. The higher the baseline threshold, the lower the counting efficiency; at high baseline levels pulses from noise and phosphorescence are suppressed, leading to a decrease in the background count rate. The selection of the discriminator height is based upon the figure of merit.

Circuitry associated with the discriminator should have good stability and sharp cutoff characteristics so that the pulse height spectrum does not experience distortion near the discriminator cutoff [112]. A sharp cutoff is particularly important in insuring the validity of the channels-ratio method for quench corrections in samples containing low radioactivity.

In many makes of spectrometer, the discriminators are frequently replaced with push-button devices which automatically select the pre-set channel width and amplification for optimum counting of either ^{14}C or ^{3}H alone or in presence of one another. The pulse height discriminator controls are usually 10-turn 1000-division potentiometers which span the full range of amplitudes of output pulses from the amplifier.

Counting Mode

Integral and differential modes of counting are practiced in liquid scintillation counting. The former relies on the use of an integral discriminator to accept all pulses above a given threshold limit, whereas the latter accepts only those appearing within the channel width set by the differential discriminators. In differential counting, the counting efficiency represents the difference between the integral efficiencies at the lower and the upper limit of the channel [113].

The interrelationships of channel width, gain and counting efficiency for samples containing ^{14}C and ^{3}H are given in Figs. 3.3 and 3.4. In integral counting, the efficiency increases rapidly at first and gradually levels off into a plateau; whereas in differential counting, it increases rapidly to a maximum and then decreases as the gain is further increased. This efficiency

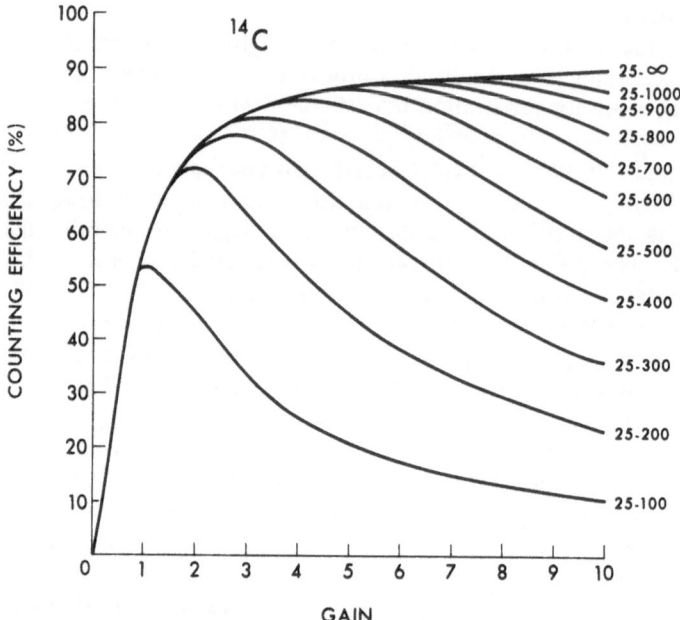

Fig. 3.3. Counting efficiency vs. gain curves at various window widths for ^{14}C.

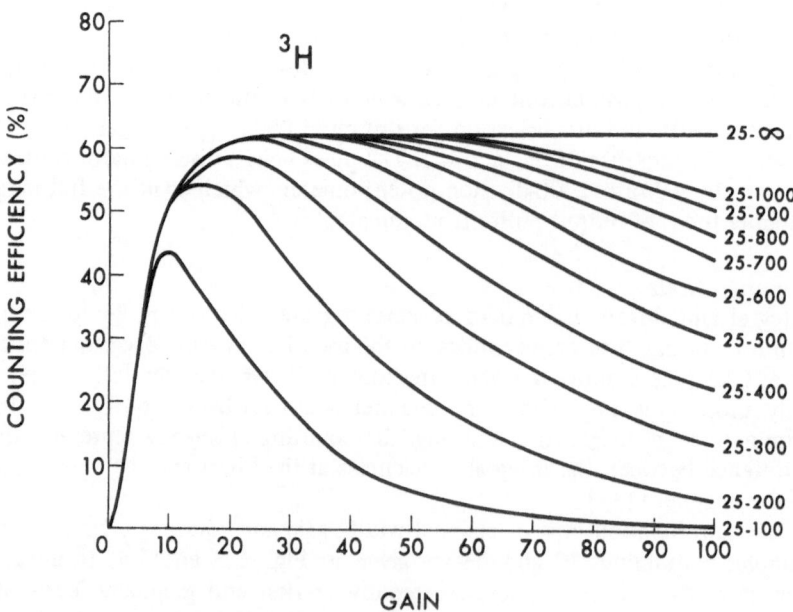

Fig. 3.4. Counting efficiency vs. gain curves at various window widths for ^{3}H.

peaking is attributed to an over-amplification of pulses at high gains to beyond the upper limit, resulting in the loss of such pulses as counts. Counting efficiency increases with an increase in the channel width but is accompanied by a shift of the balance point to a higher gain. Operation at the balance point [114] where the maximum counting efficiency occurs, is the preferred form of differential counting. In modern spectrometers the superior stability of the instrument gain has rendered infrequent the need for balance point operation.

Differential counting using a wide channel is practiced in double-isotope counting and is preferred to integral counting because large background pulses due to cosmic radiation can be eliminated to improve the figure of merit. The merits of differential vs. integral counting have been evaluated [115].

The integral count rate of a quenched sample decreases exponentially with increasing concentration of the quencher (see Section 3.5); whereas, when the same quenched sample is differential-counted at the balance point set for unquenched sample, the count rate may increase, remain constant or decrease depending upon the discriminator height and the gain. If the quenched sample is counted at its own balance point, the count rate decreases exponentially with an increase in quencher concentration [116]. Quenching causes a progressive downward shift of the pulse heights above the upper discriminator limit into the window and thence to below the detection threshold; the balance between the incoming and outgoing pulses in the window determines the change in count rate. It has been proposed that in the balance quenching method counting of all samples can be performed regardless of the extent of quenching in a narrow channel at low channel position so that the net change in count rate in the window due to quenching is essentially zero [117].

Pulse Height Spectrum

The differential pulse height spectrum is obtained by scanning with differential discriminators at a fixed window, and the integral pulse height spectrum by scanning with an integral discriminator. The integral spectrum is of approximately exponential shape and is useful for absolute standardization of the measured radionuclide (see Section 3.7). The differential spectrum represents the distribution of pulses according to amplitude or energy and its resolution is inversely proportional to the channel width. Light photons from ^{14}C and ^{3}H in an ideal liquid scintillator yield differential spectra resembling the theoretical β-energy spectra of these isotopes.

Pulse height is affected by photon yield, spectral response matching factor and baseline discriminator level, and is frequently used to evaluate scintillation solvents and solutes on a comparative basis [56]. In practical counting of samples containing ^{14}C, quenching causes a downward shift of the pulse height spectrum towards the low energy end. For ^{3}H, in addition to the shift, a decrease in pulse amplitude also occurs. Neary and Budd [96] reported a variance between the shape of the pulse height spectrum quenched by impurity and that quenched by color. Among the colors, the downward shift is more pronounced with yellow and red than with blue.

Klein and Eisler [112] observed a distortion of the pulse height spectrum near the detection threshold, especially when a high baseline discriminator level is employed. The endpoint of the spectrum of a quenched sample can be restored to that of an unquenched sample by an increase in gain. The effect of excess of gain or over-amplification on a normal spectrum is to flatten it out, thereby yielding a 'flat' spectrum [59].

3.3.3 Figure of Merit

A criterion of quality in selecting spectrometers, phototubes, sample geometry, instrument settings, etc., for precision counting is the figure of merit, defined as the ratio of the square of counting efficiency to background count rate. This ratio is derived from statistical considerations to minimize the variance of the disintegration rate of the sample for a given counting time. Wylid [118] gave the expression

$$\sigma_D^2 = \frac{1}{T(E^2/B)}$$

where σ_D is standard deviation of disintegration rate of sample, T time for gross count rate, E counting efficiency and B background count rate. If gross counts can be taken in unit time, the square root of the figure of merit becomes the reciprocal of standard deviation.

Porges [119] derived a slightly different expression for the variance

$$\sigma^2 = \left(\frac{\alpha B}{S^2}\right) \left(1 + \frac{S}{\alpha B}\right)$$

where

$$\alpha = 1 + t_S/t_B$$

where S and B are the 'true' counts of source and background; t_S and t_B refer to their counting time, respectively. This equation emphasizes that when $S/\alpha B \ll 1$, i.e., at low sample count rate, S^2/B must be maximized for precision; whereas, when $S/\alpha B \gg 1$, namely, at high sample count rate, S/B value or signal strength must be maximized regardless of S^2/B.

Putman [120] is essentially in accord with Porges but stresses that for a given counting time the relative standard deviation is proportional to $\{(s + b)^{1/2} - b^{1/2}\}$ where b is the background count rate and $(s + b)$ the signal plus background rate.

It appears that in practical counting, strict adherence to a high figure of merit in selecting instrument settings for measurement of relatively active samples may lead to no greater statistical accuracy.

3.3.4 Liquid Scintillation Spectrometers

The main function of spectrometers is to count radioactive samples with precision and accuracy with the least amount of time and labor. Liquid scintillation spectrometers can be classed as one-phototube or two-phototube instruments. The former are used for counting α-emitters; they may save on cost but suffer from lack of sensitivity and versatility in performance; the latter are highly sophisticated electronic devices and most suited for counting ^{14}C and ^{3}H.

An ideal liquid scintillation spectrometer should measure the sample activity precisely and accurately, be capable of handling a large number of samples automatically and performing a limited amount of data reduction and computation with readout presentation compatible with large computers. Thus, a liquid scintillation spectrometer becomes an integrated assembly of sample changer, coincidence light detection with two multiplier phototubes, pre-amplifiers, pulse summation circuit, amplifiers, pulse height analyzer, ultra-high speed scaler, electronic computer, and a readout device which may be a typewriter or teletype, with tape perforator attachment, or magnetic tape. Pulse summation was introduced in 1963 to improve the separation of activities of ^{14}C and ^{3}H in dual-label sample counting. A modern spectrometer has three measuring channels for multi-isotope counting or for quench correction by channels-ratio and two additional pre-set channels for external standard ratio. Gain controls and differential discriminators are provided for each of the three counting channels; factory calibrated sets for optimum counting of ^{14}C and ^{3}H alone and in the presence of each other have become standard accessories. Limited electronic computation can be carried out on cpm, standard deviation, channels-ratio and external standard ratio for print-out on typewriter or teletype. Some spectrometers have included a measuring control program which allows groups of samples to be automatically counted according to any pre-set instrument settings.

The sample changer is usually of the serpentine type, functioning dependably and efficiently; it has become the standard in all makes of modern spectrometers. In more recent models the sample chain can move forward and backward; this mechanism is very convenient for selective counting of an individual sample within a group.

All modern spectrometers are equipped with a γ-source for automatic external standardization. The selection of the source is based on its availability, half-life and the energy spectrum of its recoil electrons. The radionuclides most commonly used for this purpose are ^{133}Ba, ^{137}Cs, ^{144}Ce, ^{226}Ra and a compound source $^{226}Ra - {}^{241}Am$.

Automatic quench correction devices are newer features in spectrometer assemblies. Some manufacturers use automatic quench compensation (AQC) for restoration of quenched counts by an increase in gain [121]; others either use a small on-line computer to make quench corrections according to a pre-determined calibration curve, or employ an absolute activity analyzer (AAA) which causes a change in photoelectron optics by defocussing the multiplier phototube with a magnetic field and accommodates the quench correction of the sample to a pre-calibrated quench curve.

A simplified diagram of a liquid scintillation spectrometer system is given in Fig. 3.5.

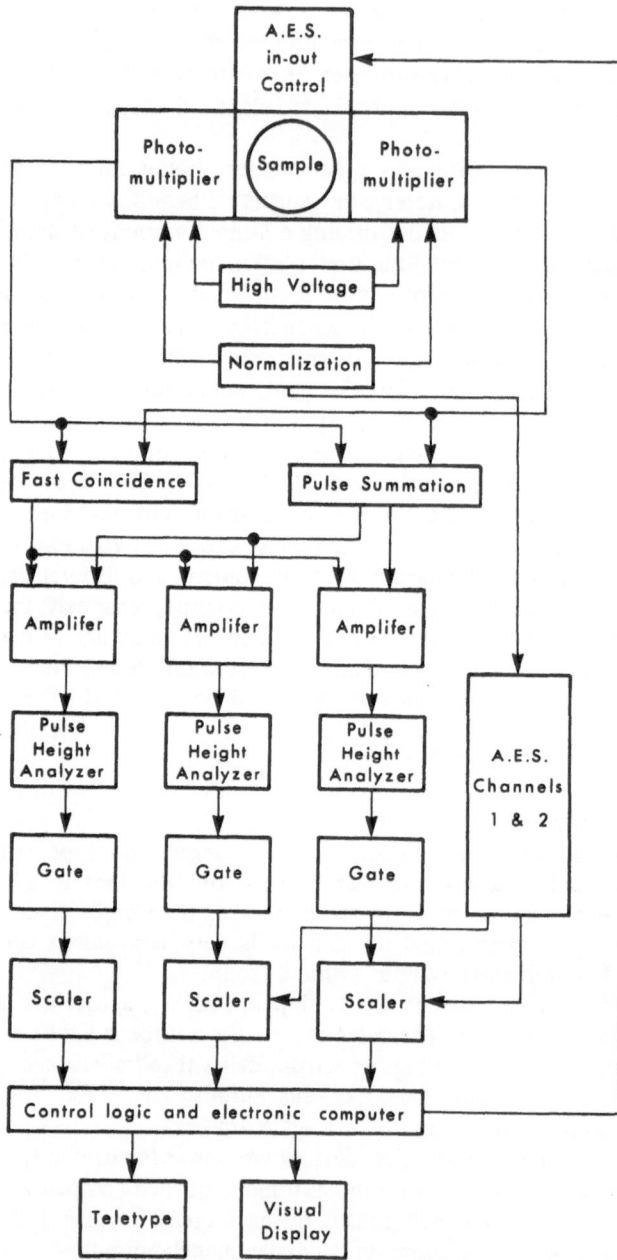

Fig. 3.5. Block diagram of a three-channel liquid scintillation spectrometer with fast coincidence, pulse summation and two pre-set channels for automatic external standardization.

3.4   SOURCES OF BACKGROUND ACTIVITY

The background count in liquid scintillation counting may arise from the following sources [111]:

- (i)   Chance coincidence. This is attributed to two noise signals each from a multiplier phototube, arriving at the gating circuit within its resolving time to be passed as a coincidence event and counted. The coincidence resolving time in a modern spectrometer is of the order of 20 nanoseconds; therefore, a noise rate of 40,000 cpm per phototube would contribute about 1 cpm in accidental count rate.
- (ii)   Cross-talk. This is a reciprocal light sensitization of the photo-cathode of one multiplier phototube by the light generated in the other phototube in the 'head-on' position in the spectrometer. Background due to noises from multiplier phototubes in low level liquid scintillation counting has been evaluated by Boyce and Cameron [122].
- (iii)   Čerenkov radiation (see Section 3.13). It arises from ^{40}K in glass sample vials from cosmic radiation, and from ^{40}K and uranium in the glass envelope of the multiplier phototubes. Plastic light guide forms another copious source of Čerenkov radiation.
- (iv)   Cosmic radiation and ambient radioactivity. Ambient radioactivity from laboratory contamination, from radium and its daughter products in the environment and in shielding material contribute to background rate.

Contribution to the background rate can be evaluated by counting a black vial for (i), an empty sample well for (i) + (ii), a solvent blank for (i) + (ii) + (iii), and a sample of liquid scintillator for the total. The difference between these successive count rates yields the individual contribution from these sources as listed.

In an impurity-quenched sample, quenching affects only the fraction of the background count due to (iv). Since signal pulses from cosmic radiation have larger amplitudes than those from ^{14}C and ^{3}H, the former can withstand a greater degree of quenching without being lost as counts. In color quenched or color-plus-impurity quenched samples both fractions due to (iii) and (iv) will be adversely affected.

The effect of quenching on background count rate can be determined by the use of a correlation curve relating the background rate (i) to the count rate or the channels ratio of an external standard [123], (ii) to the counting efficiency of the sample [68] or (iii) to the ratio of the concentration of the specimen in the sample to its half-quenching concentration [124]. The sharp difference in the background rates of samples quenched by chloroform and by methyl red [123] indicates that coloration causes additional quenching by affecting the transmission of Čerenkov light. Color quenches isotope and background solutions differently and the extent of quenching can be related directly to the reciprocal of the absorbance at 400 nm [92].

3.5   EFFICIENCY DETERMINATION

The usefulness of liquid scintillation counting as a technique for radio-activity measurement depends mainly upon the precise efficiency with which each sample is determined. For internal sample counting, the counting efficiency is affected in an individual manner. The necessity for precise and accurate efficiency determination cannot be over emphasized.

The methods for quench correction have been extensively reviewed [97, 115, 125, 126]; only a brief treatment is given in the following sections.

Internal Standard Method

The addition of a calibrated standard of the measured radionuclide to a sample that has been previously counted for efficiency determination is known as the internal standard method. If the count rate of the sample is C_1 and that of the sample plus the added internal standard C_2, then the difference in count rates $(C_2 - C_1)$ is due to that of the internal standard. This value when divided by its calibrated activity in disintegrations per unit time C_{is}, yields the counting efficiency for the particular sample; thus,

$$\text{Counting efficiency} = \epsilon = \frac{C_2 - C_1}{C_{is}}$$

The true or absolute activity of the sample is calculated as C_1/ϵ in disintegrations per unit time, e.g., as dpm.

The internal standard method is conceptually simple, long established and most widely used for quench correction in liquid scintillation counting. When properly carried out, it is the most accurate method for quench correction [127].

The accuracy of the method depends to a large extent upon the level of activity of the internal standard added. For maximum accuracy, the added internal standard should have a count rate equal to or exceeding that of the sample [46]; this requirement is verified from statistical considerations [128].

The disadvantage of the method is the loss of sample for recount and for further investigative use because of radiochemical contamination by the added internal standard. Since the count and recount of a sample are spaced in time, when a large number of samples are involved, instrument drift and the change in sample quenching properties may introduce appreciable errors. In severely quenched samples the use of an internal standard may not be justified [59] because the unequal statistical accuracy achieved with the sample count as compared to sample plus internal standard count can lead to unacceptable errors. However, the results of Roger and Moran [127] seem to indicate otherwise.

Extrapolation Method

When samples containing a radioactive specimen in increasing concentrations are integral-counted, their count rates usually do not reflect a proportionate increase with the specimen concentration as anticipated [111]. In fact, if the results are plotted, a bell-shaped curve concave toward the concentration axis is obtained which can be represented by the equation:

$$N = S_o C \exp(-qC)$$

where N is the observed count rate S_o the specific activity or count rate per unit concentration of the specimen in the absence of quenching, C the specimen concentration, and q the quenching constant. The quenching constant is equal to $0.693/C_{1/2}$ where $C_{1/2}$ is the half-quenching concentration at which the count rate is reduced to half its original value by quenching. Kowalski *et al.* [56] have derived an equation relating the $C_{1/2}$ value to energy transfer, β-radiation energy and instrument parameters.

If the logarithm of the ratio N/C, defined as the apparent specific activity, is plotted against concentration C, a straight line results which can be extrapolated to zero concentration to give the specific activity in the absence of quenching.

Although the extrapolation method can relate concentration to count rate of a labeled specimen that may quench, it has not been adopted for routine use for quench correction. The method is tedious and requires several samples for each specimen to provide enough data points for extrapolation. Nevertheless, it is inherently more accurate and dependable for efficiency determination than other methods of quench correction because statistical methods can be applied to multiple data points for extrapolation.

The method is only valid for integral counting or wide channel counting in which the pulse height spectrum of the measured radionuclide has not been truncated at the high energy end. The exponential decrease in count rate by quenching may be considered related to the approximate exponential shape of the integral pulse height spectrum; the exponential factor may be derived from other theoretical considerations [129]. Recently, Duggan and Ice [130] reported the determination of the E_{max} of various β-emitters from the relative slope of the quench plot; this parameter is identical to $C_{1/2}$ defined above.

Channels-Ratio Method

Quenching causes a decrease in photon yield and results in a downward shift of pulse height spectrum as shown in Fig. 3.6. If the spectrum is divided into two parts and their activities are measured, the ratio of the count rates will reflect the extent of quenching; its utilization for quench correction is known as the channels-ratio method.

The channels-ratio method was first introduced by Baillie [131] and improved upon by Bush [132]. As shown in Fig. 3.6, if one allows channel $(L_1 - L_2)$ to monitor the low-energy portion and channel $(L_2 - L_3)$ the high-energy portion of the spectrum and the combination $(L_1 - L_3)$ the entire spectrum, curvilinear and linear calibration curves relating counting efficiency to channels ratio can be obtained either as $(L_2 - L_3)/(L_1 - L_2)$ or as $(L_1 - L_2)/(L_1 - L_3)$ for ^3H or ^{14}C, respectively.

The channels-ratio method has the advantage of obtaining simultaneously both the sample count rate and the channels ratio. The sample is counted

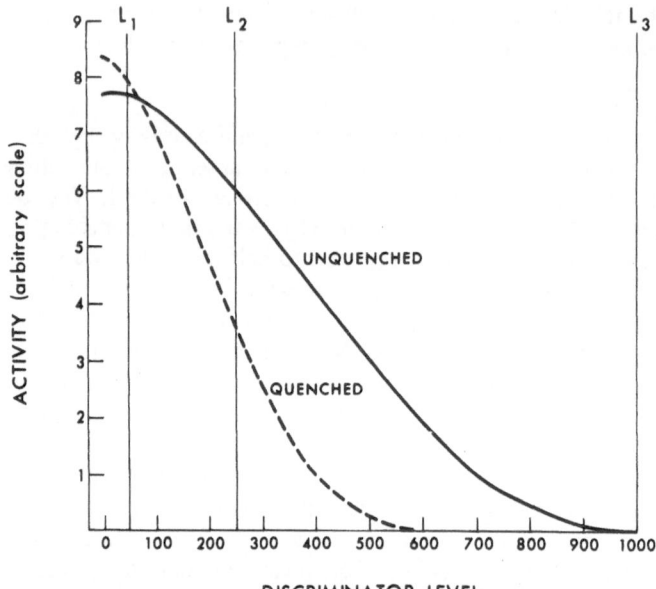

DISCRIMINATOR LEVEL

Fig. 3.6. Unquenched and quenched pulse height spectra of ^{14}C. The counting efficiency of the quenched sample is 75% of that of the argon-purged, unquenched sample. L_1, L_2 and L_3 represent discrimination levels.

once and no additional manipulation is required. The simplicity of the
method and its ease of automation and capability of handling a large number
of samples are unsurpassed.

The shortcoming of the method lies in the large errors incurred in
samples that are highly quenched or have low count rates [127]. Inaccurate
background correction and distortion of the discriminator cut-off can
adversely affect the accuracy of the channels-ratio determination.

In practice, a calibration curve correlating counting efficiency and
channels-ratio is obtained by counting a set of quenched standards of
known efficiency in two selected channels to yield the count rate ratio as
specified above. Samples that may quench are counted in the same two
channels, and the ratio of the count rates thus obtained is then related to
the counting efficiency using the calibration curve. The correlation curve for
samples containing ^{14}C quenched by impurity differs slightly from that by
color; for samples containing ^3H, no such difference exists [132]. The
correlation curve for ^3H is temperature sensitive, whereas, that for ^{14}C is
insensitive to temperature change [132].

The linearity and the slope of the correlation curve are related to channel
width and channel position [113]. Linearity allows recalibration of the
curve using only two points, and an adequate slope is essential to the sensitivity of the method [133]. The channel width is defined by

$$k = \frac{S_{o1}}{S_{o2}}$$

and the channel position or the discriminator height by

$$\gamma = \frac{\log S_{02} - \log S_2}{\log S_{01} - \log S_1}$$

where S_{01} and S_{02} are the integral count rates of unquenched sample using the lower and upper discriminator of the window at levels 1 and 2 as integral discriminators, and S_1 and S_2 the integral count rates of a slightly quenched sample at the same discriminator levels 1 and 2. From the k and γ values, the channels ratio corresponding to each counting efficiency can be calculated. From the concavity or convexity of the calculated correlation curve, it is easy to ascertain whether the channel width or the channel position should be varied to achieve approximate linearity and adequate slope [113].

The correlation curve between the counting efficiency and channels ratio can be fitted by a polynomial in which the former is expressed as a function of the latter [134, 135]; in this form, the counting data can be readily processed by machine computation (see Section 3.11).

External Standard Method

The use of either Compton recoil electrons from high energy γ-rays, photoelectrons from low energy γ-rays (< 60 keV), or X-rays [136], or a combination of these for quench correction is known as the external standard method. External standardization was introduced in 1962 independently by Higashimura *et al.* [137] and by Fleishman and Glazunov [138]. It was first used in commercial liquid scintillation spectrometers by Rapkin in 1964; since then it has become an integral part in all makes of modern spectrometers.

The advantage of the method is its ease of automation; no additional sample handling is required after the sample is prepared. In practice, the sample is counted twice, once in the presence of external standard and once alone. A correlation curve between the count rate of the γ-ray standard and the sample counting efficiency is obtained using a set of quenched standards. When a sample that may quench is counted with the external standard, the sample counting efficiency can be readily determined from its count rate using the calibration curve.

de Wachter and Fiers [139] have given detailed procedures for setting up external standardization for quench correction Since the photo- and recoil electrons generated from the external standard are more energetic than the β-particles from ^{14}C and ^{3}H, it is necessary to select proper instrument parameters to attenuate the pulse height and to ensure an approximate exponential decrease of the external standard count rate with an increase in quencher concentration. The reported phenomenon of an increase in external standard count rate with increasing quencher concentration should be avoided because it can lead to unacceptable errors [140, 141].

The disadvantage of the method is that the production of recoil electrons is dependent upon the electron density of the medium in which Compton scattering occurs. Electron density varies with sample volume, wall thickness and the presence of heavy atoms in the sample. A high electron density allows more photo- and recoil electrons to be generated, thereby enhancing the external standard count rate. The electron density of p-dioxane is 20% higher than that of toluene. A 10 per cent variation in glass wall thickness according to Higashimura et al. [137], does not introduce an error of more than 0.2% in the total count. Rogers and Moran [127] observed no correlation between the count rate of the standard and the weight of the vial. The count rate of the external standard increases with sample volume but levels off at 11 ml or beyond [142].

The use of external standard channels-ratio can avoid some of the pitfalls associated with the application of its count rate. In newer spectrometers, two special counting channels are provided for counting the external standard. Takahashi and Blanchard [141] obtained a curvilinear relationship between the external standard channels-ratio and the sample counting efficiency in which the latter can be expressed as a cubic function of the former. This form of presentation is useful for computer programming.

The accuracy of external standardization for quench correction is inferior to both the channels-ratio and the internal standard methods [127] owing to the irreproducibility of the counting geometry of the sample relative to the external source and to the dissimilarity in pulse height spectra of the external standard and the measured radionuclides (^{14}C, ^3H).

Other variations of the method include the use of an 'efficiency stick' [144, 145] and the double ratio method [146].

Miscellaneous Methods

The gain restoration method [121] involves the restoration of the endpoint of a quenched pulse height spectrum to that of the unquenched spectrum by over-amplification. This principle is used in commercial instruments for automatic quench compensation (AQC). The effectiveness of the method has been discussed by Wang [121]. Other methods proposed for efficiency determination include the gain ratio method [147], the balanced quenching method [117, 148], the dilution method [149] and a variation of the channels-ratio method [135].

3.6   MULTIPLE-ISOTOPE COUNTING

In many radiotracer studies, the use of multi-labels in one experiment can yield more meaningful data under controlled conditions than several experiments performed separately with single labels. In the isotope derivative dilution method of analysis used, for example, in the determination of the circulating concentration of aldosterone in man [150], concurrent use of ^{14}C and ^3H becomes obligatory. The feasibility of any multi-label experiment depends upon its design and the ability to measure individual labels accurately. As the energy of β-particles varies from zero to the maximum energy (E_{max}), overlap of energy spectra of β-emitters occurs. The extent of spectrum overlap is a function of the values of E_{max}. Isotope pairs can be formed from pure β-emitters such as ^{3}H, ^{14}C(^{35}S), ^{45}Ca, ^{32}P, ^{36}Cl, ^{144}Ce, etc., for double labeling, and accurate radio-assay is feasible if the endpoint energies of the paired isotopes differ from each other by a factor of about 2. Pulse summation accentuates the pulse height difference and improves the isotope separation [151].

Double label counting can be performed by the simultaneous equation method or by the isotope exclusion method [152].

According to Okita *et al.* [152] and Axelrod *et al.* [153] the discriminator ratio method is the more accurate form of the simultaneous equation method, although both are identical in theory. The former makes use of the ratios of count rates instead of counting efficiencies in the equations for calculating activities and obviates errors associated with efficiency determination. This method can be illustrated as follows using the isotope pair ^{14}C + ^{3}H; thus

$$N_1 = H_1 + C_1 \tag{3.14}$$
$$N_2 = H_2 + C_2 \tag{3.15}$$

where N_1, N_2 = net cpm of channel 1 (low-energy channel) and channel 2 (high-energy channel), respectively,

H_1, H_2 = net cpm of ^{3}H in channel 1 and 2, respectively

C_1, C_2 = net cpm of ^{14}C in channel 1 and 2, respectively

From equations (3.14) and (3.15) one can derive

$$H_1 = \frac{b\,N_1 - N_2}{b - a} \tag{3.16}$$

$$C_2 = \frac{b(N_2 - a\,N_1)}{b - a} \tag{3.17}$$

where $a = H_2/H_1$, and $b = C_2/C_1$. Therefore

$$^3\text{H dpm} = \frac{H_1}{^3\text{H counting efficiency in channel 1}}$$

$$^{14}\text{C dpm} = \frac{C_2}{^{14}\text{C counting efficiency in channel 2}}$$

The counting efficiencies for channels 1 and 2 are determined with un-quenched samples containing known standard of ^{14}C- and ^{3}H-toluene.

In the isotope exclusion method, the activity of ^{3}H is excluded entirely from the high-energy channel to simplify the computation. The sample is counted in the low-energy channel for both ^{14}C and ^{3}H and in the high-energy channel for ^{14}C alone. Such a condition is equivalent to setting a = 0 in equations (3.16) and (3.17); thus

$$H_1 = N_1 - \frac{N_2}{b} \qquad\qquad\qquad (3.18)$$

$$C_2 = N_2 \qquad\qquad\qquad (3.19)$$

Owing to its simplicity the isotope exclusion method is widely used. In practice, channel 1 is set for optimum counting of ^3H and channel 2 adjusted for that of ^{14}C. The spillover of ^3H activity in the high-energy channel is controlled by its width, or its discriminator height. According to Kobayashi and Maudsley [154], the efficiency for counting ^{14}C can be increased from 51% with no ^3H spillover to 72.5% with 1% ^3H appearing in the high-energy channel. The accuracy of the method determines the degree of spillover that can be tolerated.

The optimum width of the low-energy channel can be selected from the relative increase in ^{14}C and ^3H counting efficiencies as the channel width is increased. The lower discriminator level is set at the limit consistent with the instrument at optimum gain for ^3H; the channel width is then increased and the observed counting efficiencies for ^3H and ^{14}C at each channel width are plotted as indicated in Fig. 3.7. The point where the tangent drawn from the initial portion of the curve veers away from the curve represents the optimum width of the low-energy channel (see Section 3.10).

Another method of selecting the optimum channel width in double label counting using a log-log efficiency plot, has been described by Kobayashi and Maudsley [154].

In addition to parameters discussed from a theoretical and a practical point of view by Klein and Eisler [155], the accuracy of multi-label counting techniques also depends upon the isotope ratio, i.e., the relative activities of radionuclides in the mixture. In the isotope exclusion method, the accuracy of ^{14}C assay is unaffected by ^3H activity but the reverse is not true since appreciable ^{14}C activity appears in the ^3H channel. The isotope ratio of $^3H/^{14}C$ may vary from 100/1 to 1/20 and can still be assayed with less than 5 per cent error; but for a ratio of 1/100, the error for ^3H assay exceeds 100 per cent. Hassig and Schipper [156] noted that the only isotope ratios of $^3H/^{14}C$ giving small errors of estimate lie in the range from about 0.6 to 6.0. Very low ratios generally result in an underestimation of ^3H activity while a very high ratio yields a consistent, large over-estimation of ^{14}C.

The effect of quenching on multi-labeled samples is complex, since pulse height spectra, counting efficiencies and degree of channels overlap will change with quenching. The half-quenching concentration of a quencher increases with the β-particle energy and also with the gain of the counting channel [111]. The effect of quenching on the count rate of ^3H and ^{14}C in the low energy channel differs because the β-energies of the two nuclides differ. The effect on the count rate of ^{14}C in the low- and high-energy channels also differs owing to a variation in gain. In multi-label counting, conditions for balanced-point operation do not prevail because quenched

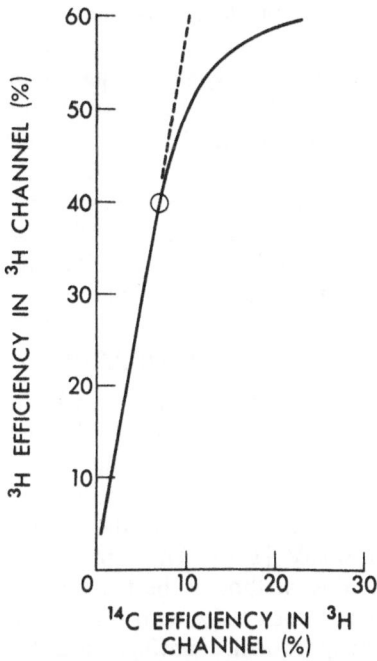

Fig. 3.7. Relative increase in counting efficiencies for ³H and for ¹⁴C in ³H channel with increasing window width. The circled region showing where the curve bends away from linearity gives the optimum channel width for ³H double-isotope counting.

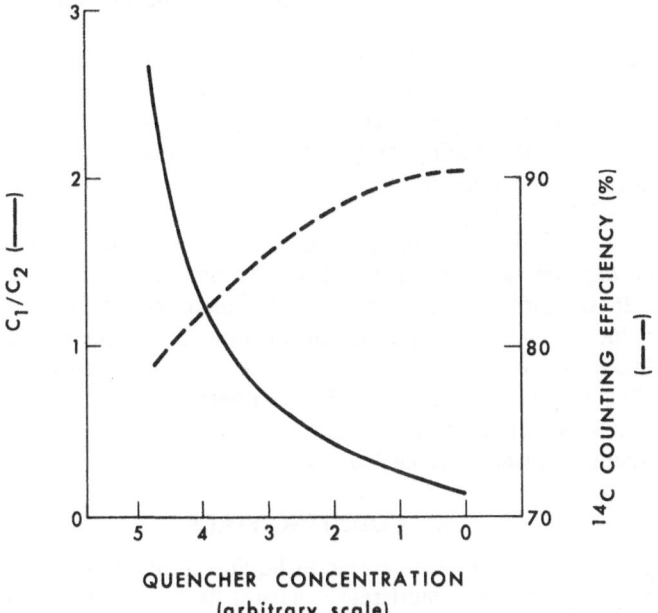

Fig. 3.8. Plots of C_1/C_2 and ¹⁴C counting efficiency vs. quencher concentration. C_1 and C_2 represent the count rates of ¹⁴C in ³H channel and in ¹⁴C channel, respectively.

samples are counted under conditions calibrated for unquenched standards. The ratio of counting efficiencies or count rates of ^{14}C in the two channels (C_1/C_2) will vary according to the degree of quenching, their relationship is shown in Fig. 3.8. It is therefore necessary to consider the change in this ratio with the degree of quenching in order to preserve the validity of equations (3.16) to (3.17).

Quench correction for multi-labeled samples can be carried out using the methods discussed in Section 3.5 and many applications have been reported [124, 134, 139, 152, 157–165]. Computer-aided data processing on multi-label counting data is discussed in Section 3.11.

3.7   ABSOLUTE STANDARDIZATION

The absolute disintegration rate of radionuclide samples can be measured by liquid scintillation counting at various discriminator or bias settings followed by extrapolation to zero bias. This method is essentially the extrapolation of the integral spectrum to zero energy and can be applied with 100 per cent efficiency to the standardization of α- as well as β-emitters with energies above 150 keV [166]. The reproducibility of the determinations carried out by this method is better than ± 0.2% and the values obtained agree within less than 0.5% with those obtained by independent methods such as 4π proportional counting, 4π α-β coincidence counting, internal gas counting or low geometry α counting [166].

The precision and accuracy of the absolute standardization depend upon the reproducibility of the extrapolation, the discriminator height, the effect of quenching and the decay scheme of the measured radionuclide [167]. The importance of preparing a stable homogeneous sample and correcting the distortion in the initial integral spectrum caused by sample absorption, pulse pile-up, increase in count rate at low energy and other distorting effects have been emphasized [166]. The necessity of reducing the high voltage supply to the multiplier phototube to diminish the interference from after-pulses when one-phototube spectrometers are used, has also been pointed out [109, 168].

On account of the high energy and monochromaticity of α-radiation, α-emitters can be determined by this method with a high degree of accuracy. In the presence of β-radiation, the distortion of the pulse height spectrum due to competing β contribution must be corrected and Vaninbroukx [169] has reported a computational method for such correction, capable of achieving an accuracy of ± 0.1% or better.

A large number of radionuclides have been standardized by this method. Some of them are listed in Table 3.6.

3.8   COUNTING VIALS

Counting vials play an important role in liquid scintillation counting; they contribute to background count rate, transmit light photons and provide a source for photo- and recoil electrons upon interaction with γ-rays from an external standard. In multiple sample counting the vials are presumed to

TABLE 3.6. *Absolute Standardization of Radionuclides by Liquid Scintillation Counting*

Type	Reference
β^--Emitters ^3H, ^{14}C, ^{32}P, ^{35}S ^{45}Ca, ^{63}Ni, ^{89}Sr, ^{91}Y, ^{99}Tc, ^{106}Ru, ^{147}Pm ^{151}Sm	166–8, 172–3
β^+-Emitter ^{22}Na	166
Electron capturers ^7Be, ^{54}Mn, ^{85}Sr	166
$\beta\gamma$-Emitters ^{56}Mn, ^{60}Co, ^{90}Sr, ^{90}Y ^{131}I, ^{137}Cs, ^{198}Au	166, 172
α-Emitters ^{210}Po, ^{233}U, ^{234}U, ^{238}Pu, ^{239}Pu ^{240}Pu, ^{241}Pu, ^{241}Am, ^{243}Am ^{242}Cm, ^{249}Bk, ^{252}Cf, ^{253}Es, Natural U and Th	168, 170

have identical optical properties and background rate but variations in the wall thickness of vials, in the type of material and in the configuration will affect this rate.

Counting vials are made of low potassium glass, polyethylene [174], nylon, quartz [175], and teflon [176]. The plastic and quartz vials yield a low background count because of the absence of ^{40}K. Among the plastic vials, both polyethylene and nylon vials are permeable to solvents; the former swell on long standing with toluene and the latter soften with water or alcohol-water mixture. Some of the vials may phosphoresce. Owing to the low background rate, the plastic vials are preferred for counting ^3H samples of low activity as they give a high figure of merit. A comparison of the background rate of the various vial materials and the figure of merit is given in Table 3.7.

Plastic vials cause more light scatter than glass vials. Light quanta internally reflected in a cell of high symmetry will be trapped within the cell if the angle of incidence is greater than the critical angle, calculated to be 42° for the air-glass interface [177]. Sand-blasting of the outside of glass vials can yield a 10–15% higher efficiency for ^3H due to increased scattering of light quanta at or near the wall. It is unnecessary to roughen the entire outside surface of the vial to achieve the desired reflection, for a strip roughening is sufficient to reduce the photon loss [178]. Roughening the inside surface shows no improvement in the escape of trapped quanta

TABLE 3.7. *Figures of Merit for Counting Vials of Different Materials*

Vial[a]	^{3}H efficiency (%)[b]	Background (counts/min)	E^2/B
Low potassium glass	31.5	25.5	39
Quartz	31.5	11.5	86
Polyethene	32.8	8.6	125
Teflon	31.7	7.0	144

[a] Data from Calf [176].
[b] Vial contains 15 ml of a dioxane scintillant.

because the close similarity in the refractive indices of toluene and glass tends to minimize the effect of asymmetry due to roughening. Opalescence and suspended objects in the sample also promote light scatter leading to an enhancement of the counting efficiency [178].

3.9   SAMPLE PREPARATION

Sample preparation is critical for the success of liquid scintillation counting as an analytical technique and has to be approached in a manner commensurate with the precision and accuracy required for the measurement. There are several excellent articles and reviews [125, 179, 180] on sample preparation to which the reader may refer. He is also well-advised to innovate and to seek optimum conditions most suitable to his purpose in sample preparation, for the introduction of a specimen into the liquid solution scintillator creates a molecular environment in which no two samples are exactly alike.

3.9.1   Homogeneous Samples

If the radioactive specimen intended for measurement can be solubilized in scintillation solvent, a homogeneous sample results. Such a sample has a uniformly distributed radioactive source with 4π counting geometry, representing one of the most ideal conditions for radioactive measurement. A homogeneous sample also allows efficiency determination to be made as a means of converting the count rate to absolute disintegration rate in order to facilitate comparison and interpretation of experimental results on an absolute scale. Unfortunately many biological tissue specimens, hydrophilic substances, macromolecules, etc., are incompatible with the solubility characteristics of the liquid scintillant; to render them soluble in the latter requires effort and ingenuity. Some of the methods for preparing homogeneous samples are discussed below.

Solubilization

A wide range of specimens including animal tissues, blood, urine, amino acids, RNA and DNA fractions in sucrose density gradients, and others, can be solubilized by the commercially available materials Bio-Solv, Digestin,

Hyamine 10-X hydroxide, NCS, Protosol and Soluene. With the exception of Bio-Solv, which has three solutions to suit the experimental conditions, these solubilizers are 0.5 to 1 M solutions of quaternary ammonium bases in methanol and toluene. Hyamine hydroxide was first introduced in liquid scintillation counting by Passmann et al. [181] as an absorbent for $^{14}CO_2$, and its potential as a solubilizing agent for amino acids and proteins was soon recognized by Vaughan et al. [182]. Since then this quaternary ammonium base has been widely used for $^{14}CO_2$ absorption and for solubilization purposes[183–185]. Hyamine solutions of animal tissues, proteins and polypeptides phosphoresce and this can interfere with accurate counting [105]. Nevertheless the success of Hyamine prompted the introduction of other organic bases as solubilizers. All the alkaline solubilizers can dissolve coarsely ground tissues and tissue homogenates but bone, tooth, cartilage and collagen fibers are not affected. During the process of solubilization, frequent agitation of the sample is required. Heating to 50° hastens solubilization but temperatures higher than those recommended cause excessive coloration and quenching.

Specific instructions in the use of each commercial solubilizer can be obtained from the manufacturer but published reports on NCS [186], Bio-Solv [187], and Digestin [188] are also available These commercial solubilizers possess high solution ability and show complete miscibility with toluene- and xylene-based scintillants, forming homogeneous samples of their digests. In order to eliminate chemiluminescence and phosphorescence (see Section 3.2.4), the alkaline sample digest is acidified either with acetic or hydrochloric acid to pH 7 or below before adding liquid scintillant. Hansen and Bush [186] pointed out that the presence of water in the digest also reduces the luminescent interferences.

The merits of NCS, Bio-Solv (BBS-1 and -3) and Hyamine in solubilizing aqueous and serum samples for solution in toluene and p-dioxane have been compared by Mueller [89] and Bray [190]. The efficiency of Hyamine as a trapping agent for $^{14}CO_2$ and its effect on enzyme systems in vitro relative to other absorbents (phenethylamine, ethanolamine and potassium hydroxide) have been investigated by Duncombe and Rising [191].

These commercial solubilizers have also been recommended for stabilizing anions for liquid scintillation counting [189] and are also used in the elution of radioactive molecules from solid absorbents or matrices [192–197]. The dissolution of acrylamide gels by solubilizers has been reviewed by Grower and Bransome [198].

The incorporation of the alkaline solubilizer in the scintillation system has an effect on the stability of the fluors. Dunn [199] found that the external standard channels-ratio decreased in systems containing Hyamine, NCS or Soluene. Among the fluors studied PPO is more stable in alkaline media than either butyl-PBD or isopropyl-PBD.

Other solubilizing agents have been tried but are less satisfactory than the quaternary ammonium bases or Bio-Solv. Petroff et al. [200] solubilized rat tissues with 2 M methanolic potassium hydroxide and incorporated the tissue digest into toluene scintillator solution with ethylene glycol monobutyl ether. Hansen and Bush [186] on the other hand, reported incomplete dissolution of tissue specimens in alcoholic KOH. High concentrations of KOH in the scintillation system enhances the background due to radiation from ^{40}K and also cause precipitation. Neujahr and Ewaldsson [201] solubilized mg quantities of bacterial cells with formamide. Francis and Hawkins [202] counted dilute labeled protein solutions in toluene scintillator containing phenethylamine.

Decolorization

The alkaline tissue digests are frequently discolored by heme and other pigments causing severe color quenching. The tissue digests can be bleached with benzoyl or hydrogen peroxide; the latter causes less chemiluminescence than the former [100]. Caution must be exercised in the routine use of peroxide oxidation to decolorize samples as those containing free amino acids or oxidizable compounds can react with peroxide resulting in a loss of radioactivity. Benevenga et al. [203] reported the loss of ^{14}C-labeled amino acids on treatment with 30 per cent hydrogen peroxide at 80° for 6 hours but observed no significant change in radioactivity when intact proteins were similarly treated. Krabisch and Bergstrom [204] reported no loss of radioactivity upon ozonization of solutions of neutral steroids. Residual oxygen from peroxide oxidation in scintillator solution causes quenching but may be removed by the use of the enzyme catalase or an oxygen scavenger, such as ascorbic acid (5—10% solution) or stannous chloride (4% in 0.1 N HCl) [103]. Mahin [205] suggested a method of oxygen removal in which a few milliliters of liquid nitrogen was allowed to evaporate from the frozen sample, the vial being then loosely capped during thawing.

Other oxidizing agents may be equally useful for decolorization. For example, chlorine water has been used by Shneour et al. [206] to bleach samples containing carotenoids and chlorophylls; Lindsay and Kurnick [207] photobleached colored NCS-treated tissue samples by long exposure to either sunlight, strong incandescent light or fluorescent light.

Oxidation Techniques

When direct solubilization cannot accommodate large amounts of specimens of low radioactivity or when solubilized biological materials are highly colored, causing severe quenching, oxidation and combustion techniques are recommended for sample preparation. The oxidation techniques discussed will include methods that either partially degrade or completely oxidize intractable and/or highly colored specimens in solution.

The use of nitric acid to digest rat skin and insect cuticle containing ^{14}C and/or ^{3}H at 70°C was introduced by O'Brien [208] neither specimen being amenable to solubilization by Hyamine or formamide. Extraneous luminescence associated with alkaline tissue solubilization and peroxide decolorization may also be avoided by acid digestion. Pfeffer et al. [209] succeeded in partially oxidizing, within 2 to 5 minutes, various minced tissues with nitric acid at 70°C, and counting the acid digest after neutralization with

Tris (*tris*(hydroxymethyl)aminomethane) buffer in a dioxane based scintillator.

Urine, feces, blood and body tissues containing ^{3}H have been oxidized by heating with a mixture of nitric and perchloric acids [210]. After complete oxidation, tritiated water mixed with acids is distilled over and the distillate neutralized and counted in a toluene solution scintillator. Samples of whole blood, plasma, tissue fluids, tissue, bone or filter discs can be oxidized directly in the counting vial with 60 per cent perchloric acid and 30 per cent hydrogen peroxide [211]. Complete digestion requires 1–2 hours at 70–80° on limited sample size (0.2 ml of tissue fluid, 100 mg of tissue or 200 mg of bone). Addition of Cellosolve and toluene scintillator to the digest completes the preparation of the sample for counting. These techniques are applicable to the preparation of samples containing ^{35}S, ^{14}C and/or ^{3}H without any loss of radioactivity or luminescent interference.

Complete oxidation of labeled materials in biological samples to ^{14}CO$_2$ and tritiated water can be performed using the Van Slyke reagent. Moore *et al.* [212] used a glass apparatus for wet oxidation and absorbed ^{14}CO$_2$ by gas diffusion directly in the counting vial containing phenethylamine. Watson and Williams [213] oxidized ^{14}C-labeled plant materials by wet combustion and absorbed ^{14}CO$_2$ in the counting vial with ethanolamine/ethylene glycol monomethyl ether by close-circuit gas circulation. Weyman *et al.* [214] used phenethylamine for in-vial absorption of ^{14}CO$_2$ immediately after the manometric measurement of total CO$_2$ in the samples.

Oxidation methods for preparing samples for counting ^{55}Fe, ^{59}Fe, ^{45}Ca and other metals in addition to ^{35}S, ^{14}C and ^{3}H have been reviewed by Jeffay [215].

Combustion Techniques
Biological specimens containing ^{35}S, ^{14}C and ^{3}H can be combusted in oxygen to ^{35}SO$_2$, ^{14}CO$_2$ and water for direct collection and incorporation into scintillation solution for counting. In this manner, impurity quenching is reduced and chemiluminescence is avoided as well.

Many variations of combustion techniques exist; they may be categorized as follows:

    (i)   oxygen flask method (Schöniger)
    (ii) vacuum-line combustion method
    (iii) oxygen train method
    (iv) flame-combustion-in-a-train method

These methods differ in the size of the sample that can be handled, in the manner of combustion, in gas collection system and in the speed of manipulation required for preparing each individual sample.

The oxygen flask combustion method, developed by Schöniger, was adapted by Kelly *et al.* [216] for biological samples containing ^{14}C and ^{3}H. The specimen, wrapped in a cellulose bag is combusted in a 2-liter oxygen-filled flask by ignition with an electrical current or spark. After completion of the combustion, ^{14}CO$_2$ is absorbed in Hyamine, phenethylamine, etc., and the tritiated water incorporated directly into a scintillation solution.

The isotope recovery is about 96 per cent on the average. The upper limit of the dry weight of samples that can be processed by this technique is 300 mg, although 500 mg samples have been safely combusted in a 2-liter rubber-stoppered flask. A similar combustion method with slight modification in the oxygen flask for a solvent well has been reported by Kalberer and Rutschman [217]. Roncucci *et al.* [218] determined simultaneously low levels of ^{35}S and ^{3}H in biological samples and found that with 6 flasks, 30 combustions could be carried out by one technician in an 8-hour day.

Several improvements to the method have been made. These include simplified construction of the flask head [218—20], side arm for trapping fluid [221, 222], safety outlet for high pressure [222], and use of infra-red ignition [223]. Dobbs [220] increased the sample weight to 1.5 g using a 3-liter oxygen flask. Gupta [224] miniaturized this technique for in-vial combustion using 2—3 mg samples containing ^{3}H, ^{14}C or ^{35}S. The collection of tritiated water was improved by Vickers [225] by a refinement of the cooling procedures. Recently, Wegner and Winkelmann [226] reported an automated system for in-vial combustion.

Although the oxygen flask method is easy to use, it suffers from inherent disadvantages: the sample size is small, purposely limited to control violent combustion and to ensure completion of oxidation; volatile samples cannot be handled without loss of radioactivity; polymeric compounds may not be completely burnt; the collection of gaseous products of combustion and the combustion itself occur in the same flask, resulting in varying degrees of oxygen quenching; and explosion sometimes occurs if the combustion flask accidentally contains residual toluene or acetone [227].

The vacuum-line combustion method introduced by Huebner and Kisieleski [228] combines the salient features of the Schöniger oxygen flask combustion with vacuum line transfer. The gaseous products of combustion are collected directly into the scintillation counting vials. The method effectively eliminates oxygen quenching. The recovery for ^{14}C and ^{3}H ranges from 97—99% and is more consistent than the recovery of the unmodified Schöniger technique. With 5 or 6 trap-vial units, 40 samples can be processed in 8 hours.

The oxygen train method is basically the classical carbon-hydrogen combustion method in which a continuous flow of oxygen carries the products of combustion through beds of catalysts to ensure complete oxidation. It was introduced by Peets *et al.* [229] for determining ^{3}H radioactivity in biological samples and was improved by Tamers and Diez [230] and by Knoche and Bell [231]. Griffiths and Mallinson [232] described the construction and operation of a furnace that burnt wet samples up to 1 g in weight and absorbed the combustion products in less than 5 minutes. The average recovery of feces spiked with ^{3}H-*n*-hexadecane, or ^{3}H-morpholine is about 99%. It is possible to prepare 8 to 10 samples in 1 hour for counting. Retention of ^{3}H radioactivity in the oxygen train is minimal because of the high flow rate and small amount of packing used.

Peterson and coworkers [233, 234] have reported an automatic oxygen train system of combustion which enables samples to be prepared for counting (^{3}H and ^{14}C separately) at the rate of one every three minutes. The merit of this system has been compared with that of the oxygen flask combustion [235].

An induction furnace has also been used by Smith [236] for combustion of biological samples containing ^{14}C.

The flame combustion-in-a-train method was developed by Kaartinen [237] and his design has been incorporated into a commercial unit. The unit carries out a flame oxidation of the sample compressed into a pellet with filter paper as added fuel material. The flame is controlled by the temperature of the igniter and the rate of oxygen flow. The tritiated water formed in the combustion is condensed and subsequently washed out with inactive water from the condenser directly into the counting vial for ^{3}H determination. The radiocarbon dioxide is absorbed in ethanolamine at 50°C in the reaction column and is delivered into a separate counting vial for ^{14}C measurement. At the conclusion of the active phase of combustion, a program is initiated in the unit which automatically flushes the system with nitrogen to eliminate oxygen quenching and delivers, in steps, the two separate liquid scintillator solutions to the two counting vials which are then removed from the sample oxidizer and capped for counting. The oxidizer can process a sample containing ^{14}C with or without ^{3}H present every 3 minutes and takes approximately one minute to prepare a sample for ^{3}H measurement. The radioisotope recovery of this method is equal or comparable to that of other methods of combustion. The performance of the commercial unit of sample oxidizer has been evaluated by Sher *et al.* [238].

Other methods of dry oxidation using sealed tube or Parr bomb have been reviewed by Jeffay [215]. The collection of gaseous products of combustion and the manipulation are time-consuming and these methods are very infrequently used.

Complexation

Radioactive cations and anions can be incorporated into liquid scintillators as salts or complexes by virtue of their inherent solubility in scintillation solvents. In absolute standardization of radionuclides and in counting of α-emitters, it is essential to maintain the radioactive ionic species in solution as stable homogeneous counting samples. When radionuclides intended for absolute counting appear as anions, they may form salts with organic bases which are soluble in aromatic hydrocarbons [239] or may be reasonably stable in dioxane-based scintillants *per se* [167]. Radioactive cations may form inorganic and organic salts or complexes that readily dissolve in scintillation solvents [240].

The early literature on liquid scintillation counting of inorganic radionuclides has been reviewed by Horrocks [240]. The formation and extraction of metal complexes is dealt with in Chapter 5 and the monograph of Marcus and Kertes treats the subject in depth [241].

A large number of radionuclides have been counted in homogeneous and

TABLE 3.8. *Liquid Scintillation Counting of Salts and Complexes of Radioactive Metal Ions*

Isotope	Chemical form	Solvent	Reference
^{45}Ca	Chloride	EtOH	243
	Chloride	DAP[a]	244
	Nitrate	Ethylene glycol/ EtOH	245
	Oxalate	HCl-EtOH	246, 247
	Perchlorate	TBP[b]	248
^{55}Fe	Phenanthroline	*iso*-Amyl alcohol/ toluene	249
	Ascorbate	Toluene/EtOH/H_2O	250
	HDEHP[c]	Toluene	251
^{63}Ni	DOP[d] complex	Xylene	109
	Pyridine dithiocyanate complex	*p*-Dioxane	252
^{95}Zr	Nitrate	TOPO[e]	253
^{99}Tc	Pertechnetate	*p*-Dioxane	167
^{106}Ru − ^{106}Rh	Nitrate	TOPO	253
	p-Toluidine complex	Toluene	109
^{144}Ce − ^{144}Pr	Nitrate	TOPO	253
^{241}Pu	HEH (EHP)[f]	Xylene	109
^{233}U, ^{234}U, ^{241}Am	TOPO	*p*-Dioxane	169, 254
	VYNS[g]	*p*-Dioxane	254
Actinides	HDEHP		170
	1-nonyldecylamine		170

[a] DAP   = dialkyl phosphate
[b] TBP   = tributyl phosphate
[c] HDEHP   = di-(2-ethylhexyl)phosphoric acid
[d] DOP   = dioctyl phosphate
[e] TOPO   = tri-octyl phosphine oxide
[f] HEH(EHP) = 2-ethylhexyl hydrogen 2-ethylhexyl phosphonate
[g] VYNS   = polyvinyl chloride-acetate copolymer

heterogeneous liquid scintillation systems. A few of those that form soluble salts and organophosphorus complexes [242] in organic solvents are given in Table 3.8.

3.9.2 Heterogeneous Samples

Suspension counting was first practiced by Hayes *et al.* [255]. Since then heterogeneous counting has become a very useful method for counting aqueous and insoluble samples in emulsion, in suspension or on solid supports. The dispersion of these specimens in liquid scintillators allows a maximum interaction between the β-radiation and the fluor molecules; it is even ideal for counting radioactive substances which are strong quenchers because the radiation can escape the phase boundary and react with the fluor while the quencher is confined to the aqueous phase and cannot exert any action on the fluor. The loss of efficiency due to self-absorption is a function of the β-radiation energy and the fineness of dispersion. For this reason, heterogeneity in the sample affects the counting efficiency for ^{3}H more adversely than that for ^{14}C. The difficulty of making valid correction for photon quenching constitutes the main drawback of heterogeneous sample counting.

Emulsion Counting

Emulsion counting, more correctly known as 'colloid counting' began with the observation of Meade and Stiglitz [256] that animal tissues suspended in Triton X-100/toluene (2/1, v/v) scintillator yielded a higher counting efficiency for ^{14}C and ^{3}H than a formamide/ethanol/toluene system. Patterson and Greene [81] investigated the emulsification of aqueous solutions in toluene/Triton and toluene/Triton/ethanol scintillation systems. In one of the systems (toluene/Triton (7/6, v/v)) studied, a fluid emulsion is formed on incorporation of 43% water, and sets to a rigid gel on cooling, in which ^{14}C can be counted with high efficiency.

The emulsion counting system is not a stable one; its appearance undergoes transition with the incorporation of increasing amounts of water, from clear to opalescent to opaque-white, corresponding approximately to the change from an apparently homogeneous state to two separate phases and to a translucent gel. The three-component phase diagrams of emulsion systems involving toluene, Triton and various aqueous solutions commonly encountered in biomedical research, have been given by Fox [257], by Chapmen and Marcroft [258] and by van der Laarse [259]; areas of homogeneity and heterogeneity are marked where a clear, a cloudy or an opaque viscous liquid or an opaque or a clear gel can be obtained. Many variables affect the phase stability and the efficiency of emulsion counting systems; these include fluor concentration, sample concentration, temperature, agitation, time of cooling, sample distribution in the vial and chemical nature of the sample.

Details of two methods for emulsion preparation have been given by Bensen [80], the higher counting efficiency being obtained with the one producing the smaller emulsion drop size.

The advantage of emulsion systems for ^{3}H counting is that they can incorporate large amounts of tritiated water of low activity. The merit of an emulsion counting system is expressed in terms of the merit number†, defined as the product of the per cent by volume of radioactive specimen present in the counting vial and the per cent counting efficiency. Moghissi *et*

†This is preferred to the 'figure of merit' used by some authors because the latter term is most frequently used to denote E^2/B.

al. [260] introduced a Y-value for comparison of low level counting systems which defines the minimum limit of detection as a function of merit number and background rate.

The merit number of an emulsion system increases with increasing amount of aqueous sample incorporated but with a concomitant decrease in counting efficiency. At high solvent-to-surfactant ratio, the merit number increases, reaches a maximum and then decreases as more water is incorporated; at an optimum ratio, the merit number maintains a gradual increase as more water is added. The efficiency of the emulsion system is dependent upon the types of surfactant used. Of 50 detergents screened for potential use in emulsion counting by Lieberman and Moghissi [261], nine detergents yielded comparable merit numbers with Triton N-101 as the best, when studied in a p-xylene/detergent (2.75/1, v/v) mixture containing 40% water.

Emulsion systems of unrevealed composition are commercially available under the trade names of Aquasol, Insta-Gel and PCS. Bohne [262] has encountered difficulties in using one of the systems for direct solubilization of biological fluids and tissue homogenates, but was able to overcome them by pre-digestion with Hyamine hydroxide. In fact, toluene/Triton/Hyamine and toluene/Triton/NCS scintillant have been used for counting neat plasma [263] and urine samples [207] containing ^{3}H. Nadarajah et al. [264] counted ^{45}Ca in serum and urine in an emulsion system containing Triton X-100. Curtiss [265] measured α-emitters in the gel region of an emulsion system. Turner [266, 267] studied the use of toluene/Triton systems for measuring ^{14}C- and ^{3}H-labeled materials and found no advantage in substituting butyl-PBD for PPO and POPOP in the emulsion. Substitution of p-xylene for toluene or using mixtures of Triton X-100 and Triton X-114 in the system increases water incorporation. Greene [268] has reported the preliminary results on a new emulsion system consisting of xylene/naphthalene/Triton X-114 which works best near 0°C. Shamoo [269] found that the addition of hygroscopic carboxymethylcellulose or its derivatives to toluene solution scintillator containing water can dramatically increase the counting efficiencies of ^{14}C and ^{3}H.

Emulsion systems are presently the only heterogeneous systems in which quench correction can be made. Turner [267] and Whyman [263] have used channels ratio and internal standard for quench correction in toluene/Triton systems and found that samples counted in emulsion and those counted in Bray's scintillant gave almost identical channels-ratio values. The accuracy and validity of quench correction by internal standard depends upon the reproducibility of formation of the emulsion after adding the standard and upon the type of standard used. For water-in-oil emulsions, it is necessary to use a tritiated water standard but for oil-in-water emulsions which correspond to the gel region, a higher efficiency is obtained with tritiated n-hexadecane. As internal standard van der Laarse [259] used only 25 microliters of tritiated water since the introduction of larger amounts of the aqueous phase may shift the relative concentration of the components in the phase diagram and result in a change of the counting characteristics of the emulsion.

Suspension Counting

Insoluble labeled materials can be counted with high efficiency as finely ground powders in a liquid scintillant. Hayes *et al.* [255] in 1956 assayed materials containing ^3H, ^{14}C, ^{35}S and ^{45}Ca including ^{14}C-labeled animal tissues and bacterial cells, by suspension counting. To prevent sedimentation of the suspension and to maintain an optimum source geometry Ott *et al.* [83] stabilized the suspension with 4% Cab-O-Sil in the solution scintillator. Addition to liquid scintillants of aluminum stearate followed by heating to 70°C and of aluminum 2-ethyl-hexanoate without heating, were used by Funt [85, 270] to effect gelation. White and Helf [84] added powdered Thixcin to a toluene scintillator to obtain a fluid gel. Recently, Bolling *et al.* [271] formed gels between toluene di-isocyanate and the branched aliphatic amines in toluene solution scintillator by the addition reaction leading to the urea-type inclusion of toluene. More recently, Benakis [272] reported the use of polyolefin resins to suspend powdered biological material, silica gel from thin layer chromatography and other powdered materials, for counting in a liquid scintillator.

In addition to its frequent use for counting $Ba^{14}CO_3$, scintillation gels containing Cab-O-Sil have been used to measure ^{14}C in animal tissues in the form of NaOH digests [273] to determine ^3H and ^{14}C on silica adsorbents from thin-layer chromatograms [274]; and to assay ^{45}Ca and ^{89}Sr in unicellular algae [275]. Eakins and Brown [276] counted ^{55}Fe and ^{59}Fe as white ferriphosphate complexes in a gel. Lloyd-Jones [277] radioassayed $Ca^{14}CO_3$ in suspension with Cab-O-Sil after obtaining a uniform precipitate with the aid of hydroxylpropyl methylcellulose. Other applications of suspension counting have been reviewed by Rapkin [179, 278].

The ratio of efficiency of suspension counting to that of homogeneous counting is known as the f-value [84, 255]. It measures the efficiency of suspension counting and varies with the particle size and the specific radioactivity of the suspended material [279]. Suspension counting with f values as low as 0.118 and as high as 0.99 has been reported [280]. The opacity of the gel also exerts an anti-quench effect by reducing the amount of light totally internally reflected. A valid quench correction for suspension counting cannot be easily made. The impracticability of quench correction using internal standards was emphasized by Bloom [281] who found a wide range of discrepancy between the $^3H/^{14}C$ ratios observed in a thixotropic gel system and a homogeneous system.

Counting on Solid Support

Radioactive solutions and suspensions containing a variety of non-volatile radioactive materials can be counted on solid supports. In practice, aliquots of the material to be measured are either spotted on or filtered through glass fiber, membrane or paper filters, which are dried and immersed in or wetted with liquid scintillator for counting.

The efficiency and reproducibility of counting on solid support depend upon the magnitude of the β-energy, the nature of the solid support, its orientation in the counting vial, the size of the sample molecule, the presence and composition of sample precipitate, and the amount of sample that becomes soluble in the scintillation solvent [282]. The thickness and absorptivity of the solid supports determine the extent of self-absorption and this affects the ^{3}H samples more markedly than the ^{14}C samples [283]. Gill [284] showed that non-absorbent glass fiber paper allows ^{3}H efficiencies of about 10%, cellulose acetate paper about 6% and standard cellulose chromatographic paper about 2% with variations depending upon the size of the molecules, for the smaller ones can readily diffuse into amorphous regions of the cellulose fibers while the large molecules remain on the surface. Self absorption and sample geometry are also affected by the amount and distribution of the sample deposited on the solid support. In addition, scintillation characteristics of the medium may alter the efficiency. If the sample dissolves partially in the scintillation solvent, the soluble fraction will be counted with an efficiency differing from that of the fraction retained on the solid support. Counting of ^{3}H on a solid phase is unreliable but for radionuclides with energies equal to or higher than that of ^{14}C, reproducible counting data can be obtained.

Heterogeneous counting on solid supports has been reviewed [278]. Johnson and Smith [285] counted $Ba^{35}SO_4$ and $Ba^{14}CO_3$ with an average of 35% efficiency on glass fiber filter paper after vigorously shaking in a dioxane scintillator to give fine division and uniform suspension. Lloyd and Rees-Evans [286] used a ring-oven to control the drying of $H_2{}^{35}SO_4$ on glass fiber filters and found less deviation among samples identically prepared under these conditions. Efficiency for $H_2{}^{35}SO_4$ on glass paper was 78% compared with a value of 82.5% obtained by homogeneous counting as an amine salt. Lithium sulfate-^{35}S was counted with less efficiency than $H_2{}^{35}SO_4$ because of the self-absorption caused by crystal packing on the paper. Malt and Miller [287] found that the TCA-precipitated radioactive fraction of RNA in sucrose density gradients can be counted after drying on glass-fiber discs with an efficiency equal to counting NH_4OH digests on Millipore filters. Cramer and Ross [288] counted ^{45}Ca in a hydrochloric acid solution of ashed rat carcass, bone and excreta on glass fiber and paper filters. Downes and Till [289] labeled wool fibers with ^{14}C-formic acid or 2-^{14}C-iodoacetic acid dissolved in a liquid scintillation solution as a means of determining the mean fiber diameter. Bull [280] counted ^{14}C-melanin on glass fiber discs with high efficiency, 74.3% for PPO + DM-POPOP and 86.7% for BBOT as fluors. Counting efficiency is also affected by the volume of the scintillator solution in the vial. Davies and Cocking [290] reported a 5–6% fall in the efficiency of counting ^{14}C-labeled tomato protein on horizontal glass-fiber discs when the volume of the scintillator was increased from 0.8 to 10 ml.

Counting on solid support has also been used to measure activities in slices of polyacrylamide gels. Helleiner and Wunner [291] treated gel slices with NH_4OH on glass fiber discs to permit much of the protein to diffuse out of the gel and counted the entire gel-glass fiber disc in a toluene scintillator. Spear and Roizman [292] hydrolyzed the gel with concentrated NH_4OH in a scintillation vial and absorbed the hydrolysate on glass fiber paper which was then dried and counted. In this manner higher efficiencies for both ^{3}H and ^{14}C were obtained than by homogeneous counting.

The anti-quench effect observed by Gordon and Curtis [178] when counting samples containing optically diffusing white materials or surfaces, is caused by a reduction of the amount of light lost through total internal reflection, thereby enhancing the counting efficiency. This increase is also observed in suspensions gelified with Cab-O-Sil. The introduction of the frosted vials for counting was an outcome of these observations (see Section 3.8).

Quench correction for counting on solid supports cannot be easily justified on a theoretical basis because the radioactive source is in solid phase and many variables in sample preparation cannot be rigorously controlled and reproduced [282]. Although channels-ratio and external standard methods have been applied, they measure only the solution efficiency and cannot ascertain the degree of self-absorption or the extent of quench reaction occurring in the solid phase.

3.9.3 Miscellaneous

Flow Monitoring
Radioactivity in chromatographic effluents can be monitored by liquid scintillation counting with a flow cell filled with anthracene or plastic scintillation beads or with one fabricated from plastic scintillator materials [293]. The effluent may also be mixed continuously with liquid scintillant so that the mixture can be counted as it flows through a coiled polyethylene tubing [294]. This technique is described in Chapter 6 and an excellent review on flow monitoring of aqueous solutions containing β-emitters has been given by Schram [295].

Counting of Radioactive Gases
Radioactive gases such as $^{14}CO_2$, ^{85}Kr, ^{14}C-ethylene and volatile labeled materials can be counted in liquid scintillation systems [296]. $^{14}CO_2$ may be absorbed in alkaline media and counted in homogeneous solution or in emulsions as aqueous solutions of sodium carbonate-^{14}C or in suspensions as barium or calcium carbonate-^{14}C. Low levels of $^{14}CO_2$ in respiratory air can be counted after conversion into benzene [297].

Volatile radioactive gases which are soluble in blood or in other physiological fluids, can be radio-assayed in a toluene/Triton emulsion scintillant. The validity of the method depends upon representative sampling and quantitative transfer of the material to be measured to the scintillant and on maintaining a gas-tight seal in the counting sample. The sample efficiency relies on the partition of the radioactive gas between the air space and the scintillant; the higher the concentration in the solution phase, the higher the efficiency. Quench correction can be carried out with the isotope channels ratio or the external standard channels-ratio method. Smith *et al.* [296] have shown that a high degree of precision can be obtained using this technique.

3.10   COUNTING STATISTICS

Optimization of sample counting conditions to achieve a high degree of statistical precision and accuracy is an ever present and desirable goal. In liquid scintillation counting, many variables contribute to the total error of a determination. For single isotope samples the error contribution from variations in instrument parameters such as channel width, efficiency performance, stability, etc., and changes in sample conditions, such as the level of activity, the degree of quenching, background rate, counting time, etc., can be predicted on the basis of propagation of errors. When an internal standard is used for quench correction, the uncertainty in the activity per unit volume of the standard solution and the error in the added volume, also contribute to the total variance of the count rate of the sample. Herberg [128, 298] evaluated the error contribution from seven quantities involved in the calculation of the absolute distintegration rate. His analysis showed that the per cent error in the calculated rate is less when conditions of high counting efficiency prevail; and that for routine counting adequate precision and accuracy can be obtained using 50,000 dpm of ^{14}C and 150,000 dpm of ^{3}H as internal standard activity and 1-minute counting time after the addition of internal standard [299].

For double isotope samples, in addition to the above considerations, one is concerned with channel selection and channel limits for each isotope, counting efficiencies of the isotope pairs in two channels, isotope ratio, isotope separation factor, performance of the counter, etc. The frequently used isotope exclusion method for double label counting, when evaluated along these lines for error contribution, was found to be less precise and less accurate than the simultaneous equation method. The counting conditions of isotope pairs for optimizing statistical accuracy in two-channel counting have been investigated by Bush [300]. As a result, the use of maximum channel width is recommended because the efficiency changes more slowly with variations in gain or high voltage or attenuation in a wide window. For the isotope pair ^{14}C and ^{3}H, the channels are selected so that the overlap of ^{14}C in ^{3}H channel is minimized, by counting ^{3}H at a gain above the balance point while the overspill of ^{3}H into the ^{14}C channel is kept below 0.6%.

Isotope separation in doubly labeled samples can be either increased or decreased at the expense of the detection efficiency for either isotope. Klein and Eisler [155] defined the separation efficiency from the ratios of the two isotopes in both counting channels and used the performance number which is the product of separation efficiency and the individual isotope efficiency at this point of optimum separation, for evaluation of scintillation counters. The factors that limit the isotope separation are the shape and overlap of the pure β-spectra themselves and the distortion introduced by the liquid scintillation processes and instrument design. According to these authors, the statistical precision and accuracy of the data depend much on the performance of the counting instrument; judging by their results on some of the commercial instruments, the counting statistics of double isotope samples can be dramatically improved by better counters yet to come.

3.11   DATA PROCESSING

Data analysis is an important aspect associated with liquid scintillation counting. The use of automatic counters can generate enormous quantities of counting data and data processing and reduction by hand calculation become extremely tedious. For example, the conversion of the gross counting data for single isotope samples to absolute radioactivity involves seven steps of arithmetic calculation. For dual-labeled samples the burden of calculation is increased many fold; to obtain individual isotope activities one must calculate the average from multiple sets of replicate counting data, ascertain their statistical accuracy, make quench corrections based on channels ratio, prepare the calibration curve using appropriate curve fitting procedures, resolve the isotope channels overlap, determine the quench effect on the spillover and on the counting efficiencies of the isotope pair in both channels, find the effect of isotope ratio on the counting accuracy, do iterative testing if required, convert the counting data into appropriate units with standard deviations and quantify the final results in relation to the whole experiment. This is a formidable task. When many dual isotope samples are counted, manual data processing to the fine degree provided by digital computers becomes impossible. Many laboratories have heretofore turned to programmed computation using programmable desk-top calculators or off-line digital computers.

Using a desk-top computer, Scott [301] reported the routine calculation of the absolute specific activities of batches of radiochemicals; Koch [302] showed a simplified method for calculating the isotope activities in double-labeled samples; and Grower and Bransome [303] calculated the absolute radioactivity of samples labeled with one or two isotopes from the liquid scintillation counting data and external standard ratio or count rate.

Because of the high speed and sophistication of large digital computers, elaborate computer programs can be written to include all pertinent analyses and statistical testings at all data points using out-putting data from the liquid scintillation counters on punched cards from interphased automatic card punch or punched tape from teletype. A computer program written in ALGOL (algebraic-English computer language) for automated testing and reduction of liquid scintillation data for single isotope samples was published in 1963. Since then, many programs have been published [305–13]. Machine computation was used by Krichevsky et al. [134] and Ninomiya [158] in the assay of double-labeled samples using channels ratio for quench correction. Spratt [311] reported computer programs written in FORTRAN IV language for calculating liquid scintillation data with quench correction by internal standard and has since modified and extended the program to accommodate quench correction by automatic external standardization [312]. Felts and Mayes [313] reported the relative errors encountered in the quantification of counting data from single- and double-labeled samples using external standardization for quench correction.

Recently, Carroll and Hauser [314] discussed a program for curve-fitting, with the objective of finding an acceptable fit at the least uncertainty using the lowest degree polynomial necessary at that uncertainty. Hassig and

Schipper [156] reported a computer program for liquid scintillation data processing that emphasizes the accuracy of the data obtained over a wide range of quench and isotope ratios. Hansen and Carroll [315] used an iterative procedure to process double isotope counting data using quench factors from the channels-ratio and the external standard ratio method, and stressed the effect of quenching on isotope overlap. Glass [316] was able by means of two FORTRAN programs, one for sample data and the other for calibration curves to detect instrument drift, contamination, misprinting, inhomogeneity, and other miscellaneous errors; Lang [317] used a computer-oriented iterative technique to correct for color and impurity quenching for more accurate determination of sample counting efficiencies.

The historical development of data handling and processing in liquid scintillation counting and the merits of on-line and off-line computations have been reviewed [318–9]. Information on many other computer programs is available from the computer library maintained by instrument manufacturers.

3.12   ČERENKOV COUNTING: INTRODUCTION

Although Čerenkov counters have been widely used in various fields of high-energy physics for detection of fast-moving charged particles and measurement of their velocity [320–3], the use of Čerenkov light as a means of measuring radioactivity in samples containing energetic β-emitters achieved importance only in recent years. The advantage of Čerenkov counting lies in the simplicity in sample preparation and sensitivity of detection. The radioactive samples can be counted directly without adulteration by scintillation chemicals and additives and they are not limited to high specific radioactivities as is necessary in liquid scintillation counting. In Čerenkov counting, large volumes of low activity samples can be measured, compensating for its inferior sensitivity compared with that of liquid scintillation counting. Owing to the nature of Čerenkov radiation, quantitative measurement is not subject to impurity quenching or self-absorption. In the following, some aspects of Čerenkov counting are considered.

3.13   ČERENKOV RADIATION

Čerenkov radiation was first observed independently by Mallet [324] in 1926 and by Čerenkov [325] in 1934 using a radium source. Čerenkov defined the nature of the radiation by further experimentation using a radium-E source [326]. A classical treatment of this phenomenon was undertaken by Frank and Tamm [327].

β-Energy Cut-off

The origin of Čerenkov radiation is attributed to the movement of a charged particle in a medium with a velocity greater than that of light in the medium. Fast-moving charged particles can travel in transparent dielectric media with a velocity v ($= \beta c$) which, being smaller than the limiting velocity of light *in vacuo*, c, is yet greater than its phase velocity c/n in that medium,

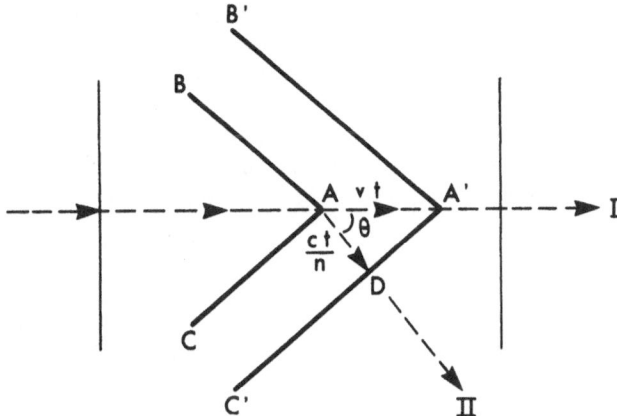

Fig. 3.9. Diagrammatic representation of conditions for emission of Cerenkov radiation. I depicts the direction of motion of the particle and II that of the light wave. BAC and B'A'C' represent cross sections of conical surfaces of the emitted light at points A and A', respectively.

n being the index of refraction of the medium. Under such circumstances, the electromagnetic field associated with the particle lags behind, and its interaction with the medium gives rise to the electromagnetic radiation known as Cerenkov radiation. This phenomenon is similar to that of a supersonic aircraft or a projectile radiating Mach's waves or of bow waves created by a moving boat on water.

Fig. 3.9 shows the conditions in which Cerenkov radiation is emitted. The electron with a velocity v moves from A to A' in a time t, traveling a distance vt along the direction of motion, while the electromagnetic radiation is propagated with a phase velocity c/n in the same time interval through a distance AD' equal to ct/n. The angle θ between the direction of motion of the particle and that of the emitted light is given by

$$\cos \theta = \frac{ct/n}{vt} = \frac{1}{\beta n} \tag{3.20}$$

This is known as the Cerenkov relation. Only when $\beta n > 1$, does Cerenkov radiation occur. The light is emitted in space along a conical surface represented in one plane in Fig. 3.9 by the Cerenkov wavefront BAC at point A and B'A'C' at point A'. If the electron is stopped after having traveled a certain distance, this conical wave will continue to propagate in the medium as a free light wave.

The Čerenkov relation sets a lower limit on the electron energy for light emission; at the cut-off limit θ equals zero or $\beta n = 1$. This determines the mutual dependency between the cut-off energy and the index of refraction of the medium. In highly dispersive media the cut-off energy is correspondingly lowered. In water (n = 1.33, d = 1.00 g/cc) and in Perspex (n = 1.50; d = 1.18 g/cc) the minimum electron energy for Cerenkov effect becomes 0.260 and 0.177 MeV, respectively. It has been reported that ^{14}C can be measured by Čerenkov counting in α-bromonaphthalene [329]. An interesting explanation of this unexpected result was given.

The half-angle, θ, of the cone of Čerenkov emission is a function of electron energy; it approaches a value of 41° with increasing energy. As the particle is decelerated in its passage through matter, the cone of emission of Čerenkov radiation contracts and vanishes at the instant when the velocity of the particle becomes identical to the phase velocity of light in the dielectric medium.

Detection

The number of photons, dN, produced in the spectral range between λ_1 and λ_2 along the path dl by the Čerenkov effect is given [322] by

$$\frac{dN}{dl} = 2\pi\alpha \left(\frac{1}{\lambda_2} - \frac{1}{\lambda_1} \right) \left(1 - \frac{1}{\beta^2 n^2} \right) \quad \text{photons/cm} \tag{3.21}$$

where α is the fine structure constant and is equal to $2\pi e^2/hc = 1/137$. The total number, N_E, of photons radiated along the track of the electron after complete dissipation of its kinetic energy in the medium can be obtained from equation (3.21) by integration between the initial and the cut-off energy; thus

$$N_E = 2\pi\alpha \left(\frac{1}{\lambda_2} - \frac{1}{\lambda_1} \right) \int_{\beta=\beta_{max}}^{1/n} \left(1 - \frac{1}{\beta^2 n^2} \right) dl$$

Using this formula Ross [328] has calculated the Čerenkov photon yield of electrons in water for selected spectral regions.

Since Čerenkov radiation is a shock wave, it contains components of all frequencies with energy flux larger towards the violet region, and in water the radiation exhibits a bluish-white appearance to the eye. When using liquid scintillation spectrometers for coincidence detection of Čerenkov radiation, quartz windowed multiplier phototubes are preferred for overall efficiency because of their greater sensitivity towards the ultraviolet portion of the emission. The head-on position of the two multiplier phototubes in the spectrometer is not considered an optimal arrangement for maximum collection for Čerenkov photons because the angular effect is manifest even in the small highly polished sample counting chamber. Ross suggested that for maximum efficiency and directional response to Čerenkov emission, the two phototubes should view the sample from the same angle.

The characteristics of pulse height spectrum of Čerenkov radiation is dependent upon particle energies and instrument parameters such as gain, discriminator height, window width, etc. These relations bear a close resemblance to those for ^{3}H counting and, as shown by Haberer [330] and Elrick and Parker [331], should be optimized for maximum counting efficiency.

3.14   PRACTICAL ASPECTS OF ČERENKOV COUNTING

Čerenkov counting has been applied to the measurement of energetic β-emitting nuclides. The efficiency is a function of β-particle energy, sample volume, the nature of dispersive medium and the presence of wavelength shifter. Table 3.9 lists data related to some of the radionuclides measured by Čerenkov counting. Other aspects of this technique are briefly dealt with below.

Choice of Counting Vials

Vials for Čerenkov counting are selected on the basis of volume, light transmission and background count rate. Haberer [330] has shown that a linear relationship exists between the sample count rate and volume in polyethylene vials. In glass vials the samples are not only counted with less efficiency, but the linear dependence between count rate and volume is only approximate. The background count rate is higher in glass than in polyethylene because of the presence of ^{40}K. Experimental results indicate that the empty vials themselves constitute a copious source of Čerenkov radiation since the addition of 20 ml sample increases the background rate only by about 20%. For a sample of 15–20 ml the glass vials usually show a background of 25–30 cpm, the plastic vials 9–15 cpm and the nylon vials 5–7 cpm. Läuchli [334] obtained the highest efficiency in nylon vials but found them inconvenient to use because of difficulties in cleaning. Polyethylene vials have a further advantage over glass vials in that for the same external dimensions they can accommodate 20% more sample [330]. For accuracy in counting, variations among the individual vials should be as small as possible [331].

Volume-Efficiency Dependence

Since the total photon yield of Čerenkov emission is increased with the length of electron path in dielectric medium, a large volume will result in a higher sample count. As the volume is increased, light transmission may become less efficient owing to the presence of optical impurities in the medium or its opacity. In polyethylene vials a maximum efficiency is obtained with a sample volume of approximately 15 ml but at larger volumes the efficiency is only slightly reduced, whereas in glass vials a greater reduction in efficiency is encountered [330].

Colored solutions adversely affect the Čerenkov counting efficiency. Ballance and Johnson [333] noted a 10- to 25-fold reduction in count rate by yellow pigments. To overcome color quenching, the samples are placed in thin-walled plastic cylinders and each cylinder is then centrally placed in a scintillation vial containing the dispersive medium. Such a device is known as the 'Čerenkov insert'. Using this device several nuclides have been counted with high efficiency in a medium of 1-chloronaphthalene [332, 333].

TABLE 3.9. *Counting Efficiency of Čerenkov Radiation from β-Emitting Nuclides*

Isotope	Efficiency (%)	Solvent	References
^{14}C	<1	Aqueous solution	332
^{22}Na	12[a], 16 (dry)[b]	α-Chloronaphthalene	333
	3.5; 6.7 (ANDA)[c]	Aqueous solution	334
^{24}Na	13.2	Aqueous solution	332
	18	Aqueous solution	331
	25[d]	Aqueous solution	335
	19[a], 52 (dry)[b]	α-Chloronaphthalene	333
^{36}Cl	1[a], 12 (dry)[b]	α-Chloronaphthalene	333
	2.3	Aqueous solution	331, 342
	7.7, 23 (ANDA)[c]	Aqueous solution	334
^{32}P	10[a], 55 (dry)[b]	α-Chloronaphthalene	333
	13	(dry sample)	336
	15	(moist sample)	336
	16	KH_2PO_4 solution	337
	25	Aqueous solution	331, 332, 342
	30.5	Na_2HPO_4 solution	338
	35.7	Chloroform/isopropanol (2/1, v/v)	339
	48	Water (in polyethylene vial)	340
^{40}K	14	Aqueous solution	331, 342
	27	Aqueous solution	330
^{42}K	42[a], 82 (dry)[b]	α-Chloronaphthalene	333
	60	Aqueous solution	331, 342
	82, 93 (ANDA)[c]	Aqueous solution	334
^{47}Ca	7.5	Aqueous solution	331, 342
^{56}Mn	40	Aqueous solution	335
^{59}Fe	<1	Aqueous solution	332
^{86}Rb	23	Aqueous solution	331, 342
	33, 61 (ANDA)[c]	Aqueous solution	334
^{90}Y	(50)[d]	Aqueous solution	330
^{90}Sr – ^{90}Y	14.2	Aqueous solution	335
^{106}Ru –	(81)[d]	Aqueous solution	330
^{106}Rh	62	Aqueous solution	331, 342
^{128}I	40–50[e], 48–60 (ANDA)[c]	Aqueous solution	341
^{137}Cs	2.1	Aqueous solution	331, 342
^{144}Ce – ^{144}Pr	(75)[d]	Aqueous solution	330
	5.4	Aqueous solution	331, 342

TABLE 3.9.   continued

Isotope	Efficiency (%)	Solvent	References
^{198}Au	5.4	Aqueous solution	331, 342
^{204}Tl	1.3	Aqueous solution	331, 342

a   Counted as the 'Cerenkov insert'.
b   Dry sample on glass fiber paper counted as the insert.
c   Counted in the presence of 7-amino-1,3-naphthalene-disulfonic acid (ANDA) as wavelength shifter.
d   Based on liquid scintillation efficiency.
e   Efficiency decreases from about 50% for iodine concentration in the range 1–3 μg to about 40% in the range 5–25 μg.

Wavelength Shift

The directional property of Cerenkov radiation causes an uneven response in the two multiplier phototubes positioned at an angle of 180°, thereby leading to a reduction of efficiency and sensitivity of coincidence detection. This angular effect can be corrected with the use of a wavelength shifter, e.g. (7-amino-naphthalene-1,3-disulfonic acid (ANDA)) which absorbs the Čerenkov radiation and then re-emits isotropically [331, 334].

Ross [343] has investigated the suitability of many fluorescent indicators as wavelength shifters. He found 4-methyl umbelliferone the best for minimizing the directional effect of Čerenkov radiation, while it exhibits a remarkable stability towards variation of pH and remains stable with time. In contrast, ANDA, although insensitive to pH changes, is much less stable. These wavelength shifters increase the counting efficiencies of low-energy nuclides much more markedly than those of high-energy ones.

Efficiency Determination

In Čerenkov counting, impurity quenching, in the sense used in liquid scintillation counting, is non-existent since no photophysical or energy transfer processes are involved. Substances which are colored or which attenuate Čerenkov emission will cause quenching. Quenching in Čerenkov counting may be avoided by decoloration of the sample by oxidation (see Section 3.9.1) or corrected by the use of an internal standard of the identical nuclide. Both methods require sample manipulation and lessen the inherent advantage of the technique. The channels-ratio method [338, 344] and the external standardization method [331, 345] have been applied independently or in conjunction with each other or with a spectrophotometric method [331, 338] for the purpose of quench correction. The pulse height spectrum of Čerenkov radiation from ^{32}P in water has an energy distribution similar to that of ^{3}H in toluene scintillator but the spectrum is more compact; addition of a yellow colored quenching agent reduces the spectrum height and concomitantly causes a downward shift. On account of this shift, Wiebe *et al.* [338] found the isotope channels-ratio effective in efficiency determination for ^{32}P in water in the presence of yellow, green and red dyes. Moir [346] showed that the channels-ratio method gives the most accurate estimation of efficiency for Čerenkov counting of ^{42}K at high counting efficiencies. On the other hand, Elrick and Parker [331] noted only a 2% change in isotope channels-ratio for a 10% change in counting efficiency. The sensitivity of channels-ratio method can be increased if high quantum efficiency photocathodes are used [342].

External standardization is useful for color quench correction but is inherently more variable and less accurate than the isotope channels-ratio method, because variations in the counting vial can cause substantial change in Čerenkov emissions from the external standard ^{226}Ra; this is demonstrated by the fact that 70% of the Čerenkov counts from a sample containing 10 ml of water came from the empty vial. Parker [342] has shown that in the efficiency correction curve different pigments at low concentrations cause only small variations but a divergence occurs at high dye concentrations.

When Cab-O-Sil is present in a sample of ^{32}P in water, it causes an initial rise followed by a decline in counting efficiency as the amounts of the thixotropic agent is increased from 100 mg/l to 500 mg/l [338]. The pulse height spectrum is attenuated at high concentrations of Cab-O-Sil but without concomitant downward shift of the spectrum towards the low energy end; so the change in counting efficiency is not accompanied by a sensitive and proportionate change in isotope channels-ratio.

REFERENCES

1. M. AGENO, M. CHIOZZOTTO and R. QUERZOLI, *Atti accad. naz. Lincei, Rend.*, **6**, 626 (1949).
2. G. T. REYNOLDS, F. B. HARRISON and G. SALVINI, *Phys. Rev.*, **78**, 488 (1950).
3. H. KALLMANN and M. FURST, *Phys. Rev.*, **79**, 857 (1950).
4. H. KALLMANN and M. FURST, *Phys. Rev.*, **81**, 853 (1951).
5. H. KALLMANN and M. FURST, *Nucleonics*, **8**(3), 32 (1951).
6. F. N. HAYES, L. C. KING and D. E. PETERSON, *J. Amer. Chem. Soc.*, **74**, 1106 (1952).
7. F. N. HAYES, B. S. ROGERS and D. G. OTT, *J. Amer. Chem. Soc.*, **77**, 1950 (1955).
8. F. N. HAYES, D. G. OTT, V. N. KERR and B. S. ROGERS, *Nucleonics*, **13**,(12), 38 (1955).
9. F. N. HAYES, D. G. OTT, and V. N. KERR, *Nucleonics*, **14**(1), 42 (1956).
10. C. L. COWAN, Jr., F. REINES, F. B. HARRISON, H. W. KRUSE and A. D. McGUIRE, *Science*, **124**, 103 (1956).
11. E. C. ANDERSON, R. L. SCHUCH, J. D. PERRINGS and W. H. LANGHAM, *Nucleonics*, **14**(1), 1 (1956).
12. E. C. ANDERSON, R. L. SCHUCH, W. R. FISHER and W. H. LANGHAM, *Science*, **125**, 1273 (1957).
13. E. C. ANDERSON, *Brit. J. Radiol. Suppl.*, **7**, 27 (1957).

14. E. C. ANDERSON, Liquid Scintillation Counting. (C. G. Bell, Jr. and F. N. Hayes, eds.) p. 211. Pergamon Press, New York, (1958).
15. W. H. LANGHAM, Liquid Scintillation Counting. (C. G. Bell, Jr. and F. N. Hayes, eds.) p. 135. Pergamon Press, New York, (1958).
16. F. N. HAYES, B. S. ROGERS and P. C. SANDERS, *Nucleonics,* 13(1), 46 (1955).
17. D. G. OTT, F. N. HAYES, V. N. KERR and R. W. BENZ, *Science,* 123, 1071 (1956).
18. D. G. OTT, F. N. HAYES, E. HANSBURY and V. N. KERR, *J. Amer. Chem. Soc.,* 79, 5448 (1957).
19. D. G. OTT, V. N. KERR, F. N. HAYES and E. HANSBURY, *J. Org. Chem.,* 25, 872 (1960).
20. F. N. HAYES and R. C. GOULD, *Science,* 117, 480 (1953).
21. R. D. HIEBERT and R. J. WATTS, *Nucleonics,* 11(12), 38 (1953).
22. L. W. PRICE, *World Medical Electronics,* 5, 3 (1967).
23. E. RAPKIN, The Current Status of Liquid Scintillation Counting. (E. D. Bransome, Jr., ed.) p. 45. Grune & Stratton, New York, (1970).
24. C. G. BELL, Jr. and F. N. HAYES (editors), Liquid Scintillation Counting. Proceedings of a Conference held at Northwestern University, August 20–22, 1957. Pergamon Press, New York, (1958).
25. G. H. DAUB, F. N. HAYES and E. SULLIVAN (editors), Proceedings of the University of New Mexico Conference on Organic Scintillation Detectors, August 15–17, 1960 (TID-7612 Instruments), U.S. Government Printing Office, Washington, D.C. (1960).
26. D. L. HORROCKS, (editor), Organic Scintillators. Proceedings of the International Symposium on Organic Scintillators, Argonne National Laboratory, June 20–22. 1966. Gordon and Breach, New York, (1968).
27. E. D. BRANSOME, Jr. (editor), The Current Status of Liquid Scintillation Counting. Grune & Stratton, New York, (1970).
28. D. L. HORROCKS, C. T. PENG (editors), Organic Scintillators and Liquid Scintillation Counting. Proceedings of the International Conference on Organic Scintillators and Liquid Scintillation Counting, University of California at San Francisco, July 7–10, 1970. Academic Press, New York, (1971).
29. Liquid Scintillation Counting. (A. Dyer, ed.), Vol. 1, Heyden & Son Ltd., London, (1971).
30. J. B. BIRKS, The Theory and Practice of Scintillation Counting. MacMillan Co., New York, (1964).
31. J. B. BIRKS, The Current Status of Liquid Scintillation Counting. (E. D. Bransome, Jr., ed.) p. 3. Grune & Stratton, New York, (1970).
32. J. H. O'DONNELL and D. F. SANGSTER, Principles of Radiation Chemistry. p. 17. Am. Elsevier Publ. Co., New York, (1970).
33. C. A. PARKER, Photoluminiscence of Solutions. p. 48. Elsevier Publ. Co., Amsterdam, (1968).
34. J. W. HUNT and J. K. THOMAS, *J. Chem. Phys.,* 46, 2954 (1967).
35. P. AUSLOOS, *Progr. React. Kinet.* (G. Porter, ed.), 5, 113 (1970).
36. F. WILKINSON, *Quart. Rev.* (*London*), 20, 403 (1966).
37. J. B. BIRKS and J. C. CONTE, *Proc. Roy. Soc.* (*London*), A303, 85 (1968).
38. S. GEORGHIOU and K. R. NAQVI, Molecular Luminescence. (E. C. Lim, ed.), p. 393. W. A. Benjamin, New York, (1969).

39. S. GEORGHIOU and I. H. MUNRO, Organic Scintillators and Liquid
    Scintillation Counting. (D. L. Horrocks, C. T. Peng, eds.), p. 339.
    Academic Press, New York, (1971).
40. R. VOLTZ, J. KLEIN, F. HEISEL, H. LAMI, G. LAUSTRIAT and A.
    COCHE, *J. Chem. Phys.*, **63**, 1259 (1966).
41. G. LAUSTRIAT, R. VOLTZ and J. KLEIN, The Current Status of
    Liquid Scintillation Counting. (E. D. Bransome, Jr., ed.), p. 13. Grune
    & Stratton, New York, (1970).
42. D. L. HORROCKS, *J. Chem. Phys.*, **52**, 1566 (1970).
43. A. SKERBELE and E. N. LASSETTRE, *J. Chem. Phys.*, **42**, 395
    (1965).
44. E. LEVIN, M. FURST and H. KALLMANN, *J. Chem. Phys.*, **54**, 2580
    (1971).
45. V. LAOR and A. WEINREB, *J. Chem. Phys*, **43**, 1565 (1965).
46. C. T. PENG, Proceedings of the University of New Mexico Conference
    on Organic Scintillation Detectors. (G. H. Daub, F. N. Hayes and
    E. Sullivan, eds.), p. 271. August 15–17, 1960 (TID-7612), U.S.
    Government Printing Office, Washington, D.C. (1960).
47. G. KALLMANN-OSTER and H. KALLMANN, *J. Chim. Phys.*, **64**, 28
    (1967).
48. Th. FÖRSTER, *Disc. Faraday Soc.* **27**, 7 (1959).
49. A. WEINREB, *J. Chem. Phys.*, **27**, 133 (1957).
50. I. B. BERLMAN, 'Handbook of Fluorescence Spectra of Aromatic
    Molecules.' Academic Press, New York. 1965; 2nd Edition, (1971).
51. G. LAUSTRIAT, *Mol. Cryst.*, **4**, 127 (1968).
52. D. L. HORROCKS, The Current Status of Liquid Scintillation Count-
    ing. (E. D. Bransome, Jr., ed.), p. 25. Grune & Stratton, New York,
    (1970).
53. P. SKARSTAD, R. MA and S. LIPSKY, *Mol. Cryst.*, **4**, 3 (1968).
54. J. W. HASTINGS and G. WEBER, *J. Opt. Soc. Amer.*, **53**, 1410
    (1963).
55. R. K. SWANK, Liquid Scintillation Counting. (C. G. Bell, Jr. and F. N.
    Hayes, eds.), p. 23. Pergamon Press, New York, (1958).
56. E. KOWALSKI, R. ANLIKER and K. SCHMID, *Int. J. Appl. Radiat.
    Isotopes*, **18**, 307 (1967).
57. M. FURST and H. KALLMANN, *J. Chem. Phys.*, **23**, 607 (1955).
58. E. GOMEZ, J. FREER and J. CASTRILLON, *Int. J. Appl. Radiat.
    Isotopes*, **22**, 243 (1971).
59. J. D. DAVIDSON and P. FEIGELSON, *Int. J. Appl. Radiat. Isotopes*,
    2, 1 (1958).
60. M. D. BARNETT, G. H. DAUB, F. N. HAYES and D. G. OTT, *J.
    Amer. Chem. Soc.*, **82**, 2282 (1960).
61. G. H. DAUB, F. N. HAYES, D. W. HOLT, L. IONESCU and J. L.
    SCHORNICK, *Mol. Cryst.*, **4**, 343 (1968).
62. H. O. WIRTH, Proceedings of the University of New Mexico Conference
    on Organic Scintillation Detectors. (G. H. Daub, F. N. Hayes,
    E. Sullivan, eds.), p. 78. August 15–17, 1960 (TID-7612), U.S. Atomic
    Energy Commission, Office of Technical Information, (1966).
63. H. O. WIRTH, F. U. HERRMANN, G. HERRMANN and W. KERN,
    Mol. Cryst., **4**, 321 (1968).
64. A. HELLER, *J. Chem. Phys.*, **35**, 1980 (1961).
65. A. HELLER, *J. Chem. Phys.*, **40**, 2839 (1964).

66. E. KOWALSKI, R. ANLIKER and K. SCHMID, Mol. Cryst., 4, 403 (1968).
67. P. LEGGATE and D. OWEN, Mol. Cryst., 4, 357 (1968).
68. B. SCALES, Int. J. Appl. Radiat. Isotopes, 18, 1 (1966).
69. W. H. MELHUISH, J. Phys. Chem., 65, 229 (1961).
70. D. L. HORROCKS and H. O. WIRTH, Mol. Cryst, 4, 375 (1968).
71. D. L. HORROCKS, J. Chem. Phys., 51, 5443 (1969).
72. I. B. BERLMAN, H. O. WIRTH and O. J. STEINGRABER, J. Phys. Chem., 75, 318 (1971).
73. D. WALKER and T. D. WAUGH, J. Heterocycl. Chem., 1, 72 (1964).
74. G. VASVARI, Int. J. Appl. Radiat. Isotopes, 16, 327 (1965).
75. M. E. ACKERMAN, G. H. DAUB, F. N. HAYES and H. A. MACKAY, Organic Scintillators and Liquid Scintillation Counting. (D. L. Horrocks, C. T. Peng, eds.), p. 315. Academic Press, New York, (1971).
76. J. W. DAVIES and T. C. HALL, Anal. Biochem., 27, 77 (1969).
77. G. GERMAI, Int. J. Appl. Radiat. Isotopes, 21, 587 (1970).
78. D. A. KALBHEN, Int. J. Appl. Radiat. Isotopes, 18, 655 (1967).
79. E. LUCATU and S. SUCIU, Proceedings of the International Conference on Luminescence. (G. Szigeti, ed.), Vol. 1, p. 503. Akadémiai Kiadó, Budapest, (1968).
80. R. H. BENSON, Anal. Chem., 38, 1353 (1966).
81. M. S. PATTERSON and R. C. GREEN, Anal. Chem., 37, 854 (1965).
82. R. C. GREEN, M. S. PATTERSON and A. H. ESTES, Anal. Chem., 40, 2035 (1968).
83. D. G. OTT, C. R. RICHMOND, T. T. TRUJILLO and H. FOREMAN, Nucleonics, 17(9), 106 (1959).
84. C. G. WHITE and S. HELF, Nucleonics, 14, (10), 46 (1956).
85. B. L. FUNT, Can. J. Chem., 39, 711 (1961).
86. C. TANIELIAN, Proceedings of the International Conference on Luminescence. (G. Szigeti, ed.), Vol. I, p. 468. Akadémiai Kiadó, Budapest, (1968).
87. G. KALLMANN-OSTER, Acta Phys. Polon., 26, 435 (1964).
88. A. HELLER, J. Chem. Phys., 36, 2858 (1962).
89. C. S. PARMENTER and J. D. RAU, J. Chem. Phys., 51, 2242 (1969).
90. H. ISHIKAWA and M. TAKIUE, Organic Scintillators and Liquid Scintillation Counting. (D. L. Horrocks, C. T. Peng, eds.), p. 387. Academic Press, New York, (1971).
91. M. D. LUMB and D. A. WEYL, Proceedings of the International Conference on Luminescence. (G. Szigeti, ed.), Vol. I, p. 477. Akadémiai Kiadó, Budapest, (1968).
92. H. H. ROSS and R. E. YERICK, Anal. Chem., 35, 794 (1963).
93. T. IWAKURA and Y. KASIDA, Proc. Symp. Radioisotope Sample Meas. Tech. Med. Biol. Vienna, p. 447. IAEA Vienna, (1965).
94. R. J. HERBERG, Anal. Chem., 32, 1468 (1960).
95. H. H. ROSS, Proc. Symp. Radioisotope Sample Meas. Tech. Med. Biol., p. 409. Vienna, IAEA. Vienna, (1965).
96. M. P. NEARY and A. L. BUDD, The Current Status of Liquid Scintillation Counting. (E. D. Bransome, Jr., ed.), p. 273. Grune & Stratton, New York, (1970).
97. C. T. PENG, The Current Status of Liquid Scintillation Counting. (E. D. Bransome, Jr., ed.), p. 283. Grune & Stratton, New York, (1970).

98. E. T. BUSH, *Int. J. Appl. Radiat. Isotopes,* **19**, 447, (1968).
99. D. A. KALBHEN, The Current Status of Liquid Scintillation Counting. (E. D. Bransome, Jr., ed.), p. 337. Grune & Stratton, New York, (1970).
100. J. WINKELMAN and G. SLATER, *Anal. Biochem.,* **20**, 365 (1967).
101. D. A. KALBHEN, Paper presented at the Tritium Symposium. Las Vegas, 1971.
102. D. A. KALBHEN, Liquid Scintillation Counting. (A. Dyer, ed.) Heyden & Son Ltd. Vol. I, p. 1. London (1971).
103. E. D. BRANSOME, Jr. and M. F. GROWER, The Current Status of Liquid Scintillation Counting. (E. D. Bransome, Jr., ed.), p. 342. Grune & Stratton, New York, (1970).
104. A. C. HOUTMAN, *Int. J. Appl. Radiat. Isotopes,* **16**, 65 (1965).
105. R. J. HERBERG, *Science,* **128**, 199 (1958).
106. P. DEBYE and J. O. EDWARDS, *Science,* **116**, 143 (1952).
107. R. A. LLOYD, S. C. ELLIS and K. H. HALLOWES, Proc. Symp. Detection Use Tritium Phys. Biol. Sci., Vol. I, p. 263. IAEA Vienna, (1962).
108. P. FODOR-CSANYI and B. LEVAY, *Int. J. Appl. Radiat. Isotopes,* **20**, 223 (1969).
109. D. L. HORROCKS and M. H. STUDIER, *Anal. Chem.,* **33**, 615 (1961).
110. R. H. BENSON and R. L. MAUTE, *Anal. Chem.,* **34**, 1122 (1962).
111. C. T. PENG, *Anal. Chem.,* **32**, 1292 (1960).
112. P. D. KLEIN and W. J. EISLER, Organic Scintillators and Liquid Scintillation Counting. (D. L. Horrocks, C. T. Peng, eds.), p. 395. Academic Press, New York, (1971).
113. C. T. PENG, *Anal. Chem.,* **41**, 16 (1969).
114. J. R. ARNOLD, *Science,* **119**, 155 (1954).
115. C. T. PENG, *Advan. Tracer Methodol.,* **3**, 81 (1966).
116. S. HELF and C. WHITE, *Anal. Chem.,* **29**, 13 (1957).
117. H. H. ROSS, *Int. J. Appl. Radiat. Isotopes,* **18**, 335 (1967).
118. G. E. A. WYLD, The Current Status of Liquid Scintillation Counting. (E. D. Bransome, Jr., ed.), p. 69. Grune & Stratton, New York, (1970).
119. K. G. PORGES, *Int. J. Appl. Radiat. Isotopes,* **19**, 711 (1968).
120. J. L. PUTMAN, *Int. J. Appl. Radiat. Isotopes,* **20**, 205 (1969).
121. C. H. WANG, The Current Status of Liquid Scintillation Counting. (E. D. Bransome, Jr., ed.), p. 305. Grune & Stratton, New York, (1970).
122. I. S. BOYCE and J. F. CAMERON, Proc. Symp. Detection Use Tritium Phys. Biol. Sci., Vol. I, p. 231. IAEA, Vienna, (1962).
123. I. T. TAKAHASHI and F. A. BLANCHARD, *Anal. Biochem.,* **29**, 154 (1969).
124. C. T. PENG, *Anal. Chem.,* **36**, 2456 (1964).
125. J. H. PARMENTIER and F. E. L. TEN HAAF, *Int. J. Appl. Radiat. Isotopes,* **20**, 305 (1969).
126. A. H. SMITH and G. W. REED, Proc. Symp. Radioisotope Sample Meas. Tech. Med. Biol., p. 427. IAEA, Vienna, (1965).
127. A. W. ROGERS and J. F. MORAN, *Anal. Biochem.,* **16**, 206 (1966).
128. R. J. HERBERG, *Anal. Chem.,* **35**, 786 (1963).
129. N. KACZMARCZYK, Organic Scintillators and Liquid Scintillation Counting. (D. L. Horrocks, C. T. Peng, eds.), p. 977. Academic Press, New York, (1971).

130. M. A. DUGGAN and R. D. ICE, Organic Scintillators and Liquid Scintillation Counting. (D. L. Horrocks, C. T. Peng, eds.), p. 1055. Academic Press, New York, (1971).
131. L. A. BAILLIE, *Int. J. Appl. Radiat. Isotopes* 8, 1 (1960).
132 E. T. BUSH, *Anal. Chem.*, 35, 1024 (1963).
133. R. J. HERBERG, Packard Technical Bulletin No. 15, December 1965.
134. M. I. KRICHEVSKY, S. A. ZAVELER and J. BULKELEY, *Anal. Biochem.*, 22, 442 (1968).
135. B. D. CADDOCK, P. T. DAVIES and J. H. DETERDING, *Int. J. Appl. Radiat. Isotopes*, 18, 209 (1966).
136. D. L. HORROCKS, *Nature*, 202, 78 (1964).
137. T. HIGASHIMURA, O. YAMADA, N. NOHARA and T. SHIDEI, *Int. J. Appl. Radiat. Isotopes*, 13, 308 (1962).
138. D. G. FLEISHMAN and V. V. GLAZUNOV, *Pribory i Tekhnika Eksperimenta*, 3, 5 (1962).
139. R. de WACHETER and W. FIERS, *Anal. Biochem.*, 18, 351 (1967).
140. P. H. SPRINGELL, *Int. J. Appl. Radiat. Isotopes*, 20, 743 (1969).
141. I. T. TAKAHASHI and F. A. BLANCHARD, *Anal. Biochem.*, 35, 411 (1970).
142. F. N. HAYES, *Advan. Tracer Methodol.*, 3, 95 (1966).
143. D. S. GLASS, *Int. J. Appl. Radiat. Isotopes*, 21, 531 (1970).
144. W. J. KAUFMAN, A. NIR, G. PARKS and R. M. HOURS, Proc. Symp. Detection Use Tritium Phys. Biol. Sci., Vol. I, p. 251. IAEA, Vienna, (1962).
145. H. E. DOBBS, *Nature*, 20, 1283 (1963).
146. E. T. BUSH, *Int. J. Appl. Radiat. Isotopes*, 19, 447 (1968).
147. N. KACZMARCZYK and I. RUGE, *Int. J. Appl. Radiat. Isotopes*, 20, 283 (1969).
148. H. H. ROSS, *Int. J. Appl. Radiat. Isotopes*, 15, 273 (1964).
149. H. TAKAHASHI, T. HATTORI and B. MARUO, *Anal. Biochem.*, 2, 447 (1961).
150. R. E. PETTERSON, *Advan. Tracer Methodol.*, 1, 265 (1963).
151. E. F. POLIC, Instrumentation in Nuclear Medicine. (G. J. Hine, ed.), p. 227. Academic Press, New York, (1967).
152. G. T. OKITA, J. J. KABARA, F. RICHARDSON and G. V. LEROY, *Nucleonics*, 15(6), 111 (1957).
153. L. R. AXELROD, C. MATTHIJSSEN, J. W. GOLDZIEHER and J. E. PULLIAM, *Acta Endocrinologica*, Suppl. 99 (1965).
154. Y. KOBAYASHI and D. V. MAUDSLEY, The Current Status of Liquid Scintillation Counting. (E. D. Bransome, Jr., ed.), p. 76. Grune & Stratton, New York, (1970).
155. P. D. KLEIN and W. J. EISLER, Jr., *Anal. Chem.*, 38, 1453 (1966).
156. B. E. HAISSIG and A. L. SCHIPPER, Jr., *Anal. Chem.*, 42, 1456 (1970).
157. R. W. HENDLER, *Anal. Biochem.*, 7, 110 (1964).
158. R. NINOMIYA, *Int. J. Appl. Radiat. Isotopes*, 17, 355 (1966).
159. J. K. WELTMAN and D. W. TALMAGE, *Int. J. Appl. Radiat. Isotopes*, 14, 541 (1963).
160. G. HETENYI, Jr. and J. REYNOLDS, *Int. J. Appl. Radiat. Isotopes*, 18, 331 (1967).
161. G. SHEPPARD and C. G. MARLOW, *Int. J. Appl. Radiat. Isotopes*, 22, 125 (1971).

162. M. B. SNIPES and F. W. LENGEMANN, *Int. J. Appl. Radiat. Isotopes,* **22,** 513 (1971).
163. J. W. DAVIES and T. C. HALL, *Anal. Biochem.,* **27,** 77 (1969).
164. R. WU, *Anal. Biochem.,* **7,** 207 (1964).
165. M. ZADUBAN, N. STALLÁROVA, Š. PALÁGYI, *J. Radioanal. Chem.,* **5,** 91 (1970).
166. R. VANINBROUKX and A. SPERNOL, *Int. J. Appl. Radiat. Isotopes,* **16,** 289 (1965).
167. G. GOLDSTEIN, *Nucleonics,* **23**(3), 67 (1965).
168. K. FRIEDRICK and M. LEISTNER, *Isotopenpraxis,* **4,** 251 (1968).
169. R. VANINBROUKX, Organic Scintillators and Liquid Scintillation Counting. (D. L. Horrocks, C. T. Peng, eds.), p. 913. Academic Press, New York, (1971).
170. W. J. McDOWELL, Organic Scintillators and Liquid Scintillation Counting. (D. L. Horrocks, C. T. Peng, eds.), p. 937. Academic Press, New York, (1971).
171. T. K. KIM and M. B. MacINNIS, Organic Scintillators and Liquid Scintillation Counting. (D. L. Horrocks, C. T. Peng, eds.), p. 925. Academic Press, New York, (1971).
172. J. STEYN, *Proc. Phys. Soc. London,* **69,** 865 (1956).
173. V. KOLAROV, Y. L. GALLIC and R. VATIN, *Int. J. Appl. Radiat. Isotopes,* **21,** 443 (1970).
174. E. RAPKIN and J. A. GIBBS, Packard Technical Bulletin No. 9, (1965).
175. B. W. AGRANOFF, *Nucleonics,* **15**(10), 106 (1957).
176. G. E. CALF, *Int. J. Appl. Radiat. Isotopes* **20,** 611 (1969).
177. E. SCHWERDTEL, *Int. J. Appl Radiat. Isotopes* **17,** 479 (1966).
178. B. E. GORDON and R. M. CURTIS, *Anal. Chem.,* **40,** 1487 (1968).
179. E. RAPKIN, Instrumentation in Nuclear Medicine. (G. Hine, ed.), p. 181. Academic Press, New York (1967).
180. Y. KOBAYASHI and D. V. MAUDSLEY, *Methods of Biochemical Analysis,* **17,** 55 (1969).
181. J. M. PASSMANN, N. S. RADIN and J. A. D. COOPER, *Anal. Chem.,* **28,** 484 (1956).
182. M. VAUGH, D. STEINBERG and J. LOGAN, *Science,* **126,** 446 (1957).
183. H. G. BADMAN and W. O. BROWN, *Analyst,* **86,** 342 (1961).
184. W. O. BROWN and H. G. BADMAN, *Biochem. J.,* **78,** 571 (1961).
185. E. RAPKIN, Packard Technical Bulletin No. 3. (Revised) (1970).
186. D. L. HANSEN and E. T. BUSH, *Anal. Biochem.,* **18,** 320 (1967).
187. D. McCLENDON, M. P. NEARY, M. GALASSI and W. STEPHENS, Organic Scintillation Counting. (D. L. Horrocks, C. T. Peng, eds.), p. 587. Academic Press, New York, (1971).
188. E. SCHAUMLOEFFEL and E. H. GRAUL, *Atompraxis,* **13,** 260 (1967).
189. E. B. MUELLER, The Current Status of Liquid Scintillation Counting. (E. D. Bransome, Jr., ed.), p. 181. Grune & Stratton, New York, (1970).
190. G. A. BRAY, The Current Status of Liquid Scintillation Counting (E. D. Bransome, Jr., ed.), p. 170. Grune & Stratton, New York, (1970).
191. W. G. DUNCOMBE and T. J. RISING, *Anal. Biochem.,* **30,** 275 (1969).

192. R. S. BASCH, *Anal. Biochem.*, **26**, 184 (1968).

193. P. N. PAUS, *Anal. Biochem.*, **42**, 372 (1971).

194. P. HELLUNG-LARSEN, *Anal. Biochem.*, **39**, 454 (1971).

195. P. V. TISHLER and C. J. EPSTEIN, *Anal. Biochem.*, **22**, 89 (1968).

196. B. MOSS and V. M. INGRAM, *Proc. Nat. Acad. Sci.*, USA **54**, 967 (1965).

197. A. V. LeBOUTON, *Anal. Biochem.*, **36**, 445 (1968).

198. M. F. GROWER and E. D. BRANSOME, Jr., The Current Status of Liquid Scintillation Counting. (E. D. Bransome, Jr., ed.), p. 263. Grune & Stratton, New York, (1970).

199. A. DUNN, *Int. J. Appl. Radiat. Isotopes*, **22**, 212 (1971).

200. C. P. PETROFF, H. H. PATT and P. P. NAIR, *Int. J. Appl. Radiat. Isotopes*, **16**, 599 (1965).

201. H. Y. NEUJAHR and B. EWALDSSON, *Anal. Biochem.*, **8**, 487, (1964).

202. G. E. FRANCIS and J. D. HAWKINS, *Int. J. Appl. Radiat. Isotopes*, **18**, 223 (1967).

203. N. J. BENEVENGA, Q. R. ROGERS and A. E. HARPER, *Anal. Biochem.*, **24**, 393 (1968).

204. L. KRABISCH and B. BORGSTROEM, *Scand. J. Gastroenterol.*, **3**, 458 (1968).

205. D. T. MAHIN, *Int. J. Appl. Radiat. Isotopes*, **17**, 185 (1966).

206. E. A. SHNEOUR, S. ARONOFF and M. R. KIRK, *Int. J. Appl. Radiat. Isotopes*, **13**, 623 (1962).

207. P. A. LINDSAY and K. B. KURNICK, *Int. J. Appl. Radiat. Isotopes*, **20**, 97 (1969).

208. R. D. O'BRIEN, *Anal. Biochem.*, **7**, 251 (1964).

209. M. PFEFFER, S. WINSTEIN, J. GAYLORD and L. INDIDOLI, *Anal. Biochem.*, **39**, 46 (1971).

210. E. H. BELCHER, *Physics in Medicine Biology*, **5**, 49 (1960).

211. D. T. MAHIN and R. T. LOFFERG, *Anal. Biochem.*, **16**, 500 (1966).

212. R. D. MOORE, C. A. CRANE and I. D. FRANTZ, Jr., *Anal. Biochem.*, **24**, 545 (1968).

213. G. R. WATSON and J. P. WILLIAMS, *Anal. Biochem.*, **33**, 356 (1970).

214. A. K. WEGMAN, J. C. WILLIAMS, A. PLENTL, *Anal. Biochem.*, **19**, 441 (1967).

215. H. JEFFAY, Packard Technical Bulletin No. 10, October, (1962).

216. R. G. KELLY, E. A. PEETS, S. GORDON and D. A. BUYSKE, *Anal. Biochem.*, **2**, 267 (1961).

217. F. KALBERER and J. RUTSCHMANN, *Helv. Chim. Acta*, **44**, 1956 (1961).

218. R. RONCUCCI, G. LAMBELIN, M. J. SIMON and W. SOUDYN, *Anal. Biochem.*, **26**, 118 (1968).

219. V. VOLF, *Int. J. Appl. Radiat. Isotopes*, **21**, 685 (1970).

220. H. E. DOBBS, *Int. J. Appl. Radiat. Isotopes*, **17**, 363 (1966).

221. R. E. OBER, A. R. HANSEN, D. MOURER, J. BAUKEMA and G. W. GWYNN, *Int. J. Appl. Radiat. Isotopes*, **20**, 703 (1969).

222. L. E. MARTIN and C. HARRISON, *Biochem. J.*, **82**, 18 (1962).

223. V. T. OLIVERIO, C. DENHAM and J. D. DAVIDSON, *Anal. Biochem.*, **4**, 188 (1962).

224. G. GUPTA, *Anal. Chem.*, **38**, 1356 (1966).

225. T. H. VICKERS, *Anal. Chem.*, **40**, 2219 (1968).
226. L. A. WEGNER and H. WINKELMANN, *Atompraxis*, **16**, 19 (1970).
227. J. D. DAVIDSON, V. T. OLIVERIO and J. I. PETERSON, The Current Status of Liquid Scintillation Counting. (E. D. Bransome, Jr., ed.), p. 222. Grune & Stratton, New York, (1970).
228. L. G. HUEBNER and W. E. KISIELESKI, *Atompraxis*, **16**, 15 (1970).
229. E. A. PEETS, J. R. FLORINI and D. A. BUYSKE, *Anal. Chem.*, **32**, 1465 (1960).
230. M. A. TAMERS and M. DIES, *Int. J. Applied Radiat. Isotopes*, **15**, 697 (1964).
231. H. W. KNOCHE and R. M. BELL, *Anal. Biochem.*, **12**, 49 (1965).
232. M. H. GRIFFITHS and A. MALLINSON, *Anal. Biochem.*, **22**, 465, (1968).
233. J. I. PETERSON, F. WAGNER, S. SIEGEL and W. NIXON, *Anal. Biochem.*, **31**, 189 (1969).
234. J. I. PETERSON, *Anal. Biochem.*, **31**, 204 (1969).
235. T. R. TYLER, A. R. REICH and C. ROSENBLUM, Organic Scintillators and Liquid Scintillation Counting. (D. L. Horrocks, C. T. Peng, eds.), p. 869. Academic Press, New York, (1971).
236. L. W. SMITH, *Anal. Biochem.*, **29**, 223 (1969).
237. N. KAARTINEN, Packard Technical Bulletin No. 18, April, (1969).
238. D. W. SHER, N. KAARTINEN, L. J. EVERETT and V. JUSTES, Jr., Organic Scintillators and Liquid Scintillation Counting. (D. L. Horrocks, C. T. Peng, eds.), p. 849. Academic Press, New York, (1971).
239. E. B. MUELLER, Organic Scintillators and Liquid Scintillation Counting. (D. L. Horrocks, C. T. Peng, eds.), p. 181. Academic Press, New York, (1971).
240. D. L. HORROCKS, Packard Technical Bulletin No. 2, Revised, November, (1962).
241. Y. MARCUS and A. S. KERTES, Ion Exchange and Solvent Extraction of Metal Complexes. Wiley, London, (1969).
242. D. F. PEPPARD, J. R. FERRARO and G. W. MASON, *J. Inorg. Nucl. Chem.*, **12**, 60 (1959).
243. T. E. F. CARR and B. J. PARSONS, *Int. J. Appl. Radiat. Isotopes*, **13**, 57 (1962).
244. J. E. HARDCASTLE, R. J. HANNAPEL and W. H. FULLER, *Int. J. Appl. Radiat. Isotopes*, **18**, 193 (1967).
245. M. SARNAT and H. JEFFAY, *Anal. Chem.*, **34**, 643 (1962).
246. L. LUTVAK, *Anal. Chem.*, **31**, 340 (1959).
247. G. PATRICK and C. STIRLING, *Int. J. Appl. Radiat. Isotopes*, **22**, 627 (1971).
248. E. R. HUMPHREYS, *Int. J. Appl. Radiat. Isotopes*, **16**, 345 (1965).
249. T. P. LEFFINGWELL, R. W. RIESS and G. S. MELVILLE, Jr., *Int. J. Appl. Radiat. Isotopes*, **13**, 75 (1962).
250. R. J. DERN and W. L. HART, *J. Lab. Clin. Med.*, **57**, 332, 460 (1961).
251. D. L. HORROCKS, *Nucl. Instrum. Meth.*, **30**, 157 (1964).
252. B. R. HARVEY and G. A. SUTTON, *Int. J. Appl. Radiat. Isotopes*, **21**, 519 (1970).
253. K. JOON and P. A. DEURLOO, *Int. J. Appl. Radiat. Isotopes*, **16**, 334 (1965).

254. H. IHLE, M. KARAYANNIS and A. MURRENHOFF, Organic Scintillators and Liquid Scintillation Counting. (D. L. Horrocks, C. T. Peng, eds.), p. 879. Academic Press, New York, (1971).

255. F. N. HAYES, B. S. ROGERS and W. H. LANGHAM, *Nucleonics,* 14(3), 48 (1956).

256. R. C. MEADE and R. A. STIGLITZ, *Int. J. Appl. Radiat. Isotopes,* 13, 11 (1962).

257. B. W. FOX, *Int. J. Appl. Radiat. Isotopes,* 19, 717 (1968).

258. D. I. CHAPMAN and J. MARCROFT, *Int. J. Appl. Radiat. Isotopes,* 22, 371 (1971).

259. J. D. VAN DER LAARSE, *Int. J. Appl. Radiat. Isotopes,* 18, 485 (1967).

260. A. A. MOGHISSI, H. L. KELLEY, J. E. REGNIER and M. W. CARTER, *Int. J. Appl. Radiat. Isotopes,* 20, 145 (1969).

261. R. LIEBERMAN and A. A. MOGHISSI, *Int. J. Appl. Radiat. Isotopes,* 21, 319 (1970).

262. F. BOHNE, *Int. J. Appl. Radiat. Isotopes* 22, 384 (1971).

263. A. E. WHYMAN, *Int. J. Appl. Radiat. Isotopes,* 21, 81 (1970).

264. A. NADARAJAH, B. LEESE and G. F. JOPLIN, *Int. J. Appl. Radiat. Isotopes,* 20, 733 (1969).

265. M. L. CURTIS, Organic Scintillators and Liquid Scintillation Counting. (D. L. Horrocks, C. T. Peng, eds.), p. 899. Academic Press, New York, (1971).

266. J. C. TURNER, *Int. J. Appl. Radiat. Isotopes,* 19, 557 (1968).

267. J. C. TURNER, *Int. J. Appl. Radiat. Isotopes,* 20, 499 (1969).

268. R. C. GREENE, The Current Status of Liquid Scintillation Counting. (E. D. Bransome, Jr., ed.), p. 189. Grune & Stratton, New York, (1970).

269. A. E. SHAMOO, *Anal. Biochem.,* 39, 311 (1971).

270. B. L. FUNT, *Nucleonics,* 14(8), 83 (1956).

271. J. N. BOLLING, W. A. MALLOW, J. W. REGISTER, Jr. and D. E. JOHNSON, *Anal. Chem.,* 39, 1508 (1967).

272. A. BENAKIS, Organic Scintillators and Liquid Scintillation Counting. (D. L. Horrocks, C. T. Peng, eds.), p. 735. Academic Press, New York, (1971).

273. R. TYE and J. D. ENGEL, *Anal. Chem.,* 37, 1225 (1965).

274. F. SNYDER and N. STEPHENS, *Anal. Biochem.,* 4, 128 (1962).

275. J. D. YARBROUGH, A. F. FENDEIS and J. C. O'KELLEY, *Int. J. Appl. Radiat. Isotopes,* 17, 453 (1966).

276. J. D. EAKINS and D. A. BROWN, *Int. J. Appl. Radiat. Isotopes,* 17, 391 (1966).

277. C. P. LLOYD-JONES, *Analyst,* 95, 366 (1970).

278. E. RAPKIN, Packard Technical Bulletin No. 5, Revised, April 1970.

279. C. T. PENG, Liquid Scintillation Counting. (C. G. Bell, Jr., F. N. Hayes, eds.), p. 198. Pergamon Press, New York, (1958).

280. A. T. BULL, *J. Label. Compounds,* 4, 181 (1968).

281. B. BLOOM, *Anal. Biochem.,* 6, 359 (1963).

282. E. D. BRANSOME, Jr. and M. F. GROWER, *Anal. Biochem.,* 38, 401 (1970).

283. C. H. WONG and D. E. JONES, *Biochem. Biophys. Res. Commun.,* 1, 203 (1959).

284. D. M. GILL, *Int. J. Appl. Radiat. Isotopes*, **18**, 393 (1968).
285. D. R. JOHNSON and J. W. SMITH, *Anal. Chem.*, **35**, 1991 (1963).
286. R. A. LLOYD and D. B. REES-EVANS, *Int. J. Appl. Radiat. Isotopes*, **16**, 393 (1965).
287. R. A. MALT and W. L. MILLER, *Anal. Biochem.*, **18**, 388 (1967).
288. C. F. CRAMER and B. H. ROSS, *Int. J. Appl. Radiat. Isotopes*, **21**, 237 (1970).
289. A. M. DOWNES and A. R. TILL, *Text. Res. J.*, **38**, 518, 523 (1968).
290. J. W. DAVIES and E. C. COCKING, *Biochim. Biophys. Acta*, **115**, 511 (1966).
291. C. W. HELLEINER and W. H. WUNNER, *Anal. Biochem.*, **39**, 333 (1971).
292. P. G. SPEAR and B. ROIZMAN, *Anal. Biochem.*, **26**, 197 (1968).
293. E. RAPKIN, Packard Technical Bulletin No. 11, February 1963.
294. J. A. HUNT, *Anal. Biochem.*, **23**, 289 (1968).
295. E. SCHRAM, The Current Status of Liquid Scintillation Counting. (E. D. Bransome, Jr., ed.), p. 95. Grune & Stratton, New York, (1970).
296. A. L. SMITH, J. W. THOMAS and H. WOLLMAN, *Int. J. Appl. Radiat. Isotopes*, **21**, 171 (1970).
297. D. S. KEARNS, *Int. J. Appl. Radiat. Isotopes*, **20**, 821 (1969).
298. R. J. HERBERG, *Anal. Chem.*, **33**, 1308 (1961).
299. R. J. HERBERG, *Anal. Chem.*, **36**, 1079 (1964).
300. E. T. BUSH, *Anal. Chem.*, **36**, 1082 (1964).
301. B. F. SCOTT, *J. Radioanal. Chem.*, **1**, 61 (1968).
302. A. L. KOCH, *Anal. Biochem.*, **23**, 352 (1968).
303. M. F. GROWER and E. D. BRANSOME, Jr., The Current Status of Liquid Scintillation Counting. (E. D. Bransome, Jr., ed.), p. 356. Grune & Stratton, New York, (1970).
304. F. A. BLANCHARD, *Int. J. Appl. Radiat. Isotopes*, **14**, 213 (1963).
305. C. MATTHIJSSEN, *Int. J. Appl. Radiat. Isotopes*, **15**, 382 (1966).
306. E. D. PLOTKA, E. G. STANT, Jr., F. A. WALTZ, V. A. GARWOOD and R. E. ERB, *Int. J. Appl. Radiat. Isotopes*, **17**, 637 (1966).
307. J. J. O'TOOLE and J. O. OSBURN, *Int. J. Appl. Radiat. Isotopes*, **19**, 821 (1968).
308. K. WINKLER, H. WOCKENFUSS, K. SIEGLER and K. ZIPPERER, *Isotopenpraxis*, **4**, 187 (1968).
309. A. W. FORREY, Organic Scintillators and Liquid Scintillation Counting. (D. L. Horrocks, C. T. Peng, eds.), p. 835. Academic Press, New York, (1971).
310. C. F. CRAMER, M. NICHOLSON, C. MOORE and K. TENG, *Int. J. Appl. Radiat. Isotopes*, **22**, 17 (1971).
311. J. L. SPRATT, *Int. J. Appl. Radiat. Isotopes*, **16**, 439 (1965).
312. J. L. SPRATT and G. L. LAGE, *Int. J. Appl. Radiat. Isotopes*, **18**, 247 (1967).
313. J. M. FELTS and P. A. MAYES, *Biochem. J.*, **105**, 735 (1967).
314. C. O. CARROLL and T. J. HOUSER, *Int. J. Appl. Radiat. Isotopes*, **21**, 261 (1970).
315. D. L. HANSEN and C. O. CARROLL, *Int. J. Appl. Radiat. Isotopes*, **22**, 677 (1971).
316. D. S. GLASS, Organic Scintillators and Liquid Scintillation Counting. (D. L. Horrocks, C. T. Peng, eds.), p. 803. Academic Press, New York, (1971).

317. J. F. LANG, Organic Scintillators and Liquid Scintillation Counting. (D. L. Horrocks, C. T. Peng, eds.), p. 823. Academic Press, New York, (1971).

318. J. L. SPRATT, The Current Status of Liquid Scintillation Counting. (E. D. Bransome, Jr., ed.), p. 349. Grune & Stratton, New York, (1970).

319. R. L. LITLE, The Current Status of Liquid Scintillation Counting. (E. D. Bransome, Jr., ed.), p. 371. Grune & Stratton, New York, (1970).

320. J. V. JELLEY, Čerenkov Radiation and its Applications. Pergamon Press, London (1958).

321. J. MARSHALL, *Phys. Rev.*, **86**, 685 (1952).

322. E. H. BELCHER, *Proc. Roy. Soc.* (*London*), **A216**, 90 (1953).

323. I. E. TAMM, *Amer. Sci.*, **47**, 169 (1959).

324. L. MALLET, *C. R. Acad. Sci. (Paris)*, **183**, 274 (1926).

325. P. A. CERENKOV, *Dokl. Akad. Nauk, SSSR*, **2**, 451 (1934).

326. P. A. CERENKOV, *Dokl. Akad. Nauk SSSR*, **14**, 101, 105 (1937).

327. I. M. FRANK and Ig. TAMM, *Dokl. Akad. Nauk SSSR*, **14**, 109 (1937).

328. H. H. ROSS, *Anal. Chem.*, **41**, 1260 (1969).

329. H. H. ROSS, The Current Status of Liquid Scintillation Counting. (E. D. Bransome, Jr., ed.), p. 123. Grune & Stratton, New York, (1970).

330. K. HABERER, *Atomwirt.*, **10**, 36 (1965). (An English translation appeared as Packard Technical Bulletin No. 16, January 1966.)

331. R. H. ELRICK and R. P. PARKER, *Int. J. Appl. Radiat. Isotopes*, **19**, 263 (1968).

332. H. BRAUNSBERG and A. GUYVER, *Anal. Biochem.*, **10**, 86 (1965).

333. P. E. BALLANCE and S. JOHNSON, *Health Phys.*, **20**, 323 (1971).

334. A. LÄUCHLI, *Int. J. Appl. Radiat. Isotopes*, **20**, 265 (1969).

335. A. de VOLPE and K. G. A. PORGES, *Int. J. Appl. Radiat. Isotopes*, **16**, 496 (1965).

336. W. HULSEN and U. PRENZEL, *Anal. Biochem.*, **26**, 483 (1968).

337. T. CLAUSEN, *Anal. Biochem.*, **22**, 70 (1968).

338. L. I. WIEBE, A. A. NOUJAIM and C. EDISS, *Int. J. Appl. Radiat. Isotopes*, **22**, 463 (1971).

339. M. K. JOHNSON, *Anal. Biochem.*, **29**, 348 (1969).

340. R. T. HAVILAND and L. I. BIEBER, *Anal. Biochem.*, **33**, 323 (1970).

341. F. L. HOCH, R. A. KURAS and J. D. JONES, *Anal. Biochem.*, **40**, 86 (1971).

342. R. P. PARKER, The Current Status of Liquid Scintillation Counting. (E. D. Bransome, Jr., ed.), p. 110. Grune & Stratton, New York, (1970).

343. H. H. ROSS, Organic Scintillators and Liquid Scintillation Counting. (D L. Horrocks, C. T. Peng, eds.), p. 757. Academic Press, New York, (1971).

344. R. D. STUBBS and A. JACKSON, *Int. J. Appl. Radiat. Isotopes*, **18**, 857 (1967).

345. J. R. BROWNELL and A. LAUCHLI, *Int. J. Appl. Radiat. Isotopes*, **20**, 797 (1969).

346. A. T. B. MOIR, *Int. J. Appl. Radiat. Isotopes*, **22**, 213 (1971).

Chapter 4

The Application of Computers in Radiochemistry

R. K. Webster

Applied Chemistry Division, A.E.R.E., Harwell, Didcot, Oxfordshire

4.1 INTRODUCTION

This chapter provides an introduction to on-line computing, and is intended as a summary of basic concepts to support the rest of the book. The aim is therefore to outline principles, and to describe a few typical examples or applications rather than to provide an extensive review of the literature. Computers now fill a very varied role in most branches of radiochemistry, and applications range from the direct use of computers as instruments (i.e. on-line computing) to complex data manipulation (off-line applications). To provide some coherence in the treatment, the major part of this chapter focusses on gamma-spectrometry and activation methods. An outline is given of the basic principles of multichannel analysers, and a simple account of the operation and programming of a small computer. This is followed by a fairly detailed description of the organisation, operation and use of a small computer as a single gamma-spectrometer, to illustrate techniques, advantages and limitations. Further sections summarise alternative methods of using dedicated computer units, for example for fully automated activation systems, and the use of medium-sized computers for simultaneous operation as several independent gamma-spectrometers. A final section summarises computer applications in some other branches of radiochemistry.

151

4.2 DATA ACQUISITION

Data acquisition in radiochemistry can be divided into two broad classes: multiscalers and multichannel analysers. A multiscaler is used to accumulate spectra which are ordered with respect to time or energy, and incorporates a store containing perhaps 512–4096 locations. Signal pulses are counted, usually for fixed time intervals, into the successive channels of the core store by synchronising channel address advance pulses with some spectral parameter, for example in Mössbauer spectrometry (chapter 11) with the rate of approach of the radiation source and the sample. By synchronising spectral sweeps and adding say n such spectra, the signal-noise ratio may generally be improved by a factor of $\sqrt{n}$.

A multi-channel analyser, however, is designed to store pulses which are generated randomly, rather than digital signals which are ordered in time and energy to suit the multi-scaler mode. The output pulse from a gamma-ray detector is fed through an amplifier to an analogue-digital converter (ADC) which can generate a train of pulses or a 12-bit number which defines the channel (e.g. up to 4096) in which the pulse is to be counted. After accumulating a series of such pulses, the gamma-spectrum can be displayed, photographed, printed, punched on paper tape, or recorded on magnetic tape or a casette. The multi-channel analyser is essentially a small computer in disguise in which the program is controlled by pre-wired knobs or switches. Both an analyser and a computer incorporate a ferrite core store, but in the computer the channel number is usually referred to as the address, and the process for counting a pulse as 'adding one to the contents of the address set by the analogue-digital converter'. On a block diagram, the only apparent difference between gamma-spectrometry systems based on a multi-channel analyser and on a small computer is that the display and ADC are shown as peripheral units in a computer system, whereas they are normally integral components of a multi-channel analyser package.

The main deficiency of a multi-channel analyser is its lack of flexibility in the face of changing requirements, for its program can only be modified with a soldering iron. As the trends in gamma-spectrometry set a demand for larger stores to provide more channels for higher resolution germanium diode systems, faster logic to meet higher resolving speeds, computer-type peripherals to handle the increasing quantity of output data, or real-time computing facilities to reduce the volume of data before output, it is worth considering whether the requirements of a multi-channel analyser could not be met more effectively by a suitably programmed small computer, with a lower risk of obsolescence. This approach is not necessarily ruled out by economics. Over the period 1956–1965, multi-channel analysers were developed to provide increased storage and performance, but at a near-exponential rise in costs, so that low cost computers began to be competitive. In more recent years the costs of high capacity multi-channel analysers and basic computers have both fallen. A generalised cost comparison is difficult as the relative costs for the two approaches will vary with the size and scope of any given installation; it should also be noted that the basic cost of a small computer may appear deceptively low, as provision must also

1	2	3	4	5	6	7	8	9	10
11	12	13	14	15	16	17	18	19	20
21	22	23	24	25	26	27	28	29	30
31	32	33	34	35	36	37	38	39	40
41	42	43	44	45	46	47	48	49	50

Fig. 4.1. Pigeon hole store.

be made for an ADC and other essential peripherals. However, in many cases the system costs can be broadly comparable, so the choice between the two approaches is controlled not so much by economics as by the requirements of any specific laboratory for gamma-spectrometry.

4.3 STRUCTURE OF A SMALL COMPUTER

This section outlines the organisation, operation and programming of a small computer, and introduces some of the terms of computer jargon such as word, bit, core store, accumulator, address, binary, octal, machine code, mnemonic, Fortran, memory, cycle time, program interrupt, cycle stealing, and data break. The account is based on a highly simplified computer structure, and is given to establish some principles for the rest of the chapter. In practice computer manufacturers have produced a wide variety of different designs, often to optimise several factors for particular applications. Some typical design factors are the range of arithmetic operations and electronic logic circuits used, the ease of expanding the size of the computer or of adding peripheral devices, and cost.

The operation sequence for a computer can be illustrated by a very simple analogy: suppose a human operator is standing in front of a set of numbered pigeon holes (Fig. 4.1), and has been 'programmed' to remove a slip of paper from pigeon hole 1, read and act upon the message written on the paper, and on completion move to pigeon hole 2 for the next message. For example, the messages in the first four pigeon holes might be as follows:

1. Look up and remember the number stored in pigeon hole 32
2. Add the number stored in pigeon hole 33
3. Store the result in pigeon hole 34
4. Go (jump) to pigeon hole 16 for the next instruction*

The information stored in pigeon holes 1–4 (and 16) represents instructions, whereas data are stored in pigeon holes 32–34, and the whole sequence is effectively a program for adding two numbers together and storing the result. This sequence will be translated into a digital computer cycle, and illustrated by programs at various programming levels.

*The instruction at 16 might, for example, be the beginning of a program routine to print the result of the addition.

In a computer, information is stored in the form of binary digits ('bits') to suit the many two-state electronic devices, e.g. lights, switches, direction of magnetisation axis. A computer word is made up as a string of bits, e.g. 8, 12, 16, 18 or 24, depending on the type and size of the computer. As a simple example of binary addition, based on 3-bit words:

011	3
010	2
101	5

The computer core store or memory provides a storage function equivalent to the nest of pigeon holes. With the advent of large scale integrated circuits, semiconductor memories are now beginning to be used, but nearly all computers to date have used magnetic core stores, which are made up from a set of ferrite rings (or cores) a few hundreths of an inch in diameter. A ferrite core can be magnetised in either the forward or reverse direction along its axis by passing a voltage pulse along a wire threaded through its centre; one direction is taken to represent '0', and the other to represent '1'. A single plane in the store consists of a set of such cores which are threaded by both 'row' and 'column' wires on a rectangular matrix. If half-size pulses are passed along a selected pair of row and column 'write' wires, the ferrite core at the intersection of these wires can have its magnetisation axis inverted, but no other location is affected. Reading from a given location is achieved in a similar manner. Combining a series of planes provides a three-dimensional store, where two dimensions can correspond to the pigeon hole addresses, and the third to the number of bits in a word.

It is also necessary to provide a series of one-word stores, for example to remember the address of the next pigeon hole to be sampled, to decode instructions, or to store data for the addition operation. These are called registers, and consist of a set of flip-flops, which are two-state devices made up from transistors or integrated circuits. The two states can be sensed from the voltage levels on two output points, and used to represent the digits 0 and 1; there are also two inputs, one of which is used to set the flop-flop to the '1-state' and the other to reset to the '0-state'. A set of these devices is combined to provide a storage register with the required number of bits.

Fig. 4.2 gives a sketch of a simplified computer layout. The block labelled memory represents the ferrite core store used for bulk storage; the arithmetic unit represents a series of electronic circuits, one for each of the functions performed by the computer such as add, store, jump, shift, etc. Of the various registers the program counter stores the address of the instruction to be followed (1–4 in the example above); the memory address stores the address of the location from which the contents are required (1–4 or 32–34 above); the memory buffer is used as a temporary store when reading the contents of any given location in the ferrite core store; the instruction register is used for the storage of instructions to permit interpretation by a decoding matrix, i.e. to switch in the required arithmetic unit for a given program instruction and to set the memory address to the required

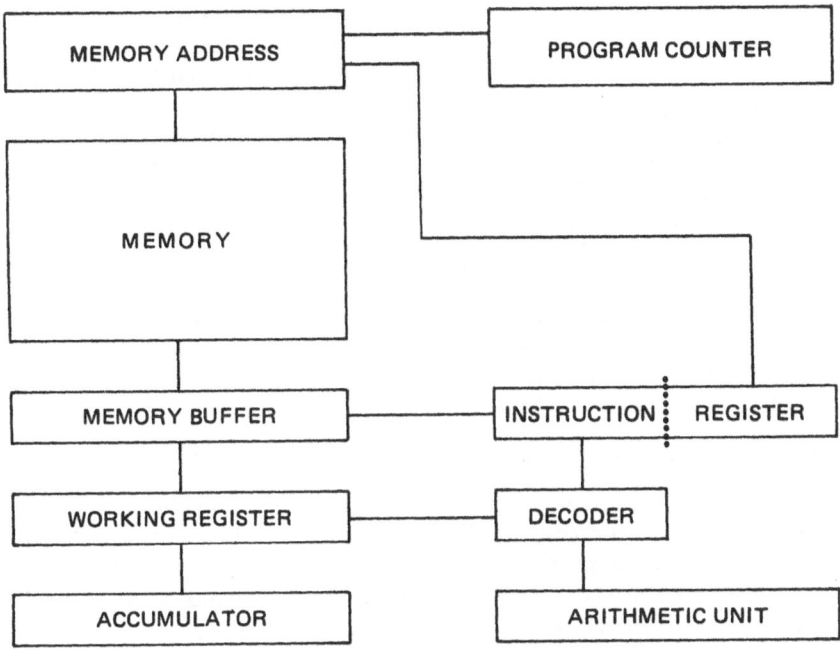

Fig. 4.2. Simplified computer layout.

data location (32–34 above); and the working register and accumulator are used in the calculation and storage of the result from each program step. This is sufficient to illustrate the principles of computer operation, but there will be many differences in detail between real computers; for example there may be more interconnections or more working registers than those shown in Fig. 4.2, or the working register may be associated directly with the arithmetic unit.

Each consecutive operation of the computer is controlled by a binary word stored as program instruction. The conventions vary, depending upon the specific computer, but suppose that a 12-bit computer is wired so that the first three bits of any program instruction are used to define the arithmetic operation, for example suppose that the binary equivalents of 1, 3 and 5 represent the 'add', 'store', and 'jump' instructions; the other 9 bits of the word can then represent the address of the location storing the data word which is to be manipulated. Two typical formats might then be as follows:

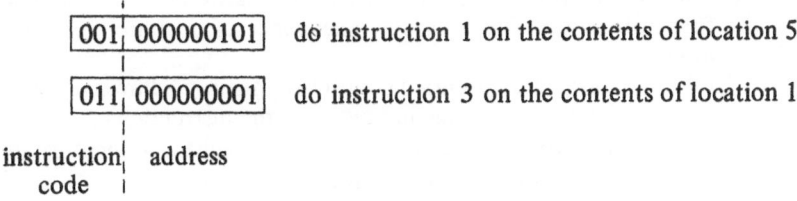

    001 000000101   do instruction 1 on the contents of location 5

    011 000000001   do instruction 3 on the contents of location 1

    instruction address
      code

In each case the status of the first three flip-flops would be sensed by the decoding matrix in order to select the corresponding arithmetic circuit; the last nine would be sensed, and the equivalent pattern transferred to the memory address register.

Referring back to the sequence, based on Fig. 4.1, which involved the storage of a program loop in locations 1–4 to add the numbers stored in locations 32 and 33 and to store the result in location 34, the corresponding binary instructions, using the above conventions, are as follows:

pigeon hole	binary bit pattern	meaning
1	001 000 100 000	instruction 1 on location 32
2	001 000 100 001	instruction 1 on location 33
3	011 000 100 010	instruction 3 on location 34
4	101 000 010 000	instruction 5 on location 16

(note: instruction 1 means ADD, 3 means STORE, 5 means JUMP).

The binary bit patterns above are those which will be stored in the first four locations of the ferrite core store. Programs can in fact be written in binary code, the most basic machine code, but in practice it is simpler to truncate them into octal numbers, or more usually to use mnemonics as part of a symbolic assembly code. These can be illustrated as follows:

pigeon hole	binary	octal	mnemonic
1	001 000 100 000	1040	TAD 40
2	001 000 100 001	1041	TAD 41
3	011 000 100 010	3042	DCA 42
4	101 000 010 000	5020	JMP 20

(mnemonics: TAD, addition; DCA, deposit and clear accumulator; JMP, jump.)

The last two columns show the octal and assembly code equivalents of the binary instructions, and to use these the computer must contain an assembly program which can translate each line of code into its binary equivalent. In higher level languages, such as Fortran, the complete sequence might be represented by a single instruction (e.g. A = B + C), and there is then no simple relation between the program instruction and the corresponding binary entry in the computer.

In the operation of a computer, an instruction is broken down into a set of sequential operations each of which is initiated by a clock-pulse. For example, referring to Fig. 4.2, at the start of the first instruction of the above sequence the program counter would be registering 'one', the address of the first instruction. The subsequent sequence of operations would then be as follows:

    (i) Copy contents of program counter to memory address register (i.e. prepare to look into pigeon hole 1 for the instruction)

    (ii) Transfer contents of address set by M.A. register to memory buffer (i.e. read the contents of pigeon hole 1)

(iii) Add '1' to the number stored in program counter (i.e. set P.C. register to be ready to point to pigeon hole 2 for the next loop)

(iv) Transfer contents of memory buffer to instruction register (to decode instruction and data address)

(v) Set memory address register for data [i.e. in above example, prepare to look into pigeon hole 32 (decimal), 40 (octal), 100,000 (binary)]

(vi) Transfer data to memory buffer

(vii) Transfer data to working register

(viii) Set circuit to bring in appropriate arithmetic unit (in this case an adder)

(ix) Add contents of working register to accumulator register

This completes one instruction cycle, and the computer now continues by transferring the contents of the program counter, incremented to 2 in the above sequence, to the memory address register to start operations for the next instruction. The time required for a complete sequence may be referred to as the instruction time, or the memory cycle time; the value varies with the instruction and with the computer but typically falls in the range 0.5–5 microseconds.

While a simple program loop could provide for regular sampling at fixed time intervals, e.g. for multiscaling, it is usually very wasteful in computer time, and other methods must be used for gamma-spectrometers where the computer has to respond quickly to pulses which occur *randomly* in time. The simplest method is to follow a Program Interrupt sequence, where the input from the gamma-spectrometer, or from any other peripheral demanding attention such as a teletype, paper tape reader, etc., 'sets a flag'; this activates electronic circuits which cause the computer (i) to stop its current program at the end of the current instruction, (ii) to jump to an interrupt servicing program which first identifies the interrupting device and then jumps to the appropriate device servicing program, and on completion (iii) to return to the next instruction of the program which was in operation when the interrupt occurred. There are many types of interrupt handling facility, but all involve connections between the peripheral devices and typically two particular locations in core memory: the first is used to store the current contents of the program counter so that the interrupted program can be reinstated later, and the second to store the starting address of the interrupt servicing program. The first action of the servicing program is to store the contents of all the processor registers in predefined memory locations so that the status of the computer is preserved; the required interrupt service routine is then carried out, the preserved contents restored to the working registers, and execution resumed at the memory location at which the interrupt occurred. If more than one external device is connected to the interrupt lines in a simple system the computer must examine sequentially a series of external signal lines, or flags, until the interrupting device is identified. An improved response time can be achieved by connecting each interrupting device to a different core location and so eliminating the identification stage. In priority interrupt schemes, if a high priority interrupt is flagged while a lower priority is being serviced, the latter is suspended until the computer has serviced the higher priority device. For high speed data transfers, special electronic hardware enables the contents of a given core store location to be modified without changing the sequence of program instructions being followed by the computer, and is referred to as a data break option, direct memory access (DMA) or autonomous data transfer. In operation, when a data transfer is to be initiated, the computer halts at the end of its current instruction, and data is transferred directly to the core store without altering the contents of the various computer registers; the transfer takes 1–4 cycles, depending on the computer, after which the computer continues with the next instruction of the program previously in operation. This does not change the sequence of instructions followed in a background program, but merely slows down its speed of execution; this method of data entry is therefore sometimes referred to as cycle-stealing or the hesitate mode. For multi-channel analysers, the external electronic hardware must specify the core store address (channel number) corresponding to the output from the ADC, and arrange for the contents to be incremented by one. For the transfer of blocks of information, for example from some external storage device, it is necessary to specify both the first location in the computer core store to which data is to be transferred ('current address') and the number of data words to be transferred ('word count'). Two registers are then set, one to the current address, and the other to minus the word count. Both registers are incremented by one each time a word is transferred, and the transfer operation is terminated when the word count register equals zero. If these registers are provided in the external device, a data word can be entered into the computer memory on each computer cycle (single cycle DMA). Alternatively, the current address and word count can be stored within the computer memory, and incremented by the computer after each data transfer. This is slower, as it requires three computer cycles per data transfer (three-cycle DMA), but involves lower interface hardware costs.

A complete computer system can range from a small unit containing a 4096 (4k) word store and equipped with only a teletype printer incorporating a slow paper tape reader and punch, to a larger system equipped with powerful peripherals such as a high speed line printer, plotting and display units, and magnetic tapes, drums or discs as backing stores for bulk storage. A disc or drum provides the fastest response time. Each word is stored at a specific point in a layer of magnetic material deposited either on a flat spinning disc or on a cylindrical rotating drum, and generally the stored information is written in tracks. Access to the starting point of a particular block of information is related to the rotation frequency, and is typically 10–40 milliseconds; given the starting point, data transfer to the computer core (by DMA) can take place at rates of up to 250,000 words per second. For the fastest response a read and write head is installed over each track so that a single rotation of the disc or drum provides access to all the stored information (fixed head system). Larger and cheaper storage can be achieved by using a single head for each surface, and scanning the head across the disc to find the particular track on which the data are stored (moving head system); however, the access time is then increased to perhaps several hundred milliseconds.

4.4   A DEDICATED COMPUTER FOR GAMMA-SPECTROMETRY

Figs. 4.3 and 4.4 illustrate the layout of a small computer organised for use as a single gamma-spectrometer. The application of two such systems to activation analysis is described by Pierce *et al.* [1], and the interfacing and basic operating system by Lewis [2]. The computer is a D.E.C. P D P-8 fitted with a 4096 (4k) 12-bit word core store, and 2 magnetic tape drives as backing store; the computer cycle time is 1.5 microseconds. The core store is organised to allocate 3k words for data storage during the accumulation of gamma-spectra leaving 1k words for the control program. The 3k words are organised by the program into 2k locations, each made up from 1½ words, to provide the equivalent of 2k channel storage and a count of up to 2^{18} in each channel. This data store can be divided, by program selection, into 1×2048, 2×1024, 4×512, 8×256 or 16×128 channel sectors as required. The program contained in the core during spectral accumulation is restricted to the routines required to handle data entry from the gamma-detector, timing operations, etc. by program interrupt, together with a background display program. After recording a gamma-spectrum, output routines to operate the Teletype, etc., can be written into the 1k program segment from one of the magnetic tapes; on completion of output the display and data acquisition program are automatically returned to the core store. Alternatively, the complete spectrum can be stored on magnetic tape in order to make the whole of the core store available to hold larger programs for data manipulation. Each magnetic tape can store a maximum of 1474 blocks of data, each block containing 128 words at fixed positions on the tape: one tape drive provides storage and an input/output system for

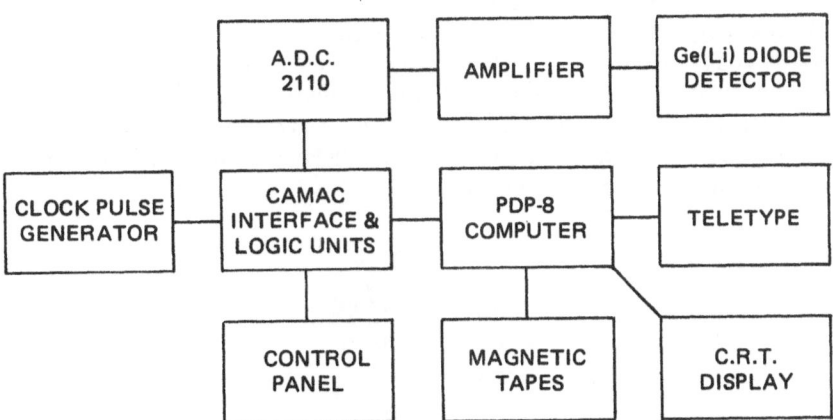

Fig. 4.3. Block diagram of PDP-8 computer system for gamma-spectrometry.

Fig. 4.4. PDP-8 computer system for gamma-spectrometry.

gamma-spectra (~90 2k channel spectra per tape); program library tapes are mounted on the second drive.

Fig. 4.5 shows the teletype record for a typical initialisation routine, and includes both output, under computer control, to prompt the operator, and the operator's responses (the latter are underlined). The first line sets a channel count limit of 2^{18} (previous setting was 2^{16}), and the next two lines set upper and lower bounds to the channels used for recording spectra; the last three entries are used to control the length of the counting period by setting either (i) a live-time limit, as indicated at 5 seconds (500 units of 10 milliseconds), or (ii) a real-time limit or (iii) a total count limit. Fig. 4.6 shows a typical output for a single peak; the area for printing is selected by moving two vertical markers on the oscilloscope by operating toggle switches on the computer console; the output gives first the range covered by the markers, then the contents of the individual channels, and finally the total count over the region selected.

$$
\begin{array}{ll}
\text{C} = 2\uparrow & 16\ \underline{18} \\
\text{ML} = & \emptyset\ \underline{5} \\
\text{MU} = & \emptyset\ \underline{2\emptyset\emptyset\emptyset} \\
\text{LT} = & \emptyset\ \underline{5\emptyset\emptyset} \\
\text{RT} = & \emptyset \\
\text{AI} = & \emptyset
\end{array}
$$

Fig. 4.5. Teletype instructions for control of gamma-spectrometer.

660–692

660	154	154	136	153
664	164	124	168	149
668	153	146	160	204
672	315	511	676	1133
676	1851	2938	4591	6406
680	7488	6725	4523	2198
684	886	226	72	15
688	8	2	5	5
692	2			

TOT: 42441

Fig. 4.6. Teletype output for part of gamma-spectrum.

Fig. 4.7 indicates the remaining parts of the total system – a control panel and modular interfacing units. The control panel provides push-button facilities (i) for starting and stopping counts, (ii) for selecting display parameters – the options include variation of the vertical gain to suit counts in the range $2^8 - 2^{18}$, choice of linear, logarithmic, or square root display modes, and facilities for displaying any part of the store, comparing any two sections, etc. The settings of the panel switches are sensed by a program loop in the computer, and action taken when any setting is changed. The control panel was included to simplify operation of the system to suit a wide range of users, although experience now suggests that control via the teletype would prove equally acceptable. The lower part of Fig. 4.7 shows interfacing units in the Harwell 7000 series, an implementation of the international ESONE standard, CAMAC [3]. It consists of a crate containing a standard data highway, and a series of modular plug-in units; the unit at the extreme right is a controller designed specifically to match the PDP-8 computer – all other modules are general purpose units chosen to suit the experiment and are independent of the computer used. Pulses from the lithium drift germanium diode detector are processed by a Harwell series type 2110 ADC; this feeds into a 2110 adapter module (Fig. 4.8) to generate program interrupt requests for accumulating the gamma-spectrum in the computer store. 100 ms, 10 ms, 1 ms, or 100 μs timing pulses, from a clock-pulse generator, are passed through the choice unit to the flag unit to generate interrupt requests; these are counted in the computer to measure or control the *total* counting time. By passing the same clock-pulses to the flag unit through an and gate module, together with inhibit pulses from the 2110 ADC system, interrupt requests corresponding to *live*-time can also be counted into the computer. A further input of 1 second pulses via the flag unit can provide a continuous record of 'universal-time' for use in the calculation of decay corrections, etc. The sense unit and drive unit modules are used to service the various switches and lamps on the control panel.

Fig. 4.9 is an expanded C.R.T. display of part of a 2k channel spectrum recorded for neutron-irradiated granite showing, for example, a central doublet corresponding to the 94.7 and 98.4 keV peaks of [233] Pa produced

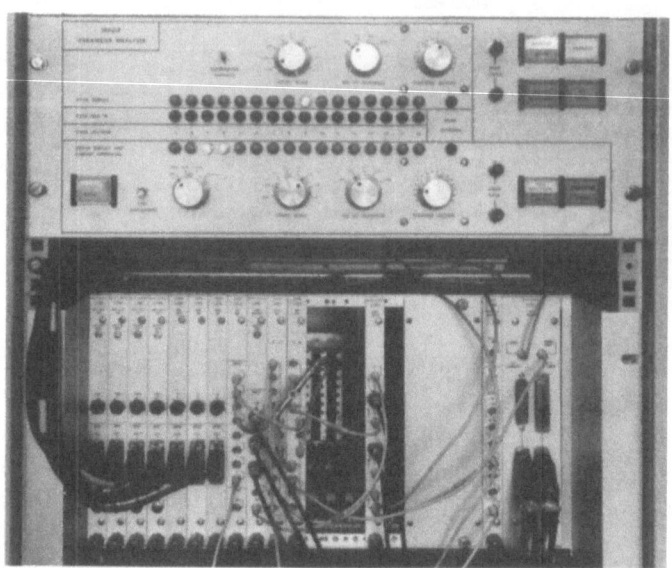

Fig. 4.7. Control panel and CAMAC modules.

from thorium (A. A. Smales and D. Mapper, unpublished data). This illustrates usage as a standard gamma-spectrometer, but as a simple example of the flexibility of a computer-based system the display can easily be converted to show three-dimensional information by introducing alternative display programs from the library tape. For example, in charged particle activation analysis, the use of collimated ion beams and a scanning system provides a two-dimensional array of data at the sample surface, and the display must cater for both the position and the count rate measured at each irradiation point. Fig. 4.10 shows an isometric display of data obtained by elastic scattering of α-particles from a transistor — the high count rates correspond to regions of high gold content. Alternatively, surface distributions can be displayed as a two-dimensional plot; thus a coloured contour map can be generated by arranging the program to display, in turn, a two-

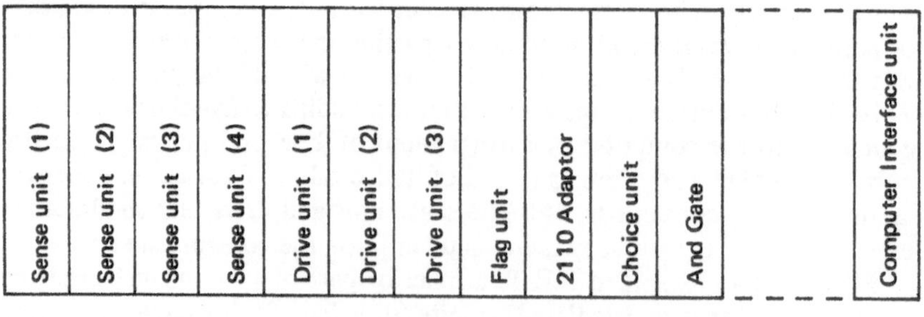

Sense unit (1)	Sense unit (2)	Sense unit (3)	Sense unit (4)	Drive unit (1)	Drive unit (2)	Drive unit (3)	Flag unit	2110 Adaptor	Choice unit	And Gate		Computer Interface unit

Fig. 4.8. Block diagram of CAMAC units.

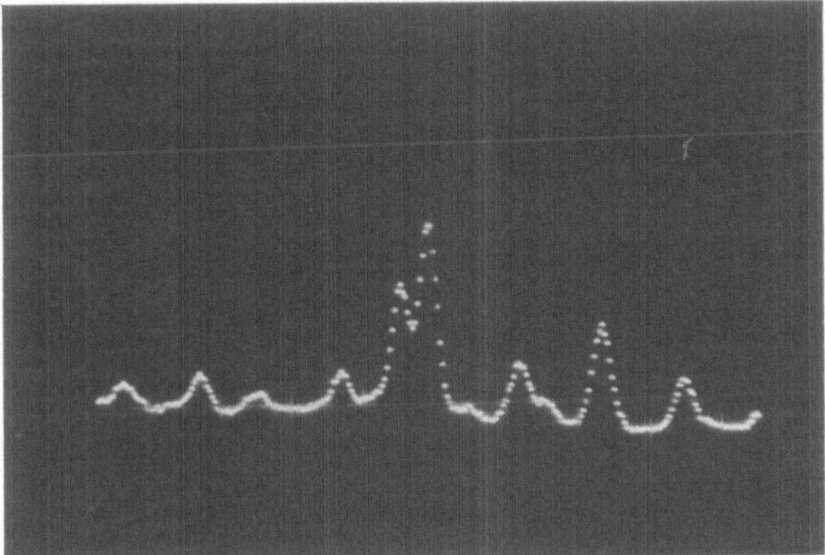

Fig. 4.9. Gamma-spectrum for neutron irradiated granite (50–160 keV).

dimensional array of results falling between various discrimination levels, and photographing each of these through different colour filters [4].

Arrangements for data processing are equally flexible, as the basic system can be modified by any individual user to suit his own requirements. For example, the counter background can be stored on the data tape and subtracted, as required, from recorded spectra; stripping routines can be used for scaled subtraction of standard spectra. These are available as options with a number of multi-channel analysers, but a computer system also permits the use of more objective calculation techniques, selected to suit the problem. For example, a short program will calculate the area under a given peak for both sample and standard, and hence their ratio, to provide almost instant analysis in simple cases; for example for the determination of manganese in aluminium alloys by neutron activation. For well-resolved spectra containing peaks for several elements, simultaneous equations can be set up and solved; for example in Table 4.1 the results calculated in this way for fast neutron activation measurements of Si and Cu in aluminium alloys show very close agreement with those obtained using more sophisticated least squares fitting procedures and an IBM 360/75 system [5]. In principle, least squares fitting procedures can also be incorporated into a small computer system, provided a disc is available as a rapid access backing store, but this type of calculation is usually handled more efficiently off-line using a larger batch processing computer. The use of a small on-line computer is better concentrated on those areas which optimise selection and recording of the gamma-spectra. For example, peaks can readily be identified by adjusting a bright spot in the display of a gamma-spectrum until it is centred

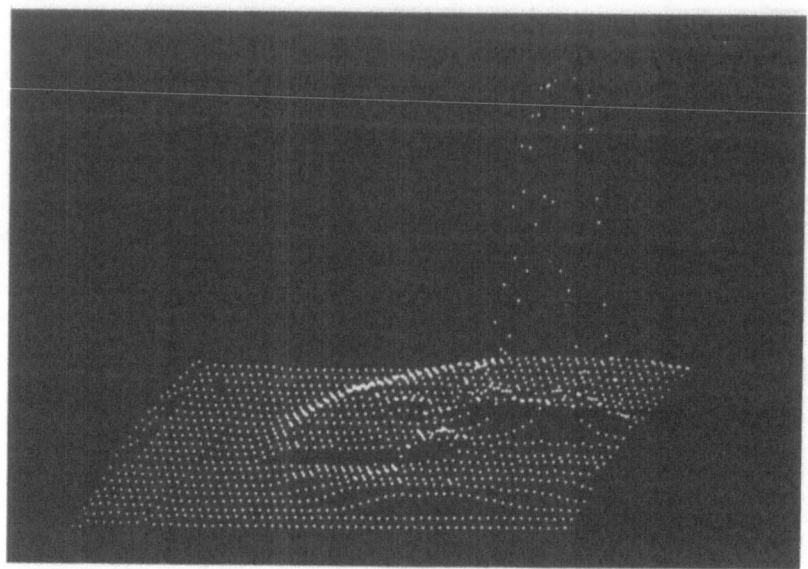

Fig. 4.10. Isometric display of results obtained by elastic scattering.

TABLE 4.1 *Comparison of Results for Si and Cu in Aluminium Alloys Determined by Fast Neutron Activation and Calculated by Simultaneous Equations (PDP-8) and Least-Squares Fitting (IBM 360/75) Methods*

Samples	Si (%)		Cu (%)	
	PDP-8	IBM	PDP-8	IBM
1	0.80	0.80	4.5	4.5
2	25.4	25.2	0.2	0.2
3	10.3	10.2	1.29	1.30
4	0.51	0.52	0.2	0.2
5	0.43	0.42	1.70	1.72

on the required peak; a program can then be called down from the library tape to provide a least squares energy calibration against spectra previously recorded for three or more energy standards.

4.5 DEDICATED COMPUTER APPLICATIONS IN NUCLEAR SPECTROSCOPY AND ACTIVATION ANALYSIS

The following paragraphs outline the use of small computers for fully-automated neutron activation systems, multi-channel analyser/computer combinations where the computer plays a more limited data processing role than above, and some alternative modes of computer operation, particularly to illustrate flexibility and changes which can be made to avoid obsolescence.

4.5.1   Automatic Neutron Activation Analysis

Section 4.4 outlines the use of a small computer for data acquisition and interpretation for gamma-spectrometry. By introducing control functions, a computer can form the centre of a fully automatic neutron activation system. Pierce *et al.* [6] have designed an activation system based on a Californium-252 neutron source and a Digico Micro-16V computer equipped with an 8k store, a 2.9 million word disc and display. Samples from a defined set of types are weighed into polythene tubes which are then introduced into a sample magazine. The operator uses a teletype to enter an identifier, the weight of material and a simple code to define the particular type of sample or elemental standard as appropriate. The computer then selects the irradiation, delay and counting times, and controls sample transfer, using either a pneumatic or a teleflex system, between the magazine, the irradiation position, the counting station and the final store for the irradiated materials. Results are then calculated using either multi-regression analysis, or simpler procedures, as appropriate. The system can be applied in either laboratory or process control applications: in the latter case the system either prints the required sample composition together with statistical data, or signals a warning and indicates how the material is out of specification. Other examples include the use of a DEC PDP-9 computer equipped with magnetic tape storage to provide automatic neutron activation systems for the analysis of steels, rocks, soils and pre-historic flints [7, 8, 9].

4.5.2   Multi-channel Analyser/Computer Combinations

As an alternative to the use of a computer as a gamma-spectrometer, a standard multi-channel analyser can be interfaced directly to the computer. This provides the data analysis capacity for energy calibration, efficiency calibration for various counting geometries, smoothing, peak location, integration, etc. without the delays of off-line computing, but avoids the real-time load placed on the computer by the direct acquisition of data [10]. This approach can simplify the programming task, and may be particularly suitable for systems where the computer is connected both to a multi-channel analyser and to other analytical equipment. Wispelaere *et al.* have described a multi-channel analyser/computer combination specifically designed for neutron activation analysis based on short-lived isotopes [11]; the technique incorporates Schonfeld's method [12] of allowing for dead-time losses in calculating the true counting rate of an isotope at the beginning of a measurement, and results are given for various impurities in iron and iron ores.

The off-line analysis of gamma-spectra is normally undertaken on large central computer systems, but a number of procedures have been described whereby complex data analysis can be made using a small laboratory computer associated with the spectrometer [13, 14, 15, 16]. Thus Op de Beeck [13] describes the use of a DEC PDP-9 equipped with magnetic tape for the analysis of data from 8 multi-channel analysers of various types and sizes. Data input is via paper tape, magnetic tape or a direct link between one of the multi-channel analysers and the computer. Various program modules smooth the data, locate peaks, calculate peak centroids, energies and areas, identify with isotopes from a library, normalise for counting, decay and irradiation times, and finally use the sample identification number and sample weight to print out concentrations in parts per million for each identified peak. DeRegge also describes the analysis of alpha-ray spectra [16].

4.5.3 Alternative Modes of On-Line Computer Operation

In addition to flexibility, a computer provides some protection against obsolescence of equipment. As detectors develop and require more channels for efficient use, the computer can be expanded by adding a further core store module. If requirements change to demand multi-parameter measurements, the mode of computer operation can be altered to match. For example, coincidence measurements between two gamma-detectors, each with 1k channel resolution, would require a million channels to record all events. One approach to the problem is to install an alternative program using associative storage, in which counting locations in the core store are assigned to those parts of the coincidence spectrum with the highest count rates. This takes advantage of the fact that much of a two-dimensional spectrum is usually empty or of little interest, and space for channels is allocated only in areas where there is significant data. As each successive pulse is detected, two locations are allocated, one to record the channel number (up to 10^6) and the other the corresponding count. As further pulses are detected, their channel numbers are compared with those already in store; if recognised, the appropriate count is increased by one, and if not, a new storage location is opened. Efficient use of storage space and computer time can be achieved by using zones, rather than individual channels. When the store has been filled, the recorded counts are examined, channel numbers with low or single counts are deleted and the process repeated. After 3–4 such interations, the computer will be storing the most significant 2k (128 zones each of 16 channels) of the million possible channels. If required, a plan or isometric display can then be used to define parts of the spectrum for further examination. This type of program can be entered into the 4k computer described in Section 4.4, with corresponding changes to the interface and control units [17].

In certain specialised cases the need for high volume storage can be avoided by undertaking event-by-event processing. For example the investigation of a reactor neutron spectrum requires a neutron spectrometer to cover the range from a few keV up to 10 MeV. An approach proposed by Wright and Silk [18] obtains the neutron energy from the angle between the triton and the alpha particle produced by the ^{6}Li(n, α)^{3}H reaction; angles from 180° to 150° correspond to incident neutron energies from about 10 keV to 10 MeV. The detection unit consists of three position-sensitive detectors: the central one is coated with lithium-6 carbonate and defines the point of interaction for an incoming neutron; the two outer detectors define the X and Y coordinates for the point of incidence of the two reaction products (α and ^{3}H). As an alternative to storing all data for all events which would require a storage capacity of over 100 million words, an on-line computer can be used to reject irrelevant events, to calculate a function of the angle between the tracks from the coordinate data for each event, to calculate the required neutron energy, and to add the count to one of 64 pre-allocated channels which cover the required range in equal energy groups on a log scale. By completing the calculation within a millisecond a very high degree of data compression can be completed in phase with the incident neutron rate.

A computer/spectrometer system can be readily modified to operate as a multi-scaler. Thus rapidly decaying spectra containing nuclides of different half-lives can be recorded according to a pre-set program of short count times, separated by wait times if required. For very short counting periods successive 128, 256 or 512 channel spectra would be recorded across the core store; for longer periods (~1 second) alternative halves of the store would be used, counting in one half while transferring the other half to a backing store.

Another trend in gamma-spectrometry is towards higher data rates. The maximum count rate is set by the response time of the complete system which depends both on the resolving time of the ADC and the detector, and on the storage time of the computer. For the system detailed in Section 4.4, the dead time is set by the ADC, and falls in the range 70–300 microseconds, depending on the channel number. With a faster ADC, the program interrupt and storage routines of the computer program would determine the maximum permissible count rate. However, the ultimate limit set by the computer is simply its access time, and by modifying the system and program to operate through the data-break channel, this time can be reduced towards 3–6 microseconds to match faster ADC's.

4.6 MULTI-ACCESS COMPUTER SYSTEMS

The discussion so far has concentrated on dedicated arrangements in which one computer takes the place of one multi-channel analyser. Larger systems can take this concept a stage further by providing for data input from several gamma-ray detectors, and for use by several independent operators. This approach is a natural extension of the dedicated computer concept, and is economically attractive when the cost of a single computer complex can be balanced against the alternative cost for up to perhaps 8 conventional multi-channel analysers. It can also provide far more powerful data storage and computational facilities than is possible with a smaller dedicated unit, particularly as more expensive peripherals can be shared between the various data acquisition stations. However, the programming and operational requirements are inevitably more complex, and the cost of software development is often overlooked when making system price comparisons. In planning the size and workload for such systems, some allowance should be made for the development and testing of software for additional applications; this activity will normally continue long after the system has been commissioned, and planning is needed to avoid unacceptable interruptions with normal operations. An analysis should also be made to ensure that there will be no interference between users when two or more have detectors running close to permitted count rate limits.

The following provides a fairly detailed account of one system to illustrate the principles involved, and then short notes on a few other systems.

Fig. 4.11 gives a block diagram for a general analytical system. It has been designed to provide for both activation analysis applications, and ultimately other analytical techniques including fluorescent X-ray spectrometry and the examination of photographic plates for optical emission spectrography [19]; this account considers only the activation applications. The system is based on a Honeywell DDP-516 computer, with a 16k word core store, 16-bit word length, and 0.96 microsecond cycle time. Peripherals include a teletype, C.R.T. display, plotter, two IBM compatible magnetic tape drives, and a fast data link channel to an IBM 370/165 computer [20] for rapid data reduction. The system operates as two independent gamma-spectrometers, based on high resolution Ge(Li) detectors, and provides six independent inputs for multiscaling from gamma-counters. These are required for gamma-activation analysis applications, where it is essential to make a series of sequential gamma-counts, from a given sample, in order to determine carbon, nitrogen and oxygen. Gamma-irradiation of all three elements leads to the formation of positron emitters (^{11}C, ^{13}N, ^{15}O); these three nuclides are therefore all detected by the 0.51 MeV gamma-ray, but they can be distinguished by resolution of the decay curve. A typical gamma-activation experiment involves the irradiation of two standards and four samples, and in view of the short half-lives involved the six irradiated specimens must be examined simultaneously using a multi-input system. A CAMAC data highway provides for interfacing all experimental stations, and also services the CRT display and the plotter.

An essential feature of a multi-user system is a supervisory program which controls the allocation of core store and peripherals between the various users, and coordinates the running of the various programs for data acquisition and processing. The system used is known as MUSTARD (Multi-User System To Acquire Real-Time Data) developed by the Harwell Electronics and Applied Physics Division [21], and allows up to eight users to share the computer, subject to core limitations. At the start of a given series of experiments, any given user must specify his requirements at the central teletype using a dialogue technique controlled by the computer program. The user types in the physical position of his units in the CAMAC crate to define a 'set'; he then enters his requirement for core store, and if sufficient is available he is allocated an 'array' and asked for a title for the experiment, and a user code to identify data subsequently collected via his set of CAMAC modules. Selection from a range of commands then allows the user to define the parameters for his particular experiment. Once this experiment definition has been entered, the operating program can be controlled either from the computer, or at the appropriate instrument station by using a small control module: this incorporates an interrupt request switch, and a

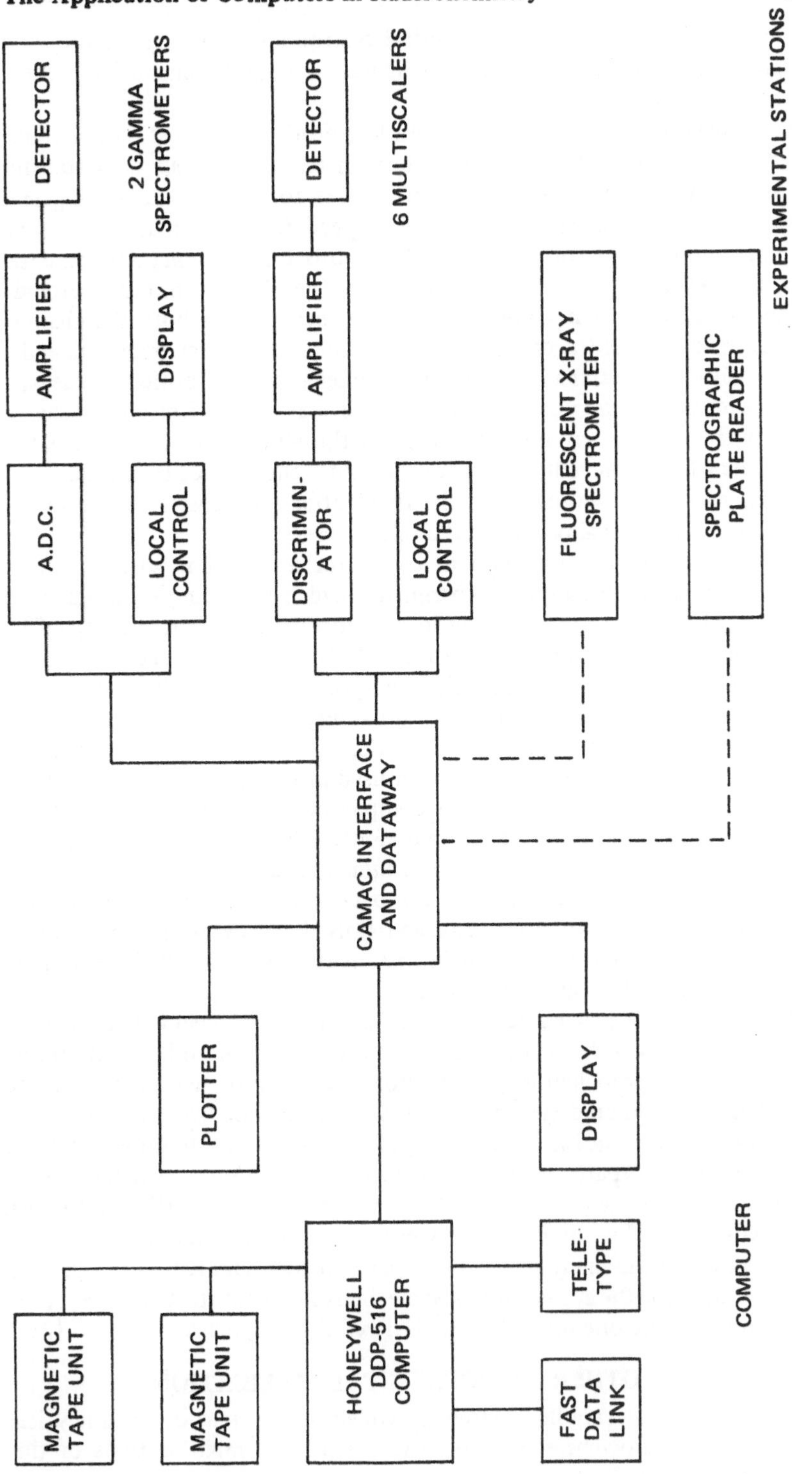

Fig. 4.11. Computer layout for general analytical system.

manual switch register to code control instructions, e.g. for starting and stopping data acquisition, or for changing count limits and other experimental parameters.

To achieve maximum efficiency and low dead times, most data is transferred to and from the core store through an autonomous data transfer channel, leaving the computer relatively free to execute monitoring and background programs, and to service slow speed devices such as the teletype by program-interrupt routines. To achieve this for incoming experimental data, for example for gamma-spectrometry, the CAMAC modules provide for the generation of a 16-bit word which incorporates both the channel address, representing the energy of the particular gamma-ray detected, and a tag to identify the particular detector. Channel addresses formed in this way are transferred autonomously into one of several data buffer areas. When a given buffer is full, an interrupt triggers the allocation of an alternative buffer area to enable data-taking to continue, and also calls in a servicing program to sort the contents of the first buffer for transfer to the appropriate store segments allocated to the various users.

The above experimental system caters for a situation where a single computer is used to support experiments which are being undertaken in different laboratories so that small local control modules are essential. The system described by Salmon and Creevy [22], for instrumental activation analysis of environmental materials, incorporates a one million word disc and is designed to accept data from up to eight gamma detectors; in this case all experimental stations are sufficiently close to permit control through a single teletype. The supervisor used provides dynamic store allocation, so that as a user logs out on completion of a run, the data area is compressed to eliminate the unused zone. This ensures that only the minimum space is committed to data acquisition at any given time, and so optimises the core store available for background processing to provide maximum flexibility for users. Sax and Daly [23] describe the use of an Interdata 3 computer to provide independent operation with four gamma detectors for radiological measurements.

This account has been restricted to providing sufficient discussion of computers to establish an appreciation of their value for on-line and off-line applications in radiochemistry. Therefore no attempt has been made to cover larger time-shared systems for nuclear physics applications, or the use of on-line data acquisition or computer control for hydrogen bubble chambers, sonic spark chambers, neutron diffraction spectrometers or accelerators. However, the use of computers at CERN (European Organisation for Nuclear Research, Geneva) has been reviewed by Bell and Øverås [24]; of the many papers on data acquisition and reduction systems, Brun et al. describe the system at the Nuclear Physics Institute, Orsay [35], and Langsford et al. the one used with the Harwell 2.8 m synchrocyclotron [26].

4.7   OTHER RADIOCHEMICAL APPLICATIONS

Computers are now applied extremely widely in all areas of the analytical sciences, and radiochemistry is no exception. The main sections of this chapter have focussed on the application of computers to the acquisition of gamma-spectra for activation analysis, but other uses noted include energy analysis for various charged particles (alpha particles, protons, etc.), and multi-scaling applications such as Mössbauer spectrometry and decay curve analyses, e.g. for mixtures of positron emitters. In general the use of computer techniques is worth considering in any area involving a significant level of data processing.

Liquid scintillation counting is a case where the calculations involved are not very complicated but can be rather laborious, particularly when many samples must be processed. Thus, the use of automatic sample changers leads to a high sample throughput, and as the activity levels and beta energies are low, it may be advisable to recycle samples in order to generate the required degree of statistical confidence in results. The calculations may involve corrections for background counts and the counter efficiency for the various radioisotopes measured; quenching corrections may be made using either internal or external standardisation, and may require the use of least squares methods for linear quench corrections. The number of calculations is greatly increased when dual labelled samples are used. The overall area is reviewed in chapter 3.

Various automatic data processing methods have been adopted. A fairly simple approach is to generate paper tape or punched card records from the counting equipment, and to submit them to a computer centre for off-line computing [27, 28, 30]. A more convenient route may be the use of a teletype in the counting laboratory, connected to a time-sharing computer service through a telephone link [31]. Alternatively off-line data reduction can be made in the laboratory using 'programmable calculators', and three different systems are reviewed by Grower and Bransome [28]. On-line systems involve the direct coupling of a small computer with the liquid scintillation counter, either to accept data as normally transmitted from the counting system to a teletype [29], or to count pulses from each detector, as they occur, directly into the computer store [32]. In both cases, four counting systems were connected to the computer; the addition of a computer is considered more economic than the use of manual systems when two or more counting units are required [32].

The format of the results given in the computer output is easily designed to suit laboratory files, and so to reduce the associated 'clerical' work. However, the output can also be selected to show results in the most advantageous way. For example a plot mode can be used to produce a bar graph for rapid scanning of samples for monitoring chromatographic column effluents [29], or to provide a histogram, for example of the radioactivity of proteins separated by gel electrophoresis [28].

Computer methods can also be applied to the analysis of data from bidimensional radiochromatograms. The chromatogram is scanned using gas flow counters [33] or a channel electron multiplier for weak beta-emitters [34] and the positional and counting data stored on paper tape; up to ten scanners can be arranged to operate with one tape punch [33]. Computer processing can then generate a number map representation of the radioactive spots on the chromatogram, together with separately plotted spot totals of absolute radioactivity.

4.8 CONCLUSIONS

The choice between a special purpose multi-channel analyser and a computer for gamma-spectrometry is not normally set by economics: costs can be broadly comparable, both for single dedicated spectrometers and for larger multi-user systems. The decision is therefore a technical one, and is set either as the optimum way of matching existing experimental requirements, or by any need to plan for future developments. A multi-channel analyser is the probable choice when the number of channels required is relatively small, or where the analysis system must be *portable*, for example for part-time use with a wide range of irradiation facilities sited in different buildings. In many other cases a computer can offer a technically attractive alternative.

In choosing a computer it is essential to ensure that (i) the size of core store is adequate to accommodate both operating programs and data, (ii) the word length is sufficient to provide the required accuracy in data storage, preferably without using multi-word working, (iii) the instruction and cycle times are short enough to meet foreseeable rates of data collection, (iv) a backing store (e.g. magnetic tape or disc) can be added to provide for bulk storage of data and program routines. It is also desirable to plan the system around a standard modular interfacing system for flexibility in setting up different experimental arrays. Given these, a computer can provide a powerful and flexible system which combines all the features of a standard multi-channel analyser with the advantages of computing facilities and peripherals. Systems can be modified quite readily to meet changes either in the technology of gamma-spectrometry, or in the objectives of a given laboratory, so the risk of obsolescence is reduced. The main penalty is the need to design and commission a suitable software system, but in practice this has not proved to be a serious limitation. Experience has shown that once set up, relatively 'unskilled' personnel quickly learn to use a computer system and to write programs for new types of data manipulation or modes of operation.

REFERENCES

1. T. B. PIERCE, R. K. WEBSTER, R. HALLETT and D. MAPPER, Proceedings of the International Conference on Modern Trends in Activation Analysis, NBS Special Publication 312 Vol. II, p. 1116, (1969).
2. A. LEWIS, U.K. At. Energy Authority Report R-5844, Harwell (1968).
3. ESONE Standard 1968: CAMAC, A Modular Instrumentation System for Data Handling. Euratom Report EUR 1831e.
4. T. B. PIERCE, P. F. PECK and K. HAINES, *J. Radioan. Chem.*, **14**, 229 (1973).

5. T. B. PIERCE, Institution of Electronic and Radio Engineers Conference on Laboratory Automation, Conference Proceedings No. 20 (1970).

6. T. B. PIERCE and D. A. NEWTON, to be published.

7. C. J. THOMPSON, *Nuclear Applications,* 6, 559 (1969).

8. C. J. THOMPSON, AECL Report CPSR-235 (1969).

9. M. DeBRUIN and P. M. J. KORTHOVEN, *Anal. Chem.,* 44, 2382 (1972).

10. J. A. KEENAN and G. B. LARRABEE, *Chemical Instrumentation,* 3, 125 (1971).

11. C. De WISPELAERE, J. OP De BEECK and J. HOSTE, *Anal. Chem.,* 45, 547 (1973).

12. E. SCHONFELD, *Nucl. Instr. and Methods,* 42, 213 (1966).

13. J. P. OP De BEECK, *J. Radioan. Chem.,* 11, 283 (1972).

14. P. J. M. KORTHOVEN, Interuniversity Reactor Institute, Delft, IRI Report 133-72-09 (1972).

15. E. ACHTERBERG, F. C. IGLESIAS, A. E. JECH, J. A. MORAGUES, M. PÉREZ, J. J. ROSSI, W. SCHEUER and J. A. SUÁREZ, IEEE, *Trans. on Nucl. Sci,* NS-19 No. 53 (1972).

16. P. DE REGGE, *Nucl. Instr. and Methods,* 102, 269 (1972).

17. G. C. BEST, *IEEE Trans. on Nucl Sci.,* NS-13 566 (1966).

18. S. B. WRIGHT and M. G. SILK, AFIPS Conf. Proc. 33, Pt. 2, 1968 Fall Joint Computer Conference, San Francisco, p. 1099 (1968).

19. J. W. McMILLAN and J. W. HAYNES, *Proc. Soc. Anal. Chem.,* 7, 202 (1970).

20. J. G. AUSTIN, R. C. M. BARNES and P. J. B. FERGUS, U.K. At. Energy Authority Report R-5529 (1967).

21. G. C. BEST and I. N. HOOTON, Proc. ISPRA Nuclear Electronics Symp., Euratom Report EUR 4289e, 305 (1969).

22. L. SALMON and M. G. CREEVY, Proc. I.A.E.A. Symp. on Nuclear Techniques in Environmental Pollution, 147 (1971).

23. N. I. SAX and J. C. DALY, *Nuclear Applications and Technology,* 8, 516 (1970).

24. R. T. BELL and H. ØVERAS, *IBM J. of Res. and Dev.,* 13 (1) 104 (1969).

25. J. C. BRUN, F. PICARD, R. SELLEM and G. VERROUST, *IEEE Trans. on Nucl. Sci.,* NS-19 No. 1 654 (1972).

26. A. LANGSFORD, O. N. JARVIS and C. WHITEHEAD, U.K. At. Energy Authority Report R-6832 (1971).

27. J. L. SPRATT, The Current Status of Liquid Scintillation Counting. (E. D. Bransome, Jr., ed.), p. 349 Grune and Stratton, New York, (1970).

28. M. F. GROWER and E. D. BRANSOME, The Current Status of Liquid Scintillation Counting. (E. D. Bransome, Jr., ed.), p. 356 Grune and Stratton, New York, (1970).

29. R. L. LITLE, The Current Status of Liquid Scintillation Counting. (E. D. Bransome, Jr. ed.), p. 371 Grune and Stratton, New York, (1970).

30. G. F. CRAMER, M. NICHOLSON, C. MOORE and K. TENG, *Intern. J. Appl. Radiation Isotopes,* 22, 17 (1971).

31. A. R. REICH, *J. Radioan. Chem.* 6, 437 (1970).

32. Y. ASHKENAZI and I. CARMI, *Nucl. Instr. and Methods,* 89, 125 (1970).

33. E. B. CHAIN, A. E. LOWE and K. R. L. MANSFORD, *J. Chromatog.*, **53**, 293 (1970).
34. T. O. SEIM and S. PRYDZ, *J. Chromatog.*, **73**, 173 (1972).

Chapter 5

Separation Methods for Inorganic Species

D. I. Coomber

Laboratory of the Government Chemist (Department of Industry), Cornwall House, Stamford Street London, S.E.1

5.1   INTRODUCTION

This chapter gives an account of the principal methods used in separating radioactive elements prior to measurement by methods depending on their radioactivity. This is followed by some examples of separations of elements of particular interest in radiochemistry and applications of these methods to sequential separations.

In many cases radiochemical separation procedures are based on those used in non-radiochemical work though there is some change of emphasis. For example, in precipitations the need to achieve the complete recovery necessary in gravimetric analysis disappears, as corrections for an incomplete yield can be applied. However, a trace of an impurity which would be unimportant in gravimetric analysis may contribute a large unwanted activity to the counting source and several re-precipitations in the presence of hold-back carriers may be needed. Some techniques, e.g. ion exchange, extraction chromatography, can be carried out under near ideal conditions for separation as the amounts of material involved are small. Some methods are used more widely in radiochemical separations, e.g. co-precipitation, while others such as those based on recoil or isotopic exchange have no counterpart in non-radiochemical work.

A number of factors determine the choice of method apart from the specific properties of the elements themselves. These include the nature of the matrix, the level of activity of the nuclide being determined and of the other radionuclides present, their half-lives and other nuclear properties which determine the method of measurement, and the accuracy required. The problems vary from that presented in the separation of a component from highly active material, where the initial separation stages may need remote handling behind shielding, to the case where low levels of activity have to be separated sequentially from a limited amount of sample prior to low level counting [1].

Advances in instrumentation in recent years have reduced the amount of separation needed in many cases. The development of solid state detectors has brought about high resolution gamma spectrometry and a limited chemical separation may be all that is required, though some separation into groups, the removal of one or two nuclides which complicate the gamma spectrum, e.g. ^{24}Na or ^{32}P, or a separation of the bulk of the matrix material may be an advantage. Similarly the silicon diode has simplified alpha spectrometry and the analysis of alpha emitters. Complete separations are still required in the analysis of nuclides which emit only beta particles or for measurements of low levels of activity.

Information on the aspects of radiochemistry dealt with in this chapter is widely spread in the chemical literature. Special mention can be made of the series of monographs covering the radiochemistry of most of the elements, and a number of specialised techniques, published during the early 1960's by the Subcommittee on Radiochemistry of the National Academy of Sciences — Science Research Council, U.S.A. Some of these have been subsequently revised and others are in course of revision. A complete list will be found in the recently revised monograph on the Radiochemistry of Mercury [2]. Many methods of separation of the rarer metals are given by Korkisch [3]. Separations pertaining to activation analysis are dealt with by Bowen and Gibbons [4]. Translations of a series of monographs on the Analytical Chemistry of the Elements, prepared by the Vernadskii Institute of Geochemistry and Analytical Chemistry, Academy of Sciences of the USSR, which include many references to radiochemical separations, are being published, e.g. Ref. [254].

5.2 PRECIPITATION AND ADSORPTION METHODS

5.2.1 The Use of Carriers in Radiochemical Separations

In many cases the mass of a radionuclide is very small and a quantity of the inactive element is added as carrier. For precipitations, milligram amounts of carrier are used while in other separations smaller amounts may be added, e.g. to limit adsorption effects or reduce oxidation or reduction by traces of impurities. The recovery of the added carrier can be determined by weighing a suitable compound when milligram amounts are used, though this is limited in accuracy, particularly when the duplicate filter paper method is employed. Other methods of chemical analysis are used [5, 6] which means the destruction of the counting source or taking an aliquot of a solution at the final stage.

It is essential for the carrier and the radioactive species to be in the same chemical form to avoid fractionation in the separations. Steps to bring about exchange are particularly necessary when the element is capable of existing in several valency states, e.g. iodine or ruthenium, or when hydrolysis and polymer formation can occur, e.g. zirconium or plutonium. In the former case the radioelement and carrier should be taken through all the possible valency states to ensure exchange, while in the latter, polymers should be destroyed by heating with strong acids after addition of carrier.

Where a non-isotopic carrier has to be used, e.g. barium for radium, the behaviour of the two elements will not be identical and the yield of radium cannot be assumed to be the same as that of the barium. Even with the closely similar rare earths Denechaud et al. [7] found the yield changed through the series and it was necessary to determine it for several members. In some cases another radioisotope of the element can be used to measure the chemical recovery, e.g. polonium-208 for polonium-210 or plutonium-236 for plutonium-239, the isotopes being distinguished in the counting procedure.

5.2.2   Precipitation and Co-precipitation

Precipitation and co-precipitation were the earliest techniques used in radio-chemical separations and their use has been described in many standard texts [8, 9]. The amounts of added carrier used in precipitation steps may be a disadvantage when subsequent procedures, e.g. ion exchange, are most satisfactory with small amounts of material.

Co-precipitation is a process in which a tracer ion is carried by another precipitate, though the tracer concentration is too low for it to be precipitated on its own. Several mechanisms are involved; compound formation, the formation of solid solutions, mechanical inclusion and surface adsorption. In co-precipitation the carrier ion is usually added before the precipitating ion so that the former is incorporated into the precipitate and not merely adsorbed on the surface of a pre-formed precipitate. One advantage of co-precipitation is that the wanted ion may be separated from the carrier and subsequent steps continued with smaller amounts of material, e.g. carrier iron separated by solvent extraction.

Co-precipitation is frequently used to remove other ions from solution by scavenging and for this purpose hydrous oxides, especially ferric hydroxide, are used. Although their carrying properties are non-specific, separations are possible by adjusting the conditions. Examples are the use of ferric hydroxide for separating trace yttrium from strontium [10], hafnium from tantalum targets [11] and the rapid separation of molybdenum, zirconium and niobium in investigations on the $^{99}Mo \rightarrow ^{99}Zr \rightarrow ^{99}Nb$ fission chain [12].

Co-crystallisation is the carrying of traces of metals on organic precipitates [13]. A number of papers have been published by H. V. Weiss and co-workers. For example they investigated the carrying of several metal ions on α-nitroso β-naphthol at different pH values [14]. Cobalt was carried at a lower pH than several other metals. Kuznetsov et al. [15] carried traces of uranium oxinate (down to concentrations as low as 1 in 2×10^{11} on a number of organic precipitates which they described as 'indifferent co-precipitants'.

The Weisz ring oven technique [16] is included here as it depends on some ions being fixed in the centre of a paper by precipitation or co-precipitation while others are washed into the ring zone. Examples of its use are the separation of molybdenum-99 from irradiated uranium for the measurement of uranium-235 depletion [17], a procedure which only takes five minutes, and the separation of long-lived fission products [18, 19].

5.2.3 Adsorption on Pre-formed Precipitates

The removal of ions from solution by passing through pre-formed precipitates is a useful alternative to carrier precipitation particularly for rapid separations [20, 21]. The processes involved include ion exchange, isotope exchange, precipitation, mixed crystal formation by recrystallisation, etc.

Girardi *et al.* [22] determined the retention behaviour of some 66 radioactive ions towards 11 materials from 11 acid solutions following a standardised procedure. Two of the materials were ion exchange resins and the remainder ionic precipitates, e.g. MnO_2, SnO_2 and antimony pentoxide. The 'clear cut' data, i.e. $> 95\%$ retention or $< 5\%$ retention, was recorded on punched cards, for rapid assessment of possible separations, while a computer programme was devised for dealing with more elaborate separation sequences [23]. The disadvantage of the method, recognised by the authors, was that the data, obtained from experiments with single components, might not apply exactly to samples containing other substances.

5.2.4 Residue Adsorption

Kirby [24] used the adsorption of traces of radionuclides for separations. When a tracer mixture of strontium and yttrium in dilute acid was evaporated on any surface and subsequently re-evaporated, after addition of dilute ammonium hydroxide, the strontium was removed by washing with distilled water while the yttrium remained fixed. He also separated the members of the $^{227}Ac \rightarrow {}^{227}Th \rightarrow {}^{223}Ra$ series, though radium-223 allowed to grow in after 'fixing' was only partially removed, due to recoil penetration [25].

5.3   ION EXCHANGE

5.3.1   Introduction

Ion exchange on synthetic resin exchangers is very widely used in radiochemical analysis. The scope of these materials has been greatly extended over the past 25 years with the use of complexing agents in cation exchange, eluents of high ionic strength in anion exchange and the use of mixed solvent systems. More limited use has been made of chelating resins and synthetic inorganic exchangers. The two procedures used most frequently are ion exchange elution chromatography and selective sorption. In the former, near equilibrium conditions are maintained and the process is fairly slow, though rapid separations have been carried out with resins of small particle size operating under high pressure. In selective sorption, one species is much more strongly retained by the exchanger than the others, equilibrium conditions need not be maintained, and high flow rates can be employed. The theory and use of ion exchange is described in a book by Samuelson [26], which includes many radiochemical applications, and in one by Rieman and Walton [27].

5.3.2   Synthetic Organic Exchangers

Introduction

When a cation exchange resin with its counter ion A^{-m}, represented by RA, reaches equilibrium with a solution containing ions, B^{-n}, the selectivity co-efficient is defined by

$$k_{A,B} = \frac{[B]_r^m \; [A]^n}{[B]^m \; [A]_r^n} \tag{5.1}$$

where the brackets denote concentrations, r indicates the resin phase and the signs of the ionic charges are omitted for clarity. Equation (5.1) is usually expressed so that $k_{A,B} > 1$.

The thermodynamic exchange constant, or corrected selectivity co-efficient, $k_{A,B}^a$ is given by

$$k_{A,B}^a = k_{A,B} \frac{\gamma_{B_r} \gamma_A}{\gamma_B \gamma_{A_r}} \tag{5.2}$$

where γ_A etc. are the relevant activity coefficients.

By choosing reference states so as to make $k_{A,B}^a = 1$

$$k_{A,B} = \frac{\gamma_B \gamma_{A_r}}{\gamma_{B_r} \gamma_A} \tag{5.3}$$

$k_{A,B}$ is not a true constant but will be approximately so during a chromato-graphic separation, in which the ion A is present in excess, while the exchanging ion, B, is present only in traces, so that the activity coefficients are unaltered during the exchange process. This condition applies in many radiochemical separations. Analogous considerations apply to exchange of anions on anion exchangers.

The adsorbability of the ion, B, is given by a weight distribution coeffi-cient, D_g (weight of B per Kg dry resin/weight of B per litre of solution), or by a volume distribution coefficient, D_v (weight of B per litre of resin bed/weight per litre of solution). They are related by $D_v = D_g \rho$ where ρ is the bed density.

From equation (5.1)

$$D_{g_B} = \frac{[B]_r}{[B]} = \left(k_{A,B} \frac{[A]_r^n}{[A]^n} \right)^{1/m} \tag{5.4}$$

Under the conditions stated above $k_{A,B}$ and $[A]_r$ are constant and

$$D_{g_B} \propto \frac{1}{[A]^{n/m}} \tag{5.5}$$

and a plot of log D_{g_B} versus log A gives a straight line of slope $-n/m$. Distribution coefficients may be determined batchwise, by equilibrating a quantity of resin with a solution containing the ion in question and measuring the amounts in the two phases, or by elution measurements from columns [28].

When two trace ions B and C are present the ratio of their distribution coefficients give the separation factor,

$$\alpha_{B,C} = \frac{D_B}{D_C} \tag{5.6}$$

In elution chromatography the ions to be separated are adsorbed on the top of the column and an eluent containing acid, base, salts or complexing agents passed through the column. The adsorbed ions are eluted in succession, those with the lowest distribution coefficients appearing first. If the loading of the column is small, i.e. the amounts of the trace ions are only a few per cent of the column capacity, and the flow rate is sufficiently low, the elution curve will be approximately Gaussian as in Fig. 5.1. In practice the peak may tail, due to variation of $k_{A,B}$ with concentration of ions in the elution band [29].

The peak elution volume, v, is related to the total column volume, V, and to D_v by

$$\ddot{D}_v = \frac{v - iV}{V} \tag{5.7}$$

where i is the void fraction of the column.

Martin and Synge [30] applied the theoretical plate concept from distillation theory to chromatographic columns and Mayer and Tomkins [31] and Glueckauf [32] extended it to ion exchange chromatography. The number of equivalent theoretical plates in a column is given by [32]

$$N = \frac{h}{L} = 8\left(\frac{v}{\beta}\right)^2 \tag{5.8}$$

where L is the length of the column, h the height of an equivalent theoretical plate and β is defined in Fig. 5.1.

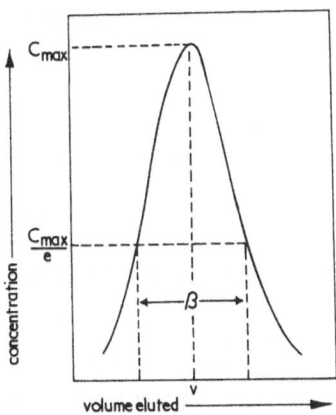

Fig. 5.1. Theoretical elution curve of a single component on an ion exchange column.

Glueckauf calculated the relation between h and the parameters of the column, i.e. particle radius, flow rate and the diffusion coefficients of the ion in the resin and interstitial phases. One term in his equation, which includes the diffusion coefficient of the ion in the resin, is approximately inversely proportional to D_v so that h will vary for a series of ions.

The efficiency of separation of two kinds of ions eluted successively from an ion exchange column depends on the overlap of the elution peaks. This is determined by the separation factor, α, and by the widths of the peaks, which are related to h by equation (5.8). α will generally increase with resin cross linking, while h increases with cross linking, particle size and flow rate and decreases with temperature. Fig 5.2 shows some elution curves of the rare earths with a fairly high cross-linked cation exchange resin [33]. A column of the same resin with larger particles (10–75 μm compared with 10–25 μm) gave approximately the same elution curve but with a flow rate five times slower. Cornish [34] calculated the column parameters required

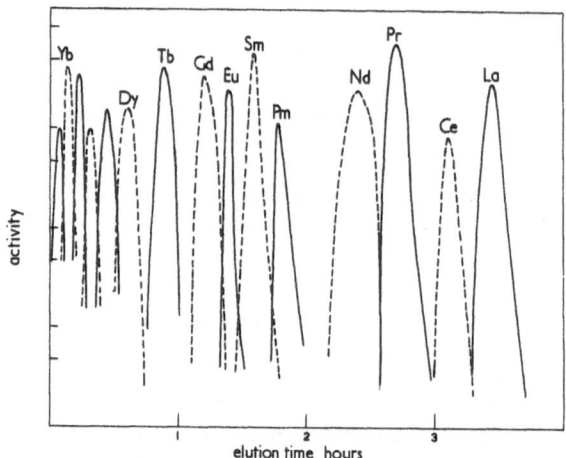

Fig. 5.2. Elution of rare earth elements from a cation exchange column. Legend: Dowex 50-X10 cation exchanger, column dimensions 6 mm × 500 mm, particle diameter 10–25 μm, flow rate 1.1 ml/minute. (After Aubouin and Laverlochere [33])

for separating two ions, based on the theory of Glueckauf, taking into account the acceptable cross contamination, the selectivity coefficients, the solution concentrations of the two substances, the resin size and the flow rate. A method of calculating the cross contamination occurring when a parent-daughter pair is being separated, the time of separation being an appreciable fraction of the half-life of the daughter, has been recently published [35].

When a series of ions is eluted, e.g. the rare earths, the elution peaks are not evenly spaced, the distance between peaks being greater for ions eluted later in the series. More even spacing and a reduction in time taken to elute the series is possible by changing the composition of the eluent stepwise or continuously (gradient elution) [36]. The use of an eluent of increasing concentration also reduces the tailing of the peaks referred to previously. Many devices have been used for delivering an eluting solution of changing composition [27], p. 115). With an eluent of changing composition equation (5.7) does not apply and methods of calculating plate heights under these conditions have been published [37].

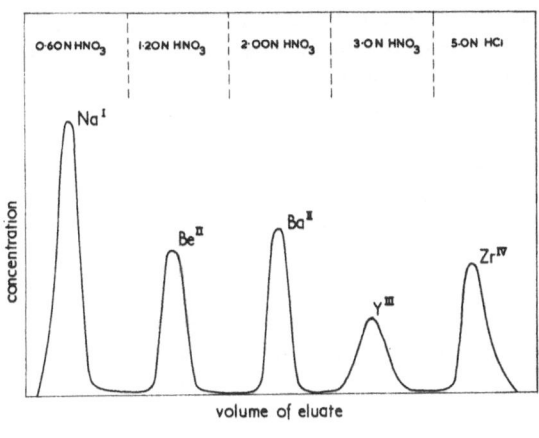

D_v *values*

Na	12.0	5.8	3.4	2.0	–
Be	33.0	12.0	6.6	4.5	2.0
Ba	200.0	45.0	13.0	6.0	7.0
Y	700.0	130.0	35.8	13.9	5.0
Zr	10^4	4000.0	652.0	112.0	5.0

Fig. 5.3. Chromatographic separation of ions of increasing valency on a cation exchanger (AG 50 W-X8). (After Strelow, Rethemeyer and Bothma [38])

Fig. 5.3. shows the elution curves of a series of ions of increasing charge [38]. The values of the distribution coefficients in the various eluting acids are shown below the peaks. The eluents were chosen so that the ion being eluted had a value of D_v of about 10 (italicised in Fig. 5.3) while those being retained had higher values [28].

Cation Exchange Separations
The affinity of cations for a cation exchanger increases with ionic charge and, for a group of elements of the same charge, with atomic number. In the lanthanide series, however, this order is reversed due to the 'lanthanide contraction'. With a non-complexing eluent the separation factor for two ions of the same charge is small. For example $\alpha_{Sr,Ca}$ on Dowex 50-X8 is about 1.25 while $\alpha_{La,Lu}$ (the first and last members of the lanthanide series) is only 2.15 [39]. The separation of cations may be greatly increased by eluting with complexing agents, making use of the differences between formation constants.

The separation factor is then modified and becomes

$$\alpha'_{B,C} = \alpha_{B,C} \, \frac{k'_B}{k'_C}$$

where k'_B and k'_C are the formation constants of the ions B and C with the complexing agent. For example, in the case of calcium and strontium with ethylenediaminetetraacetic acid (EDTA) as complexing agent the ratio of the formation constants is about 100 so that $\alpha'_{Sr,Ca}$ is approximately 125. An example of the use of complexing agents in eluting a series of metal ions from a cation column is shown in Fig. 5.4.

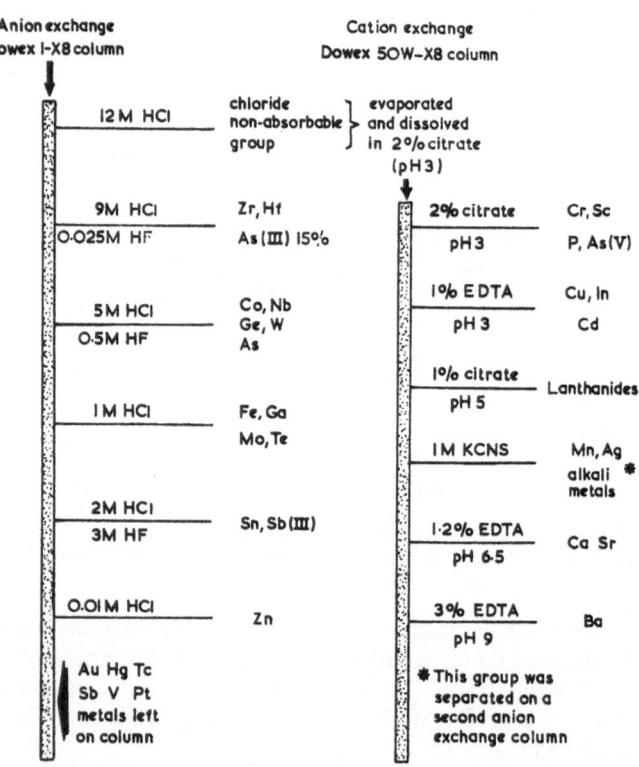

Fig. 5.4.   Sequential separation scheme using anion exchange from strong acids and cation exchange with complexing agents. (After Jervis and Wong [274])

Anion Exchange Separations

Much of the early work on the anion exchange of metal ions, undertaken for the separation of fission products, was due to Kraus and co-workers. Metal ions adsorbed from hydrochloric acid solutions included not only those, e.g. iron and gallium, which would be expected to form chloride complexes, but most of the metals apart from the alkali metals, the alkaline earths and lanthanides, aluminium, nickel and thorium. Exact theoretical treatment was difficult and Kraus and Nelson [40] measured the values of D_v for most of the elements in the periodic table in a range of molarities of hydrochloric acid. Their results were given in the form of a periodic table display, part of which is shown in Fig. 5.5. They also published data for hydrochloric acid solutions containing small amounts of hydrofluoric acid [41] which prevented hydrolysis of some metals at low acid concentrations. Although these results were based on one make of resin they give a useful guide to possible separations with strongly basic anion exchangers and these, and other data published on other systems, have been widely used in planning separations, e.g. Fig. 5.4. Table 5.1(a) gives some of the systems published.

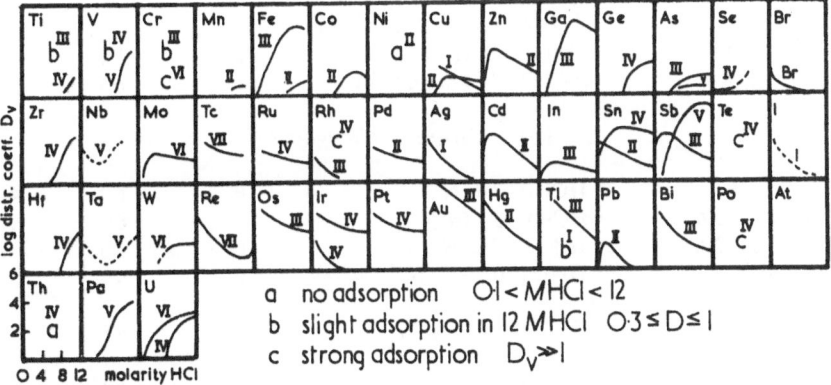

Fig. 5.5. The adsorption of elements on an anion exchanger (Dowex 1 X 10) from hydrochloric acid solutions. (After Kraus and Nelson [40])

Adsorption on Cation Exchangers from Strong Acids

Djurfeldt and Samuelson [42] found that ferric ions were strongly adsorbed on a cation exchange resin from hydrochloric acid solutions and attributed this to adsorption by the polystyrene matrix. Nelson *et al.* [44] measured the adsorption of a large number of elements from hydrochloric and perchloric acid solutions. The most striking observation was the marked adsorption of scandium, yttrium, the lanthanides and trivalent actinides from 9 M perchloric acid. The order of elution was the reverse of that in dilute acid solutions, e.g. barium eluted before calcium. Table 5.1(b) gives references to some of the systems published.

Chelating Resins [57]

The active groups attached to the resin matrix in chelating resins have a high selectivity for certain metals, differing from the exchangers described previously where the affinity depends on physical properties such as ionic charge and ionic radius. The best known chelating resin is the commercial Dowex Chelating Resin A-1 in which the iminodiacetate group is incorporated into

TABLE 5.1. *Ion Exchange Data*

Exchanger	Range of eluents	Mode of data presentation	Number of elements	Ref.
	(a) *Anion Exchangers*			
Dowex 1-X10	HCl 0.1–12 M	P	61	40
DIAION SAN 1X8	HCl 1–9 M	P	42	43
Dowex 1-X10	HF 1–24 M	P	50	45
Dowex 1-X10	HNO_3 0.1–14 M	P	70	46
Bio-Rad AG1-X8	H_2SO_4 0.01–4 N	T	27	47
Dowex 1-XB	dil. H_2SO_4	D	26	48
Dowex 1-X4	HNO_3 1–12 M + 0.2, 1.0 and 5 M HF	P	19	49
Dowex 1-X8	HBr 1–8 N	D	14	50
Dowex 1-X8	2–17.4 M acetic acid	P	65	51
	(b) *Cation Exchangers.*			
Dowex 50-X4	HCl 0–14 N	P	68	44
Dowex 50-X4	$HClO_4$ 0–14 N	P	60	44
Dowex 50-X4	HBr 0–12 N	P	70	52
Bio-Rad AG 50W-X8	HNO_3 0.1–4 N	P	49	38
Bio-Rad AG 50W-X8	H_2SO_4 0.1–4 N	P	45	38
Bio-Rad AG 50W-X8	HCl 0.1–3 N acetone 0–95%	T	54	53
Bio-Rad AG 50W-X8	HCl 0.1–3 N ethanol 0–95%	T	45	54
	(c) *Inorganic Exchangers*			
Hydrous ZrO_2 Zr phosphate Zr tungstate Zr molybdate	HNO_3 pH 1–5	P	60	55
MnO_2	HNO_3 0–5 N	P	61	56

P = Periodic table of D_v versus acid concentrations

T = Table of figures

D = Distribution curves versus acid concentration

a styrene divinylbenzene matrix. This resin has a low degree of cross linking and exhibits considerable volume change with change in ionic strength of the medium. The selectivity for divalent metals is high [58]. This resin has the advantage over ordinary ion exchange resins that metals can be extracted from solutions of high ionic strength and it has been used in the extraction of cobalt from seawater [59]. A resin prepared by Savvin [60] incorporating the chelating group in Arsenazo I was used in the separation of americium and curium from plutonium and fission products. The difficulty with commercially produced resins possessing groups of high specificity and restricted applications is that they may not remain in production [61].

5.3.3 Inorganic Ion Exchangers
The ion exchange properties of certain inorganic substances including the hydrous oxides of zirconium and tin, acid salts such as zirconium phosphate, and salts of heteropolyacids, e.g. ammonium molybdophosphate, have been described [62–64]. Inorganic exchangers are much more resistant to high temperature and to radiation than organic exchangers and are not subject to swelling and shrinking with change in the ionic strength of the medium. They show high selectivities for certain ions and the high affinity for caesium has been of great importance in the nuclear energy field. Stannic molybdate has been prepared with a high selectivity for lead, and stannic phosphate with a high selectivity for copper [65]. Baetsle and Huys [66] prepared a polyantimonic acid with a high selectivity for strontium over both calcium and barium. The nearly specific adsorption of sodium on a specially prepared hydrated antimony pentoxide [67] has been used to remove sodium-24 from irradiated samples.

5.3.4 Ion Exchange in Organic and Mixed Solvents
The use of organic solvents and mixed organic-aqueous solvents has been systematically investigated by Korkisch [68]. Replacement of water by solvent molecules, e.g. alcohols or ketones, lowers the dielectric constant, favours ion association and decreases ion hydration. The swelling of the resin phase is usually lower in organic solvents, giving the same effect as increased cross-linking. The result is that selectivities and separation coefficients are greater in mixed solvents than in aqueous solutions. Fig. 5.8 shows the effect of addition of methyl alcohol in the cation exchange separation of the alkali metals [186]. Korkisch and Arrhenius [69] proposed anion exchange from acetic acid – nitric acid solution to adsorb trace metals (U, Th, lanthanides, Y, Cd, Pb, Bi) from bulk constituents (Fe, Al, Ca) as a means of separating them in the analysis of marine sediments. Morrow [70] used methanol solutions of HCl to separate actinides from bomb debris and methanolic nitric acid solution for separating lanthanides and actinides.

5.3.5 High Pressure Ion Exchange Chromatography
High pressure ion exchange was developed by Scott [71] for biochemical separations and, following him, Campbell and Buxton [72, 73] used fine particle resins (Dowex 50-X12, 10–20 μm diameter) in high resolution preparative columns for the separation of lanthanides and trivalent actinides on a milligram scale. The system has the advantages that, with the high flow rates possible, radiation damage to the resin is reduced while gases formed

by radiolysis remain dissolved. A similar method was used by Farrar *et al.* [74] in the analytical separation of berkelium from highly radioactive solutions.

5.4   ISOTOPE EXCHANGE

In isotope exchange a radioisotope in one phase, e.g. an aqueous solution, exchanges with stable isotopes present in considerable excess in another phase so that a high proportion of the activity is transferred to the second phase. For example, traces of silver in solution exchange rapidly with inactive silver in a silver chloride precipitate and most of the silver activity goes into the solid phase [75].

Kuroda and Oguma [76] separated ^{90}Sr and ^{90}Y on the tracer scale by thin layer chromatography on a mixture of strontium sulphate and silica gel. On running the chromatogram the yttrium moved close to the solvent front while the strontium activity was removed by isotope exchange near the starting line. Tang and Maletskos [77] used isotope exchange to remove sodium-24 and potassium-42 activities from irradiated tissue samples. An acid digest of the tissue mixed with acetone was passed through columns of potassium and sodium chlorides. Isotope exchange of iodide on an 'iodinated' resin was used by Heurtebise and Ross [78] in separating iodine from irradiated biological fluids

A special case of isotope exchange is amalgam exchange in which the second phase is an amalgam of the metal corresponding to the radioactive ion which is to be separated. After exchange, brought about by shaking the amalgam with the solution, the activity is measured directly in the amalgam, e.g. by measuring gamma emission, or after stripping into aqueous solution. De Voe *et al.* [79] suggested that interferences could be removed by a preliminary extraction with an amalgam of the element immediately below the desired element in the electromotive series. Silker [80] used the method in determining zinc-65 in reactor cooling water, copper-64 being removed by a preliminary treatment with a cadmium amalgam. Qureshi and Nagi [81] investigated the optimum conditions for zinc separation and gave references to earlier work. Sometimes special conditions have to be found; the first attempts to exchange gallium were unsuccessful but Ruch [82] found that exchange took place from an acid medium of high ionic strength in presence of thiocyanate.

5.5   CHROMATOGRAPHIC METHODS

5.5.1   Introduction

Most chromatographic procedures have been applied to inorganic radio-chemical separations though ion exchange chromatography (Section 5.3) and reversed-phase partition chromatography (Section 5.6.8) have been used the most widely.

5.5.2 Paper and Thin-Layer Chromatography

Many separations have been carried out on filter paper. R_f data for a number of systems including adsorption chromatography from aqueous solution, partition chromatography, papers loaded with ion exchange resins and modified celluloses, published in the previous decade have been given by Lederer and Majani [83]. An interesting investigation on the use of modified and treated papers for the separation of technetium in different valency states has been described by Shukla [84]. Thin-layer chromatography has the advantage that supports other than cellulose, which may interfere with carrier-free separation of radionuclides due to reducing substances present in the paper, can be used. In a review of inorganic thin-layer chromatography Lederer [85] states that, provided the optimum conditions are chosen, thin-layer methods have no advantage over those with paper as regards speed of separation.

5.5.3 Gas Chromatography

Gas chromatography has been used in the concentration of tritium [86], the separation of krypton and xenon fission products [87, 88] and in the separation of fission produced halogens using alkyl halides formed by recoil reaction in methane [89].

A large number of volatile metal compounds could be used for the gas chromatographic separation of metals [90] but, as yet, work has been limited to volatile halides [91] and fluorinated diketone compounds. The earlier use of the latter required a lengthy preliminary separation of the chelate compounds [92] but Sievers [93] found that volatile chelates could be formed directly from a number of metals and metal compounds by direct reaction at a high temperature in sealed tubes. Zvarova and Zvara separated the higher lanthanides [94] and trivalent actinides [95] on a glass capillary column using the volatile double chlorides formed with aluminium trichloride, e.g. $YbAlCl_6$.

5.6 LIQUID-LIQUID EXTRACTION AND EXTRACTION CHROMATOGRAPHY

5.6.1 Introduction

Liquid-liquid extraction is widely used in radiochemical work as it can be carried out rapidly with simple equipment which is capable of automation, it can be made very selective and can be applied to a wide range of concentrations. The principles have been described by Morrison and Freiser [96] and by De, Khopkar and Chalmers [97]. Separations usually depend on the distribution of material between an aqueous phase and an organic phase, the two being practically immiscible. The extracted material may consist of simple molecules, e.g. iodine, ruthenium tetroxide, or it may be rendered soluble in the organic phase by chelate formation, solvation, ion pairing or ion exchange processes. A classification of extraction systems was given by Morrison and Freiser [98]. The special case of substoichiometric extraction is described in Chapter 8. The less exact term 'solvent extraction' is used through the remainder of this chapter for convenience.

5.6.2   Chelate Extraction [99]

When metals form uncharged inner complexes with large organic molecules and the complexes are not appreciably hydrated these can be extracted into immiscible organic solvents. The extraction can be represented by

$$M_w^{n+} + nHL_o = ML_{n_o} + nH_w^+ \tag{5.9}$$

The suffixes, w and o, refer to the aqueous and organic phases. The extraction constant is given by

$$K_{ex} = \frac{[ML_n]_o}{[M^{n+}]_w} \; \frac{[H^+]_w^n}{[HL]_o^n} \tag{5.10}$$

The concentration distribution ratio, D_c (or distribution or extraction coefficient) is the ratio of the total concentration of the metal in the two phases and is given by

$$D_c = \frac{[ML_n]_o}{[M^{n+}]_w} = K_{ex} \frac{[HL]_o^n}{[H^+]_w^n} \tag{5.11}$$

When two species are present the ratio of the D_c values is the separation factor. Equations (5.10) and (5.11) apply where the only metal complex formed is the uncharged ML_n. If the concentration of metal ions in the aqueous phase is reduced by the formation of charged complexes, e.g. ML_{n-p}^+ by hydrolysis giving hydroxy complexes or by the presence of water soluble complexes with a masking agent, then D_c will be reduced. In most radiochemical separations the ligand is in considerable excess, to make the extraction as complete as possible, hydrolysis is avoided, and masking agents are chosen so as to form complexes only with interfering ions. It is also assumed that the activity coefficients, considered in both phases, remain constant. Poskanzer and Foreman [100] give an account of these effects in the case of 2-thenoyltrifluoroacetone (TTA) complexes.

If K_{ex} is constant a plot of D_c versus $\log HL_o$ at constant pH will be a line of slope n (the extractant dependency) and one of $\log D_c$ versus pH at constant chelating agent concentration will also be a line of slope n.

The recovery factor, R, is the fraction of a substance extracted. For extraction into the organic phase

$$R = \frac{D_c}{1 + D_c} \tag{5.12}$$

for equal phase volumes, or $R = 1 - (1 + rD_c)^{-m}$ when m successive extractions are performed and r = volume of organic phase/volume of aqueous phase. The ratio of R values for two substances gives the enrichment factor.

5.6.3   Acid Esters Derived from Phosphorus

The extractive properties of di-(2-ethylhexyl) orthophosphoric acid (HDEHP) were discovered by Peppard *et al.* [101] following observations by several workers on the effects of acid esters present as impurities in the neutral extractant, tributyl phosphate (TBP). A number of other acid esters derived from phosphoric and phosphonic acids have been used [102] but HDEHP has found the widest applications.

While extraction of cations by solutions of these acid esters in diluents bears some resemblance to exchange on cationic exchangers the analogy is not exact. For instance, the separation obtained with adjacent members of the lanthanides is far greater than is obtained on a cationic exchanger with a non-complexing agent. They can best be considered as chelating agents and Peppard *et al.* [101] refer to HDEHP as the 'high-acid' analogue of TTA.

HDEHP is present as a dimer in hydrocarbon solvents and as a monomer in alcohols. The reaction with a number of ions can be represented as follows for extraction by HDEHP in toluene:

$$La^{3+} + 3(HL)_2 = La(HL_2)_3 + 3H^+$$

$$UO_2^{2+} + 2(HL)_2 = UO_2(HL_2)_2 + 2H^+$$

The extractant dependency and the pH dependency are both numerically equal to the charge on the ions in aqueous solution. In other cases the stoichiometry is more complex. In the case of calcium

$$Ca^{2+} + 3(HL)_2 = Ca(HL_2)_2 . (HL)_2 + 2H^+$$

while with thorium the stoichiometry varies with the nature of the acid and the pH [103].

The presence of small amounts of certain impurities in HDEHP may have a marked effect on its extractive properties. Fig. 5.6 shows the effect of the presence of about 1% of the mono-ester (H_2 MEHP) in HDEHP. The extraction of the high atomic number lanthanides is not affected but that of the lower members is greatly increased. The presence of the neutral ester increases the extraction of uranium by synergism (Section 5.7.6) while 2-ethyl hexanol reduces the extraction of some metals. Purification of HDEHP is generally carried out by partition between solvents [101, 104]. A review of the uses of HDEHP in extractions from strong acids is given by Qureshi *et al.* [105].

5.6.4   Ion Association Extraction

Ion pairs containing large organic groups can be extracted into organic solvents, e.g. tetraphenylarsonium perrhenate into chloroform, caesium dipicrylaminate into nitrobenzene. The solubility of uranium nitrate in ether and the extraction of ferric chloride into ether have been known for many years. Many other oxygenated solvents including ethers, ketones and esters have been used in the extraction of metal ions. Ether was used in the separation of uranium in fuel processing before the introduction of tributyl phosphate, first used by Warf [106] as a substitute for ether is the extraction of the ceric ion.

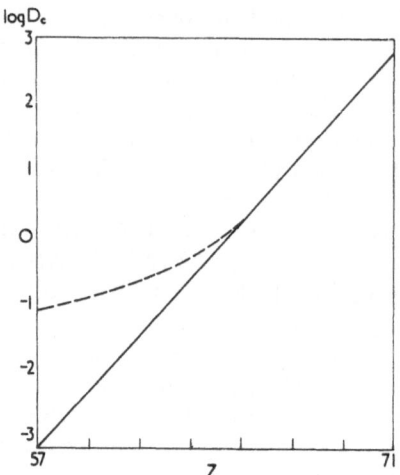

Fig. 5.6. The effect of an impurity on the extraction properties of di-(2-ethylhexyl) orthophosphoric acid. Legend: The continuous line shows the relation between the distribution constant and Z for the rare earth series when the extractant was highly purified HDEHP in toluene. The broken line shows the effect of 1% of the di-acid ester (H_2MEHP) in the HEDHP. The effect is negligible for the high Z members but increases markedly for the members of the series below gadolinium. (After Peppard, Mason, Maier and Driscoll [101])

The mechanism of these extractions is more complicated than chelate extraction though the extraction of uranium nitrate, for instance, can be formally represented by [107]

$$UO_2^{2+} + 2NO_3^- + 2TBP = UO_2(NO_3)_2 . 2TBP$$

The extraction of the trivalent lanthanides and actinides is given by

$$M^{3+} + 3NO_3^- + 3TBP = M(NO_3)_3 . (TBP)_3$$

and thorium by

$$Th^{4+} + 4NO_3^- + 2TBP = Th(NO_3)_4 . (TBP)_2$$

The extractive power of other neutral phosphorus compounds increases in the order

$$(RO)_3 . PO < (RO)_2 R.PO < (RO)R_2 . PO < R_3 . PO$$

One representative of the last group, n-octyl phosphine oxide, (TOPO), is a very important extractant and has been reviewed by White and Ross [108].

Data for a number of neutral extractants given in the form of periodic table displays of D_c versus acid strength have been published [109–111].

5.6.5 Extraction with Long Chain Amines

A general account of extraction with long chain amines has been given by Moore [112]. They are often referred to as liquid anion exchangers, due to the analogy between amine extraction and anion exchange with solid exchangers, though there are differences: the organic phase, unlike the solid phase in resin exchangers, contains little water. However, extraction curves have the same general form as those of solid exchangers.

The wide choice of amines available; primary, secondary, tertiary and quaternary ammonium compounds; and the variations in organic diluents and in the aqueous medium make amine extraction systems very versatile. Because separations can be performed rapidly amine extractants have an advantage over solid exchangers, i.e. for work in high radiation fields. Many of the amine extractants used most frequently are not single substances but commercial mixtures of closely related compounds.

Data for several amine extractants is included in the compilations referred to at the end of Section 5.6.4. Information on the extraction of many metals by tetrapropyl-, tetrabutyl- and tetrahexyl-ammonium salts from mineral acids and sodium hydroxide have been given by Maeck *et al.* [113].

5.6.6 Extraction with Mixed Ligands

The term synergism was applied by Blake *et al.* [114] to describe the marked enhancement in the extraction of the uranium(VI)-HDEHP complex brought about by the presence of small amounts of neutral compounds, e.g. Tributyl phosphate (TBP), trioctyl phosphine oxide (TOPO), though the effect had been observed previously by Cuninghame *et al.* [115]. Synergism has since been found to be much more general [116]. The explanation of the effect is that water molecules which complete the co-ordination of the metal complex are replaced by molecules of the neutral compound, increasing the solubility of the complex in the organic phase. Alimarin *et al.* [117] explained, in a similar way, the fact that the TTA chelate of NpO_2^+ (co-ordinatively unsaturated) is soluble in isobutanol but not in benzene, while the Ce(III) chelate (co-ordinatively saturated and requiring no water molecules to complete the co-ordination) is soluble in both solvents. The behaviour of small amounts of compounds in synergism underlies the possible effect of trace impurities in solvents used in solvent extraction. The reverse effect, anti-synergism, has been observed with greater concentrations of the neutral compound. Interaction between the chelating agent and the neutral compound is probably involved [118].

5.6.7 General Practical Considerations

In separating a substance from one or more others by solvent extraction the following points have to be considered:

(i) The solubility of the extracted compound should be high and the separation factors should be as large as possible. This may be brought about by the choice of extractant, the choice of solvent, the use of masking agents and pH control.

(ii) The solvent should be immiscible or nearly immiscible with the aqueous phase and its specific gravity should differ as much as possible from that of the aqueous phase for rapid separation of the phases and to minimise the formation of emulsions. In some cases, e.g. for automatic separation systems, the choice of solvent may be determined by the need for it to be lighter or heavier than the aqueous phase. If a substance is added to prevent emulsion formation, e.g. an aliphatic alcohol, it should not have any effect on the chemistry of the system.

(iii) The substance being extracted may be an interfering nuclide only intended for rejection. If, however, it is the wanted species the subsequent separation stages will have to be considered. The extraction may be the final stage in the separation and the organic phase may be counted direct, e.g. by gamma spectrometry, it may be incorporated into a medium for liquid scintillation counting (see Chapter 3) or it may be prepared for planchet counting by evaporation and ignition to destroy organic matter. Alternatively the separated species may be re-extracted into an aqueous solution for subsequent treatment. The re-extraction may involve valency changes using oxidising or reducing agents which may operate in either phase.

Where the separation factors are large one or two extractions followed by washing steps and re-extraction may suffice and the equipment used will be quite simple. Where the separation factors are small, multiple extractions and multiple scrubbing steps may be required. This subject is dealt with in detail by Peppard ([102], p. 7) and his account includes a simplified countercurrent distribution method ('push through' method) carried out with a few separating funnels. In general the equipment for countercurrent distribution separations is complicated [119, 120] and multiple extractions for analytical purposes can be carried out more easily by extraction chromatography.

5.6.8 Reversed-Phase Extraction Chromatography

In reversed-phase extraction chromatography the organic phase is maintained stationary while the aqueous phase is mobile. A recent review by Cerrai and Ghersini [121] gives a full account of the technique with many radiochemical applications. The organic phase may be supported on paper or on particulate material used in thin layers or columns. Supporting materials include cellulose powder, glass powder or micro-beads, silica gel, kieselguhr, polythene granules, styryl divinylbenzene polymers, polytrifluoroethylene or polytetrafluoroethylene, e.g. teflon. Many of the substances used in solvent extraction have been applied to extraction chromatography including long chain amines, acidic esters, e.g HDEHP, and neutral substances such as TBP, TOPO and methyl isobutyl ketone.

In columns the parameters are related to the distribution coefficient of a substance by the equation

$$D_c = \frac{V_{max} - V_o}{V_s}$$ (5.13)

where V_{max} is the volume of eluent required to reach the peak concentration in the effluent, V_o is the void volume of the column and V_s is the volume of the organic extractant in the stationary phase.

The plate theory dealt with in the section on ion exchange (Section 5.4.2) can be applied to extraction columns and the number of equivalent theoretical plates calculated from the equation

$$N = 8 \left(\frac{V_{max}}{\beta} \right)^2$$ (5.14)

which is equivalent to equation (5.8).

As with ion exchange columns, when large separation factors are involved, one substance can be retained on a column while the others pass through. The plate height is not very critical in this case and fast flow rates can be used. For separations of several metals, e.g. the lanthanides, where separation factors between adjacent members are small, step elution or gradient elution is used and the number of theoretical plates in the column is important. The factors affecting column performance have been described by several workers [122, 123].

Fig. 5.7, taken from Horwitz et al. [123], shows the effect of flow rate

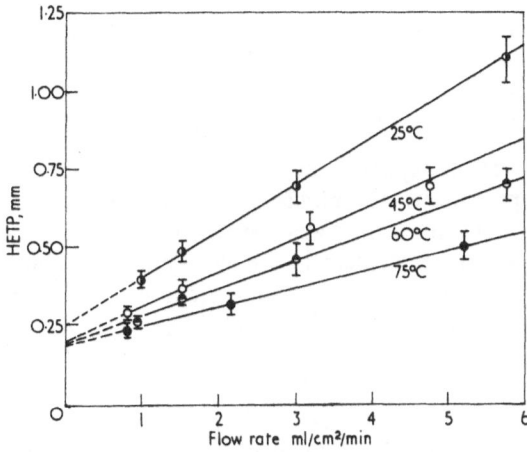

Fig. 5.7. The effect of flow rate and temperature on the height of an equivalent theoretical plate of an extraction column. Legend: Column material HDEHP on celite (0.27 equiv. per Kg dry bed). Dimensions 0.06 cm² X 10 cm. Bed density 0.4 g/ml. V_o = 0.72. Curium (III) eluted with 0.15 N nitric acid. (After Horwitz, Bloomquist and Henderson [123])

and temperature on the plate height of an extraction column. The experimental conditions are fairly typical. It is seen that the column capacity is an order lower than those of ion exchange resin columns.

In paper or thin layer applications R_f values and distribution coefficients are related by the equation

$$D_c = A\left(\frac{1}{R_f} - 1\right) \qquad (5.15)$$

where A is the ratio of the areas of cross section of the aqueous and stationary phases in the paper or thin layer. The constant A may not be accurately defined, or easy to determine, but for practical purposes the separation factor for two components a and b

$$\alpha_{a,b} = \left(\frac{1}{R_{fa}} - 1\right)\Big/\left(\frac{1}{R_{fb}} - 1\right) \qquad (5.16)$$

can often be used. The inert support may take part in a chromatographic process [124] but this is less likely with a fluoride polymer support than with cellulose or kieselguhr.

Combining equation (5.15) with equation (5.11) gives

$$\log\left(\frac{1}{R_f} - 1\right) = -\log A + \log K_{ex} + n \log [HL]_o + n \text{ pH} \qquad (5.17)$$

With HDEHP as extractant Kimura [125] found for a large number of metals $\log\left(\frac{1}{R_f} - 1\right)$ gave straight line plots versus pH and Pierce et al. [126] obtained n $\sim$ 3 for the rare earths. They did not attempt to relate $\log\left(\frac{1}{R_f} - 1\right)$ and $[HL]_o$ as the effective value of $[HL]_o$ was not known.

5.7 DISTILLATION, SUBLIMATION AND GAS SWEEPING METHODS

Many elements can be separated by methods depending on their volatility. These include preparative techniques such as the sublimation of metals in the carrier-free state, from irradiated targets, which are not dealt with here.

The halogens can be distilled in the elemental state (fluorine as hydrofluoric acid), ruthenium, osmium, manganese and technetium as the higher oxides and chromium as chromyl chloride. The following elements can be distilled as halides: As, Sb, Sn, Se, Re, Hg and Ge. By varying the conditions, e.g. oxidation state, separations can be effected, e.g. Tc from Ru, [127] Ru from Os [128] and Hg from As [129].

Several elements can be separated as their volatile hydrides. Greendale and Love [130] found a rapid evolution of AsH_3 and SbH_3 was achieved by adding an acid solution containing the element in question to an excess of hot zinc. The method was extended to selenium [131] and tellurium ([20], p. 18). The separation of radon isotopes from their precursors by gas sweeping methods is described in Section 5.9.1.

5.8 ELECTROCHEMICAL METHODS

5.8.1 Electromigration Techniques

The separation of substances by migration in an electric field has been used less with inorganic species than with organic substances. The differences in mobility of simple ions with the same charge depends only on their radii [132] and are small, but as in cation exchange chromatography, the differences can be increased by the use of complexing agents. For example, paper electrophoresis of 63 metals using N-(2-hydroxyethyl)iminodiacetate as complexing agent was investigated by Jokl et al. [133]. However, separation factors may not be so great as in the case of ion exchange with the same complexing agents [134]. The technique has the advantage of speed, many separations requiring less than 30 minutes. The disadvantages are the need to take precautions in the use of high voltages and the problems of heat dissipation. Reviews of electrophoresis on paper have been given [135, 136] and only a few applications are mentioned here. Adloff et al. [137] separated natural radioelements of short half-life using apparatus similar to that described by Gross [138]. The lanthanides have been separated on acetyl cellulose [139] with α-hydroxyisobutyric acid (α-HIBA) as complexing agent and the method has been used for determining these elements in meteorites by activation analysis [140].

In continuous electrophoresis (electrochromatography) the supporting electrolyte flows in a direction at right angles to the applied field. When the supporting medium is paper the path of each ion is determined by its mobility and its chromatographic behaviour and the separated ions leave the bottom of the paper at different points. It was first used in the separation of fission products by Strain et al. [141] and Pučar et al. [142] used it in rare earth separations.

Electrophoretic ion focussing, described in a series of papers by Schumacher et al. [143, 144] consists of electrophoresis on paper, the complexing agent being introduced into the cathode compartment and a complex destroying agent, e.g. acid, into the anode compartment. The concentration of complexing agent and the pH vary along the paper and the complexes of the different ions form narrow zones at their iso-electric points.

Counter current electrolysis, first used by Brewer et al. [145] in isotopic separations, has been applied to separations in activation analysis [146, 147]. Electrolysis takes place in a long cell with porous barriers to limit mixing. The electrolyte flows through the cell in opposition to the movement of ions. Ions of different mobility concentrate in distinct zones, due to the decrease in ion velocity with increase of ion density.

Enrichment of tritium, relative to hydrogen, in water is frequently carried out by electrolysis, the tritium being concentrated in the residue [148, 149].

5.8.2 Electrodeposition

Electrodeposition techniques have been used less for separations of ions than as a means of producing thin sources for counting alpha [150] emitters, or low energy beta emitters, although separations have been carried out by controlled potential deposition at a mercury cathode, e.g. for the separation of technetium from fission products [151] and for group separations of fission products [152].

For source preparation the actinides have been deposited as hydrated oxide films from alkaline solution [153] or from dilute acid containing ammonium salts [154]. A more rapid deposition was obtained using uranium as a carrier [155]. Parker *et al.* [156] deposited many metals from solutions in organic solvents, e.g. iso-propanol, and called the process 'molecular plating' because the deposition involved positively charged molecular clusters rather than ions. A potential of several hundred volts was required.

Spontaneous electrodeposition (internal electrolysis) is frequently used as a final step in the separation of polonium [157]. A number of metals, e.g. Co, Ni, Ag, Hg, Zn, have been deposited on thin metallised films by internal electrolysis using a magnesium anode [158]. Millard [159] determined polonium-210 and lead-212 in minerals for checking radioactive equilibrium by spontaneous deposition of the former and subsequent electrodeposition of the latter.

Electrodeposition of metals may be sensitive to the presence of other substances. For example, the deposition of polonium on silver is inhibited by the presence of iron unless a reducing agent is present [159], separation of carrier-free copper from irradiated zinc by spontaneous deposition on platinum black was found to be very sensitive to the presence of gold and silver [160], and in presence of fluoride traces of rare earths can interfere with the deposition of americium [161]. In many cases uncertainties of yield can be corrected by the use of another radioisotope as an internal standard.

5.9   SEPARATIONS OF SOME ELEMENTS OF SPECIAL INTEREST

5.9.1   Natural Radioelements in the Thorium and Uranium Series
Measurements of the natural radioactive elements are of importance from the point of view of environmental contamination and in studies in the geochemical and cosmochemical sciences, where many dating methods are based on the ratios of nuclides in these series.

Separations of thorium, protactinium and uranium have been carried out by many methods [3] but those based on ion exchange and solvent extraction procedures are used most frequently [162–164].

Where a nuclide is a member of a decay series it is often an advantage to measure another member of the series, e.g. a daughter nuclide, when radioactive equilibrium, or a measured fraction of this equilibrium, can be established. Such a separation may give additional decontamination, the daughter may be easier to measure than the parent and it may enable one nuclide to be distinguished from its other isotopes.

The separation and measurement of radon-222 as an indicator for radium-226 is very specific. The radon is removed by gas sweeping from a solution containing the radium, after equilibrium or a known fraction of equilibrium has been reached, purified in a gas train and counted by means of an ion chamber [165, 166] or by alpha scintillation counting [167, 168]. Very low levels of radium have been determined in this way [169]. A similar technique has been used to measure nuclides in the thorium and actino-uranium series [170]. In this case the two radon isotopes have short half-lives (radon-220, $t_{1/2}$ 54 sec and radon-219, $t_{1/2}$ 4 sec) and a system in which the counting gas is recirculated was employed. This technique has been used in measuring thorium-232 (via radon-220) and protactinium-231 (via radon-219) in dating deep sea cores [171].

Radium-228 is a low energy beta emitter and is most easily determined via its actinium-228 daughter. Methods include a prior separation of radium-228 and subsequent removal of the actinium daughter, after in-growth, by solvent extraction or ion exchange, or direct separation of the actinium-228 from the sample [172]. Similarly, lead-210, a low energy beta emitter, can be determined via its bismuth-210 daughter, an energetic beta emitter [173], or its polonium-210 descendant, an alpha emitter [174]. Thorium-228 has been determined by separating the radium-224 daughter [175]. Radium-226, radium-224 and radium-223 have been determined by solvent extraction and counting of the radioactive lead and bismuth descendants after various ingrowth periods [176]. These isotopes have also been determined by co-precipitation on barium sulphate, after removing interfering nuclides [177, 178], and distinguished from one another by growth-decay measurements [177].

However, this method needs caution, e.g. the presence of radium-228 can cause a positive error in the determination of radium-226 due to the growth of its alpha emitting daughters [179]. Smith and Mercer [180] separated the radium isotopes from barium carrier by ion exchange and measured them by alpha spectrometry using the non-natural radium-225 as a tracer for determining the chemical yield of radium.

Polonium-210 is generally separated by co-precipitation on tellurium [181], ion exchange [52, 182] or solvent extraction [183]. The final step usually involves spontaneous deposition on silver [157] or on other metals, e.g. nickel [184].

5.9.2 Caesium

Considerable attention has been given to the separation of caesium because of the importance of fission product caesium-137. A review of the precipitants used for caesium has been given by Yamagata [185]. Cation exchange separation of all the alkali metals has been investigated extensively [186] (Fig. 5.8). Many inorganic exchangers have a high selectivity for caesium. Ammonium-12-molybdophosphate (AMP) was first investigated by Smit [187] and has since been used widely for caesium separations. A full account of the use of this, and other materials, has been given by Folsom and Sreekumaran [188]. Several solvent extraction systems have been used for caesium.

.   Leaf [189] used sodium tetraphenylborate in amyl acetate for the

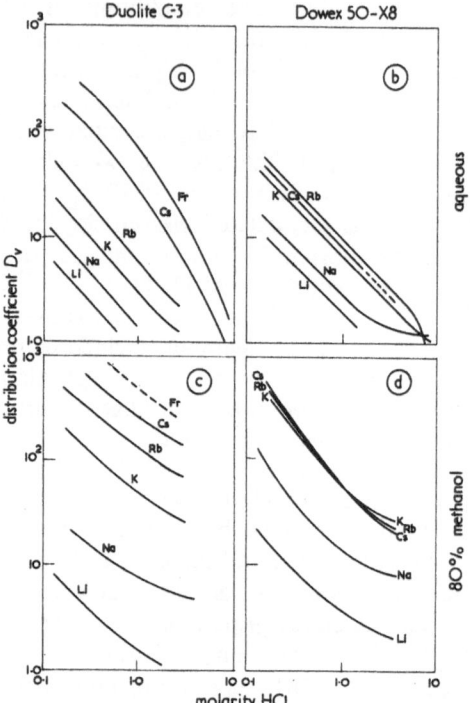

Fig. 5.8. Adsorption of alkali metal ions from hydrochloric acid solutions on two resins from aqueous and aqueous-methanol media. (After Nelson, Michelson, Phillips and Kraus [186])

extraction of caesium-137 from irradiated uranium for burn-up determination, EDTA being used to complex zirconium, etc. Tetraphenylborate has been used in an extractive chromatographic method of separation [190]. Extraction of caesium with 4-sec-butyl-2-(a-methylbenzyl)phenol (BAMBP) in cyclohexane from strongly alkaline solution has been used in the analysis of fission product solutions [191].

5.9.3 Alkaline Earths

Strontium-89 and strontium-90 are important fission products and, because they both only emit beta radiation, separation from other elements is frequently required. Separation and measurement of yttrium-90 is used to distinguish strontium-90 and strontium-89 (the yttrium-89 daughter is stable).

The low solubility of strontium and barium nitrates in nitric acid is the usual method of separating these elements from calcium and many other metals [192, 193]. Barium is usually separated from strontium by the method of Sunderman and Meinke [194] which uses the low solubility of barium chromate at pH 4. The separation was improved by the use of EDTA as a complexing agent for strontium [195].

Many separations of these elements from one another have been achieved by cation exchange, eluting with complexing agents; references to many of which are given in a paper by Strelow *et al.* [196]. In their method they used ammonium malonate in separating the beryllium-barium series. This complexing agent had the advantage of being easily destroyed subsequently. Stoeppler [197] separated radiostrontium from mixed fission products on an anion exchanger in the hydroxide form, the only fission products eluting with the strontium being barium and caesium. Cation exchange separations of barium and radium using EDTA have been described by Duyckaerts *et al.* [198] and by Nelson [199].

Many methods have been used for separating yttrium-90 from strontium-90 including co-precipitation of yttrium on ferric hydroxide [10], cation exchange with complexing agents [198] and solvent extraction with TTA [200], TBP [201] and HDEHP [202, 203].

5.9.4 Yttrium, the Lanthanides and Actinides

With the development of nuclear energy the analytical chemistry of the lanthanides and actinides, which presented similar problems, became important and many elegant separation procedures were devised. The early work on the chemistry of the actinides has been described by Katz and Seaborg [204] and the analytical chemistry has been reviewed by Metz and Waterbury [205]. Table 5.2 lists all the elements in this group. The numbers beneath the element symbols are the principal valency states exhibited in solution, the most stable states being italicised.

TABLE 5.2. *Valency States of the Lanthanides and Actinides Exhibited in Solution*

Sc														
3														
Y														
3														
La	Ce	Pr	Nd	Pm	Sm	Eu	Gd	Tb	Dy	Ho	Er	Tm	Yb	Lu
3	*3*,4	3	3	3	2,*3*	2,*3*	3	3	3	3	3	3	2,*3*	3
Ac	Th	Pa	U	Np	Pu	Am	Cm	Bk	Cf	Es	Fm	Md	No	Lw
3	4	4,5	2.4 5,6	3,4 5,6	3,4 5,6	*3* 5,6	3	3,4	*3*,4	3	3	2,*3*	2,*3*	3

The Lanthanides

Precipitation, ion exchange or solvent extraction are usually used in the separation of the lanthanides as a group from other elements [140]. Separations within the group based on valency changes are limited, as most of the lanthanides only exist in the trivalent state. Cerium can be separated from the other members, after oxidation to Ce(IV), by precipitation as iodate [206], anion exchange from nitric acid [207] or by solvent extraction [106]. Europium, ytterbium and samarium have been separated by reduction to the divalent state [208, 209].

The intergroup separation of the trivalent members has generally been carried out by cation exchange chromatography, eluting with complexing agents, or by extraction chromatography. A number of complexing agents have been used in the ion exchange method. Choppin *et al.* [210] found alpha hydroxy iso-butyrate (*a*-HIBA) to be the best complexing agent with a highly cross linked resin (Dowex 50-X12) at 87° with gradient elution. Smith and Hoffman [211] obtained equally good separations with a lower cross linked resin at room temperature, the average separation factor for adjacent lanthanide pairs being 2.0. Their procedure has been followed by most workers using ion exchange subsequently [212–214]. Many separations have been based on extraction chromatography. With HDEHP as extractant average separation factors of 2.5 have been obtained [215, 216]. The highest average separation factor (2.8) was obtained with 2-ethyl-hexylphenyl phosphonic acid (HEHϕP) [217]. In many systems yttrium appears with the heavy rare earths, close to holmium, which makes separation from the fission product lower lanthanides relatively simple.

Actinide Separations Based on Valency Changes

The elements uranium to americium, and berkelium, can exist in more than one oxidation state and the stability of the higher states decreases with atomic number, so that separations from other elements and from one another can be made by differential oxidation followed by co-precipitation, ion exchange or solvent extraction. However, the solution chemistry of these elements is complicated, e.g. Np and Pu may co-exist in several valency states, and hydrolysis and polymerisation takes place at low acidities [218]. The different valency states are carried more or less specifically by different precipitants, e.g. LaF_3 and $BiPO_4$ carry the 3- and 4-valent states while $ZrPO_4$ and zirconium phenylarsonate carry the 4-valent states. Co-precipitation of 4-valent plutonium on LaF_3 and $BiPO_4$ was used in the original extraction process [219] but the main use in analysis is for initial separations before using more selective methods. However, Sill and Williams [220, 221] have used a scheme for separating all the actinides which form 5- and 6-valent states by sequential oxidation and co-precipitation of the lower valency states on barium sulphate in presence of potassium sulphate. High recoveries and decontamination were obtained under carefully standardised conditions.

The higher oxidation states of the actinides are absorbed from concentrated HCl solutions on anion exchangers while the 3-valent states are unabsorbed. Selectivities increase in the order U < Np < Pu while thorium is not absorbed at all. An example of the use of this is the separation of U, Np and Pu [222, 223]. The mixture (Np and Pu being in the 4-valent state) was added to the anion exchange column in 9M HCl. Elution with 9M HCl removed the 'chloride non-unadsorbable' group; 9M HCl + ammonium iodide eluted Pu, after reduction to Pu(III), 4M HCl eluted Np and 0.5M HCl eluted U.

The stability of the complexes of the actinides increases in the order $MO_2^+ < M^{3+} < MO_2^{2+} < M^{4+}$ and in the same valency state with increase of atomic number:

$$Th \; < \; U \; < \; Np \; < \; Pu$$
$$0.5 \quad\quad -0.3 \quad\quad -0.34 \quad\quad\quad -0.85$$

The figures below the element symbols are the $pH_{50\%}$ values for extraction of the 4-valent states into TTA in benzene [100]. Moore has published methods based on TTA extraction for separating neptunium [224], berkelium [225] and americium [226]. Berkelium was oxidised to Bk(IV) with dichromate when, in the absence of sulphate, Np and Pu were oxidised to the non-extractable 6-valent states. The separation of Am depended on the non-extraction of Am(V) from an acetate buffer (pH 5) when all the other actinides and lanthanides were extracted.

The 4-valent actinides are also extracted from strongly acid solutions with HDEHP. Fardy and Chilton [227] extracted Pu(IV) from 6M HCl with HDEHP in diethyl benzene when the co-extraction of many elements, e.g. iron, the rare earths, was at a minimum. Co-extracted zirconium was removed by stripping the plutonium in the trivalent state, a reducing agent soluble in the organic phase being required. An interesting sequential method for separating a number of elements from urine samples was proposed by Peppard et al. [228]. The lanthanides and actinides were all extracted from the acidified sample with a solution of mono-ethylhexyl phosphoric acid (H_2MEHP). The 3-, 6- and 4-valent species were then sequentially re-extracted into the aqueous phase by adding increasing amounts of 2-ethyl hexanol to the organic phase. The higher alcohol acted as an anti-synergist.

Americium is the highest atomic number actinide capable of existing in the 6-valent state. Holcomb [229] found a rapid oxidation was possible with potassium perxenate (K_4XeO_6). The Am(VI) and Ce(III) were separated by adsorbing the latter on a specially prepared CaF_2 column.

Separation of the Trivalent Actinides from the Lanthanides

The trivalent actinides and the lanthanides cannot be separated from one another by the methods available for separations within the two groups, e.g. ion exchange with complexing agents, as the elution peaks overlap. For example, americium and promethium are eluted close together, their ionic radii being nearly identical. Street and Seaborg [230] separated americium and curium from the lanthanides by cation exchange from 13.5 M HCl, the actinides being more strongly complexed by chloride than the lanthanides and eluted earlier. Hulet et al. [231] used concentrated LiCl as eluent in an anion exchange system, the lanthanides being eluted first, in this case, while Moore and Mullins [232] used a similar system with a liquid anion exchanger. The trivalent actinides form more stable complexes with thiocyanate than the lanthanides and this has been used to separate them by anion exchange [233, 234] and by extraction with long chain quaternary ammonium salts [235, 236]. Weaver and Kappelman [237] extracted the lanthanides preferentially with HDEHP from a solution containing lactic acid and diethylenetriamine pentaacetic acid. The separation factor between the least extractable lanthanide (neodymium) and the most extractable actinide (californium) was ten. Separations by anion exchange using mixed solvents have been referred to on page 187.

Berkelium-249 decays mainly by emission of a low energy beta-particle and high decontamination factors are required [74]. The separation of berkelium from cerium is complicated by the similar values of their III—IV oxidation potentials. Moore has proposed several methods of separation based on anion exchange and extraction chromatography depending on the more stable complexes formed by Ce(IV) with nitrate [238]. High pressure liquid chromatography has been used to separate berkelium from all other elements except europium [74], TTA extraction being used for the subsequent separation of europium and berkelium.

Separation of the Trivalent Actinides from One Another [239]
This is a similar problem to that of separating the trivalent lanthanides. Cation exchange, eluting with citrate, was used in the discovery of berkelium and californium ([219], p. 155). More recently cation exchange with *a*-HIBA [210] and extraction chromatography with HDEHP have been used [240] the latter giving the higher separation factors [241].

5.9.5   Zirconium, Niobium and Hafnium
Nuclides of zirconium and niobium are important fission products and many methods have been used for separating them from other fission products and from one another. The classical precipitation techniques for separating zirconium used barium fluozirconate and either cupferron or mandelic acid [242]. Anion exchange from hydrochloric acid has been used for determinations in fallout [243] and in uranium alloys [244].

Zirconium-TTA chelate can be extracted from strongly acid solutions [100] and this has been used in its separation [245, 246]. The presence of hydrogen peroxide reduces the co-extraction of niobium in TTA extractions [245] and in extractions with acid phosphate esters [202, 247].

A highly selective method for separating zirconium from irradiated fuels, depending on extraction with N-benzoyl-N-phenylhydroxylamine (BPHA) in benzene from a neutral solution containing a number of masking agents, gave a decontamination factor for niobium of 10^6 and, with the exception of EDTA, many of the usual complexing agents did not interfere [248]. A rapid method of separating niobium from commercial radioactive solutions was used by Moore *et al.* [249] in which the fluoride complex of niobium was extracted from H_2SO_4/HF solution by di-isobutyl carbinol. Zirconium was not extracted and the method had the advantage that the fluoride ion outcomplexed any oxalate often present in such solutions.

5.9.6   Molybdenum and Technetium
Molybdenum-99 is formed in fission in high yield and is often used as a standard in the determination of the relative yields of other fission nuclides [250]. Molybdenum has been separated by solvent extraction, e.g. chloride and thiocyanate extractions with oxygenated solvents, but the most specific extractant is the chelate reagent, benzoin-*a*-oxime, e.g. in the

separation of molybdenum-99 from other fission products [251]. Technetium-99 has been separated from its molybdenum-99 parent by extraction of tetraphenylarsonium pertechnetate into chloroform [252] when the molybdenum is not extracted. Because the pertechnetate ion is carrier-free, and easily reduced by impurities present in the organic reagent, a holding oxidant (H_2O_2) should be present. An interesting ion exchange separation was achieved by passing a solution of molybdate and pertechnetate through an anion exchanger in the phosphate form when the molybdate was preferentially absorbed [253].

No stable isotope of technetium exists and rhenium is used as a non-isotopic carrier. Technetium and rhenium are both extracted in their highest valency states from alkaline solution by oxygen containing solvents, e.g. methyl ethyl ketone. By reduction of the technetium to the 4-valent state with hydrazine [254] or hydroxylamine [255] only rhenium is extracted.

5.9.7   Ruthenium

Ruthenium is of special importance in radiochemistry because two of its isotopes are fission products of fairly long half-life and high yield (^{103}Ru and ^{106}Ru) and its complicated chemistry created many problems in nuclear technology. Separations of ruthenium usually involve distillation or solvent extraction of the tetroxide, precipitation of RuO_2 after reduction by alcohol and precipitation of the metal by reduction with magnesium [256].

5.9.8   Bromine and Iodine

Many bromine and iodine nuclides are formed in fission. Iodine-131 is of special importance being one of the most hazardous fission products. Some of the shorter-lived bromine and iodine fission products are of interest as they are delayed neutron precursors. Separations of bromine and iodine are usually based on the use of the selective oxidation-reduction cycles of Glendenin [257] and distillation or solvent extraction. These oxidation-reduction cycles ensure complete interchange between the active species and added carrier and they can be carried out rapidly and repeated as many times as necessary to achieve the required degree of decontamination [258]. An apparatus which enables these separations to be carried out automatically, with some sacrifice of chemical yield, in about a minute has been described [259].

5.10   MANUAL AND AUTOMATIC MULTI-ELEMENT SEPARATIONS

5.10.1   Introduction

It is frequently useful to determine a number of elements in the same sample of material. Schemes have been devised for sequential separation of nuclides from irradiated material and from environmental samples and many such separations have been used for multi-element determination in activation analysis.

5.10.2   Fission Products

Several sequential schemes for separating fission products have depended mainly on classical precipitation techniques [260–262] Crouch and Cook [222] included ion exchange and solvent extraction steps and Boni [263] based a separation of fission and activation products on anion exchange from strong acids with some precipitation stages, both systems being capable of dealing with samples containing considerable amounts of extraneous matter, e.g. in biological samples, soils. de Wet and Crouch [264] included ion exchange, solvent extraction and extraction chromatography for carrier-free separations prior to mass spectrographic analysis.

Natsume *et al.* [265] separated groups of elements on a single cation exchanger with subsidiary cation and anion exchange columns for further separations. Very small amounts of carrier were used (10–100 μg) and the elements were either recovered quantitatively or the chemical yields determined polarographically. Herrmann *et al.* [266] used extraction columns (TBP and HDEHP) in a sequential scheme but evaporations stages were necessary between the different columns.

5.10.3 Separations in Activation Analysis

The methods used in separations for activation analysis depend on the elements of interest and their number, the matrix, the half-lives of the nuclides being determined and the counting procedures. A number of schemes are listed in Table 7.6 (page 286) and many can be found in the Proceedings of the Conferences on Modern Trends in Activation Analysis [267, 268].

Several separations use distillation as an initial step [269–271] while some are based entirely on ion exchange. Aubouin *et al.* [272] separated some 38 elements, most of them completely and the others into groups of two or three, using five anion exchange columns and two cation exchange columns. Ricq [273] modified this scheme using interconnecting columns for rapid separations. In both cases columns could be left out for dealing with more limited numbers of elements. Jervin and Wong [274] separated 40 elements into 14 groups on three columns (see Fig. 5.4). Forster and Schwabe [275] proposed a scheme for use in the activation analysis of trace impurities in highly pure materials, in which 60 cations were separated into groups by extraction from solutions of decreasing acidity with solutions of two chelating agents, diethylammonium-diethyldithiocarbamate and N-benzoyl-N-phenylhydroxylamine in chloroform. The groups were subdivided by re-extractions with different aqueous reagents.

In the analysis of lunar material Allen *et al.* [276] separated 39 elements into 12 groups in a series of precipitation and solvent extraction steps, carried out by three workers acting in a programmed co-operation. The order of separation was determined by the half-lives of the nuclides being measured and the last group was ready for counting seven hours after the end of the irradiation.

In spite of the high resolution of Ge(Li) detectors there is still an advantage in some prior chemical processing in order to reduce the complex spectra and limit dead-time losses in the multi-channel analyser due to high gross counts. Filby *et al.* [277] determined 32 elements in rocks by gamma spectrometry after removal of sodium and tantalum with HAP. For the determination of some 45 elements in rock samples Morrison *et al.* [278] separated them, after irradiation, into six groups before high resolution gamma spectrometry, the germanium solid state detector system including a coincidence-anticoincidence arrangement for Compton suppression. Arsenic and bromine were first removed from the prepared sample by distillation, sodium (and tantalum) by treatment with HAP, and the remaining elements divided into three groups by anion exchange from strong hydrochloric acid. The 'chloride non-adsorbable' group was divided into two by a TBP extraction. In a similar method used for dealing with biological samples [279] the TBP extraction step was replaced by one using cation exchange.

5.10.4   Automatic Separations

Automatic procedures have been devised for the separation of short-lived nuclides where a manual process would be too slow, or for the handling of a number of samples. A semi-automated ion exchange method of separating long-lived fission products was described by Mathers *et al.* [280] and automatic ion exchange equipment for separating fission products for analytical process control has been described [281]. Fourcy *et al.* [282] used automatic ion exchange in which a peristaltic pump distributed a series of eluting solutions to eight columns in parallel, the collection of fractions being synchronised with the change of eluents.

Samsahl [270, 283] described a system for activation analysis of biological materials in which a large number of elements were separated into groups by a distillation stage followed by selective sorption on a series of ion exchange, chelating resin and extraction chromatographic columns. The system, a development of work described in earlier publications, was designed to separate strongly radioactive elements singly, while the others were isolated into groups of short-lived and long-lived, high and low gamma energy nuclides giving maximum resolution with a simple sodium iodide gamma spectrometer. The flow of sample solution, and the reagents added between columns, was maintained by a single pump operating a series of syringes. The required flow rates were determined by the dimensions of the syringes and columns. This gave more precise and reproducible flows than was given by peristaltic pumps. The principle is illustrated in Fig. 5.9 which shows the separation of Hg, Sb, As and Se after their removal from the sample by a preliminary distillation [270]. The contents of the columns were counted without elution and with nine columns the separation time was only five minutes.

A phase-separating centrifuge for solvent extractions described by Vallis and Perkin [284, 285] is illustrated in Fig. 5.10. The liquids are mixed in a centrifuge cup and then separated by spinning. The top of the cup is an annular porous barrier which may be hydrophilic (sintered glass) or hydrophobic (PTFE). With the hydrophilic barrier the aqueous layer is ejected at low speed of rotation while the organic layer, which may be either lighter or heavier than the aqueous layer, requires a higher speed of rotation. In the case of a hydrophobic barrier the organic layer is ejected at the lower speed

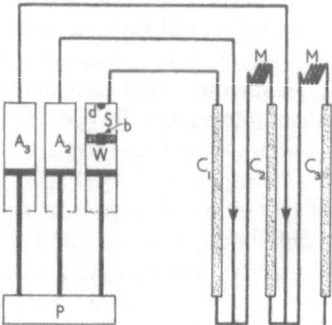

Fig. 5.9. Automatic ion exchange. Legend: C_1, C_2 and C_3 are three anion exchange columns (Dowex 2). The piston P operates three syringes containing sample and eluents. The sample, S, passes through column C_1 when mercury is retained. Wash liquid from W follows the sample when the bung, b, is forced out by the stop, d. The effluent from C_1 passes through column C_2 via a mixing coil, M, after being joined by eluent from A_2. C_2 removes antimony. The effluent from C_2 passes through C_3 mixed with eluent from A_3 and As and Se are retained. (After Samsahl, Wester and Langström [270])

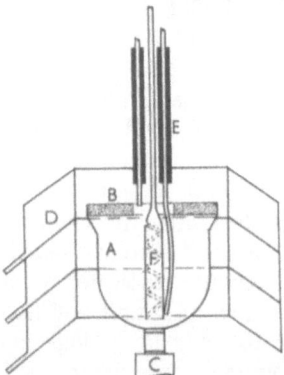

Fig. 5.10. Phase-separating centrifuge. Legend: A, glass centrifuge cup; B, porous barrier; C, centrifuge motor shaft; D, collectors; E, support for delivery tubes; F, stirrer. (After Sutton and Vallis [285])

of rotation. The stirrer and the tubes for introducing samples, reagents, etc. pass through the centre of the annular top. Fig. 5.11 shows a commercial unit incorporating the device. Another phase-separating device due to Steed and Trowell [286] was used for mixing and separating an aqueous layer from a more dense organic layer. The interface between the layers is sensed by the change in conductivity between two electrodes. Goode *et al.* [287] used this separator in an automatic equipment in which a neutron activated sample was extracted by a series of complexing agents, all in carbon tetrachloride. Reagents were added to the aqueous solution between extractions.

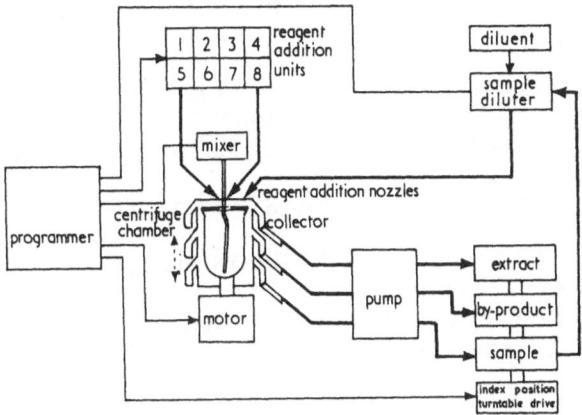

Fig. 5.11. Centrichem automatic separating equipment incorporating the Vallis and Perkin phase-separating centrifuge. (After Sutton and Vallis [285])

The whole operation was controlled by a programming unit using signals recorded on tape. Some 25 elements were separated into six groups for gamma spectrometry in about 1½ hours.

Comar and Le Poec [288] developed an automatic continuous system for separating iodine-128 from irradiated biological samples. This included sample taking, dilution with carrier, ion exchange, solvent extraction, distillation and counting. The samples were separated in the flow line by air bubbles but there was approximately 2% contamination between samples and those differing appreciably in activity could not be allowed to follow one another.

REFERENCES

1. M. HONDA and J. R. ARNOLD, *Geochim. et Cosmochim. Acta,* **23,** 219 (1961).
2. J. ROESMER, The Radiochemistry of Mercury. Nuclear Science Series NAS-NS-3026, U.S. National Academy of Sciences, (1970).
3. J. KORKISCH, Modern Methods for the Separation of the Rarer Metal Ions. Pergamon Press, New York, (1969).
4. H. J. M. BOWEN and D. GIBBONS, Radioactivation Analysis. Clarendon Press, Oxford, (1963).
5. ELINOR F. NORTON, Chemical Yield Determinations in Radiochemistry. Nuclear Science Series NAS-NS-3111, U.S. National Academy of Sciences, (1967).
6. R. G. MONK and J. HERRINGTON, *Anal. Chim. Acta,* **24,** 481 (1961).
7. E. B. DENECHAUD, P. A. HELMKE and L. H. HASKIN, *J. Radioan. Chem.,* **6,** 97 (1970).

8. N. A. BONNER and M. KAHN, Radioactivity Applied to Chemistry. (A. C. Wahl and N. A. Bonner, eds.) p. 102. John Wiley, New York, (1951).

9. R. T. OVERMAN and H. M. CLARK, Radioisotope Techniques. p. 352. McGraw Hill, New York, (1960).

10. G. A. WELFORD and E. L. CHIOTIS, *Anal. Chim. Acta*, **31**, 376 (1964).

11. A. I. NOVIKOV, T. M. ZAKREVSKAYA and G. K. RYANZANOVA, *Radiokhimiya*, **10**, 368 (1968).

12. D. E. TROUTNER, R. L. FERGUSON and G. D. O'KELLEY, *Phys. Rev.*, **130**, 1466 (1963).

13. F. H. FIRSCHING, Chelates in Analytical Chemistry. (H. A. Flaschka and A. J. Barnard, eds.) Vol. 2, p. 117. Marcel Dekker, New York, (1969).

14. H. V. WEISS, M. G. LAI and A. GILLESPIE, *Anal. Chim. Acta*, **25**, 550 (1961).

15. V. I. KUZNETSOV and V. V. GORSHKOV, *Radiokhimiya*, **5**, 93 (1963).

16. H. WEISZ, Microanalysis by the Ring Oven Technique. 2nd ed. Pergamon Press, Oxford, (1970).

17. D. A. HILTON and D. REED, *Analyst*, **89**, 599 (1964).

18. H. WEISZ and D. KLOCKOW, *Talanta*, **12**, 55 (1965).

19. F. LOLEY and H. MALISSA, *Anal. Chim. Acta*, **34**, 278 (1966).

20. G. HERRMANN and H. O. DENSCHLAG, *Ann. Revs. Nucl. Science*, **19**, 4 (1969).

21. B. D. ERDAL and A. C. WAHL, *J. Inorg. Nucl. Chem.*, **33**, 2763 (1971).

22. F. GIRARDI, R. PIETRA and E. SABBIONI, *J. Radioan. Chem.*, **5**, 141 (1970).

23. F. GIRARDI, G. GUZZI and G. DICOLA, *J. Radioan. Chem.*, **6**, 359 (1970).

24. H. W. KIRBY, *J. Inorg. Nucl. Chem.*, **25**, 481 (1963).

25. H. W. KIRBY, *J. Inorg. Nucl. Chem.*, **31**, 3375 (1969).

26. O. SAMUELSON, Ion Exchange Separations in Analytical Chemistry. John Wiley, New York, (1963).

27. W. RIEMAN and H. WALTON, Ion Exchange in Analytical Chemistry. Pergamon Press, Oxford, (1970).

28. K. A. KRAUS and F. NELSON, ASTM Special Technical Publication No. 195 (1958).

29. J. E. SALMON, *Pure and Applied Chem.*, **25**,(4) 707 (1971).

30. A. J. P. MARTIN and R. L. M. SYNGE, *Biochem. J.*, **35**, 1358 (1941).

31. S. W. MAYER and E. R. TOMPKINS, *J. Amer. Chem. Soc.*, **69**, 2866 (1947).

32. E. GLUECKAUF, *Trans. Farad. Soc.*, **51**, 34 (1955).

33. G. AUBOIN and J. LAVERLOCHERE, *J. Radioan. Chem.*, **1**, 123 (1968).

34. F. W. CORNISH, *Analyst*, **83**, 634 (1958).

35. P. J. KAROL, *Anal. Chem.*, **43**, 1383 (1971).

36. W. E. NERVIK, *J. Phys. Chem.*, **59**, 690 (1955).

37. F. MOLNAR, A. HORVATH and V. A. KHALKIN, *J. Chromatog.*, **26**, 215 (1967).

38. F. W. E. STRELOW, R. RETHEMEYER and C. J. C. BOTHMA, *Anal. Chem.*, **37**, 106 (1965).
39. J. P. SURLS and G. R. CHOPPIN, *J. Amer. Chem. Soc.*, **79**, 855 (1957).
40. K. A. KRAUS and F. NELSON, Proc. 1st Intern. Conf. Peaceful Uses At. Energy, **7**, 113 (1956).
41. F. NELSON, R. M. RUSH and K. A. KRAUS, *J. Amer. Chem. Soc.*, **82**, 339 (1960).
42. R. DJURFELDT and O. SAMUELSON, *Acta Chem. Scand.*, **4**, 165 (1950).
43. S. YOKOTSUKA, E. AKATSU and K. UENO, *J. Nucl. Sci. Tech. (Tokyo)*, **8**, 622 (1971).
44. F. NELSON, K. MURASE and K. A. KRAUS, *J. Chromatog.*, **13**, 503 (1964).
45. J. P. FARIS, *Anal. Chem.*, **32**, 521 (1960).
46. J. P. FARIS and R. F. BUCHANAN, *Anal. Chem.*, **36**, 1158 (1964).
47. F. W. E. STRELOW and C. J. C. BOTHMA, *Anal. Chem.*, **39**, 595 (1967).
48. L. DANIELLSON, *Acta, Chem. Scand.*, **19**, 670 (1965).
49. E. A. HUFF, *Anal. Chem.*, **36**, 1921 (1964).
50. T. ANDERSON and A. B. KNUTSEN, *Acta. Chem. Scand.*, **16**, 849 (1962).
51. P. van den WINKEL, F. de CORTE and J. HOSTE, *Anal. Chim. Acta*, **56**, 241 (1971).
52. F. NELSON and D. C. MICHELSON, *J. Chromatog.*, **25**, 414 (1966).
53. F. W. E. STRELOW, A. H. VICTOR, C. R. van ZYL and CYNTHIA ELOFF, *Anal. Chem.*, **43**, 870 (1971).
54. F. W. E. STRELOW, C. R. van ZYL and C. J. C. BOTHMA, *Anal. Chim. Acta*, **45**, 81 (1969).
55. W. J. MAECK, M. E. HUSSY and J. E. REIN, *Anal. Chem.*, **35**, 2087 (1963).
56. C. BIGLIOCIA, F. GIRARDI, J. PAULY, E. SABBIONI, S. MELONI and A. PROVASOLI, *Anal. Chem.*, **39**, 1635 (1967).
57. B. TREMILLON, *Z. anal. Chem.*, **236**, 472 (1968).
58. R. ROSSET, *Bull. Soc. Chim. France*, (1966) 59.
59. M. G. LAI and H. A. GOYA, U.S.N.R.D.L. 67-11 (1968).
60. S. B. SAVVIN, *J. Radioan. Chem.*, **2**, 369 (1969).
61. T. E. GREEN, S. L. LAW and W. J. CAMPBELL, *Anal. Chem.*, **42**, 1749 (1970).
62. K. A. KRAUS, H. O. PHILLIPS, T. A. CARLSON and J. S. JOHNSON, Proc. 2nd. Intern. Conf. Peaceful Uses At. Energy, **28**, 3 (1958).
63. C. B. AMPHLETT, Inorganic Ion Exchangers. Elsevier, New York, (1964).
64. M. J. FULLER, *Chromatog. Revs.*, **14**(1), 45 (1971).
65. M. QURESHI and J. P. RAWAT, *J. Inorg. Nucl. Chem.*, **30**, 305 (1968).
66. L. H. BAETSLE and D. HUYS, *J. Inorg. Nucl. Chem.*, **30**, 639 (1968).
67. F. GIRARDI and E. SABBIONI, *J. Radioan. Chem.*, **1**, 169 (1968).
68. J. KORKISCH, Progress in Nuclear Energy, Series IX 6, (H. A. Elion and D. C. Stewart, eds.) p. 1. Pergamon Press, Oxford, (1966).
69. J. KORKISCH and G. ARRHENIUS, *Anal. Chem.*, **36**, 850 (1964).

70. R. J. MORROW, *Talanta,* **13,** 1265 (1966).
71. C. D. SCOTT, *Anal. Biochem.,* **24,** 292 (1968).
72. D. O. CAMPBELL and S. R. BUXTON, U.S.A.E.A. Report ORNL-TM-1876 (1967).
73. D. O. CAMPBELL and S. R. BUXTON, *Ind. Eng. Chem., Process Design and Development,* **9,** 89 (1970).
74. L. G. FARRAR, J. H. COOPER and F. L. MOORE, *Anal. Chem.,* **40,** 1602 (1968).
75. W. ECKHART, G. HERRMANN and H. D. SCHÜSSLER, *Z. anal. Chem.,* **226,** 71 (1967).
76. R. KURODA and K. OGUMA, *Anal. Chem.,* **39,** 1003 (1967).
77. CHUNG-WAI TANG and C. J. MALETSKOS, *Science,* **167,** 52 (1970).
78. M. HEURTEBISE and W. J. ROSS, *Anal. Chem.,* **43,** 1438 (1971).
79. J. R. DeVOE, C. K. KIM and W. W. MEINKE, *Talanta,* **3,** 298 (1960).
80. W. B. SILKER, *Anal. Chem.,* **33,** 233 (1961).
81. I. H. QURESHI and F. I. NAGI, *J. Inorg. Nucl. Chem.,* **29,** 2879 (1967).
82. R. R. RUCH, *Anal. Chim. Acta,* **47,** 425 (1969).
83. M. LEDERER and C. MAJANI, *Chromatog. Reviews,* **12,** 239 (1970).
84. S. K. SHUKLA, *J. Chromatog.,* **21,** 92 (1966).
85. M. LEDERER, *Chromatog. Reviews,* **9,** 115 (1967).
86. H. PERSCHKE, M. B. A. CRESPI and G. B. COOK, *Intern. J. Appl. Radiation Isotopes,* **20,** 813 (1969).
87. R. C. KOCH and G. L. GRANDY, *Anal. Chem.,* **33,** 43 (1961).
88. D. W. OCKENDEN and R. H. TOMLINSON, *Can. J. Chem.,* **40,** 1594 (1962).
89. H. O. DENSHLAG and A. A. GORDUS, *Z. anal. Chem.,* **226,** 62 (1967).
90. B. I. ANVAER and Y. S. DRUGOV, *Zh. Analit. Khim.,* **26,** 1180 (1971).
91. J. TADMOR, *J. Gas Chromatog.,* **2,** 385 (1964).
92. G. P. MORIE and T. S. SWEET, *Anal. Chem.,* **37,** 1552 (1965).
93. R. E. SIEVERS, J. W. CONNOLLY and W. D. ROSS, *J. Gas Chromatog.,* **5,** 241 (1967).
94. T. S. ZVAROVA and I. ZVARA, *J. Chromatog.,* **44,** 604 (1969).
95. T. S. ZVAROVA and I. ZVARA, *J. Chromatog.,* **49,** 290 (1970).
96. G. H. MORRISON and H. FREISER, Solvent Extraction in Analytical Chemistry. John Wiley, New York, (1957).
97. A. K. DE, S. M. KHOPKAR and R. A. CHALMERS, Solvent Extraction of Metals. Van Nostrand, London, (1970).
98. G. H. MORRISON and H. FREISER, *Anal. Chem.,* **36,** 93R (1964).
99. J. STARÝ, The Solvent Extraction of Metal Chelates. Pergamon Press, Oxford, (1964).
100. A. M. POSKANZER and B. M. FOREMAN, *J. Inorg. Nucl. Chem.,* **16,** 323 (1961).
101. D. F. PEPPARD, G. W. MASON, J. L. MAIER and W. J. DRISCOLL, *J. Inorg. Nucl. Chem.,* **4,** 334 (1957).
102. D. F. PEPPARD, Advances in Inorganic Chemistry and Radiochemistry. (H. J. Emeleus and A. G. Sharp, eds.) Vol. 9, p. 25. Academic Press, New York, (1966).

103. D. F. PEPPARD, G. W. MASON and S. McCARTY, *J. Inorg. Nucl. Chem.*, **13**, 138 (1960).
104. D. C. STEWART and H. W. CRANDALL, *J. Amer. Chem. Soc.*, **73**, 1377 (1951).
105. I. H. QURESHI, L. T. MACLENDON and P. D. La FLEUR, *Radiochim. Acta*, **12**, 107 (1969).
106. J. C. WARF, *J. Amer. Chem. Soc.*, **71**, 3257 (1949).
107. F. S. MARTIN and R. J. W. HOLT, *Quarterly Reviews, Chem. Soc.*, **13**, 327 (1959).
108. J. C. WHITE and W. J. ROSS, Separations by Solvent Extraction with Tri-n-Octyl Phosphine Oxide. Nuclear Science Series NAS-NS-3102, U.S. National Academy of Sciences, (1961).
109. T. ISHIMORI and E. NAKAMURA, Data of Inorganic Solvent Extraction, Part 1, JAERI-1047 (1963).
110. T. ISHIMORI, E. AKATSU, WEN-PIN CHENG, K. TSUKEUCHI and T. OSAKABE, Data of Inorganic Solvent Extraction, Part 2, JAERI-1062 (1964).
111. T. ISHIMORI, E. AKATSU, K. TSUKUECHI, T. KOBUNE, Y. USUBA, K. KIMURA, G. ONAWA and H. UCHIGAMA, Data of Inorganic Solvent Extraction, Part 3, JAERI-1106 (1966).
112. F. L. MOORE, Liquid-liquid Extraction with High Molecular Weight Amines. Nuclear Science Series NAS-NS-3101 (1960).
113. W. J. MAECK, G. L. BOOMAN, M. E. KUSSY and J. E. REIN, *Anal. Chem.*, **33**, 1775 (1961).
114. C. A. BLAKE, C. F. BAES, K. B. BROWN, C. F. COLEMAN and J. C. WHITE, Proc. 2nd Intern. Conf. Peaceful Uses At. Energy, **28**, 289 (1958).
115. J. G. CUNINGHAME, P. SCARGILL and H. H. WILLIS, AERE Report C/M 215 (1956).
116. H. M. N. H. IRVING, Solvent Extraction Chemistry. (D. Dyrssen, J. O. Liljenzin and J. Rydberg, eds.) p. 91. North Holland, Amsterdam, (1967).
117. I. P. ALIMARIN and Y. A. ZOLATOV, *Talanta*, **9**, 891 (1962).
118. R. A. TOURNIER and M. W. DAVIS, *Separation Science*, **7**, 159 (1972).
119. L. C. CRAIG and O. POST, *Anal. Chem.*, **21**, 500 (1949).
120. Y. ITO and R. L. BOWMAN, *J. Chromatog. Science*, **8**, 315 (1970).
121. E. CERRAI and G. GHERSINI, Advances in Chromatography. (J. Calvin Giddings and R. A. Keller, eds.) Vol. 9, p. 3. Marcel Dekker, New York, (1970).
122. S. SIEKIERSKI and R. J. SOCHACKA, *J. Chromatog.*, **16**, 374 (1964).
123. E. P. HORWITZ, C. A. A. BLOOMQUIST and D. J. HENDERSON, *J. Inorg. Nucl. Chem.*, **31**, 1149, 3255 (1969).
124. E. CERRAI and C. TESTA, *J. Chromatog.*, **7**, 112 (1962).
125. K. KIMURA, *Bull. Chem. Soc. Japan*, **34**, 63 (1960).
126. T. B. PIERCE, P. F. PECK and R. S. HOBBS, *J. Chromatog.*, **12**, 81 (1963).
127. R. J. MEYER, R. D. OLDHAM and R. P. LARSEN, *Anal. Chem.*, **36**, 1975 (1964).
128. A. D. WESTLAND and F. E. BEAMISH, *Anal. Chem.*, **26**, 739 (1954).
129. B. SJÖSTRAND, *Anal. Chem.*, **36**, 814 (1964).

130. A. E. GREENDALE and D. L. LOVE, *Anal. Chem.*, **35**, 632 (1963).
131. L. TOMLINSON and M. H. HURDUS, *J. Inorg. Nucl. Chem.*, **30**, 1995 (1968).
132. V. JOKL, *J. Chromatog.*, **13**, 451 (1964).
133. V. JOKL, M. UNDEUTSCH and J. MAJER, *J. Chromatog.*, **26**, 208 (1967).
134. J. STARÝ, *Talanta*, **13**, 421 (1966).
135. R A. BAILEY and L. YAFFE, *Chromatog. Revs.*, **3**, 159 (1961).
136. D. I. RYABCHIKOV, E. K. KORCHEMNAYA and V. I. NAUMOVA, *Zh. Analit. Khim.*, **23**, 741 (1968).
137. J. P. ADLOFF and R. BERTRAND, *J. Electroan, Chem.*, **5**, 461 (1963).
138. D. GROSS, *J. Chromatog.*, **10**, 221 (1963).
139. K. AITZETMÜLLER, K. BUCHTELA, F. GRASS and F. HECHT, *Mikrochim. Acta*, (1966) 1101.
140. F. GRASS and R. KITTLE, *Mikrochim. Acta*, (1971) 371.
141. T. R. SATO, W. P. NORRIS and H. H. STRAIN, *Anal. Chem.*, **26**, 267 (1954).
142. Z. PUČAR and Z. KONRAD-JAKOVAC, *J. Chromatog.*, **9**, 106 (1962).
143. E. SCHUMACHER, *Helv. Chim. Acta*, **40**, 221 (1957).
144. E. SCHUMACHER and W. FRIEDLI, *Helv. Chim. Acta.*, **43**, 1706 (1960).
145. A. K. BREWER, S. L. MADORSKY, J. K. TAYLOR, V. H. DIBELER, P. BRADT, O. L. PARHAM, R. J. BRITTEN and J. G. REID, *J. Res. Nat. Bur. Stand.*, **38**, 137 (1947).
146. D. BEHNE, *Kerntechnik*, **12**, 112 (1970).
147. D. BEHNE, *Radiochem. Radioan. Lett.*, **6**, 39 (1971).
148. R. A. ALLEN, D. B. SMITH, R. L. OTLET and D. S. RAWSON, *Nucl. Instr. Meth.*, **45**, 1 (1966).
149. P. METSON, *Analyst*, **94**, 1122 (1969).
150. L. YAFFE, *Ann. Revs. Nucl. Science*, **12**, 153 (1962).
151. A. E. GREENDALE and D. L. LOVE, *Anal. Chem.*, **32**, 780 (1960).
152. A. BESSON, Y. PRIGENT and F. VAN-KOTE, CEA-R-3788 (1969).
153. H. W. MILLER and R. J. BROWNS, *Anal. Chem.*, **24**, 536 (1952).
154. R. F. MITCHELL, *Anal. Chem.*, **32**, 326 (1960).
155. M. Y. DONNAN and E. K. DUKES, *Anal. Chem.*, **36**, 392 (1964).
156. W. PARKER and R. FALK, *Nucl. Instr. Meth.*, **16**, 355 (1962).
157. S. C. BLACK, *Health Phys.*, **7**, 87 (1961).
158. Y. UJIHIRA and J. C. ROY, *Can. J. Chem.*, **46**, 1221 (1968).
159. H. T. MILLARD, *Anal. Chem.*, **35**, 1017 (1963).
160. J. MILLER, *J. Radioan. Chem.*, **4**, 35 (1970).
161. K. W. PUPHAL and D. R. OLSEN, *Anal. Chem.*, **44**, 284 (1972).
162. K. A. KRAUS, G. E. MOORE and F. NELSON, *J. Amer. Chem. Soc.*, **78**, 2692 (1956).
163. J. N. ROSHOLT, B. R. DOE and M. TATSUMOTO, *Geol. Soc. of America Bulletin*, **77**, 987 (1966).
164. TEH-LUNG KU, *J. Geophys. Res.*, **73**, 2271 (1968).
165. R. B. JACOBI, *J. Chem. Soc.*, (1949) S 314.
166. R. D. EVANS, *Rev. Sci. Inst.*, **4**, 223 (1933).
167. P. E. DAMON and H. I. HYDE, *Rev. Sci. Inst.*, **23**, 766 (1952).
168. H. F. LUCAS, *Rev. Sci. Instr.*, **28**, 680 (1957).

169. A. A. POMANSKI, S. A. SEVERNIJI and E. P. TRIFONOVA, *Soviet J. At. Eng.*, **27**, 36 (1969).
170. C. GIFFIN, A. KAUFMAN and W. BROECKER, *J. Geophys. Res.*, **68**, 1749 (1963).
171. D. HEYE, *Geochim. et Cosmochim. Acta*, **34**, 389 (1970).
172. J. O. JOHNSON, Geol. Survey Water-Supply Paper 1696-G U.S. Printing Office, Washington, (1971).
173. W. H. HENRY and B. A. LOVERIDGE, A.E.R.E.-R 3795 (1961).
174. R. B. HOLTZMAN, *Health Phys.*, **9**, 385 (1963).
175. H. G. PETROW, A. COVER, W. SCHIESLE and E. PARSONS, *Anal. Chem.*, **36**, 1601 (1964).
176. E. J. BARATTA and A. C. HERRINGTON, U.S. At. Energy Comm. Report, WIN-118 (1960).
177. J. N. ROSHOLT, *Anal. Chem.*, **29**, 1398 (1961).
178. A. S. GOLDIN, *Anal. Chem.*, **33**, 406 (1961).
179. Report of the Government Chemist. p. 153. H.M. Stationery Office, London, (1968).
180. K. A. SMITH and E. R. MERCER, *J. Radioan. Chem.*, **5**, 303 (1970).
181. H. V. MOYER, U.S. At. Energy Comm. Report, TID-5221 (1956).
182. T. ISHIMORI, *Bull. Chem. Soc. Japan*, **28**, 432 (1955).
183. C. T. BISHOP, U.S. At. Energy Comm. Report, MLM-1721 (1970).
184. R. L. BLANCHARD, *Anal. Chem.*, **38**, 189 (1966).
185. N. YAMAGATA, *Bull. Inst. Publ. Health (Tokyo)* **14**, 59 (1965).
186. F. NELSON, D. C. MICHELSON, H. O. PHILLIPS and K. A. KRAUS, *J. Chromatog.*, **20**, 107 (1965).
187. J. R. VAN SMIT, *Nature*, **181**, 1530 (1958).
188. T. R. FOLSOM and S. C. SREEKUMARAN, Reference Methods for Marine Radioactivity Studies. p. 129, IAEA Tech. Reports Series No. 118, Vienna, (1970).
189. A. C. LEAF, U.S. At. Energy Comm., Rept. HW-72199 (1962).
190. C. CESARANO, G. PUGNETTI and C. TESTA, *J. Chromatog.*, **19**, 589 (1965).
191. W. J. ROSS and J. C. WHITE, *Anal. Chem.*, **36**, 1998 (1964).
192. H. H. WILLARD and E. W. GOODSPEED, *Ind. Eng. Chem., Anal. Ed.*, **8**, 414 (1936).
193. E. J. BARRATA and F. E. KNOWLES, *Anal. Chem.*, **43**, 1138 (1971).
194. D. N. SUNDERMAN and W. W. MEINKE, *Anal. Chem.*, **29**, 1578 (1957).
195. H. A. C. MONTGOMERY, *Analyst*, **85**, 524 (1960).
196. F. W. E. STRELOW, C. R. van ZYL and C. R. NOLTE, *Anal. Chim. Acta*, **40**, 145 (1968).
197. M. STOEPPLER, *Z. anal. Chem.*, **250**, 237 (1970).
198. G. DUYCKAERTS and R. LEJEUNE, *J. Chromatog.*, **3**, 58 (1960).
199. F. NELSON, *J. Chromatog.*, **16**, 403 (1964).
200. A. S. GOLDIN, R. J. VELTEN and G. W. FISHKORN, *Anal. Chem.*, **31**, 1490 (1959).
201. A. S. GOLDIN and R. J. VELTEN, *Anal. Chem.*, **33**, 149 (1961).
202. T. V. HEALY, *Radiochim. Acta*, **2**, 52 (1963).
203. H. G. PETROW, *Anal. Chem.*, **37**, 584 (1965).
204. J. J. KATZ and G. T. SEABORG, The Chemistry of the Actinide Elements. Methuen, London, (1957).

205. C. F. METZ and G. R. WATERBURY, Treatise on Analytical Chemistry. (I. M. Kolthoff and P. J. Elving, eds.) Vol. 9 Part II, p. 189. Interscience, New York London, (1962).
206. H. G. PETROW, *Anal. Chem.*, **26**, 1514 (1954).
207. D. MACDONALD, *Anal. Chem.*, **33**, 1807 (1961).
208. F. R. LAWLESS and M. A. WAHLGREN, *J. Radioan. Chem.*, **5**, 11 (1970).
209. L. N. MOSKVIN, *Radiokhimiya*, **5**, 747 (1963).
210. G. R. CHOPPIN, B. G. HARVEY and S. G. THOMPSON, *J. Inorg. Nucl. Chem.*, **2**, 66 (1956).
211. H. L. SMITH and D. C. HOFFMAN, *J. Inorg. Nucl. Chem.*, **3**, 243 (1956).
212. K. WOLFSBERG, *Anal. Chem.*, **34**, 519 (1962).
213. L. WISH and S. C. FOTI, *J. Chromatog.*, **20**, 2927 (1965).
214. M. THEIN, M. N. RAO and P. K. KURODA, *J. Inorg. Nucl. Chem.*, **30**, 1145 (1968).
215. J. W. WINCHESTER, *J. Chromatog.*, **10**, 502 (1963).
216. T. B. PIERCE and P. F. PECK, *Nature*, **195**, 597 (1962).
217. I. FIDELIS and S. SIEKIERSKI, *J. Chromatog.*, **17**, 542 (1965).
218. F. L. MOORE and J. E. HUDGENS, *Anal. Chem.*, **29**, 1767 (1957).
219. G. T. SEABORG, The Transuranium Elements. Methuen, London, (1958).
220. C. W. SILL and R. L. WILLIAMS, *Health Physics*, **17**, 89 (1969).
221. C. W. SILL and R. L. WILLIAMS, *Anal. Chem.*, **41**, 1625 (1969).
222. E. A. C. CROUCH and G. B. COOK, *J. Inorg. Nucl. Chem.*, **2**, 223 (1956).
223. F. NELSON, D. C. MICHELSON and J. H. HOLOWAY, *J. Chromatog.*, **14**, 258 (1964).
224. F. L. MOORE, *Anal. Chem.*, **29**, 1941 (1957).
225. F. L. MOORE, *Anal. Chem.*, **38**, 1872 (1966).
226. J. R. STOKELY and F. L. MOORE, *Anal. Chem.*, **39**, 994 (1967).
227. J. J. FARDY and J. M. CHILTON, *J. Inorg. Nucl. Chem.*, **33**, 3247 (1969).
228. D. F. PEPPARD, G. W. MASON and C. M. ANDREJASICH, *J. Inorg. Nucl. Chem.*, **25**, 1175 (1963).
229. H. P. HOLCOMB, *Anal. Chem.*, **37**, 415 (1965).
230. K. STREET and G. T. SEABORG, *J. Amer. Chem. Soc.*, **72**, 2790 (1950).
231. E. K. HULET, R. G. GUTMACHER and M. S. COOPS, *J. Inorg. Nucl. Chem.*, **17**, 350 (1961).
232. F. L. MOORE and W. T. MULLINS, *Anal. Chem.*, **37**, 419 (1965).
233. J. P. SURLS and G. R. CHOPPIN, *J. Inorg. Nucl. Chem.*, **4**, 62 (1957).
234. J. S. COLEMAN, L. B. ASPREY and R. C. CHISHOLM, *J. Inorg. Nucl. Chem.*, **31**, 1167 (1969).
235. F. L. MOORE, *Anal. Chem.*, **36**, 2158 (1964).
236. E. A. HUFF, *J. Chromatog.*, **27**, 229 (1967).
237. B. WEAVER and F. A. KAPPELMAN, *J. Inorg. Nucl. Chem.*, **30**, 263 (1968).
238. M. T. KELLEY, *Pure and Applied Chem.*, **26**, 65 (1971).
239. J. STARÝ, *Talanta*, **13**, 421 (1966).
240. F. L. MOORE and A. JURRIAANSE, *Anal. Chem.*, **39**, 733 (1967).

241. K. A. GAVRILOV, E. G. VUZDZ, J. STARY and W. T. SENG, *Talanta*, 13, 471 (1966).
242. E. P. STEINBERG, The Radiochemistry of Zirconium and Hafnium. Nuclear Science Series NAS-NS-3011, U.S. Academy of Sciences, Washington, (1960).
243. G. A. WELFORD, W. R. COLLINS, R. S. MORSE and D. C. SUTTON, *Talanta*, 5, 168 (1960).
244. J. O. KARTTUNEN, *Anal. Chem.*, 35, 1044 (1963).
245. F. L. MOORE, *Anal. Chem.*, 28, 997 (1956).
246. W. J. MAECK, S. F. MARSH and J. E. REIN, *Anal. Chem.*, 35, 292 (1963).
247. J. KRTIL and A. MORAVEK, *Coll. Czech. Chem. Comm.*, 35, 2915 (1970).
248. R. VILLARREAL, J. O. YOUNG and J. R. KRSUL, *Anal. Chem.*, 42, 1419 (1970).
249. A. JURRIAANSE and F. L. MOORE, *Anal. Chem.*, 38, 964 (1966).
250. R. J. AMANI, Proc. I.A.E.A. Symposium on Standardisation of Radio-nuclides. p. 613 Vienna (1967).
251. L. WISH, *Anal. Chem.*, 34, 625 (1962).
252. S. TRIBALAT, *Anal. Chim. Acta*, 3, 113 (1949).
253. D. I. RYABCHIKOV and L. V. BORISOVA, *Zh. Analit. Khim.*, 13, 155 (1958).
254. A. A. POZDYNAKOV and B. Y. SPIVAKOV: quoted in Analytical Chemistry of Technetium, Promethium, Astatine and Francium. A. K. Lavrukhina and A. A. Pozdynakov, Ann. Arbor-Humphrey, London, (1970).
255. A. D. MATTHEWS and J. P. RILEY, *Anal. Chim. Acta*, 51, 455 (1970).
256. E. I. WYATT and R. R. RICKARD, The Radiochemistry of Ruthenium. Nuclear Science Series NAS-NS-3029, U.S. National Academy of Sciences, Washington, (1961).
257. J. KLEINBERG and G. A. COWAN, The Radiochemistry of Fluorine, Chlorine, Bromine and Iodine. Nuclear Science Series NAS-NS-3005, U.S. National Academy of Sciences, Washington, (1960).
258. H. R. VON GUTEN, K. F. FLYNN and L. E. GLENDENIN, *J. Inorg. Nucl. Chem.*, 31, 3357 (1969).
259. P. DEL MARMOL, P. FETTWEIS and D. C. PERRICOS, *Radiochim. Acta*, 16, 4 (1971).
260. E. L. GEIGER, *Anal. Chem.*, 31, 806 (1959).
261. R. G. D. OSMOND, T. W. EVETT, J. W. ARDEN, M. B. LOVETT and B. SWEENY, U.K. Atomic Energy Authority Report AM-84, Harwell, (1961).
262. R. WOOD and L. A. RICHARDS, *Analyst*, 90, 606 (1965).
263. A. L. BONI, *Anal. Chem.*, 32, 599 (1960).
264. W. J. DE WET and E. A. C. CROUCH, *J. Inorg. Nucl. Chem.*, 27, 1735 (1965).
265. H. NATSUME, H. UMEZAWA, T. SAZUKI, F. ICHIKAWA, T. SATO, S. BABA and H. AMANO, *J. Radioan. Chem.*, 7, 189 (1971).
266. G. HERRMANN, R. DENIG and N. TRAUTMAN, *J. Radioan. Chem.*, 6, 331 (1970).
267. Proceedings of the International Conference on Modern Trends in Activation Analysis, Texas A and M University, College Station, (1965).

268. Proceedings of the International Conference on Modern Trends in Activation Analysis, NBS Special Publication 312 (1969).
269. W. J. ROSS, *Anal. Chem.*, **36**, 1114 (1964).
270. K. SAMSAHL, P. O. WESTER and O. LANGSTRÖM, *Anal. Chem.*, **39**, 1480 (1967).
271. F. HERMANN, W. KIESL. F. KLUGER and F. HECHT, *Mikrochim. Acta,* (1971) 225.
272. G. AUBOUIN, J. DIEBOLT, E. JUNOD and J. LAVERLOCHERE, Proceedings of the International Conference on Modern Trends in Activation Analysis, Texas A and M University, College Station, (1965).
273. J. C. RICQ, *J. Radioan. Chem.*, **1**, 443 (1968).
274. R. E. JERVIS and K. Y. WONG, Proceedings I.A.E.A. Symposium on Nuclear Activation in the Life Sciences, p. 137. Vienna, (1967).
275. H. FORSTER and K. SCHWABE, *Anal. Chim. Acta*, **45**, 511 (1969).
276. R. O. ALLEN, L. A. HASKIN, M. R. ANDERSON and O. MUELLER, *J. Radioan. Chem.*, **6**, 115 (1970).
277. R. H. FILBY, W. A. HALLER and K. R. SHAP, *J. Radioan. Chem.*, **5**, 277 (1970).
278. G. H. MORRISON, J. T. GERARD, A. TRAVESI, R. L. CURRIE, S. F. PETERSON and N. M. POTTER, *Anal. Chem.*, **41**, 1633 (1969).
279. G. H. MORRISON and N. H. POTTER, *Anal. Chem.*, **44**, 839 (1972).
280. W. G. MATHERS and C. W. K. HOELKE, *Anal. Chem.*, **35**, 2064 (1963).
281. O. BOBLETER, H. P. FORSTER and A. G. REUSCHEL, 4th Int. Conf. Peaceful Uses of Atomic Energy, A/Conf. 49/P-211 (1971).
282. A. FOURCY, M. NEUBERGER, G. GARREC, A. FER and J. P. GARREC, Proc. of the International Conference on Modern Trends in Activation Analysis, N.B.S. Special Publication 312 (1969).
283. K. SAMSAHL, Aktiebologet Atomenergi Report AE-389 (1970).
284. D. G. VALLIS and J. L. PERKIN, *J. Inorg. Nucl. Chem.*, **22**, 1 (1961).
285. D. W. SUTTON and D. G. VALLIS, *J. Radioan. Chem.*, **2**, 377 (1969).
286. K. C. STEED and F. TROWELL, U.K. Patent Appl. 17329/67.
287. G. C. GOODE, C. W. BAKER and N. M. BROOKE, *Analyst*, **94**, 728 (1969).
288. D. COMAR and C. LE POEC, Proc. Int. Conference on Modern Trends in Activation Analysis, p. 351. Texas A and M University, College Station, (1965).

Chapter 6

Radiochromatography and Radioelectrophoresis

W. G. Duncombe and P. Johnson

The Wellcome Research Laboratories
Langley Court, Beckenham, Kent

6.1 INTRODUCTION

The application of chromatographic principles and the use of radionuclides in combination have produced a wide series of analytical techniques which have led to major advances in many fields of scientific endeavour.

The research worker has open to him various choices in two main areas of methodology; on the one hand, the choice of the type of chromatography or electrophoresis which is to be used in separating the radioactive components of a mixture and, on the other hand, the choice of suitable systems for detecting and measuring the radioactivity present in those components.

Throughout this chapter we have assumed knowledge by the reader of the general practice of various types of chromatography and electrophoresis, and have emphasized only those aspects which are particularly relevant to the use of radioactive compounds.

In selecting apparatus for the measurement of chromatographically or electrophoretically separated radioactive substances the choice of detection

TABLE 6.1. *Methods of Detection of β-Particles*

Type of system	Typical background (cpm)	Approximate efficiency (%)	
		^{3}H	^{14}C
A. *General Qualitative Methods*			
Autoradiography			
Spark chamber			
B. *Counting Systems*			
G.M. end window	10	0	10
Windowless G.M.	4	2	10–15
Glass spiral + CaF$_3$ (Eu) crystals for organic solutions	85*–180	1.5	18 (liquid)
Plastic scintillator disc cell	90*		5 (liquid) 40* (CO$_2$)
Ionization chamber	300	75–90	50
Plastic scintillator tubing	50	0.06* 2 (water)	46* (liquid) 60–80 (CO$_2$)
Powdered anthracene	40–50	10	65–80
Liquid scintillator	50 100–200*	20–40	80–95
Gas-flow proportional counter	Lead shield 20	70	90
	Scintillator shield 2	65	95

*With single PM tube.

system will depend not only on the type of nuclide to be detected, the efficiency of the detector for that nuclide and the background response, but also on the particular needs of the operator and the compatibility of the radioactivity detector with the chosen chromatographic system.

Various methods of detecting β-particles are listed in Table 6.1, as it is with β-emitting nuclides that most chromatographers are concerned and where problems of detection are often considerable. Detectors exclusively for γ-emitting nuclides, such as solid crystal scintillator counters, have therefore been omitted.

Table 6.1 shows both qualitative and quantitative types of detection system. Because the use of carbon-14 and tritium greatly outweighs that of all other nuclides, particularly in the biomedical field, orders of efficiency are given for measuring these in various counting systems. Some of the systems will be seen to give very poor counting efficiency, particularly for tritium, but the research worker is often faced with the use of a relatively inefficient counting system as the most convenient, sometimes the only, means of detecting radioactivity on chromatographic or electrophoretic media.

Many of these detection systems, both quantitative and qualitative, will be referred to throughout the text as appropriate to the chromatographic system under consideration. Although the many different types of chromatography and electrophoresis bring special problems or advantages which must be taken into account in the detection and measurement of radioactivity, in many practical situations paper chromatograms, electrophoresis papers, flexible thin-layer sheets and impregnated glass fibre papers may be considered as identical media.

6.2 DETECTION AND MEASUREMENT OF RADIOACTIVITY AFTER CHROMATOGRAPHY AND ELECTROPHORESIS ON THIN MEDIA

Radioactivity on solid media can be detected by both of the general methods A and B referred to in Table 6.1, although not all of the instrumental methods listed under B are applicable. This section can be further classified into methods in which the medium may remain intact or be subdivided. In the former the complete chromatogram is typically scanned by moving it beneath a radiation detector; in the latter, sections of paper or scrapings from thin-layer plates may be counted directly, e.g. in a heterogeneous liquid scintillation system, or the radioactive material may be extracted from the medium before counting.

Each method has its own advantages and disadvantages for specific applications. No single technique is ideal and the most appropriate one must be chosen in the light of experimental requirements. Some of the relevant factors will be discussed later, but it is necessary first to discuss the peculiar features of chromatography of radioactive substances.

6.2.1 Special Factors Affecting Chromatography and Electrophoresis of Radioactive Materials

Although in principle any chromatographic or electrophoretic technique may be used for radioactive samples, certain features require particular attention to avoid misleading results. These may be related either to the intrinsic radioactivity of the sample or to various chromatographic and chemical effects.

The radioactivity of a compound can give rise to a number of effects causing errors of interpretation. For practical chromatographic purposes these are of two main types: (1) primary radioactive disintegration of atoms in labelled molecules; and (2) reactions in which labelled molecules interact with nuclear radiation produced by the sample, or with free radicals or other excited species produced by the radiation.

The first type is unimportant with most nuclides of biological interest if the labelled molecule contains one radioactive atom only. When this decays the only products are non-radioactive. If the molecule contains more than one radioactive atom, however, then decay of one of these will result in one or more radioactive fragments chemically different from the original molecule. This effect becomes more important with increasing specific activity and molecular size and with decreasing half-life. Thus macro-molecules of high specific activity can give rise to a high proportion of radioactive impurities. A rather similar effect is produced with molecules labelled with a nuclide that decays to a radioactive daughter (e.g. scandium-47 from calcium-47). A very clear account of such primary decay effects is given by Evans [1].

In the second type of radiation effect the chemical decomposition of a labelled molecule is caused by the energy associated with the radioactive disintegration of atoms in other molecules and leads to the appearance of radioactive impurities. This is of particular importance with high specific activity tritium-labelled compounds and can cause serious trouble in the storage of such compounds. This possibility should also be considered when purifying such materials by preparative TLC, when radioactive impurities may arise on the plate in the 'purified' sample during autoradiographic exposure. Catch [2] and Evans [1] give very good accounts of these effects and of methods for reducing decomposition during storage.

Differences in behaviour of two compounds identical except for isotopic composition can also occur. Such isotope effects may be primary, resulting from formation or breaking of bonds involving the different isotopes, or secondary, when such bonds are not directly involved. The former are of interest in studying chemical reactions, while the latter are of the type which might be observed in chromatographic separations. Such effects have indeed been reported in the chromatography of compounds containing hydrogen or carbon isotopes and are discussed by Catch [2], Evans [1] and Raaen et al. [3]. With organic compounds the effects are usually very small and the general conclusion is that isotope effects are likely to be insignificant in normal biomedical applications of chromatography.

A number of effects are due to the extremely small amount of material commonly handled. These include loss of radioactivity by volatilization in paper and thin-layer chromatography, which becomes proportionately more significant as the quantity decreases. It can occur with compounds such as glycerol and oxalic acid and by hydrolysis of salts of weak acids and bases (e.g. sodium acetate, aniline hydrochloride [2]). These effects may sometimes be reduced by the addition of non-radioactive carrier, provided the volatility is not too great. The presence of an unexpected volatile compound must also be considered, e.g. as an impurity in a compound whose radiochemical purity is being checked, when a false impression of 100% purity could be obtained. Losses by volatilization can be checked by relating the total activity of all spots to that of the sample originally applied [2].

Artefacts may also be caused by adsorption of minute amounts of radioactive material, e.g. at the origin of a paper or thin-layer chromatogram, in a gas-liquid or liquid chromatographic column or from dilute solutions on the walls of sample containers. This can give rise to spurious 'impurities', to contamination of subsequent column runs or to disappearance of labelled materials. Again these effects can be minimised by the addition of carrier.

Small samples are more susceptible to chemical, microbial or photodecomposition in very dilute solution or when supported on high surface area materials such as paper or TLC media.

Decomposition may occur at any stage of the chromatographic processing, and is particularly likely if the developed chromatogram is stored or subjected to a long autoradiographic exposure. If decomposition occurs after the development of a one-dimensional chromatogram it is usually of little importance unless a volatile product is formed or it is desired to extract the detected spots for further study. Decomposition during a run will give misleading results, however, and it is necessary to know whether radioactive spots represent compounds genuinely present in the initial sample or decomposition products. Useful information can be obtained by carrying out two-dimensional chromatography, using the same solvent system and length of run in both directions, followed by autoradiography. Radioactive compounds originally present in the sample applied will appear as a series of spots lying on a diagonal, but if decomposition has taken place at the origin after application, or during the first development, it is likely that more will take place during the intermediate drying or the second development, and radioactive spots will appear that do not lie on the diagonal, as in Fig. 6.1.

A purely chromatographic effect to be considered is the variation in migration of compounds on thin-layers and papers with sample loading. A minute radioactive sample may thus have a different R_f from a larger chemical standard run alongside, and identity or non-identity of compounds on this basis may be misleading. Co-chromatography of the radioactive sample with added carriers of possible compounds may help to resolve this difficulty. It is also vital to guard against loss of radioactivity from thin media caused by chemical visualisation techniques. Visualisation of samples is often required in conjunction with autoradiography or scanning, and some of the standard methods are very likely to cause trouble through such

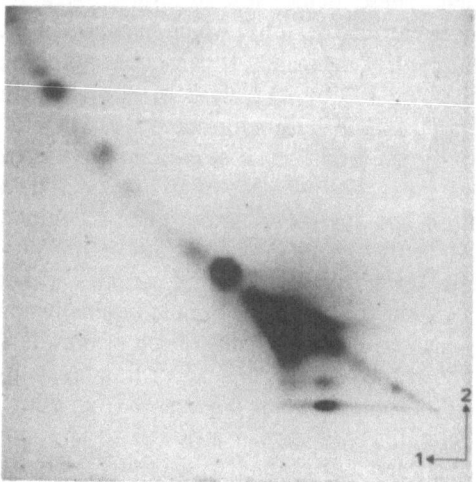

Fig. 6.1. Checking for decomposition during chromatography by two-dimensional runs in the same solvent system.

loss. It need hardly be emphasized that sulphuric acid spraying and charring of organic samples on TLC plates should not be carried out before detection of radioactivity, but the chemistry of apparently more innocuous reactions should also be considered carefully. Thus ninhydrin will cause loss of radio-activity from amino acids labelled with carbon-14 in the carboxyl group. Reagents may also cause positive or negative artefacts on autoradiographs by chemical interaction with the photographic emulsion (see p. 227), and reagents, or resultant coloured derivatives, may cause quenching problems in liquid scintillation counting.

In most circumstances it will be sufficient to carry out visualisation after detection of radioactivity when any technique may be used. However, it may sometimes be necessary to locate the sample first, and then it is import-ant to use a visualisation method that does not give rise to any interference. Iodine is a useful general purpose reagent for many organic compounds on TLC plates, and it has been reported [4] that it does not quench in liquid scintillation counting. The use of ultra-violet absorption or fluorescence is ideal, where applicable, and is often very sensitive. A very useful spray for lipids and lipophilic materials, used in our laboratory, consists of the scintil-lator dimethyl POPOP ref. Table 4.3 (0.005%, w/v in toluene/ethanol 1:1). This allows sensitive localisation under long-wave UV light, causes no arte-facts in autoradiography and does not interfere in scintillation counting; in the latter respect it is much superior to the widely-used rhodamine or dichlorofluorescein.

Artefacts arising during the drying of chromatograms will be discussed in Section 6.2.2.

6.2.2 Autoradiography of Paper and Thin-Layer Media

The preparation of an autoradiograph by contact with a photographic emulsion is applicable to all types of thin-layer adsorbent, paper and impregnated paper used in chromatography and electrophoresis. The subject has been discussed in detail by Rogers [5] and has received particularly wide usage with thin-layer chromatography [6].

General Principles

Autoradiography is based on the fact that photographic emulsions are darkened when exposed to ionising radiation. Factors such as type of nuclide, efficiency, resolution, background, type of film, temperature and method of exposure all play a part in the success or failure of the technique. With β- and γ-emitting nuclides grain density autoradiography is used, in which the exposed film is examined visually for blackening in areas corresponding to the presence of radioactive material.

Differential Autoradiography

Beta-particles are emitted with a wide range of energies, the maximum energies increasing approximately by factors of ten in the order tritium, carbon-14 and phosphorus-32.

Such energy differences can be exploited in differential autoradiography. Carbon-14 and phosphorus-32 chromatograms exposed to a double layer of X-ray film show blackening of the first film due to both isotopes whereas only the phosphorus β-particles have sufficient energy to pass through the first film and blacken the second [7]. A similar technique has been applied to sulphur-35 and phosphorus-32 labelled substances [7, 8]. Use has been made of X-ray film sensitive on both sides when the lower energy beta-particles only interact with the emulsion side in contact with the chromatogram [9]. Iodine-131 has been detected in the presence of soft beta emitters by interposing aluminium foil between the chromatogram and X-ray film [10–13]. Presumably this differentiation could also be achieved by the method of Jackson and Kahn [14], who exposed Polaroid film in its pack to iodine-131 and obtained a rapid positive print in the usual way. The difference in half-lives of nuclides can also be exploited for differential purposes by allowing short-lived activity to decay before autoradiography.

Choice of Nuclide

Autoradiography of γ-emitters is not ideal, as the efficiency of response is poor; but γ-rays can be a source of unwanted background in the autoradiography of beta-emitters.

It can be seen that as the maximum energies of the β-particles increase, film blackening will occur at greater distances from the source, with a decrease in resolution. From this point of view, tritium is a more attractive choice for autoradiography than is carbon-14, and where an element has more than one isotope, other factors being equal, the lowest energy isotope would be the one of choice for maximum autoradiographic resolution. On the other hand, the low energy of tritium β-particles results in a high degree of absorption in a typical thin-layer chromatogram coating, which in consequence necessitates a much longer exposure time than with carbon-14. However, the exposure time for tritium can be greatly reduced by the technique of low temperature scintillation autoradiography.

The choice of tritium or carbon-14 is therefore likely to depend on cost, ease of synthesis, the specific activities attainable and the possibility of biological exchange.

Exposure Conditions

The most popular method for autoradiography of chromatograms and electrophoretograms is undoubtedly simple apposition to a 'no-screen' X-ray film, which may or may not have double-sided emulsion. Such films have a high sensitivity owing to large grain size which causes a greater interaction with β-particles, they are easy to cut to size and manipulate, and give a low and uniform background. All X-ray films, however, deteriorate on storage, and are sensitive to light. Practical details of the exposure of such films are described on p. 231.

Although simple contact with X-ray film works well for the autoradiography of carbon-14 compounds, it is a slow and relatively insensitive method for tritium. Rogers [15] and Markman [16] attempted to overcome this by application to the chromatogram of a liquid photographic emulsion, in order to obtain a more intimate contact between the nuclide and the emulsion. Such impregnation methods suffer from the disadvantages of possible chemical interaction between the chromatographic medium and its compounds and the photographic emulsion, and of possible migration of compounds. Some steroids, for example, have been found to densensitize X-ray film [17] and Markman [16] found that pretreatment of chromatographic paper was necessary to prevent desensitization of the sprayed-on nuclear emulsion. The method offered no gain in sensitivity for tritium over simple contact with an X-ray film but was further modified by Chamberlain and his colleagues [18] to include the use of more sensitive X-ray emulsions.

This procedure resulted in a gain in sensitivity by a factor of ten for tritium, but offered no advantage for carbon-14.

Using normal contact with a fast X-ray film, carbon-14 at an activity of 0.01 nCi/cm^2 as an infinitely thin film has been stated to give a perceptible image in 24 hours ([2], p. 102), although Randerath [19] gives the lower limit of detection five times this. Under similar conditions Evans [1] gives about 5 μCi/cm^2 as the equivalent figure for tritium, although 1 mCi/cm^2 is readily detected in less than an hour. Chamberlain et al. [18] quoted 0.3 μCi/cm^2/day as a minimum sensitivity for tritium either by contact autoradiography on no-screen X-ray film or by impregnation of the chromatogram with a photographic emulsion, and 0.03 μCi/cm^2/day by impregnation with X-ray emulsion.

The best method for tritium is low-temperature scintillation autoradiography [19] (sensitivity 2–3 nCi/cm^2/day) as described later in this section.

Hais and Macek [20] give approximate limits of detection for the following nuclides separated by paper chromatography (no spot sizes given) and exposed for six-days to no-screen X-ray film: carbon-14, 0.004 μCi; phosphorus-32, 0.0005 μCi; sulphur-35, 0.003 μCi and iodine-131, 0.0005 μCi.

Chemical Interference
Chemographic effects can result from chemical interaction of compounds on the chromatogram with the photographic emulsion, by direct contact or by sublimation, giving rise to false 'radioactive' areas. Thus 'autoradiography' of identical unlabelled samples should always be investigated before accepting blackening of a photographic emulsion as positive proof of the presence of radioactivity. Negative chemography can also arise either from a reduction in sensitivity of the emulsion or from accelerated fading of the latent image. The two effects can be independent and in many cases the concentration threshold of a compound for loss of sensitivity is much less than that required to produce latent image fading.

The first effect can be checked by preparing an 'autoradiograph' of a chromatogram of an identical non-radioactive sample exposed for the same time as the autoradiograph of the radioactive sample and then subjected to controlled uniform fogging of the whole film area, followed by photographic processing. Ideally the fogging should be done by exposure to radiation from the same nuclide as the experimental samples, though in practice fogging with visible light will usually be satisfactory. The aim is to see whether fogging is reduced in those areas in contact with the samples. The second effect can be checked by a similar technique, but fogging the film *before* 'autoradiography' of the non-radioactive sample.

The above effects can be lessened, if high energy isotopes are being used, by placing a very thin sheet of plastic film between the chromatogram and the X-ray film. Light spraying with a plasticizer such as polyvinyl propionate [21] (e.g. 'Neatan': Merck) may also help, a technique which has also been used to prevent 'flaking' of thin layers during autoradiography from plates spread in the laboratory [18]. Such spraying can, however, lead to variable background, spreading of the spots and even interaction with the photographic emulsion and is to be avoided if possible. In our experience the incidence of chemical interference is very infrequent and with layers containing binder and particularly with commercial pre-coated plates precautions to prevent flaking are not normally required.

Chemical interference leading to a variable high background or reduced response can also arise from the use of chromogenic sprays or from residual chromatographic solvents and the use of any spray reagent may lead to diffusion of spots on the chromatogram.

Artefacts might also be caused by chemiluminescence (e.g. from oxidation) occurring on the chromatogram during exposure, or from phosphorescence effects produced by light used to examine the chromatogram although such effects have not been reported as major sources of trouble in autoradiography.

Chromatogram-Drying Effects

Several workers have reported migration of radioactive components to the surface of a developed chromatogram during drying [22–24]. A study of the effect of drying one side of a developed paper chromatogram with a hot-air blower showed unequal migration of radioactive components to that side of the paper, resulting in a greater or lesser autoradiographic response depending on which side of the paper was chosen to be in contact with the film. Such an effect may be considered as chromatography within the paper perpendicular to the normal direction of development and is probably greater with solutes having a higher R_f value [23].

For qualitative autoradiography therefore, the most efficient exposure will be obtained from a paper which has been dried as rapidly as possible on the side which is placed in contact with the photographic emulsion. The same will be true of a thin-layer plate which has been hot-air dried on the

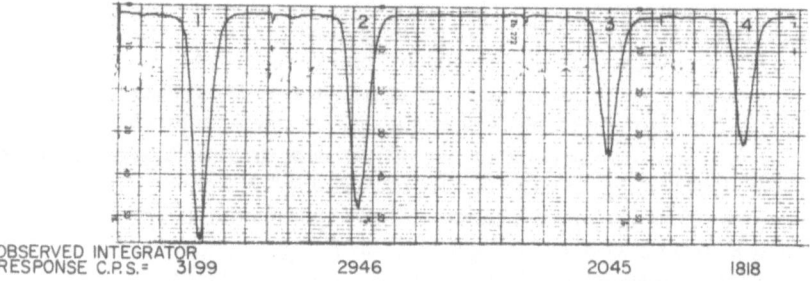

OBSERVED INTEGRATOR
RESPONSE C.P.S.=   3199   2946   2045   1818

Fig. 6.2.   Effect of different methods of drying a T.L.C. plate on the response of a windowless scanner to identical amounts of applied carbon-14.
(1)   Hot air blower on adsorbent.
(2)   Cold air blower on adsorbent.
(3)   Hot plate on underside of glass.
(4)   Standing at room temperature.

surface, and Fig. 6.2 shows various scans with a thin-layer scanner [25] of identical chromatograms dried in different ways. For autoradiography, therefore, it is obviously pointless to strip the plate with 'Neatan' and adhesive tape in order to expose the less efficient underside to the X-ray film. Such stripping is best avoided also because a rigid thin-layer coating on glass provides a better surface for uniform contact with X-ray film.

Quantitative Autoradiography

Autoradiography has been referred to so far as an essentially qualitative technique but it is possible by using suitable standards and controlled conditions to obtain approximately quantitative results from an autoradiograph either by densitometric scanning or by visual inspection.

If autoradiographs exposed at different times are to be compared subsequently, it is obvious that for a given amount of radioactivity the film must have received the same effective exposure time, and therefore with short-lived isotopes decay corrections must be made. To solve this problem with iodine-131 ($t_{1/2}$ = 8 days), calculation of exposure times was facilitated by the construction of a nomogram which allowed the required time to be simply read off the scale [26] and for calculating the exposure of one autoradiograph required to differ from another by a certain factor. A special-purpose slide rule for maintaining a constant exposure in successive autoradiographs with iodine-131 has also been reported [27].

One of the difficulties associated with quantitative autoradiography is latent-image fading, particularly in biological work when long exposures are required to obtain a response from low activity constituents in the presence of high activity ones. A latent image formed early in an autoradiographic exposure will not necessarily remain for the whole of a long exposure. Ray and Stevens [28] have investigated ways of preventing fading, for example in an inert gas atmosphere.

A further difficulty in quantitative autoradiography arises from the non-linear response of a photographic emulsion to ionising radiation. Long exposures required to obtain a response from minor components will reach the limit of blackening of the film for high activities present ([2], p. 86).

Sublimation

The phenomenon of chemography, both positive and negative, referred to earlier in this section, introduces obvious problems in quantitative autoradiography. Losses due to volatility are also a hazard, and further difficulties were underlined by Wilson and Spedding [29], who reported unsuspected sublimation of compounds into the emulsion during autoradiography. Sublimation may have uses in the autoradiography of certain types of compound, but the problems would be great and the technique needs detailed investigation. It is certain that variations in sensitivity of response attributable to this phenomenon must always be kept in mind in quantitative autoradiography, as must also the general likelihood of sublimation and interaction with the emulsion of a volatile non-radioactive component.

Scintillation Autoradiography

In 1958, Wilson [30] described the impregnation of a paper chromatogram with a scintillator to facilitate the detection of tritium by the response of X-ray film to the light thus produced. He called the technique 'scintillation autoradiography', and the terms 'fluorography' and 'scintillation fluorography' [31] have since been used. Randerath [19] showed that, in the presence of scintillator, the response to tritium was primarily due to the interaction of light photons with the emulsion (fluorography) whereas with very low activities of carbon-14 direct interaction of β-particles with the emulsion was involved (autoradiography).

Randerath [19] has reported an excellent systematic study of film detection methods for tritium and other low energy β-emitting isotopes. The method of choice is one in which the chromatogram is impregnated with a critical amount of scintillator solution (PPO in ether) and exposed by contact with a screen-type X-ray film over solid carbon dioxide ($-78°$). Kodak RB-54 Royal Blue Medical X-ray film, a blue sensitive screen-type film, was the best film of several tested. When the final image was chemically intensified, sensitivities for tritium and carbon-14 were 2–3 nCi/cm² /day and 0.05–0.06 nCi/cm² /day respectively, although below 2–3 nCi/cm² of carbon-14 the scintillator effect was negligible, as was the effect of temperature. Although Randerath found the scintillation method to be superior to normal autoradiography for carbon-14 levels greater than 10 nCi/cm², he was unable to confirm an earlier report [32] of a more than 30-fold increase in sensitivity for carbon-14 at levels less than 0.1 nCi/cm².

We have found with Randerath's technique that even with no-screen X-ray film the method is clearly superior to that of normal autoradiography for tritium (cf. Chamberlain et al. [18]).

Colour Film Detection Methods
The possible response of colour film to radiation was first investigated by Buckaloo and Cohn in 1956 [33]. Filter paper containing spots of radioactivity was placed directly in contact with Ektachrome film, and the resulting images after processing were compared with colour standards. The results suggested that the type of colour produced was related to the energy of emitted β-particles for various nuclides, permitting a differentiation between widely differing pairs such as phorphorus-32 (green-yellow) and carbon-14 (blue-purple).

Recently Tio and Sisenwine [34] reported the use of Polaroid colour film and showed different colour responses to carbon-14 and tritium.

Autoradiography of Electrophoretograms on Gel Media
The techniques described so far have been concerned with thin layer or paper chromatograms or electrophoretograms. It is also possible to autoradiograph slabs of polyacrylamide gels, but it is difficult to obtain these in a dry form suitable for autoradiography due to problems of cracking and shrinkage which cause distortion of the bands. Daniels and Wild [35] overcame these problems by bonding the gel to a piece of hardwood with resin glue and allowing it to dry for a few days, after which normal contact autoradiography was successfully employed. Another method described for cylindrical columns of polyacrylamide gels [36] uses longitudinal slices prepared by means of a special support and cutting wire and dried on to filter paper on a vacuum bed under infra-red heating. This method prevents shrinking and produces an easily handled section in one to two hours; we have used it successfully for thin (2–3 mm) slabs of polyacrylamide gel.

Lambiotte [37] has reported the use of X-ray film directly for electrophoresis, immersing the film in a buffered solution and removing excess liquid between sheets of filter paper. Electrophoretic separations of several amino acids and other compounds were achieved in the emulsion, and on subsequent storage in the dark gave autoradiographs which could detect 1 nCi of tritium in six days.

Practical Autoradiography

Many laboratories have developed their own working procedure for practical autoradiography, and much of the finer detail becomes a matter of personal preference. The following procedure is used routinely in the authors' laboratory for paper chromatograms or electrophoretograms and thin layer plates.

Preparation of the Chromatogram

The chromatogram (or electrophoretogram) after development is dried rapidly on one side only with a hot or cold air drier (unless volatile or labile compounds are involved) and spots of radioactive marker ink (promethium-147 in India ink [38]) are applied as required.

Exposure of the Chromatogram

All procedures are carried out under safe-light working at a distance of at least five feet from a bench level (Patterson orange 15 W) or ceiling mounted (Wratten series OB, indirect illumination) safe-light for no longer than 15 minutes.

Agfa-Gevaert Curix RP X-ray film, size 20.3 cm × 25.4 cm, Kodirex No-screen X-ray film or Kodak RB-54 Royal Blue Medical screen-type X-ray film is stored in bulk at $4°$, the pack being well wrapped in black polyethylene. Individual sheets within their outer wrapper are removed in the dark room and guillotined or cut with scissors to the required size, taking care to avoid introduction of fingerprints at any stage. Wearing of disposable gloves is recommended.

All solvents are removed completely from chromatograms before exposure. Paper or thin layer plates up to 20 cm × 20 cm are housed during exposure in individual wooden boxes which are stored within a cabinet (except for high energy β- or γ-emitting isotopes) in the dark room for the required period of time. In the case of thin layer chromatograms, the plate is placed in a box with the adsorbent layer uppermost and the freshly cut sheet of X-ray film is removed from its outer wrapper and placed with care directly in contact with the adsorbent. An uncoated glass plate which has been scrupulously cleaned is laid carefully on top to form a sandwich, followed by a piece of foam rubber and the lid of the box which slots under a lip at one edge and presses down firmly when clipped into position. The tray is labelled and placed in the exposure cabinet, the door of which closes firmly on to rubber foam and is held in place with two hook-and-eye catches.

Papers larger than 20 cm × 20 cm are fixed to the film with paper clips or staples, taking care that the side of the paper dried in the air stream is in direct contact with the film. Location after exposure is assisted by radioactive ink markers or by making an asymmetric pattern of two or three holes with a small hand punch. The paper/film pair is then put in a light-tight envelope or box and left during exposure under an even pressure between foam rubber pads. Alternatively, the paper/film pair is sandwiched between sheets of plate glass and wrapped in black paper or polythene.

It should be emphasized that chromatograms with high energy β- or

γ-emitting nuclides should not be stacked on top of one another during autoradiographic exposure.

Development of Autoradiographs

Developer: Kodak DX-80 diluted one in five, preferably just before use, to give a final volume of 500 ml.

Fixer: Kodak FX-40 diluted one in five to give a final volume of 300 ml. The fixer need not be fresh and can be re-used.

Hardener: Kodak HX-40 which is added to the fixer (50 ml per litre).

Method: The X-ray film is removed from the exposure box (safe-light), taking care not to damage the absorbent layer of a thin layer plate, and is placed in developer in a shallow tray which is gently rocked for a period of four minutes. The film is then removed from the developer, rinsed by passing through a tray of water and placed in fixer plus hardener in a shallow tray. Thorough wetting is achieved by rocking the tray and the clearing time noted, the film being kept in the fixer for twice the clearing time. On removal from the fixing bath, the film is washed in running tap water for half an hour, given a final rinse in water containing a little liquid detergent and allowed to drip dry.

Practical Scintillation Autoradiography for Tritium on Thin-Layer Chromatograms

Autoradiography of tritium labelled compounds on thin-layer plates is carried out in our laboratory as described by Randerath [19, 31]. A 7% (w/v) solution of 2,5-diphenyloxazole (PPO) in diethyl ether (16 ml per 20 cm X 20 cm plate) is rapidly but carefully poured over the entire plate and distributed evenly by tilting. The plate is allowed to dry in a vertical position for several minutes, and then sandwiched between glass plates in contact with Kodak RB-54 Royal Blue Medical screen-type X-ray film, as described for conventional autoradiography. The sandwich may be bound with surgical tape or housed in a thin paper envelope, but an exposure box which gives good thermal insulation should not be used. The plate is laid flat in a polystyrene-insulated chest (ice chests sold for camping are ideal for the purpose) and covered with solid carbon dioxide which is replenished as necessary. The exposed film is developed as already described, except that longer developing times of up to twenty minutes can give increased sensitivity, and chemical intensification may also prove useful [19].

6.2.3   Scanning Methods for Radioactivity on Thin Media

General Principles

The usual system involves a transport mechanism for moving the chromatogram at a suitable speed close to some radiation detector. For one-dimensional paper scans the paper, usually in the form of a strip up to about 2 inches wide, is moved at constant speed by a capstan. Many chromatograms can be taped together to form a long strip which is stored on one spool and taken up on another after passage through the detector. Some paper scanners can be adapted for use with narrow glass thin-layer plates (20 × 5 cm). For standard 20 × 20 cm plates, special thin-layer scanners have been developed, in which the plate rests on a flat bed and is moved under the detector. The latter can also traverse the plate, manually or automatically, so that several one-dimensional lanes can be scanned in sequence.

The radiation detector is shielded from the chromatogram by a thin metal plate provided with a collimating slit at right angles to the direction of movement of the chromatogram, so that only a narrow band of the radio-active sample is exposed to the detector at any one time.

The output from the detector is fed into a ratemeter with chart recorder or into a scaler with digital print-out on paper tape. The print-out tape or recorder chart is usually synchronized with the scanning speed, so that the chromatogram can be exactly aligned with the record of its radioactivity. The scanning speed and slit width are variable, enabling optimum counting conditions to be selected.

Two-dimensional paper chromatograms can be cut into strips which are scanned one-dimensionally. However for large papers this results in a great volume of data for examination and, if the cuts have divided any of the radioactive zones so that parts appear on two or more strips, interpretation can be difficult. As far as we are aware no instrument for two-dimensional scanning of large papers is now commercially available, presumably because of the greatly increased use of TLC. A well-tried design is described by Lowe *et al.* [39, 40], in which the output data can be processed by computer.

Thin layer plates and smaller papers can be scanned two-dimensionally on the standard TLC scanners, the detector traversing device being arranged to operate automatically and scan successive lanes a few mm apart. Again there is a large output of data. For qualitative purposes a useful accessory is the 'dot-printer' [41] which prints out a pattern of short lines, the number per unit area being proportional to the counting rate. A simulated autoradio-graph is thus obtained, which is claimed (for carbon-14) to give in an over-night scan a picture similar to an autoradiographic exposure of two weeks. (Fig. 6.3.)

Detectors

The simplest detector system consists of a thin-end-window Geiger counter in contact with a paper chromatogram. Greater sensitivity is obtained with two counters, one on each side of the paper ('4π geometry'), which also helps to average out variations in the surface radioactivity on the two sides of the paper. This is adequate for moderate activities of carbon-14 or nuclides of higher energy but such counters are insensitive to tritium. With the widespread use of tritium it is now usual to have an opposed pair of windowless gas-flow counters for scanning papers. One wall of each counter, containing the collimating slit, is in contact with the paper and a suitable counting gas mixture passes through each counter. The paper acts as a reasonably good gas seal and the gas consumption is small. The counting

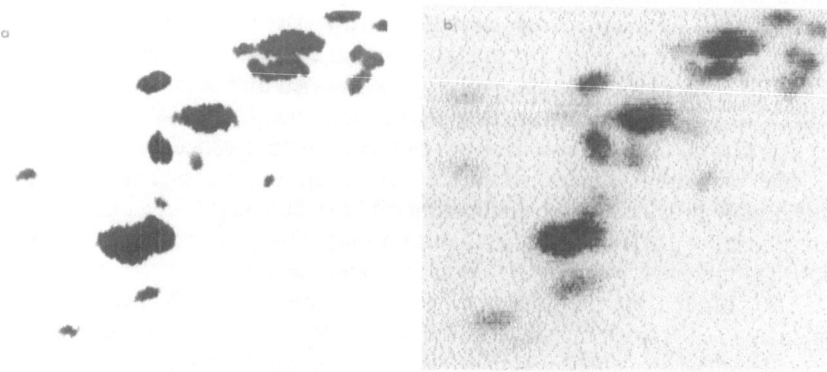

Fig. 6.3. (a) Autoradiograph of a two-dimensional thin-layer chromatogram (exposure
2 weeks) (b) 'Dot-printer' scan of the same chromatogram (overnight run).
(Courtesy Berthold/Frieseke, GmBH).

efficiency of a pair of such counters is up to 2—3% of the activity in the
paper for tritium and 20—35% for carbon-14 depending on the paper thickness. A particular problem in the use of such counters is the development of
electrostatic charges on the paper, which affect the electrostatic field set up
by the counter anode wire, resulting in unpredictable changes in background
and counting efficiency. This is minimised by avoiding high scan speeds and
wide collimator slits. The use of a fine gauge wire mesh in the collimator slit
is also helpful.

This problem can also arise when adhesive tape is used for joining
together strips of chromatography paper for automatic scanning. Johnson *et
al* [42] found that various adhesive tapes gave rise to considerable peaks
when passed through a 4π windowless gas-flow counter and only a commercial masking tape gave a satisfactory background.

Gas-flow counters may be designed for operation in either the Geiger or
proportional mode. The former requires comparatively simple circuitry but
the dead time of the counter causes appreciable loss of counts at high
counting rates [43]. The latter mode requires more complex circuitry but
has a much shorter dead time.

When scanning thin-layer plates only one detector is needed and this
cannot be in contact with the surface. It must, however, be quite close for
two reasons: (1) with nuclides of low energy, particularly tritium, the range
of the beta particles in gases is very small, and the efficiency of detection
falls off rapidly as the detector/plate distance increases: (2) as the detector/
plate distance increases the detector senses an increasing number of radioactive particles from areas of the sample not immediately below the collimator, and resolution is impaired. This effect is greater with nuclides of
higher energy. Wood [44] has discussed these effects further and has
suggested 0.5 mm as a reasonable compromise for detector/plate distance
although on our scanner [25] with precoated plates we work with
clearances as low as 0.1 mm. The construction of the plate support and

transport mechanism must be such as to ensure that the detector/plate distance remains constant throughout the scan. It is also important that the TLC plates should be plane and have glass of uniform thickness. Counting efficiency for thin-layers is about 0.5—2% for tritium and 15—30% for carbon-14, depending on the layer thickness.

Detection by scintillation counting can also be used, for instance a windowless anthracene crystal scintillator [45] operated in darkness, which is also useful for gamma-emitting nuclides, to which gas-flow counters are rather insensitive. Chromatograms can also be impregnated with scintillating material and scanned by a photomultiplier tube [46—49] or cut up and the sections counted in empty vials in a liquid scintillation counter [50]. In the latter method efficiencies were 10—20% for tritium and 55—58% for carbon-14.

Chromatographic Materials

For quantitative measurements attention must be paid to uniformity of the chromatographic media, so that absorption of the radiation in the medium is constant over the whole area, particularly for tritium. Chromatographic paper is reasonably consistent, but the thickness of the layer on a TLC plate can vary considerably. Thin papers or layers are advantageous with tritium, since self-absorption becomes a problem with thicker layers.

The commercially available precoated glass TLC plates have much to commend them. The adsorbent layer is more uniform than on many laboratory-coated plates and is somewhat more resistant to abrasion thus minimising contamination of the detector with dust. For the latter purpose, spraying the chromatogram with an aqueous dispersion of polyvinyl propionate [21] is also helpful but may lower the counting rate, particularly with tritium, and can also cause spreading of the spots.

Metal foil-backed TLC 'plates' are not in general suitable for use with scanners; it is difficult to make them lie sufficiently flat. The commercial precoated 'plates' on a plastic backing usually lie quite flat and can be scanned satisfactorily with a TLC scanner. The 5 cm wide strips can also be used with some paper scanners if a leading strip of paper is taped on, as the adsorbent layer is quite resistant to abrasion. The latter can be minimised if the faces of the detectors in contact with the strip are covered with self-adhesive PTFE film, except over the area of the collimator slit.

Layers from glass plates can also be used in a paper scanner when sprayed with a plastic dispersion, dried and stripped off with adhesive tape. The prepared strip may then be scanned, but it may be necessary also to spray the back lightly to avoid loose dust. This treatment reduces the counting efficiency, and with any such treatment the plastic spray, or soaking in water to strip off the film, may cause diffusion of radioactive materials on the thin layer. A non-aqueous spray, based on polystyrene [51], has been recommended for use with water-soluble materials. Chemical visualisation methods cannot be applied after these treatments.

With all thin media it is convenient to apply a radioactive ink mark (see page 231), beyond the chromatographic solvent front, so that the chromatogram and recorder or printer paper can be precisely correlated.

Recording the Detector Output

The different methods of recording, already mentioned, have their own advantages and disadvantages.

The ratemeter system gives an analogue output on a chart recorder, enabling the pattern of radioactive zones on the chromatogram to be seen as a series of peaks on the trace, but it suffers from the inherent property of such differentiating devices that the signal is averaged over a finite time period so that the system does not respond instantaneously to changes in count rate. The time constant of the ratemeter circuit (a measure of this averaging time) can be varied, and needs to be adjusted according to the counting rate and to other parameters of the complete system. It can be shown [52] that the statistical fluctuation of the randomly arriving radioactive pulses, expressed as a fraction of the mean counting rate R, is proportional to $1/\sqrt{2TR}$, where T is the time constant of the ratemeter circuit. This fluctuation manifests itself as 'noise' on the recorder trace, causing uncertainty as to the true value of the mean count rate. It is reduced by increasing R, so that, other parameters being equal, a smoother trace is obtained with samples of higher activity. The relative noise can also be reduced by increasing the time constant T, but this causes an increase in the time lag between a change in count rate and the corresponding response of the ratemeter, resulting in a peak that is displaced and distorted (see Fig. 6.4). Mathematical treatments for correcting ratemeter readings for lag due to the time constant have been given by Pilgrim [53] and by Holm [54]. However, these methods are rather tedious to apply and are seldom justifiable for use in chromatograph scanning.

Other parameters of the system that need to be optimized are the width of the collimator slit and the scanning speed. It is evident that a wide slit will increase the counting rate but reduce the resolution of two closely-spaced peaks. A very narrow slit will give good resolution but will decrease the count rate so that statistical fluctuations will be larger. A very fast scan

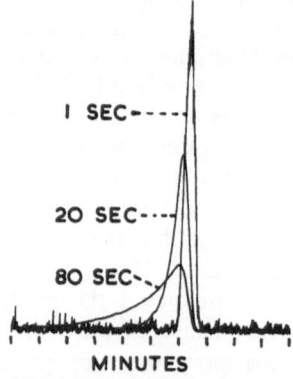

I SEC----

20 SEC---

80 SEC-.

MINUTES

Fig. 6.4.   Effect of altering the time constant of a ratemeter. A radioactive spot was scanned in a paper chromatogram scanner and the output recorded (time direction from right to left). Increasing the time constant reduces statistical fluctuations but causes a lag in the appearance of peak activity and lowers and distorts the peak.

speed will result in few counts being collected, again giving large fluctuations, and will also distort the peak shape, while a very slow speed, though giving good sensitivity and resolution, will limit the throughput of the machine. Some commercial instruments [55] now save time by scanning radioactive zones automatically at optimum speed and inactive zones at a higher speed. A detailed discussion of operating conditions is given by Johnson [56], together with a graphical method for their selection. Other authors have suggested formulae relating these variables so that optimum operating conditions may be chosen [44, 57].

Although such expressions may be useful guides, other factors may be more important in practice. Thus, if only semi-quantitative results are required it may be better to accept less than the optimum resolution or sensitivity in the interest of speed. Some typical operating conditions used with Panax Thin-Layer Chromatogram Scanners are shown in Table 6.2.

TABLE 6.2. *Typical Operating Conditions for Panax Thin-Layer Scanner*

Nuclide	Activity in spot (dpm)	Scan speed (cm/hr)	Counts/sec F.S.D.	Time constant (sec)
Carbon-14	5×10^3	12	10	10
	2×10^4	60	30	3
Tritium	6×10^4	6	3	30
	3×10^6	180	100	1

Collimator slit width: 2 mm
Samples applied as 5–10 mm diameter spots

If the chromatogram contains peaks of widely differing activity the conventional linear ratemeter is inconvenient. It is necessary to adjust its sensitivity range manually during the scanning process so that low activity peaks are detectable and high activity peaks do not exceed the recorder chart width. This difficulty can be overcome by using a logarithmic ratemeter with a sufficiently wide range, which is very useful for qualitative purposes. Such instruments are less accurate than those with linear response however and may not be suitable for quantitative measurements.

Another system for recording scans employs a scaler and digital print-out on paper tape synchronized with the scanning speed. The chromatogram may be advanced either continuously or in discrete steps, the scaler being arranged to integrate and print out the counts collected during a pre-set time or distance of scan. A limiting factor here is the physical size of the printer type face, which controls the closeness of the print-out. Thus in one

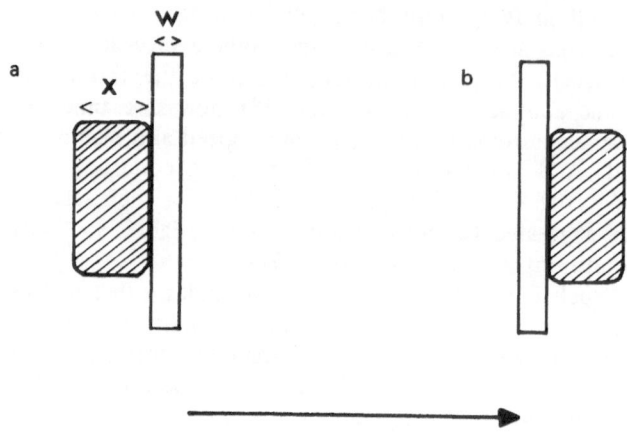

CHROMATOGRAM MOVEMENT

Fig. 6.5.   Effect of scanner slit width on resolution. Detection of radioactive zone (hatched) starts at (a) and ceases at (b), during which time the zone has travelled a distance x + w. The apparent width of a radioactive zone is thus greater than its true width by the slit width.

paper chromatogram scanner [58] this distance is 0.125 inch, which is a measure of the best resolution attainable, no matter how the other operating conditions are optimized.

The advantage of such systems over the ratemeter is that the scaler responds to changes in counting rate with no time lag, so that the time constant effects of the ratemeter, particularly troublesome with low activity samples, are avoided. However, the statistics of counting radioactive disintegrations [52] still apply, i.e. a mean value of N counts has a standard deviation of $\pm\sqrt{N}$. Other parameters of the scanning system already discussed, i.e. slit width and scan speed, have the same effect in the scaler system as with the ratemeter, and must be adjusted for optimum results.

The resolution obtainable with a scanning system can never be as precise as that given by an autoradiograph. The activity in a radioactive zone is first detected when the leading edge of the zone reaches the near side of the detector slit (Fig. 6.5a) and ceases as the trailing edge leaves the far side (Fig. 6.5b). The base width of the recorded peak is thus the true length of the radioactive zone plus the slit width. This is theoretically the most

Fig. 6.6.   Loss of resolution between adjacent radioactive zones (hatched) caused by chromatographic distortion. In neither case does the collimator slit (open rectangle) observe a non-radioactive zone.

favourable result and other factors such as the ratemeter time constant degrade the response further.

Another possibility of error arises if the shape of the separated zones on the chromatogram is distorted, so that the detector slit covers part of two adjacent zones (Fig. 6.6); no clear-cut separation will then be apparent, though the separation would be visible on an autoradiograph.

The chief advantage of scanning over autoradiography is that of speed. Besides the saving of time this may also be of value if the samples, should they be required for further use after elution, tend to undergo chemical decomposition on the chromatogram.

Another instrumental method for visualising chromatograms is a development of the 'spark-chamber', described by Pullan, Howard and Perry [59]. The original instrument consisted of a planar array of parallel wires supported close to another array running at right angles to the first. In an atmosphere of a suitable quench gas a high voltage was applied between the two sets of wires beneath which was placed the thin-layer chromatogram. Ionisation caused by a radioactive sample gave rise to sparking between the wires, which was recorded photographically on Polaroid film. Improvements in the physical arrangement of the electrodes were made later [60], but the resolution was still severely limited by the spatial separation of the electrode elements. This has now been overcome in a commercially available instrument [175] by a small amplitude rotary movement of the electrode system over the TLC plate, so that every part of the radioactive zone is 'scanned'.

Fig. 6.7 shows a typical result obtained in our laboratory. This method is particularly valuable in that visualisation of the whole area is carried out at the same time. The result is similar to an autoradiograph, but much shorter

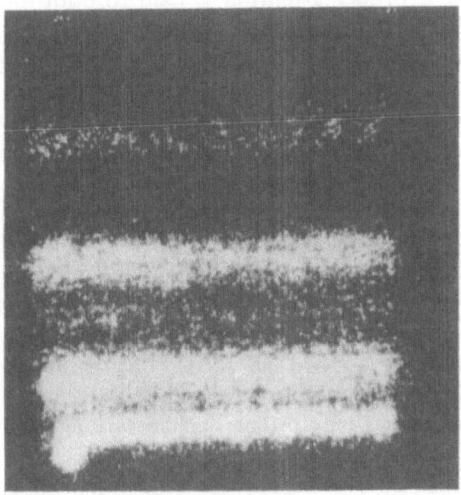

Fig. 6.7. Spark chamber photograph of a thin-layer chromatogram of ^{14}C-labelled lipids. Exposure time 10 minutes.

exposure times are needed, ranging from a fraction of a minute for high activity preparative plates to an hour or more for analytical levels. No quantitative record of radioactivity is obtained but quantitation can be effected by scraping off the radioactive zones and counting, as described in Section 6.2.4.

A further method permitting examination of the whole area of a chromatogram is the use of an array of small counters operating simultaneously. A commercial version [174] consists of 1622 gas flow detectors covering an area of 8 × 8 inches. Counts from the detectors are displayed as a pattern of dots on a cathode ray tube where they can be photographed with a Polaroid Land camera. Total counts accumulated during timed intervals either over the whole area or in selected radioactive zones can also be integrated and displayed on a scaler.

An interesting technique recently reported [61] makes use of a zonal furnace to oxidize organic materials on a thin-layer chromatogram prepared on the inner surface of a glass tube. The resulting CO_2 is monitored for mass by a thermal conductivity detector and for radioactivity by a proportional gas flow counter.

Quantitative Scanning Methods

The foregoing discussion has been concerned mainly with obtaining a reasonably accurate representation of the distribution of radioactivity on the chromatographic medium. With suitable apparatus this can be extended so that the radioactivity in the various separated zones can be measured quantitatively. For this purpose the whole system must be operating in such a way that the final observed measurement (e.g., chart recorder deflection) is a linear function of the absolute amount of radioactivity in that part of the sample zone 'seen' by the collimator slit. This overall linear response will only be obtained if proper attention is paid to both the chromatographic and instrumental aspects of the whole system.

The first prerequisite is good chromatographic technique. Bands should be well separated and not skewed or distorted. Equally important is the elimination of factors discussed previously which can cause variations in the ratio: radioactivity detectable at the surface of the chromatogram/absolute radioactivity in the chromatographic medium.

It is also desirable that the collimator slit be long enough to 'see' the whole width of the zone being scanned. If this is not so the proportion not 'seen' will vary with the zone width (Fig. 6.8) and varying errors will arise with different spots on the same chromatogram, since spots spread in both dimensions with increasing R_f. A typical collimator length of 15 mm will usually cover samples applied as spots, but band application, though preferred for its increased resolution, may lead to errors of this sort.

Secondly, instrumental conditions must be arranged so that two requirements are met:

(i)   The recorded output from the detector must be proportional to the radioactivity at the surface of the chromatogram. This needs, among other factors, a detector of stable counting efficiency,

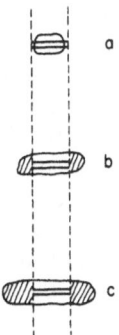

Fig. 6.8   Relationship between width of radioactive zone and collimator slit width.
(a) Whole of zone scanned by collimator. (b, c) Varying proportions of
zones not scanned (hatched areas).

constant detector/chromatogram distance, samples of such activity
that there is no significant loss of counts due to dead time in
detector or circuitry and a recording system showing linear
response.

(ii)   The distribution of radioactivity on the chromatogram must be
represented as accurately as possible. This depends on the variable
instrumental parameters already discussed, for which values must
be chosen appropriate to the sample being examined. If justified,
further corrections can be made, e.g., as already mentioned for
ratemeter outputs [53, 54].

Given an approximation to these ideal conditions it is possible for
reasonably quantitative measurements to be made.

For a ratemeter/chart recorder system the total radioactivity in a chromatographic zone is proportional to the area under the corresponding
recorded peak for any one set of operating conditions. If alterations are
made, e.g., to recorder sensitivity to accommodate zones of widely differing
activity, then appropriate corrections will have to be made.

Peak areas may be measured by any of the standard methods. Using peak
dimensions the most reliable value for the area is given by [height × width
at half height], provided the peak is not markedly asymmetric. The use of
peak height alone as a measure of relative area cannot be recommended since
the same amount of radioactivity may produce a high, sharp peak with one
substance and a lower, broader peak for another on the same chromatogram.

The foregoing considerations apply to the comparison of radioactivities
in different zones on the same chromatogram. Although in principle this
could be extended to comparisons between zones on different chromatograms it would be difficult in practice to obtain quantitative results,
particularly with tritium, since small variations in drying, layer thickness,
detector/chromatogram distance, etc., would produce appreciable changes

in counting efficiency. Quantitative scanning is therefore best restricted to comparisons on one chromatogram, and even this requires careful attention to detail. However, such comparisons form one of the most common applications of radiochromatography, and the rapidity with which results can be obtained make this a most valuable technique.

Scanning techniques in most cases leave the radioactive fractions unchanged on the chromatographic medium, so that they may be extracted and subjected to further chemical procedures, etc. Other favourable characteristics of scanning are its minimum sample preparation and automatic operation, both saving valuable operator time. Against these must be set the rather low efficiency for tritium and the limitations on quantitative work discussed above.

6.2.4   Determination of Radioactivity in Sections and Extracts of Thin Media

In contrast to scanning, the methods described here involve subdivision of the chromatogram, the individual fractions being dealt with separately. More handling of radioactive material thus results, and the avoidance of cross-contamination of samples is important. It is desirable to wear disposable (e.g. vinyl) gloves, and even then the actual radioactive zones on papers or thin layers should not be touched. Paper chromatograms should be laid on sheets of clean paper for all marking and cutting operations, cut sections being handled with forceps. Thin-layer plates may be handled over a tray in a fume cupboard, to minimize the spread of radioactive dust.

Sectioning of the Chromatogram

This can be approached in two ways. In the first method precise localization of the radioactive zones must be carried out. This is best achieved by autoradiography, though the exposure time needed may be inconvenient. The exact limit of the radioactive zone may be difficult to define since the concentration of sample diminishes towards the periphery of the zone and if the exposure has not been sufficient the low activity areas may not cause blackening of the film. With increasing exposure this effect becomes of less importance, and the autoradiograph represents more faithfully the radioactive area. Caution must therefore be exercised when using autoradiography to delineate the radioactive zones and a rather larger area than indicated should be removed if possible. Difficulties arise if the separation between zones is small, and if one is trying to measure adjacent low and high activity zones there may be considerable errors in the former, e.g. when measuring a small amount of impurity in a labelled compound.

Alternatively it may be possible to visualize the zones of interest by spraying with a suitable chromogenic reagent. Precautions to be taken when using such reagents with radioactive materials have been discussed in Section 6.2.1. In all cases relying on chemical visualization there is the possibility that radioactive zones below the limit of chemical detectability may be missed. As with autoradiography the degree of sensitivity of the visualization method may make it difficult to delineate the radioactive area precisely. An advantage of visualization is that only areas of interest are examined, thus limiting the number of samples to be counted. This makes it practicable for use with two dimensional chromatograms. Qualitative scanning, as described, may also be used for localizing radioactive zones.

Having located the areas of interest these may be outlined with pencil (for papers) or a fine point (for TLC plates). A sheet of opal glass, strongly illuminated from below, has been found very useful for outlining areas located by autoradiography, and is conveniently built into a bench top as a permanent viewing panel. The chromatogram is placed on top of the auto-radiograph (precise location being obtained by the use of radioactive ink markers; see p. 231) and viewed by transmitted light. The autoradiographic spots can usually be seen sufficiently well, even when separated from a TLC layer by the thickness of the glass plate; if not, the spots on the film can first be ringed with a felt pen. A record should now be made of the marked chromatogram, e.g. by photography or tracing, showing the position of the areas to be removed for counting and of any duplicate samples and reference compounds that may have been run, together with the origin and solvent front. The chromatogram may now be sectioned for counting.

When cutting samples from paper it is desirable to surround the radio-active part by an adequate non-radioactive area to permit the paper to be cut to a suitable shape if subsequent chromatographic elution of the radio-active material is required.

Areas from thin-layer plates may be removed most simply by scraping with a chisel-ended spatula and tapping the powder on to clean glazed paper, care being taken to clean the scraper after removal of each area. If the relative radioactivities of the various areas are known (e.g. by inspection of an autoradiograph) the lower activities should be dealt with first, to minimize contamination. Scraping TLC plates is rather a dust-producing operation, however, and it is better to use some sort of suction collector. Many varieties of these have been suggested, some of the more useful being described in standard texts [62]. Most of them involve collection of the powder in a soxhlet thimble or sintered glass filter, and a number are available commercially [63]. An advantage of all these methods is that a non-labile sample may be recovered unchanged and uncontaminated by any other zone.

The second method, involving 'blind' sectioning, must be used if location methods are undesirable or if time is insufficient for an autoradiographic exposure. The main disadvantage of this method is that if the chromat-ographic zones are distorted then a section taken at right angles to the direction of separation may include material from more than one component and resolution will be impaired. The effect is similar to that obtained when scanning distorted zones (see Fig. 6.7). On the other hand the whole of the chromatogram is examined, so that no radioactive zone can be missed. Provided the zones are not distorted the resolution obtainable by this method is only limited by the size of the fractions taken, and can approach that given by autoradiography, but the higher the resolution desired the greater the number of samples to be counted. Limitations on sample counting usually make this method impracticable for examining the whole area of a two-dimensional chromatogram.

As with the sample location method, before final sectioning a record should be made of the chromatogram and of the position of the cuts to be made. If a duplicate sample lane has been run it should be visualized and recorded in some way, as should lanes of reference compounds, for correlation with subsequent counting results.

For sectioning, papers are cut into strips at right angles to the direction of chromatography, at a suitable width for the resolution required. Bands from thin-layer preparations are removed similarly, either by a suction collector or by scraping. Such bands usually need to be quite narrow (1–2 mm) to obtain adequate resolution, however, and accurate removal by manual methods of large numbers of fractions is tedious. To simplify this work Snyder [4, 64] has developed an automatic zone scraper which removes narrow bands from TLC plates and transfers them directly to scintillation vials for counting. A subsequent paper [65] describes a sophisticated system for computer analysis of a series of such samples.

Zonal scanning probably shows most benefit when dealing with quite low levels of radioactivity. Snyder [66] found that two peaks having a total activity of only 280 d.p.m. of carbon-14 could be shown up quite well by zonal scanning while with conventional windowless counter scanning the peaks were indistinguishable in the general noise of the recorder trace.

In practice, both methods can be combined by first removing located areas and then zone sectioning the remainder.

Sample Preparation and Counting

Either method of sectioning of the chromatogram produces one or more samples for counting, which can be handled in various ways. The most accurate result is obtained by quantitative extraction of the radioactive material from the medium, followed by homogeneous liquid scintillation counting. If quantitative extraction is impracticable, portions of the medium (paper, TLC scrapings, etc.) may be combusted and the products counted. Often a semi-quantitative result is sufficient, which may be obtained by immersing portions of the solid material in a scintillator solution either directly or in gel suspension.

Counting after Extraction

There is a considerable literature on the extraction of non-radioactive samples from paper and thin-layer chromatograms and the principles are equally applicable to radioactive materials [62, 67]. In brief, elution can be carried out either chromatographically or by extraction either directly in solvent or in a reflux extractor, taking precautions appropriate to the substance being extracted (e.g. cold extraction for thermolabile materials). Chromatographic elution of thin-layer materials is conveniently done in a micro-column [68].

Whatever method is adopted, the extract is counted after addition of a scintillator solution with which it is miscible. Normal liquid scintillation counting is carried out, with standardization by any of the usual methods, and with a suitable instrument dual labelled samples may be counted (e.g. carbon-14 and tritium). Preliminary evaporation of the extract may be necessary if the volume is large or if the solvent is a powerful scintillation quencher. It is important that the extracted radioactive material is completely soluble in the final solution for counting. For instance, water-soluble compounds may not remain dissolved when their aqueous solutions are added to dioxan-based scintillator solutions. Usually, however, the weight of material in the extract is so small that such effects are not likely to occur. Recently the use of detergents has been introduced to improve the scintillation counting of aqueous solutions [69–72].

The completeness of extraction should be checked when dealing with a new system by heterogeneous counting of the solid medium after extraction, which will indicate whether any significant amount of radioactivity remains.

Counting after Combustion

Sometimes extraction of the radioactive material from the chromatographic medium may be impossible, or undesirable because of quenching. Quantitative results may still be obtained if the medium is combusted and the products taken up in a suitable absorber and mixed with a compatible scintillator solution. The method seems to have been very little used for direct combustion of chromatographic media, one application being the oxygen-flask combustion of paper segments by Baxter and Senoner [73]. Parmentier and ten Haaf [74] give references to many modifications of the basic oxygen-flask method; a very simple and inexpensive arrangement is combustion in plastic bags as described by Gupta [75]. Commercial combustion apparatus is now available [76] and the authors have used a furnace designed by Griffiths and Mallinson [77]. Scrapings from TLC cellulose plates can be combusted in a suitable sample holder, and with added fuel, such as cellulose, even samples on non-combustible chromatographic supports (glass-fibre paper, silicic acid, alumina, etc.) can be satisfactorily combusted [78].

Suspension Counting

A much simpler method of dealing with paper or TLC fractions is to immerse them directly in a liquid scintillator and count as a heterogeneous system (Section 3.9.2). As far as chromatographic samples are concerned, the possibility of obtaining quantitative results depends on knowing to what extent the radioactive material is extracted from the solid medium into the solution.

The most favourable situation occurs with complete extraction, the presence of a small amount of settled TLC powder or of a piece of paper in the counting vial having little effect. Normal standardization methods can be used and dual label experiments carried out. Completeness of extraction must of course be checked. Use of a scintillator solution containing water [65, 80] may help by deactivating TLC media, while strong organic bases (e.g. Hyamine, NCS [81], Bio-Solv [82]) are useful for solubilizing biological materials such as proteins and nucleic acids that are insoluble in scintillator solutions. These strongly alkaline solutions may exhibit chemi-luminescence, which can be counteracted with acid [85, 86].

If the radioactive sample remains completely adsorbed on the solid medium the counting rate will be affected by absorption of the radiation in the solid phase and by the orientation of the latter, and in general it will be impossible to apply any of the normal methods for measuring counting efficiency in homogeneous systems. Heterogeneous counting is thus of chief value for the comparison of activities, e.g. of fractions from one chromato-gram, and even here considerable variations may arise, particularly with tritium. One advantage is that chemical or colour quenching may be less than with solution counting, and it may be possible to wash out the scintil-lator solution and extract the radioactive material for further chemical work [94].

The effects of orientation of paper segments have been widely investi-gated and some useful reviews published [87, 74, 88], the consensus of opinion being that orientation is of importance only with very low energy nuclides such as tritium. Davidson [87] recommends placing paper strips in scintillator solution in a narrow tube supported axially in a counting vial.

Fig. 6.9. Counting small strip cut from a paper chromatogram. From left to right: Packard plastic counting vial; the same modified by removal of neck; screw-topped glass tube containing sample strip damped with scintillation fluid; the same supported in modified vial. This assembly is placed in the liquid scintillation counter. Using the low volatility scintillator described in the text it is not necessary to cap the glass tubes.

We have used this arrangement (see Fig. 6.9), treating the paper with just enough non-volatile scintillator solution (butyl-PBD [89] (0.5%w/v) in isopropylbiphenyl [90]) to wet it without excess dripping off. This is very economical of materials and prevents evaporation. Table 6.3 shows results obtained with paper sections containing tritiated uridine. Repeat counts were made, either *in situ* in the counting chamber or recycling between counts in the sample conveyor belt of a Packard Tri-Carb Liquid Scintillation Spectrometer, in which the axial orientation of the vials tends to change during transit. It is seen that variations in orientation cause a considerable increase in the standard deviation of the counting rate; this method should therefore only be used for semi-quantitative work, e.g. for scanning a complete chromatogram.

TABLE 6.3. *Effect of Orientation on Liquid Scintillation Counting of Paper Sections*

Sample number	Counting method	Obser-vations	Mean counts	S.D.	S.D. as % of mean counts
1	In situ	22	1739	38.7	2.2
	Recycling	23	1772	252.5	14.2
2	In situ	28	4858	143.5	3.0
	Recycling	23	4789	532.6	11.1

In the case of radioactive material adsorbed on particles of TLC medium the counting efficiency can be improved by keeping the powder suspended in the scintillator. Various methods have been proposed for this [91], a widely used one [85] in current use being a thixotropic suspension of finely divided silica [92] in the scintillator solution. This also helps to minimise adsorption of radioactive materials on to the walls of glass counting vials [93]. The use of polythene vials has also been recommended for the latter purpose [85].

The least satisfactory situation for heterogeneous counting occurs when the radioactive material is only partly extracted into the scintillator solution. Even the relative activities of different samples will be uncertain since the counting efficiencies for the solid and liquid phases will be different and the distribution of activity between them unknown. Suggestions [64, 95] that the absolute activity of such a system may be obtained from the individual counting rates and efficiencies for the two phases are hardly practicable, particularly for tritium.

Counting Samples from Electrophoresis Gels
The methods of sample preparation described above are not generally applicable to polyacrylamide gels, etc., owing to the insolubility of the gel and the difficulty of extracting samples. A number of special techniques have thus been developed for use with beta-emitters, gamma-emitters being more conveniently dealt with by direct counting of sections of gel.

Homogenizing of the gel and dispersion in the scintillator solution [96] may be adequate for comparative purposes. If absolute standardization of counting is required it will be necessary to extract the labelled material, for instance by heating with a strong organic base [97, 98]. Ward *et al.* [99] found that 99% of labelled protein was extracted from fragments of polyacrylamide gels on shaking with a scintillator solution containing ammonia and NCS [81] for two hours. The ammonia reduced chemiluminescence caused by the NCS.

Another approach is to modify the gel formulation. Thus Alpers and Glickman [100] and Anker [101] have used cross-linking agents which make the gel soluble in dilute alkali [100] or perchloric acid [101] and compatible with scintillation fluid.

Apart from these methods many techniques have been used for the elution (electrophoretic and otherwise) of non-radioactive materials from electrophoretic gels [96]. In general these are adaptable for radioactive work, the extracts being concentrated if necessary and counted by a suitable technique.

6.3   GAS-LIQUID RADIOCHROMATOGRAPHY

In detecting radioactivity in the component peaks emerging from a gas chromatograph, there is a primary choice between discontinuous, off-line counting and continuous on-line flow-detection systems. Where one already has a suitable counter, the former system is cheap and adequate for many purposes and does allow counting of low-activity samples for long periods of time. However, apart from advantages which could be grouped under the general heading of 'convenience', such as simultaneous detection of mass and radioactivity, it should be emphasized that the compelling reason for using a continuous-flow system is that it obviates the danger, inherent in any discontinuous system, that high specific activity peaks of low mass, which would not be recorded by a mass-detector, may be completely missed by off-line counting. Several reviews of the subject have appeared [6, 102, 103, 104].

6.3.1   Discontinuous Off-Line Detection of Radioactive Gas Chromatographic Effluent

The descriptive name given to this type of operation implies that the effluent is to be trapped or condensed in some way, peak by peak, the radioactivity of each peak being counted separately. Table 6.4 lists some trapping methods which can be used in this type of application. Obviously in type 1, scintillation vials could be used directly as trapping vessels, containing a suitable trapping fluid compatible with a scintillator mixture for direct liquid scintillation counting. The effluent may also be bubbled directly into cold scintillator solution [105] which allows very sensitive differential liquid scintillation counting of carbon-14 and tritium [106]. The limitation of such a system is the need to ascertain for each type of compound that the chosen trapping fluid will be efficient under the given conditions of temperature and flow rate. In types 2 and 3, direct counting of γ-emitters could be achieved if the trapping device were compatible with a counter assembly, but compounds labelled with β-emitting isotopes would have to be isolated, except where scintillator crystals were used. It is usual to wet both scintillation crystals and absorbent cotton wool with a suitable non-volatile organic solvent and again the possible variations of trapping

TABLE 6.4. *Condensation of GLC Effluent for Counting of Radioactivity*

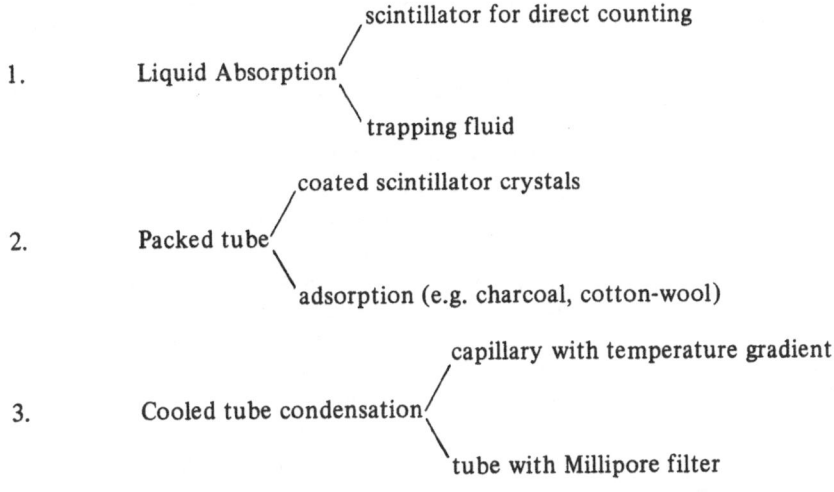

1. Liquid Absorption
    - scintillator for direct counting
    - trapping fluid

2. Packed tube
    - coated scintillator crystals
    - adsorption (e.g. charcoal, cotton-wool)

3. Cooled tube condensation
    - capillary with temperature gradient
    - tube with Millipore filter

efficiency under various conditions must be taken into consideration. Fatty acid methyl esters have been condensed in this way in a glass tube containing methanol-moistened cotton wool [107].

Hajra and Radin [108] trapped carbon-14 labelled fatty acid methyl esters quantitatively using a modified Swinny hypodermic Millipore filter adapter [109] attached to a glass tube. Any fog which did not condense in the glass tube was effectively trapped in the filter and the whole assembly was easy to wash out with solvent for subsequent liquid scintillation counting.

In one well-known design of fraction collector [110], obtainable commercially [76], samples in the effluent from the gas chromatograph are trapped on silicone-coated scintillator crystals in glass tube cartridges. In the original description anthracene was used, and the cartridges were mounted in special holders and counted directly in a liquid scintillation counter. However the efficiency of tritium counting was very low and a considerable improvement was made by substituting coated *p*-terphenyl crystals for the anthracene and immersing the whole cartridge in a PPO-toluene scintillator solution after collection of the sample [101, 111].

Greater efficiencies can be obtained by extruding the contents of the cartridge into a counting vial of scintillator solution. For fairly non-volatile samples it is possible to use as the trapping agent toluene-damped cellulose cigarette filters [112] or even dry glass wool [113].

The flame ionisation detector has been used as a combined mass detector and combustion chamber, the CO_2 resulting from carbon-14-labelled fractions being trapped in various ways [114–116].

Whatever trapping method is used it is essential that no condensation of a peak occurs between its leaving the exit port of the gas chromatograph and entering the trapping system which normally means the use of a heated transfer line between the exit port of the detector block or splitter and the trapping system.

6.3.2 Continuous On-Line Detection of Radioactive Gas Chromatographic Effluent

In using a flow detection system there are three choices:

(i)   The effluent vapour can be continuously condensed in the counting system for continuous monitoring of the total level of radioactivity, giving an 'integral' record.

In this category of detection one must inevitably refer to the elegant pioneering work of Popjak and his colleagues [117, 118] who constructed a system in which the effluent gases from their gas chromatograph could be continuously trapped in liquid scintillator and recorded as described by James [6].

Whatever type of trapping and counting is used, a point will be reached when full scale deflection is achieved on the recorder, such as might occur with early peaks of high specific activity, making it necessary to change the range or wash out the system. A small peak of radioactivity may be masked by a preceding large one, making it necessary to wash out and measure the small peak alone. Nevertheless, such a system can be extremely sensitive.

(ii) The counting system can be operated at the same temperature as the gas chromatograph which normally means a heated counter, measuring effluent radioactivity in the vapour phase. A heated flow-through proportional counter has been offered commercially [119] but there are many problems associated with such counters [6]. For example, difficulties associated with short operating voltage plateaux at elevated temperatures must become more acute as gas chromatographers use even higher temperature stationary phases. Welch *et al.* [120] used a proportional flow counter with a thin internal window, directly on line to a gas chromatograph, but rather low efficiency was obtained and the upper temperature limit was 125–150°. The use of ionization chambers at moderately high temperatures (240 and 300°) for direct measurement of gas chromatographic effluent has been described [121, 122].

(iii) The effluent from the gas chromatograph can be converted to a permanent gas mixture and counted in the gas phase at room temperature in an ionization chamber or proportional counter or by passing through plastic scintillator tubing.

Following the use by Wolfgang and Rowland [123] of a proportional counter on line to a gas chromatograph, James and Piper [124, 125] designed a system which gained wide acceptance and formed the basis for an early commercial model [126]. Effluent from the gas chromatograph was led into a combustion furnace and passed over heated copper oxide for conversion of carbon-14 to carbon dioxide and over iron filings for conver-

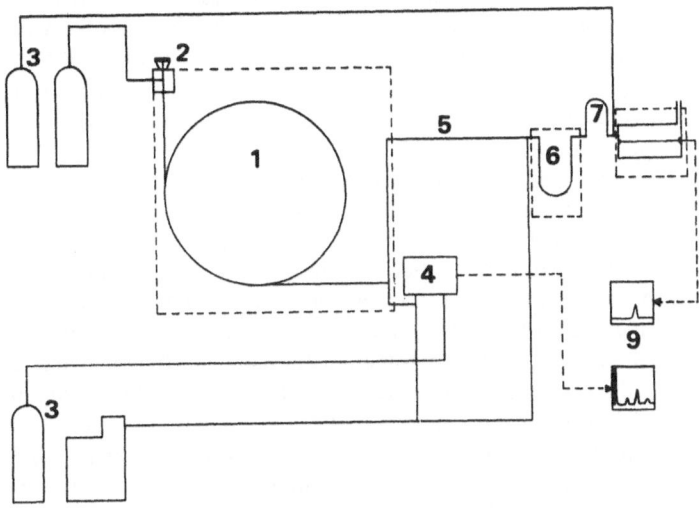

Fig. 6.10. Equipment for gas-liquid radiochromatography.

(1) G.L.C. oven.
(2) Injection ports.
(3) Gas supplies and flow controllers.
(4) Mass detectors.
(5) Heated transfer line.
(6) Combusion furnace.
(7) Drying tube.
(8) Gas-flow proportional counter.
(9) Electronics and recorders.

sion of tritiated water to hydrogen for subsequent counting as permanent gases in a proportional counter. A similar system was described by Karmen [127].

The use of a combustion method was found by Winkelman and Karmen [128] to be necessary for the successful operation of an ionization chamber on line to a gas chromatograph as the ionization chamber was found to respond to unlabelled material in the gas stream. A similar combustion method and ionization chamber has also been used with interrupted elution gas chromatography [129].

Karmen *et al.* [130] converted the gases leaving the column to carbon dioxide or hydrogen which passed through a cartridge containing anthracene crystals mounted in a scintillation counter to give a continuous monitoring of radioactivity. Benson and Maute [131] have used a plastic scintillator cell for counting isolated gas samples which might be applicable to gas chromatography and a plastic scintillator gas flow cell is commercially available [132]. A flow cell for counting alkali-trapped carbon dioxide has been described [115].

In our laboratory [133] an F and M 5750 Research Gas Chromatograph [134] has been linked by means of a heated transfer line to a Berthold RGC 170 gas-flow proportional counter [41, 135]. (Figs. 6.10a and b). The transfer line consists of a length of 1/8″ stainless steel tubing bound with a Variac-controlled Electrothermal heating tape with an outer wrapping of asbestos cord. The furnace inlet and proportional counter assembly have since been modified to allow easy coupling and uncoupling from the gas chromatograph by sliding the counter assembly sideways, and to ensure the shortest possible transfer line [136].

The effluent peaks can be converted to permanent gases either by the well-tried method of oxidative combustion or by the more novel method of hydrogenative cracking. The latter method, introduced by Simon *et al.* [137] in 1965, has particular advantages for compounds which show pseudo-activity or memory effects when using the oxidative process.

The detection of components of a mixture of ^{14}C fatty acid methyl esters (separated gas-chromatographically on diethyleneglycol succinate columns) shows a higher efficiency of detection of radioactivity by oxidation than by hydrogenation (Fig. 6.11). A similar loss in efficiency using hydrogenation has been observed in the analysis of carbon-14 labelled basic drugs separated on OV17 columns. In order to explain this observation it is necessary to examine the equation for overall efficiency ($\bar{E}$) of the system:

$$\bar{E} = \frac{kEv}{f_t} \tag{6.1}$$

where

$$
\begin{array}{lll}
k & = & \text{a constant} \\
E & = & \text{the efficiency of the counter} \\
v & = & \text{the volume of the counter} \\
f_t & = & \text{gas flow rate}
\end{array}
$$

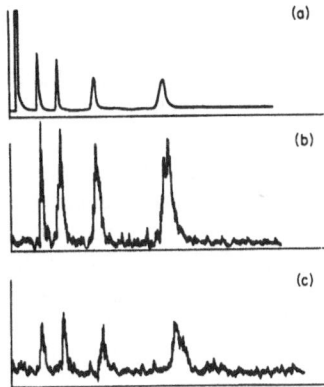

Fig. 6.11. Gas radiochromatographic detection of a mixture of carbon-14 fatty acid methyl esters. (a) Mass response (F.I.D.). (b) Radioactivity response after oxidative combustion. (c) Radioactivity response after hydrogenative cracking.

The operation of a gas-flow proportional counter requires the introduction of a suitable quench gas, in our case methane, which must form a suitably high proportion of the total gas flowing through the system. Normally, therefore, the gases entering the proportional counter consist of combustion products, carrier gas and methane. In the cracking process, a considerable amount of hydrogen must be introduced, which necessitates a corresponding increase in methane flow to preserve the proportion of quench gas. Thus with added hydrogen and increased methane flow, the f_t term of the equation is increased with a corresponding fall in $\bar{E}$ if other conditions such as the counter tube volume are kept constant. Nevertheless the cracking process is a useful alternative to oxidative combustion both for carbon-14 and for tritium in particular and has further possible applications with compounds which may be difficult to combust in oxygen.

When a non-destructive mass detector is used on the gas chromatograph, the effluent gases may be passed directly from the detector to the combustion furnace and counting system. Martin [138] passed the whole of the column effluent into a combustion furnace and measured total carbon dioxide in a microthermistor detector and radioactive carbon dioxide in a proportional counter. When a destructive detector such as flame ionisation is used, it is necessary to split the stream at the end of the gas chromatographic column, so that the bulk of the effluent is carried to the combustion furnace with a fixed small proportion entering the mass detector. Under normal conditions of gas chromatographic analysis, a split ratio of 1:20 may be obtained by using a splitter of equal bore in both arms with lengths of coil in the ratio 20:1. However, the back pressure created by the combusion system is such that with the equipment in our laboratory we find it is necessary to use splitter arms of bore b with length y and bore b/2 with length 60 y in order to obtain a 20:1 split ratio when the combustion furnace is in use.

Our system has a normal background of about 20 c.p.m. which slowly increases with prolonged usage apparently due to accumulation of non-volatile material in the counter chamber. The background can be returned to normal by using the convenient 'baking-out' facility of the built-in heater for the proportional counter.

Realistic minimum measurable quantities would be 0.1 nCi of carbon-14 per peak, and slightly higher for tritium depending on the method used. Integration can be achieved by use of a scaler. A commercially available combustion/detection system [140] based on a design by Simpson [139] uses an anticoincidence shield of plastic scintillator and massive lead shielding round the counter tube to reduce the background to 1–3 cpm so that smaller radioactive peaks can be detected which, over three years, the authors found had a stable background.

6.3.3   Radioactive Labelling as a Means of Gas Chromatographic Detection

Just as neutron activation analysis of chromatograms is not strictly radio-chromatography, so the introduction of a radioactive label after vapours have passed through a column is not gas-liquid radiochromatography in the normally accepted sense but is most conveniently mentioned in this section. Behrendt [141] has given a preliminary account of this method in which exchange-labelling of compounds can be achieved after their separation on a gas chromatographic column. The resulting radioactivity can then be determined, giving mass detection at picogram levels or even lower. In spite of considerable technical difficulties the method has great potential, and Behrendt [141] described not only a detector using carbon-14 exchange after combustion of effluent to carbon dioxide, but also extension of the method to many other nuclides.

6.3.4   Special Factors to Be Considered in Gas-Liquid Radiochromatography

The efficiency of recovery of radioactivity and measurement by a proportional flow counter after separation by gas chromatography has been seen to be governed by various factors expressed in equation (6.1). Thus, the inherent efficiency of the counter itself, the volume of the tube, the flow rate, the type of carrier gas, the temperature of the counter tube and its operating voltage, will all affect the overall efficiency.

As in any counting system, when using a proportional counter one must take account of background, quenching and ghost peaks or pseudo-activity. In addition, in using a continuous flow system with peaks which may follow one another closely there is the problem of memory effects.

The background of a proportional flow-counter operating at room temperature will remain fairly constant, but can be increased by bleed from septa and from stationary phases, from condensation of radioactivity and also from compounds which are incompletely or only slowly converted to permanent gases during any catalytic process and may therefore leach out during subsequent operations. If the counter tube can be heated it is often possible to return to low background conditions by flowing gas through the tube for a period at elevated temperature. If this fails then the counter must be stripped down and cleaned.

Simon *et al.* [137] studied the problems associated with the fact that most compounds can either quench or exhibit pseudo-activity if counted directly. They found that compounds containing halogen or nitro groups showed quenching with all gas mixtures and that aromatic compounds could show very strong pseudo-activity.

These problems can be overcome by combusting carbon-14 labelled compounds to carbon dioxide which can be counted at room temperature. However, although tritium-labelled compounds are converted to water and transformed to hydrogen or acetylene, Karmen *et al.* [111] and Cacace and Perez [129] reported memory effects with some tritium-labelled compounds which could cause problems with closely-spaced peaks. James [6] also observed bad tailing with tritium if the gases came into contact with cold metal surfaces before entering the counter.

A further problem in the analysis of tritium by oxidative combustion arises when iron filings are used to convert the water formed to hydrogen. The ferric oxide formed strongly absorbs water, and the iron must therefore be kept in the reduced form. Bleeding hydrogen into the system at a rate of 2–3 ml per minute will accomplish this, and prevent release of 'trapped' tritium by unlabelled compounds which exchange with tritiated water held by the ferric oxide [102, 111].

Using hydrogenative cracking of compounds, mainly to methane, Simon *et al.* [137] reported that no interference was obtained from unlabelled compounds containing halogens or sulphur. Their results did suggest, however, that chains of 16 or more carbon atoms may be incompletely cracked, giving a lower efficiency for carbon-14 labelled compounds, although this effect was not seen for similar compounds labelled with tritium. It may be that longer chains partially form hydrocarbons higher than methane which may lower the counting efficiency of the system.

A particular problem arises in the chromatography of materials of high specific activity if they react on the column so as to cause contamination of other components in the mixture. Pascaud [142] reported that transesterification of fatty acid methyl esters occurred on a polyester stationary phase, resulting in contamination of a linoleate fraction by methyl palmitate of very high specific activity. Such contamination, which may be insignificant in terms of mass, can be very misleading in terms of radioactivity.

From a consideration of the large number of factors which can affect the response of any flow counter on line to a gas chromatograph, it will be appreciated that for quantitative work calibration of the system will be necessary. In some cases injection into the gas chromatograph of known amounts of labelled standards similar to the test compounds will suffice, but there may be problems due to different patterns of adsorption or efficiency of combustion. Where oxidative conversion to carbon dioxide is to be used, standardisation by the introduction of known amounts of labelled carbon dioxide is a good method, but often difficult to achieve in practice. Karmen [143] has discussed these problems and has described a simple generator of carbon-14 labelled carbon dioxide for the calibration of flow detectors.

6.4   DETECTION AND MEASUREMENT OF RADIOACTIVITY IN LIQUID CHROMATOGRAPHY EFFLUENTS

The general principles of radioactivity measurement for any technique producing a liquid stream are similar to those for gas-liquid chromatography given in Section 6.3 and measurements may be made on discrete fractions or by a continuous on-line flow-detection system. Whichever method is used it is desirable to have some method for measuring and recording the mass peaks in the effluent to correlate mass with radioactivity. Such detection methods may be almost any of those normally used for non-radioactive work, and the physical arrangement of mass and radioactivity detectors will depend on the types being used.

In many cases column chromatography is used as a preparative method and some form of fraction collector is required, so that if a flow-type radioactivity detector is used it must obviously precede the fraction collector. With the development of high pressure liquid chromatography and of sensitive mass detectors there is an increasing use of columns for high-resolution analytical work in which fraction collection is not required, giving, in principle, more flexibility in the relative positions of mass and radioactivity detectors. These will still often be determined by their characteristics; for instance, it is essential for a flow-type radioactivity detector to precede the mass detector in an amino-acid analyser, since the samples undergo chemical decomposition in the mass detector with possible loss of radioactivity. Indeed this sequence is desirable with most mass detectors, as it avoids the possibility of radioactive contamination or hold-up in the latter, causing spurious radioactive effects following an active peak.

Whatever systems are used it is essential that the mass and radioactivity detectors should be close together in terms of liquid passage time, so that correlation between the two measurements is as good as possible. Other necessary provisions include adequate radiation shielding between chromatography column and radioactivity detector if gamma or hard beta emitters are being used and a hood over the apparatus if escape of any volatile radioactive material from the system is possible.

6.4.1   Counting Discrete Liquid Samples

Perhaps the simplest system is the use of a normal fraction collector. The fractions can be examined for the presence of eluted sample by a suitable chemical or physical method and for radioactivity by counting aliquots. In the first instance it may only be necessary to do this semi-quantitatively as a rough indication of which fractions are radioactive; for this purpose samples for counting may be taken quickly with a Pasteur pipette, washed between samples with eluting solvent. Appropriate fractions representing a single radioactive peak may then be combined for quantitative examination.

For liquid scintillation counting it may be necessary to evaporate the sample in the counting vial if the eluting solvent causes excessive quenching or if the volume has to be large because of low radioactivity. If the samples are so active that a few drops of solution are adequate for counting then it is possible to spot samples on small pieces of filter paper which are dried, damped with isopropylbiphenyl-based scintillator and counted, as described in Section 6.2.4; or successive samples could be spotted at intervals on a strip of chromatography paper, dried and counted in a paper strip scanner (see Section 6.2.3). This technique has been described for the quantitative measurement of radioactivity in samples of cells containing labelled macromolecules [144], and should be applicable to chromatography fractions.

Samples may also be dried on planchettes and counted with a windowless gas-flow counter or with a thin-window Geiger counter for suitable nuclides. Loss of volatile radioactive material should be borne in mind.

These general methods of counting discrete liquid samples can of course be used with any system producing such samples. Thus instead of using a fraction collector one might collect effluent containing a single peak as indicated by some form of mass detector.

Thin-layer and paper chromatography and electrophoresis are valuable adjuncts to fraction collecting. For instance a number of fractions can be run on one chromatogram, subjected to autoradiography or scanning and visualisation and checked for chemical and radiochemical homogeneity against standards run simultaneously.

6.4.2   Continuous Flow Counting of Liquid Samples

The use of a continuous flow detector for radioactivity has advantages over counting discrete samples as it is non-destructive, less handling of samples is involved, and with suitable apparatus a continuous record is obtained of the changing radioactivity in the effluent. On the other hand the volume of liquid being counted at any instant is limited (see below) and the counting efficiency may be less than for discrete samples in a liquid scintillator. In a flow system it is impracticable to concentrate the sample for counting, so that flow counting may not be suitable for samples of low radioactivity.

The relationship between various parameters of the system have been presented by Piez [145]. Suppose we have a solution containing a peak of radioactivity of total activity A (dpm) flowing at a constant rate v (ml/min) through a flow cell with a fluid volume V (ml) in a counting system of efficency E (%). The total net sample count collected, C, is given by

$$C = A\frac{VE}{100\,v} \tag{6.2}$$

The sensitivity of the system can be increased by increasing the cell volume V, but this is determined by the resolution required, since closely spaced radioactive peaks will not be separated unless the cell is small enough. A typical cell volume would be about 1 ml. Reducing the flow rate v will also increase the sensitivity, but the flow is often determined by the chromatographic conditions, so that in practice the sensitivity is controlled by the counting efficiency and the amount of radioactivity. The effect of background on counting precision is also discussed by Piez [145] who gives the equation:

$$\sigma = \frac{100(C_T + B.t)^{\frac{1}{2}}}{C} \tag{6.3}$$

where

σ	=	the standard error of counting (%)
C_T	=	the total count
B	=	the background counting rate (cpm)
t	=	time required for the radioactive peak to traverse the cell (min)
C	=	total net sample count

Flow Detectors

A flow cell of any type must fulfil certain requirements for efficient operation. As stated above, the cell volume should be as large as possible (for maximum sensitivity) compatible with the chromatographic resolution, but must be designed so that there is no mixing and no dead spaces in which hold-up could occur. The shape of the cell, including surface/volume ratio, may also influence the counting efficiency, and the material of which the cell is made may impose restrictions on its use.

The simplest flow detection system uses some form of Geiger or proportional counter. Dobbs [146] has described a simple adaptation of an end-window counter used to count carbon-14 in organic solutions at a specific activity down to 2.5 nCi/ml. Bangham [147] designed a counter of increased sensitivity with a large area thin window, used for carbon-14 and sulphur-35. For phosphorus-32 and other high energy beta emitters Geiger counters incorporating a thin-walled glass tube flowing samples are available [148].

For flow counting of tritium and other low energy beta emitters, it is necessary to use scintillation counting in which the sample stream flows through a transparent cell containing some scintillating material insoluble in the liquid, the cell being viewed by one or two photomultiplier tubes.

Two main types of scintillation cell have been used. In one the cell is made of plastic incorporating organic scintillator. The liquid flows in a shallow channel machined in the face of a block of the plastic scintillator and covered with a sealed-on sheet of the same material; or through a flat spiral of plastic scintillator tubing [149]. The efficiency of counting is of the order of 5% for carbon-14 but is very low for tritium. Because the plastic material used is attacked by organic solvents such as chloroform or acetone such cells are restricted to the use of aqueous and lower alcoholic solutions.

In the second type of cell a channel or tube contains a solid scintillator, usually anthracene, in a finely divided form, the sample liquid flowing in the interstices between the scintillator particles. This improved geometry results in an increased counting efficiency. With a typical commercially available cell [150] carbon-14 can be counted at 65–80% efficiency and tritium as high as 10%. In this type also the materials preclude the use of many organic solvents but for these inorganic scintillators may be used. One detector of this type [151] consists of a glass spiral filled with europium-activated calcium fluoride crystals, mounted in a plastic cell. It is sensitive to carbon-14 and tritium with counting efficiencies about the same as with anthracene. Glass scintillator powders are also available for this purpose [152].

The efficiency of conversion of beta particles to photons depends on the particle size of the solid scintillator. The mean free path should be as small as possible, particularly when using tritium, but very fine material may cause an undesirable back pressure in the chromatographic system and, though probably of minor importance, reduce the light collection efficiency from the cell. For practical purposes the particle size does not seem to be very important. Schram and Lombaert [153] recommended anthracene of 0.3 mm particle size for counting carbon-14 and 0.15 mm for tritium, while Piez [154] used normal crystalline material from which the fines had been removed.

A number of factors can affect counting results. The solid scintillator filling of the cell may pack down under the chromatographic flow altering the counting geometry and efficiency. Efficiency may also be reduced by quenching due to colour in the chromatographic eluate, yellow being a particularly efficient absorber of scintillation photons. Efficiency also depends on the range of the beta particles in the liquid and thus on the density of the liquid, but this is not usually of practical significance. A change in efficiency could, however, occur if gradient elution chromatography involved considerable changes in solvent density, but the importance of the effect would depend also on the scintillator particle size and the energy of the beta particles. As far as we are aware this has not been investigated quantitatively. Finally, radioactive contamination of the cell must be considered. For instance, some organic compounds may be adsorbed by anthracene leading to an increase in the background counting rate.

Another type of scintillation flow detector has been described by Hunt [155]. In this, part of the aqueous chromatographic effluent passes through a spectrophotometer for mass detection (of nucleotides, etc.) and part is mixed with a stream of dioxan scintillator solution which passes through a flow cell in a liquid scintillation counter. Efficiencies of up to 15% for tritium and 70% for carbon-14 were obtained, and the apparatus would detect peaks containing 0.9 nCi of tritium or 0.2 nCi of carbon-14 per ml of column effluent.

Measuring and Recording the Detector Output

Whether using Geiger or scintillation detection the output from the detector is fed to a scaler/timer/printer combination, to a ratemeter/recorder, or to both so that simultaneous analogue and integrated results are obtained. The major manufacturers of liquid scintillation counters supply flow cells that can be inserted into the counting chamber in place of the usual liquid scintillation counting vial. This is a convenient arrangement in that the versatile electronics can be used to give integrated count print-out for various time intervals, double-label counting, etc. In addition the scaler output can usually feed a ratemeter. Alternatively bench top scintillation flow-monitoring systems are available.

Apart from the nature of the detector the use of a flow counter resembles in many ways the scanning of chromatograms already described in Section 6.2.3, the rate of flow of the sample being analogous to the scanning speed. In flow counting, however, the geometry of the detector/sample combination is fixed and the counting efficiency should be stable, so that errors of the type introduced into quantitative scanning by variable sample/detector distance, chromatogram thickness, etc. are avoided.

In liquid chromatography it is often sufficient to measure the relative radioactivities of the various peaks in one run when the absolute counting efficiency of the system and other variables, such as flow rate, need not be known and the required results are given by the integrated counts or the relative peak areas, exactly as described in Section 6.2.3. It should be noted, however, that even here uniformity of flow rate is vital and is best achieved by the use of a peristaltic pump to feed the chromatography column.

For absolute measurements a factor relating observed counts (or peak area) with absolute radioactivity must be determined. This is most reliably done by running a sample of known radioactivity through the detector under exactly the same operating conditions, which may change if the same cell is cleaned and repacked with scintillator.

Equations relating various parameters of a flow detector system and the sample and background counts have already been given [equations (6.2) and (6.3)] and these can be used to calculate the sensitivity and precision of measurements.

6.5   SPECIAL TECHNIQUES

6.5.1   Neutron Activation Analysis

Analysis of substances on chromatograms or electrophoretograms by neutron activation is limited to compounds which contain atomic species which will result in radioactive isotopes of sufficiently long half-life following exposure to thermal neutrons. Apart from this limitation, the worker has a choice of irradiating a mixture and then chromatographing the resulting products, or of irradiating the chromatogram after chromatography and scanning the resulting radioactive spots. The former method suffers from the disadvantage that even comparatively mild doses of thermal neutrons can lead to degradation and rearrangement of compounds contained in the sample [156, 157] and the latter has received the wider application.

The need to present the sample for irradiation as a fairly compact specimen has limited the method to rolled-up strips of paper after chromatography or electrophoresis. Even the use of very small glass thin-layer plates (e.g. microscope slides) was precluded by the high background which arises when glass is bombarded with neutrons. The general availability of flexible sheets of pre-coated thin layer media may lead to wider use of the technique in thin-layer chromatography if flexible materials with low activation backgrounds can be used, but binders such as calcium sulphate must be avoided owing to the activation of inorganic atoms such as calcium and sulphur by neutrons. Whatever medium is used, the resulting radioactive components must then be detected or measured by methods described elsewhere in this chapter.

In the biomedical field, neutron activation of chromatograms has found particular application in the analysis of phospholipids [42, 157, 158].

Two of the major problems to be overcome in the use of the method are first, background interference arising from materials present in the chromatographic medium, which can be sufficient to mask completely the desired radioactive product, and secondly, the fragility of the chromatographic strip after exposure to the neutron flux. Methods of overcoming these problems in the paper chromatographic analysis of phospholipids by careful selection of irradiation conditions and pre-treatment of selected chromatography paper have been described by Johnson et al. [42]. It was found that the high background radioactivity obtained from normal chromatography paper after bombardment with neutrons could be reduced by 70% by pre-treatment of the paper with formaldehyde followed by overnight washing with water.

Robinson [159] criticised the technique of neutron activation analysis of chromatograms as a routine method on the grounds that the severe neutron bombardment caused disintegration of the chromatography paper and made subsequent handling difficult but such problems can be eliminated by the choice of suitably mild conditions [42].

The technique of neutron activation has received limited and mainly qualitative application in some fields, undoubtedly due in part to the lack of access of most research workers to irradiation facilities. Nevertheless, the method has sufficient possibilities to warrant further study of its uses both qualitatively and quantitatively in many fields.

6.5.2 Chromatography on Tritiated Media

The well-known ease of exchange of hydrogen in many molecules makes it possible to introduce a hydrogen isotope into a compound during its passage through a chromatographic medium. Originally used as a means of introducing deuterium for mass spectrometric studies by the use of alumina columns or by gas liquid chromatography [160–163], the method was extended by many workers to the radioactive labelling of compounds by the use of chromatographic media which contained exchangeable tritium [162, 164–167].

More recently, the idea of exchange-labelling has been used in the technique of self-labelling chromatographic analysis. Klein and Erenrich [168] described the use of radioactive alumina, prepared by exchange of its surface hydroxyl groups with tritiated water, in the quantitative analysis of mixtures of keto-steroids. The method appears to be applicable to submicrogram quantities, but depends on certain requirements such as the ability to prepare a uniform adsorbent of known specific activity, the knowledge of exchange capability and its linearity with sample weight and information about the behaviour of exchangeable groups in the molecule. The method also necessarily depends on efficient chromatographic resolution of the compounds in a mixture, but seems capable of greater application than it has received so far.

6.5.3 Miscellaneous Counting Techniques

So far we have discussed the use of conventional liquid scintillation counting for discrete solid and liquid samples, but other methods may be advantageous for particular purposes. With chromatograms of gamma-emitting nuclides, sample preparation for counting may be much simplified if a gamma scintillation counter is available. Dry portions of paper, TLC media, etc may then be counted directly, and it will only be necessary to ensure that the geometry of different samples is approximately the same.

For hard beta emitters in aqueous solution Čerenkov counting (see Chapter 3) has been found useful. Phosphorus-32, sodium-24 and potassium-42 are among the nuclides of biological interest that have been counted in this way [74, 169, 170].

In some cases it may be advantageous to use a beta scintillation counting system in which a solid phosphor is suspended in an aqueous or alcoholic solution of a radioactive sample [74, 88, 171], when the sample can be recovered unchanged after counting.

6.5.4 Combination of Various Types of Chromatography

It is not our intention to refer to the many uses of discontinuous sequential or parallel chromatography, in which the various methods may be used complementary to each other, as such uses will vary widely between different laboratories and types of work. Janak [172] refers to many combinations of chromatographic techniques which have been applied and are equally applicable to radioactive or unlabelled materials.

It is possible to combine directly certain types of chromatography (and electrophoresis), one of the more obvious examples being the condensation of gas chromatographic effluent onto a paper or thin-layer plate for subsequent separation by electrophoresis or chromatography [172, 173]. The main advantages of such a combination stem from the different separating factors involved in the various types of method. For instance, gas chromatography depends primarily on volatility, and temperature is one of the most important factors in the separation process while being of minor significance in paper chromatography. If thin-layer or paper chromatography is used before gas-liquid chromatography, the method is discontinuous and is merely a combination of complementary principles as discussed above. It can be advantageous however, to apply gas chromatographic effluent directly to a thin-layer plate or paper situated at the exit port of the gas chromatograph, particularly where emergent radioactive peaks are too close to be collected separately or to be resolved in a flow counter. The plate or paper may be moved under the exit port of the gas chromatograph in three basic modes: at constant speed, at a logarithmic rate or stepwise. Detailed and comprehensive accounts of the combination of gas-liquid chromatography and thin layer chromatography have been written by Janak [172] and by Kaiser [173].

Effluent from a liquid chromatography column can be similarly treated, though the method has little real application to the broad bands from conventional columns and may be more applicable to the relatively sharp bands of small amounts of material obtained from narrow bore high pressure liquid chromatography columns.

In all cases where effluent from a column technique, either gas or liquid, is applied to a thin medium, the new chromatogram can be later scanned or assayed by the methods described in Section 6.2.

Acknowledgements

The authors wish to acknowledge their colleagues, Dr. R. H. Nimms-Smith, Mr. E. J. Kentish of the Wellcome Research Laboratories and Dr. G. W. W. Stevens of Kodak Ltd., for advice on some of the aspects of autoradiography.

REFERENCES

1. E. A. EVANS, Tritium and its Compounds. Butterworths, London, (1966).
2. J. R. CATCH, Carbon-14 Compounds. Butterworths, London, (1961)
3. V. F. RAAEN, G. A. ROPP and H. P. RAAEN, Carbon-14. McGraw Hill, New York, (1968).
4. F. SNYDER and H. KIMBLE, *Anal. Biochem.*, 11, 510 (1965).
5. A. W. ROGERS, Techniques of Autoradiography. Elsevier, London, (1967).
6. A. T. JAMES, New Biochemical Separations (A. T. James and L. J. Morris eds.), Chapter 1. Van Nostrand, London, (1964).
7. E. S. KEMPNER and J. H. MILLER, *Science*, 135, 1063 (1961).
8. M. T. GILLIES, *Nature*, 182, 1683 (1958).
9. W. G. DUNCOMBE, *Nature*, 183, 319 (1959).
10. A. S. KESTON, S. UDENFRIEND and R. K. CANNAN, *J. Am. Chem. Soc.*, 68, 1390 (1946).
11. A. S. KESTON, S. UDENFRIEND and M. LEVY., *J. Am. Chem. Soc.*, 72, 748 (1950).
12. S. UDENFRIEND, *J. Biol. Chem.*, 187, 65 (1950).
13. S. UDENFRIEND, T. C. CLARK and E. TITUS, *Experientia*, 8, 379 (1952).
14. D. D. JACKSON and M. KAHN, *Intern. J. Appl. Radiation Isotopes*, 20, 742 (1969).
15. A. W. ROGERS, *Nature*, 184, 721 (1959).
16. B. MARKMAN, *J. Chromatog.*, 11, 118 (1963).
17. G. S. RICHARDSON, I. WELIKY, W. BATCHELDER, M. GRIFFITH and L. L. ENGEL, *J. Chromatog.*, 12, 115 (1963)
18. J. CHAMBERLAIN, A. HUGHES, A. W. ROGERS and G. H. THOMAS, *Nature*, 201, 774 (1964).
19. K. RANDERATH, *Anal. Biochem.*, 34, 188 (1970).

20. I. M. HAIS and K. MACEK (Editors), Paper Chromatography, p.210. Academic Press, New York, (1963).
21. e.g. "Neatan"; Merck.
22. A. J. TOMISEK and B. T. JOHNSON, *J. Chromatog.*, **33**, 329 (1968).
23. W. G. DUNCOMBE, *J. Chromatog.*, **36**, 557 (1968).
24. R. F. PHILIPS and W. R. WATERFIELD, *J. Chromatog.*, **40**, 309 (1969)
25 Radio-TLC Scanner, Type RTLS − 1A., Panax Equipment Limited; Redhill, England.
26. W. G. DUNCOMBE, *Intern. J. Appl. Radiation Isotopes*, **10**, 212 (1961).
27. E. C. STOWELL, *Nature*, **198**, 318 (1963).
28. R. C. RAY and G. W. W. STEVENS, *Brit. J. Radiol.*, **26**, 362 (1953).
29. A. T. WILSON and D. J. SPEDDING, *J. Chromatog.*, **18**, 76 (1965).
30. A. T. WILSON, *Nature*, **182**, 524 (1958).
31. K. RANDERATH, *Anal. Chem.*, **41**, 991 (1969).
32. S. F. CONTRACTOR and B. SHANE, *J. Chromatog.*, **41**, 483 (1969)
33. G. W. BUCKALOO and D. V. COHN, *Science,* **123**, 333 (1956).
34. C. O. TIO and S. F. SISENWINE, *J. Chromatog.*, **48**, 555 (1970).
35. M. J. DANIELS and D. G. WILD, *Anal. Biochem.*, **35**, 544 (1970).
36. G. FAIRBANKS, C. LEVINTHAL and R. H. REEDER, *Biochem. Biophys. Res. Commun.* **20**, 393 (1965).
37. M. LAMBIOTTE, *Atomlight, No.* **45**, 10 (1965).
38. J. E. GARDINER, *Nature*, **197**, 414 (1963).
39. A. E. LOWE, Quantitative Paper and Thin Layer Chromatography (E. J. Shellard ed.), Chapter 10. Academic Press, London and New York, (1968).
40. E. B. CHAIN, A. E. LOWE and K. R. L. MANSFORD, *J. Chromatog.*, **53**, 293 (1970).
41. Laboratorium Prof. Dr. Berthold, Wildbad, W. Germany.
42. P. JOHNSON, E. J. WEBER, H. E. CARTER and M. S. KROBER, *J. Lipid Res.*, **6**, 425 (1965).
43. R. G. DAVIS, *Biomed. Eng.*, March, 154, (1966).
44. B. A. WOOD, Quantitative Paper and Thin Layer Chromatography (E. J. Shellard ed.), Chapter 9. Academic Press, London and New York (1968).
45. System D.0212/XGD-01, Panax Equipment Limited, Redhill, England.
46. H. H. SELIGER and B. W. AGRANOFF, *Anal. Chem.*, **31**, 1607 (1959).
47. S. PRYDZ and K. S. SKAMMELSRUD, *J. Chromatog.*, **32**, 732 (1968).
48. T. B. MELO and S. PRYDZ, *Anal. Chem.*, **42**, 1093 (1970).
49. Photoscanner; Societe d'Applications Industrielles de la Physique, Paris.
50. M. M. NAKSHBANDI, *Intern. J. Appl. Radiation Isotopes*, **16**, 157 (1965).
51. R. A. SCHWANE and R. S. NAKON, *Anal. Chem.*, **37**, 315 (1965).
52. W. J. WHITEHOUSE and J. L. PUTMAN, Radioactive Isotopes Chapter V, Section 7 Clarendon Press, Oxford, (1953).
53. D. H. PILGRIM, *Intern. J. Appl. Radiation Isotopes*, **16**, 461 (1965).
54. L. W. HOLM, *Rev. Sci. Instr.*, **27**, 370 (1956).
55. e.g. Telefunken A. G., Ulm, W. Germany.

56. M. J. JOHNSON, *J. Chromatog.*, **20**, 100 (1965).
57. R. R. WILLIAMS and R. E. SMITH, *Proc. Soc. Exp. Biol. Med.*, **77**, 169 (1951).
58. Ekco Electronics Limited, Southend-on-Sea, England; Chromatogram Scanner, type N 679.
59. B. R. PULLAN, R. HOWARD and B. J. PERRY, *Nucleonics*, **24**, 72 (1966).
60. B. R. PULLAN, Quantitative Paper and Thin-Layer Chromatography (E. J. Shellard ed.), Chapter 11. Academic Press, London and New York, (1968).
61. E. HAAHTI, R. VIHKO, I. JAAKONMÄKI and R. S. EVANS, *J. Chromatog. Sci.*, **8**, 370 (1970).
62. E. STAHL (Editor), Thin Layer Chromatography 2nd Ed. pp. 101, 148 George Allen and Unwin Limited, London, (1969).
63. e.g. TLC Spot Remover; Quickfit Instrumentation, Stone, Staffs., England.
64. F. SNYDER, *Anal. Biochem.*, **9**, 183 (1964).
65. F. SNYDER and D. SMITH, *Separation Sci.*, **1**, 709 (1966).
66. F. SNYDER, *Separation Sci.*, **1**, 655 (1966).
67. I. M. HAIS, Paper Chromatography, (I. M. Hais and K. Macek eds.) p. 185 Academic Press, New York and London, (1963).
68. e.g. Microcolumn Chromatography Apparatus; Quickfit Instrumentation, Stone, Staffs., England.
69. J. C. TURNER, *Intern. J. Appl. Radiation Isotopes*, **19**, 557 (1968).
70. J. C. TURNER, *Intern. J. Appl. Radiation Isotopes*, **20**, 499 (1969).
71. B. W. FOX, *Lab. Pract.* **17**, 595 (1968).
72. R. LIEBERMAN and A. A. MOGHISSI, *Intern. J. Appl. Radiation Isotopes*, **21**, 319 (1970).
73. C. F. BAXTER and I. SENONER, *Anal. Biochem.*, **7**, 55 (1964).
74. J. H. PARMENTIER and F. E. L. TEN HAAF, *Intern. J. Appl. Radiation Isotopes*, **20**, 305 (1969).
75. G. N. GUPTA, *Microchem. J.*, **13**,4 (1968).
76. Packard Instrument Co. Inc., Downers Grove, Illinois, U.S.A.
77. M. H. GRIFFITHS and A. MALLINSON, *Anal Biochem.*, **22**, 465 (1968).
78. H. E. DOBBS, *Intern. J. Appl. Radiation Isotopes*, **19**, 155 (1968).
79. D. M. GILL, *Intern. J. Appl. Radiation Isotopes*, **18**, 393 (1967).
80. F. SNYDER, *Atomlight*, No. 58 (1967).
81. "NCS Solubilizer" – Amersham/Searle, Des Plaines, Illinois, U.S.A.
82. Beckman Instruments Inc., Fullerton, California, U.S.A.
83. H. TAKAHASHI, T. HATTORI and B. MARUO, *Anal. Biochem.*, **2**, 447 (1961).
84. D. L. HANSEN and E. T. BUSH, *Anal. Biochem.*, **18**, 320 (1967).
85. F. SNYDER and N. STEPHENS, *Anal. Biochem.*, **4**, 128 (1962).
86. W. F. BOUSQUET and J. E. CHRISTIAN, *Anal. Chem.*, **32**, 722 (1960).
87. E. A. DAVIDSON, Packard Tech. Bull., No. 4 (1962).
88. E. RAPKIN, *Intern. J. Appl. Radiation Isotopes*, **15**, 69 (1964).
89. Ciba (A.R.L.) Limited, Duxford, Cambridge.
90. W. L. BUCK and R. K. SWANK, *Rev. Sci. Instr.*, **29**, 252 (1958).
91. E. RAPKIN, Packard Tech. Bull. No. 5 (1961).
92. e.g. "Aerosil" (Bush, Beach and Segner Bayley, London); or "Cab-O-Sil" (Cabot Corp., Kokomo, Indiana, U.S.A.).

93. F. A. BLANCHARD and I. T. TAKAHASHI, *Anal. Chem.*, **33**, 975 (1961).
94. N. J. H. MERCER and J. F. HENDERSON, *Anal. Biochem.*, **13**, 559 (1965).
95. J. WILLENBRINK, *Intern. J. Appl. Radiation Isotopes*, **14**, 237 (1963).
96. A. H. GORDON, Electrophoresis of proteins in polyacrylamide and starch gels, North-Holland Publ. Co., Amsterdam and London, (1969).
97. R. S. BASCH, *Anal. Biochem.*, **26**, 184 (1968).
98. M. ZAITLIN AND V. HARIHARASUBRAMANIAN, *Anal. Biochem.*, **35**, 296 (1970).
99. S. WARD, D. L. WILSON and J. J. GILLIAM, *Anal. Biochem.*, **38**, 90 (1970).
100. D. H. ALPERS and R. GLICKMAN, *Anal. Biochem.*, **35**, 314 (1970).
101. H. S. ANKER, *F.E.B.S. Letters*, **7**, 293 (1970).
102. A. KARMEN, *J. Am. Oil Chemists' Soc.*, **44**, 18 (1967).
103. P. G. W. SCOTT, *Process Biochem.*, **16**, (1967).
104. A. KARMEN, Packard Technical Bulletin, No. 14, (1965).
105. F. A. IDDINGS and J. T. WADE, *J. Gas Chromatog.*, **1**, (11) 31 (1963).
106. P. J. THOMAS and H. J. DUTTON, *Anal. Chem.*, **41**, 657 (1969).
107. H. MEINERTZ and V. P. DOLE, *J. Lipid Res.*, **3**, 140 (1962).
108. A. K. HAJRA and N. S. RADIN, *J. Lipid Res.*, **3**, 131 (1962).
109. Millipore Filter Corporation, Bedford, Mass., U.S.A.
110. A. KARMEN, L. GIUFFRIDA and R. L. BOWMAN, *J. Lipid Res.*, **3**, 44 (1962).
111. A. KARMEN, I. McCAFFREY, J. W. WINKELMAN and R. L. BOWMAN, *Anal. Chem.*, **35**, 536 (1963).
112. W. G. DUNCOMBE and T. J. RISING, *Biochem. J.*, **109**, 449 (1968).
113. M. BENNETT and E. COON, *J. Lipid Res.*, **7**, 448 (1966).
114. W. A. CRAMER, J. P. W. HOUTMAN, R. O. KOCH and G. J. PIET, *Intern. J. Appl. Radiation Isotopes*, **17**, 97 (1966).
115. W. A. CRAMER, J. P. W. HOUTMAN, G. J. PIET and E. DE BOER, *Intern. J. Appl. Radiation Isotopes*, **20**, 129 (1969).
116. J. D. ROBBINS and J. E. BAKKE, *J. Gas Chromatog.*, **5**, 525 (1967).
117. G. POPJAK, A. E. LOWE, D. MOORE, L. BROWN and F. A. SMITH, *J. Lipid Res.*, **1**, 29 (1959).
118. G. POPJAK, A. E. LOWE and D. MOORE, *J. Lipid Res.*, **3**, 364 (1962).
119. Model 4998 Gas Radiochromatograph System, Nuclear Chicago, Des Plaines, Illinois 60018, U.S.A.
120. M. WELCH, R. WITHNELL and A. P. WOLF, *Anal. Chem.*, **39**, 275 (1967).
121. L. H. MASON, H. J. DUTTON and L. R. BAIR, *J. Chromatog.*, **2**, 322 (1959).
122. D. C. NELSON, P. C. RESSLER and R. C. HAWES, *Anal. Chem.*, **35**, 1575 (1963).
123. R. WOLFGANG and F. S. ROWLAND, *Anal. Chem.*, **30**, 903 (1958).
124. A. T. JAMES and E. A. PIPER, *J. Chromatog.*, **5**, 265 (1961).
125. A. T. JAMES and E. A. PIPER, *Anal. Chem.*, **35**, 515 (1963).
126. W. G. Pye Limited, Cambridge, England.
127. A. KARMEN, *J. Gas Chromatog.*, **5**, 502 (1967).
128. J. WINKELMAN and A. KARMEN, *Anal. Chem.*, **34**, 1067 (1962).

129. F. CACACE and G. PEREZ, *Anal. Chem.*, **41**, 368 (1969).
130. A. KARMEN, I. McCAFFREY and R. L. BOWMAN, *J. Lipid Res.*, **3**, 372 (1962).
131. R. H. BENSON and R. L. MAUTE, *Intern. J. App. Radiation Isotopes.* **17**, 488 (1966).
132. Type NE 802 gas flow cell, Nuclear Enterprises Limited, Edinburgh, Scotland.
133. P. JOHNSON and H. R. HAZELTON, *Proc. Soc. Anal. Chem.*, **6**, 162 (1969).
134. Hewlett-Packard Corp., Avondale, Pennsylvania 19311, U.S.A.
135. F. BERTHOLD and M. WENZEL, Instrumentation in Nuclear Medicine, Vol. 1, p. 251, Academic Press, New York, (1967).
136. H. R. HAZELTON and P. JOHNSON, *Brit. J. Pharmacol.*, **39**, 236P (1970).
137. H. SIMON, G. MUELLHOFER and R. MEDINA, Proc. I.A.E.A. Symp. Radioisotope Sample Measurement, Tech. Med. Biol., p. 317. Vienna, (1965).
138. R. O. MARTIN, *Anal. Chem.*, **40**, 1197 (1968).
139. T. H. SIMPSON, *J. Chromatog.*, **38**, 24 (1968).
140. Low-background radiodetector F 2111, Panax Equipment Limited, Redhill, England.
141. ST. BEHRENDT, *Z. Physik. Chem.*, **20**, 367 (1959).
142. M. PASCAUD, *J. Chromatog.*, **10**, 125 (1963).
143. A. KARMEN, *J. Lipid Res.*, **8**, 61 (1967).
144. M. WENZEL and W. STOHR, *Anal Biochem.*, **37**, 282 (1970).
145. K. PIEZ, Tech. Bull. No. 15 Nuclear-Chicago Corp., Des Plaines, Illinois, U.S.A. (1964).
146. H. E. DOBBS, *J. Chromatog.* 2, 572 (1959).
147. D. R. BANGHAM, *Biochem. J.* **62**, 552 (1956).
148. e.g. Type F10; 20th Century Electronics Limited, New Addington, Surrey, England.
149. e.g. Plastic scintillator Spiral Flow Cell, Type NE 801; Nuclear Enterprises Limited, Edinburgh, Scotland.
150. Model 316 Liquid Flow Cell; Packard Instrument Co. Inc., Downers Grove, Illinois, U.S.A.
151. Flow Cell for Organic Media, Type NE 808; Nuclear Enterprises Limited, Edinburgh, Scotland.
152. Glass Scintillator Powder, Type NE 901; Nuclear Enterprises Limited, Edinburgh, Scotland.
153. E. SCHRAM and R. LOMBAERT, *Anal. Biochem.*, **3**, 68 (1962).
154. K. A. PIEZ, *Anal. Biochem.*, **4**, 444 (1962).
155. J. A. HUNT, *Anal. Biochem.*, **23**, 289 (1968).
156. P. A. SELLERS, T. R. SATO and H. H. STRAIN, *J. Inorg. Nucl. Chem.*, **5**, 31 (1957).
157. J. HOLZI, *Natürwissenschaften*, **51**, 241 (1964).
158. F. NAKAYAMA and R. BLOMSTRAND, *Acta Chem. Scand.*, **15**, 1595 (1961).
159. J. R. ROBINSON, *Can. J. Biochem. Physiol.*, **40**, 1460 (1962).
160. K. MISLOW, M. A. W. GLASS, H. B. HOPPS, E. SIMON and G. H. WAHL, *J. Am. Chem. Soc.*, **86**, 1710 (1964).
161. W. J. RICHTER, M. SENN and A. L. BURLINGAME, *Tetrahedron Letters*, 1235 (1965).

162 M. SENN, W. J. RICHTER and A. L. BURLINGAME, *J. Am. Chem. Soc.*, **87**, 680 (1965).
163. G. STÖCKLIN, F. SCHMIDT-BLEECK and W. HERR, *Angew. Chem.* **73**, 220 (1961).
164. H. ELIAS in Proc. Conf. on Methods of Preparing and Storing Marked Molecules, p. 531 (Brussels, Nov. 13th-16th 1963) Euratom (1964).
165. H. ELIAS, K. H. LIESER and F. SORG, *Radiochim. Acta,* **2**, 30 (1963).
166. P. D. KLEIN and J. C. KNIGHT, *J. Am. Chem. Soc.*, **87**, 2657 (1965).
167. F. SCHMIDT-BLEECK, G. STÖCKLIN and W. HERR, *Angew, Chem.*, **72**, 778 (1960).
168. P. D. KLEIN and E. H. ERENRICH, *Anal. Chem.*, **38**, 480 (1966).
169. P. J. GARRAHAN and I. M. GLYNN, *J. Physiol.*, **186**, 55P (1966).
170. R. T. HAVILAND and L. L. BIEBER, *Anal. Biochem.*, **33**, 323 (1970).
171. D. STEINBERG, *Nature,* **182**, 740 (1958).
172. J. JANAK, Progress in Thin Layer Chromatography and Related Methods (A. Niederwieser and G. Pataki eds.), Vol. 2, p. 63. Ann Arbor — Humphrey, Michigan, (1971).
173. R. KAISER, Ancillary Techniques of Gas Chromatography, (L. S. Ettre and W. H. McFadden eds.), p. 299. Wiley-Interscience, New York, (1969).
174. Model 6000 Beta Camera; Baird-Atomic, Cambridge, Mass., U.S.A.
175. Beta-graph Radiochromatogram Imaging System; Panax Equipment Ltd; Redhill, England.

Chapter 7

Activation Analysis

H. J. M. Bowen

Department of Chemistry, The University
Whiteknights Park, Reading

7.1   INTRODUCTION

Activation is a technique of elementary analysis involving transmutation of elements by nuclear reactions. A sample is exposed to a flux ϕ of transmuting particles for a time period t, so that most of the elements in it become radioactive. The specific activity s induced in any parent nuclide may be calculated from the formula

$$s = 6.02 \times 10^{26}\, \phi \sigma f A^{-1}\, (0.5)^{t'/t\frac{1}{2}}\, [1 - (0.5)^{t'/t\frac{1}{2}}] \tag{7.1}$$

where

    s = specific activity in disintegrations s^{-1} kg^{-1}
    ϕ = flux of particles in m^{-2} s^{-1}
    σ = cross-section for particle interaction with parent nuclide in m^2
    f = fractional abundance of parent nuclide in element concerned
    A = atomic weight of parent nuclide
    t' = time between activation and counting ⎫
    t = activation period   ⎬ in the same units
    $t_{\frac{1}{2}}$= half life of daughter nuclide ⎭

In practice, the count rate, a, from a sample is compared with the count rate, a_0, from a standard containing a mass, m_0, of the element sought, after sample and standard have been exposed to the same flux of transmuting particles. The unknown mass m of element in the sample is given by

$$m = m_0 a/a_0$$

The transmuting particles are usually neutrons, though they may be high energy photons, protons or other charged particles. A major technical problem is that of sorting out the different radioactive nuclides produced, either by radiochemistry, gamma spectrometry or a combination of these techniques.

Equation (7.1) shows that in order to obtain maximum sensitivity from the technique, the following conditions must be satisfied:

(i)   The flux of particles must be as high as possible. In practice, the highest available sustained neutron or proton fluxes are of the order of $10^{18} - 10^{19}$ m^{-2} s^{-1}, though higher transient fluxes may be obtained.

(ii) The cross-section of the nuclear reaction must be as high as possible. The fact that cross sections of (n, γ) reactions with thermal neutrons are relatively large accounts for much of the popularity of these particles for activation analysis. On the other hand, the low cross sections for reactions with high-energy photons have led to a neglect of gamma activation techniques. In the literature cross sections are often quoted in 'barns' where a barn = 10^{-24} cm^2 = 10^{-28} m^2 = 100 fm^2.

(iii)   The fractional abundance of the parent nuclide is usually fixed, but may be adjusted in some controlled experiments. The atomic weight of the parent and the half-life of the daughter nuclide cannot be altered.

(iv)   The time between activation and counting should be as short as possible, especially for short-lived daughter nuclides. However, in some cases it pays to increase this time to allow shorter-lived impurities to decay away.

(v)   The activation period is generally chosen to be of the same order of magnitude as the half-life of the daughter nuclide, unless this is very long. If $t = t_{1/2}$, 50% of the maximum possible specific activity is obtained, and the activation of long-lived impurities is minimised.

7.1.1   History of the Technique

From a historical point of view, three stages in the development of the technique may be distinguished. The first stage, during which the technique was a scientific curiosity, lasted from 1936 until about 1948. The discovery that neutrons were produced by mixing radium and beryllium [1] was soon followed by the first applications of neutron activation analysis [2]. Applications of deuteron activation [3] and alpha activation [4] soon followed. However, applications were limited partly by the primitive state of radiation detection (scintillation counting was not discovered until 1948) but still more by the difficulty of producing large fluxes of neutrons and other particles. Only 34 papers on activation analysis appeared in the literature during this stage. The construction of the first nuclear reactor in 1942, and the development of cyclotrons for the acceleration of charged particles, helped to initiate the second stage, which was one of consolidation.

The consolidation phase lasted about ten years from 1949 until 1959. During this phase a large number of applications of the technique were developed, mostly by nuclear scientists who were interested in other fields such as geochemistry or biochemistry. The expansion of interest during this stage resulted in the publication of about 600 research papers on a great variety of applications. Notable technical developments during this period included the development of scintillation counting and multichannel analysers and the production of commercial neutron generators. These developments ushered in the third, or current, stage of activation analysis.

The current phase is characterised by the utilisation of the technique by scientists whose primary interest is not in nuclear science: that is, activation analysis is taking its place among the armoury of modern tools for elementary analysis. The number of research papers on the topic is still increasing exponentially, with a doubling period of about 3 years (Fig. 7.1). In addi-

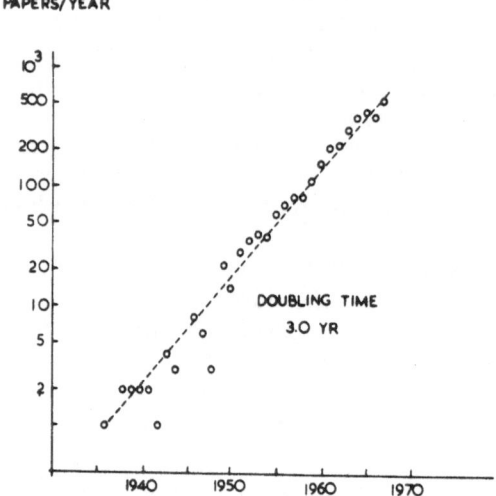

Fig. 7.1.　Rate of increase of activation analysis literature.

tion conferences, reviews, bibliographies, books and radiochemical journals have proliferated at an alarming rate during the last decade. It is impossible to mention all the latest technical developments, but among the most important are the automation of activation analysis and gamma spectrometry [5], the automation of radiochemical techniques [6], the development of pulsed reactors [7] and the use of germanium semiconductor detectors [8]. There are at present no signs of any declining interest in the technique.

7.2　ADVANTAGES AND DISADVANTAGES

7.2.1　Nature of Sample and Its Destruction

Activation analysis can be carried out for any type of sample, solid, liquid or gaseous, but is mostly used for solid samples. Liquid samples, or hydrated salts, may generate considerable amounts of gas when exposed to radiation, so that they may explode when irradiated in poorly sealed ampoules. In theory, activation analysis is a non-destructive technique, as is illustrated by its *in vivo* applications. In practice, activation analysis always slightly changes the ratios of nuclides present in a sample, and destroys the sample if radiochemistry is employed.

7.2.2　Range of Elements Determinable

Activation analysis is a technique for determining individual nuclides or elements, and gives no information on the chemical state of these elements. This is its main limitation. For example, activation analysis may be used to determine total iron in a soil sample, but cannot distinguish between iron(II), iron(III) or iron humate compounds.

On the other hand it has been used to determine all the 83 elements between hydrogen and bismuth, including technetium and promethium, as well as the radioactive elements radium, actinium, thorium, protoactinium, uranium and plutonium. Its most common form, neutron activation analysis, has been used for 77 of the 83 elements lighter than bismuth, but is unsuitable for boron, beryllium, carbon, hydrogen, helium-4 and lithium. Activation analysis is a multi-element technique, but it has not yet proved possible to determine all 77 elements in a single run. However Samsahl has been able to determine 40 elements in biological material [9], and Morrison *et al.* have determined 45 elements in rocks [10].

7.2.3 Sensitivity
The sensitivity of the technique may be worked out for any nuclear reaction from equation (7.1). Thus for the reaction

$$^{55}Mn(n, \gamma)^{56}Mn$$

using a flux of 10^{18} thermal neutrons m^{-2} s^{-1}, we know that $\sigma = 13.6 \times 10^{-28}$ m^2, $f = 1$ and $A = 55$. Putting $t = t'$, $= t_{1/2} = 2.6$ hours in order to simplify the arithmetic, we find

$$S = \frac{6.02 \times 10^{16} \times 13.6 \times 0.5 \times 0.5}{55} \text{ disintegrations } s^{-1} \text{ kg}^{-1}$$

$$= 3.7 \times 10^{15} \text{ disintegrations } s^{-1} \text{ kg}^{-1}$$

Since 3.7 disintegrations s^{-1} are readily detectable, we can determine less than 10^{-15} kg, or 1 picogram of manganese. The detectable amounts of many elements using thermal neutron activation analysis are given in Table 7.1. It can be seen that, apart from the light elements, the technique has a sensitivity which is equal or superior to most other available analytical methods.

TABLE 7.1. *Sensitivity of Thermal Neutron Activation Analysis*

10 – 100 ng	Fe, S
1 – 10 ng	Zr
0.1 – 1 ng	Ag Ca*Cr Mg*Ni Rb Se Te Ti*Zn
10 – 100 pg	Cd, Ce, Cl, Cs, Gd, Ge, Mo, Nd, Os, Pt, Ru, Sn, Ta, Tb, Th, Tm
1 – 10 pg	Al*,Ar, As, Ba, Br, Co, Cu, Er, Ga, Hg, I*, In, Na, Pd, Pr, Sb, Sc, U, W, Yb
0.1 – 1 pg	Hf, Ho, Ir, La, Re, Rh*, Sm, V*
10 – 100 fg	Au, Dy, Eu, In, Mn

Assumptions: Flux of neutrons $= 10^{18}$ $m^{-2}s^{-1}$. Irradiation time = ten half-lives or 100 hr, whichever is shorter. Zero time between irradiation and counting. *$t_{1/2} < 30$ min.

The main factors limiting the sensitivity are the available flux ϕ, σ, t, t′, and $t_{1/2}$. These factors are discussed in Section 7.4, but it should be noted that experimental difficulties are most serious for product nuclides of half-life shorter than about 30 minutes. In practice, this includes the six elements mentioned in Section 7.2.2 and also Al, Cl, F, I, Mg, N, Nb, Ne, O, Rh, Ti, and V when activated by thermal neutrons. Such elements are best determined in laboratories adjacent to the reactor, neutron generator or cyclotron used to activate them.

Note that the number of atoms of an element converted to a radio-nuclide is very small: it is equal to $n\phi\sigma t$, where n is the number of atoms irradiated. Putting in the figures for the manganese irradiated described above, we find that only one atom in 10^5 is converted to ^{56}Mn under these conditions.

7.2.4 Selectivity

The selectivity of the technique depends on the efficiency of radiochemical separation, and/or of gamma spectroscopy employed. Radiochemical separations have been worked out for all nuclides of sufficiently long half-life, and in skilled hands these can lead to excellent selectivity. Since a separated radionuclide has several independent properties which may be measured, such as its half-life, beta spectrum and gamma spectrum, there are several checks that the separation is indeed complete. The resolution of gamma spectrometry is often not adequate for counting one radionuclide in the presence of large amounts of others. For example, both ^{27}Mg and ^{56}Mn have gamma rays at 0.84 MeV, and all positron-emitters have a gamma ray at 0.51 MeV. In such cases the half-life of the gamma peak may give an indication that two or more nuclides are responsible for it.

7.2.5 Interferences

A major advantage of activation analysis is that inactive impurities introduced after activation do not affect the determination. Thus no special care is needed in purifying acids for dissolving active samples, or in the preparation of reagents for radiochemistry. Only radioactive contaminants can affect the result.

On the other hand trouble may be caused by interfering nuclear reactions. In practice these are only serious between elements differing in atomic number by one or two units. An example is the mutual interference of the three elements phosphorus, sulphur and chlorine.

When these elements are activated in a reactor, the following reactions may occur:

Reaction	Cross section (fm^2)
^{31}P(n,γ)^{32}P	19
^{34}S(n,γ)^{35}S	1.1
^{32}S(n,p)^{32}P	30
^{35}Cl(n,γ)^{36}Cl	3300
^{35}Cl(n,p)^{35}S	30
^{35}Cl(n,α)^{32}P	19

It can be seen that both sulphur and chlorine will seriously interfere with the analysis of phosphorus by activation. This is a particularly bad case and it is the exception rather than the rule for interfering nuclear reactions to have comparable cross-sections.

7.2.6   Accuracy and Precision

Accuracy measures the closeness of a result to the truth: precision measures the scatter of a series of determinations. Both accuracy and precision have been tested using standard materials [11–16] and by comparison with other techniques. The tests have shown that activation analysis is an accurate technique in almost all cases, that is, results obtained by activation agree with those obtained by other techniques within the observed scatter. In a very few cases, notably in the determination of potassium by neutron activation followed by gamma spectrometry, consistently low results may be obtained; but it seems that these are caused by technical faults in the gamma spectrometry.

The precision of the technique as usually practised is around ±5%, inter-laboratory precision is between 5% and 10% while extremely careful work can produce precisions of ±1% [17].

7.2.7   Costs and Hazards

The cost of capital equipment for activation analysis may be quite high, but the cost per analysis is low, comparable to that of other techniques. Few laboratories can afford to buy a nuclear reactor, neutron generator, or cyclotron solely for activation analysis. However, space in existing reactors can readily be hired. In Britain, a volume of 30 cm^3 can be exposed to a flux of 10^{16} neutrons m^{-2} s^{-1} for up to a week at a cost of £10, and it is often possible to pack 100 samples into this volume. Sample preparation and radiochemical separations require skilled technicians and may be expensive. Counting may be carried out using relatively simple detection equipment (costing about £300), or with multichannel analysers costing up to £7,000. Additional costs may be incurred if the results are worked out by computers.

Activation analysis can usually be carried out in an ordinary laboratory provided with a fume hood, metal trays, rubber gloves and a few lead bricks for shielding. Only rarely does an activated sample becomes a radiation hazard, as when it contains large amounts of sodium or chlorine (e.g. sea water). The amounts of radioactivity to be handled usually lie between picocuries and a few millicuries, and the usual precautions must be taken to avoid ingestion of radioactive material and to minimise exposure to radiation.

7.3   PREPARATION OF SAMPLES AND STANDARDS

7.3.1   Sample Collection

The biggest experimental difficulty in activation analysis is often the collection of the sample without contaminating it. The technique is so sensitive that it can determine traces of contaminants which are undetectable by most other techniques. Techniques of handling samples which have been proved to involve neither contamination nor loss of material are rare or non-existent.

A primary problem is dust control, since a clean sample will pick up dust if left uncovered on a laboratory bench, and some samples may have been accumulating dust for years. The rate of dust accumulation depends on the surface area exposed [18], and is usually of the order of $2-20$ mg m^{-2}s^{-1}. Samples should therefore be handled in dry boxes or in clean rooms with laminar air flow.

Drying of samples is often used to reduce their volume and hence economise in reactor space. It should be avoided where mercury is to be determined, since this element is volatile at room temperature. Some forms of arsenic, bromine, chlorine, iodine, selenium and sulphur are volatile below $100°C$, and their rate of loss under the particular drying conditions used should be tested.

Samples used for activation are usually small, and may weigh between a few μg and 1 g or more. Care must be taken that such small samples are truly representative of the material being analysed, especially when this is inhomogeneous.

7.3.2 Solid Samples

Hard samples, such as rocks, are frequently brittle, and may then be wrapped in polyethylene and pulverised by impact. Uncontaminated particles with fresh surfaces can then be selected for analysis.

Soft samples, such as biological material, can be cut with home-made knives made of silica or hard plastic to avoid metal contamination. Both these materials may leave fragments of the blade in the sample, and new materials, such as boron nitride or pure titanium, may prove more suitable. Stainless steel blades and forceps should be avoided in view of their high content of chromium, manganese, molybdenum, vanadium, tungsten etc. Biological material is often handled in the frozen state, to avoid leakage of elements in the fluids exuding when the tissue is cut.

7.3.3 Liquid Samples

Liquids are liable to contamination from their container walls, and to losses by adsorption. Samples should be collected in clean polyethylene or silica ampoules, and blanks should be run. Few containers can be cleaned so thoroughly that the sodium and manganese blanks are immeasurably small after activation. Many silica containers have a gamma spectrum after activation indicating the presence of antimony-124. Adsorption losses may be reduced by boiling the container with acids containing suitable carriers after activation.

7.3.4 Containment for Activation

Activation data for some elements commonly used for containment are given in Table 7.2.

TABLE 7.2. *Activation Data for Elements Used in Containers*

Element	n absorption cross section (fm^2)	Half-life of product	Specific* activity of product $(mCi\ g^{-1})$	Proton reaction cross section (fm^2)
Al	23	168 s	128	1.0
Be	1.0	2.7×10^6 years	9×10^{-8}	1.2
C	0.33	5760 years	1.6×10^{-8}	20
H	33	12.3 years	7×10^{-7}	0
O	0.02	29 s	4×10^{-4}	50
Si	13	2.6 hours	2.0	18

*Neutron flux 10^{-18} m^{-2} s^{-1}. Irradiation time ten half-lives or 100 h, whichever is the shorter.

In practice, the most widely used canning materials for neutron activation are polyethylene, aluminium and silica. Polyethylene is convenient for cleaning and heat sealing, but becomes brittle and degrades after exposure to $10^{21} - 10^{22}$ neutrons m^{-2}. Hence it cannot be used for long activation times. Aluminium foil is convenient for packaging solids, but becomes very radioactive on exposure to neutrons and gives rise to sodium-24 by the (n, α) reaction. Silica ampoules are mostly used for activating liquid samples, and may develop a high pressure during activation, making them somewhat hazardous to open. Transfer of activities from containers to samples has been discussed by Brune [19].

Thin tantalum foil has been used for containment of samples for proton activation, since this metal has a high threshold energy for reaction with protons.

7.3.5 Preparation of Standards

The ideal standard should be as nearly as possible identical with the sample in size, shape and chemical composition. In this way any errors due to flux inhomogeneities or self-shielding during the activation period are likely to be the same for both sample and standard and will tend to cancel out. In most cases the analyst will have to prepare his own standard and cannot rely on commercially available materials other than pure chemicals.

Standards are usually prepared by dissolving known masses of pure elements or their oxides in a suitable acid, so that the primary standardisation is by weight. Small weighed aliquots of the diluted solution are then taken as standards. This can be done in several ways.

(a) The liquid aliquot is sealed into a polyethylene or silica ampoule.
(b) The liquid aliquot is adsorbed onto a small square of ashless filter paper, which is then dried and packaged.
(c) The liquid aliquot is adsorbed onto a known weight of sample, which is then dried and packaged.

Yet another technique for preparing standards consists of adsorbing a known weight of substance from solution onto a known weight of ion-exchange resin. Individual resin beads can then be used as standards in the nanogram to picogram range [20, 21].

Girardi et al. [22] have suggested using a single comparator standard when many elements are to be determined in a sample by thermal neutron activation. They used equal lengths of wire containing 90% Al, 9% Mg and 1% Co as single comparator standards: each standard contained 211 μg Co. The relative activities induced in 16 pure elements relative to cobalt were first determined experimentally. The ^{60}Co activity induced in the single comparator standard was then used to calculate the amounts of 16 elements present in activated samples. This method appears to be satisfactory for routine analyses in a reactor with a constant flux of neutrons, a high proportion of which have thermal energies.

Leliaert et al. [23] have suggested an internal standard technique to determine an unknown element x as follows. Take two samples of masses m_1 and m_2, and add a known mass m of element x to the second of these. After activation, select an element y, present in both samples, which has a much longer half-life than x.
Then

$$\frac{a_1(y)}{a_2(y)} = \frac{m_1}{m_2} \alpha$$

where a_1 and a_2 are activities in counts s^{-1} and α is a factor, nearly equal to 1, to allow for any difference in neutron flux reaching samples 1 and 2. It is easy to show that the fraction of x in the sample is given by

$$\frac{ma_1(x)}{\alpha m_1 a_2(x) - m_2 a_1(x)}$$

This method should be useful where large flux inhomogeneities or neutron self-shielding effects are anticipated, but has not been widely employed.

7.4   THE ACTIVATION PROCESS

7.4.1   Isotopic Neutron Sources

Several isotopic neutron sources have been suggested for activation analysis, as shown in Table 7.3. None of these have found any major applications, owing to the low neutron fluxes available, but they are useful for demonstrating activation to students [24]. Ricci & Handley [25] have recently suggested using ^{252}Cf as a spontaneous fission source of neutrons, as its theoretical neutron output is 2.34×10^{12} neutrons $s^{-1}g^{-1}$. Unfortunately this is not a commercial proposition at the current cost of this nuclide, which is 10^9 per gram (about 600 Ci).

TABLE 7.3. *Isotopic Neutron Source Data*

Source of neutrons	Half-life	Yield: available neutrons (s^{-1})	Mean neutron energy (MeV)
^{124}Sb + Be (γ,n)	60 days	1.6×10^6 Ci^{-1}	0.024
^{241}Am + Be (α,n)	470 years	2.2×10^6 Ci^{-1}	4.0
^{226}Ra + Be (α,n)	1620 years	1.0×10^7 Ci^{-1}	4.0
^{252}Cf	2.46 years	2.3×10^{12} g^{-1}	0.7

7.4.2 Neutron Generators [26]

A number of neutron generators are commercially available and may be installed in any laboratory with suitable shielding. The commonest reaction used to produce neutrons is the bombardment of zirconium tritide targets with 0.1–0.5 MeV deuterons. This gives rise to a point source of neutrons with energies of about 14 MeV, by the reaction:

$$^2H + {}^3H \longrightarrow {}^4He + {}^1n$$

Large linear accelerators producing beams of electrons with energies of 1.5 MeV upwards have also been used to produce neutrons. The electrons strike a platinum target which gives rise to gamma radiation, and this reacts with beryllium foil to give neutrons

$$^9Be + \gamma \longrightarrow 2\,{}^4He + {}^1n$$

provided the incident gamma rays have energies exceeding 1.66 MeV. These neutrons have energies in the range 0.5–5 MeV.

Fluxes available from neutron generators depend on a number of factors such as beam currents, target geometry, etc. It appears that the rate of neutron production is of the order of $10^{10} - 6 \times 10^{13}$ s^{-1} from the deuteron accelerators, and $10^{11} - 2 \times 10^{14}$ s^{-1} from the electron accelerators. The corresponding neutron fluxes available (in m^{-2}s^{-1}) are about 100 fold higher. The main practical problems are those of deterioration of targets, non-uniformity of neutron flux and biological shielding.

Despite these snags, sensitivities of detection have been published for 60 elements [27, 28], and fast neutron activation is routinely used for the determination of oxygen in several matrices by the (n,p) reaction. 1 μg of oxygen can be determined quite rapidly. It has also been suggested as a means of determining nitrogen in hydrocarbons [29], to determine Al, Fe, Mg and Si in rocks [30, 31]. Activation analysis has been carried out with pulsed neutrons to determine nuclides with half-lives down to 4 ms [32].

7.4.3 Neutrons from Nuclear Reactors [33]

The bulk of neutron activation work is carried out using nuclear reactors. These consist of uranium dispersed in a moderator, whose function is to slow down the neutrons produced by fission. Natural uranium contains 99.27% ^{238}U and 0.72% ^{235}U, but when used in reactors it is frequently enriched in the lighter isotope. Commonly used moderators include water, heavy water or graphite, all of which need to be extremely pure. The amount of uranium required varies from 0.6 kg ^{235}U to 40,000 kg natural uranium used in the extinct Harwell reactor BEPO. In the former case the moderator need only be a few tens of kg of D_2O, while in the latter case 8 $\times$ 10^5 kg graphite was used. Reactors are usually large, immoveable objects and are also very expensive. The cheapest cost £80,000 ($200,000), and large reactors may cost 1—5 million dollars.

The neutrons produced in reactors vary in energy between 0.02 eV and about 25 MeV, with a mean energy of about 2 MeV. Their energies are highest in the vicinity of uranium fuel elements, and lowest in regions occupied by the moderator. Fig. 7.2 shows the relative fluxes of thermal (E

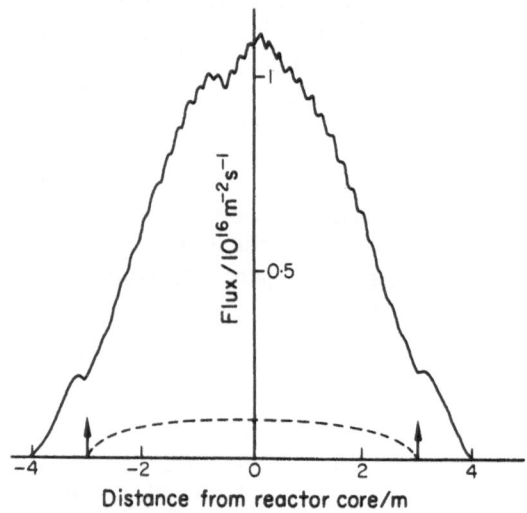

Fig. 7.2. Flux distribution of thermal (————) and fast (- - - - - - -) neutrons across the core of the reactor BEPO.

= 0.025 eV) and fast (E > 1MeV) neutrons in a typical reactor. The fluxes of thermal neutrons obtained from research reactors are usually of the order of 10^{15} — 10^{18} m^{-2} s^{-1}, and the volume of sample which can be exposed to these fluxes is usually about 50 cm^3. Such samples will also be exposed to fluxes of fast neutrons of the same order as those of the thermal neutrons, and to a flux of gamma photons (energies between 0 and 7 MeV) which deposit 10—100 W kg^{-1} in the sample. This seldom causes overheating problems except in samples with high neutron cross sections, which absorb extra energy from reactor neutrons. The main effect of the gamma photons is to cause decomposition in most organic materials exposed to them; for example, polyethylene ampoules will become brittle and break up after exposure to > 10^{21} neutrons m^{-2}. If overheating is a problem, samples can be cooled before or during activation [34, 35].

Since thermal neutrons are more often used for activation than are fast neutrons, activation is best carried out in tubes surrounded by moderator. Light water gives a lower flux ratio of thermal to fast neutrons than do D_2O

or graphite. If fast neutron activation is specifically requested, the sample should be surrounded with 1 mm thick cadmium foil. Cadmium has an enormous cross section for thermal neutrons (0.25 fm^2) but a small cross section for fast neutrons. Attempts to increase the fast neutron flux by surrounding samples with ^{235}U are expensive and have not been much used. Resonance neutrons, with energies between 1 and 10^3 keV, have only rarely been used for activation [36–38].

Reactors with a negative temperature coefficient of reactivity can be 'pulsed' by conversion to a supercritical state for short periods. For example, the Triga (^{235}U/water) reactor can be pulsed reproducibly for 6 ms, so that the peak neutron flux in the pulse is $2 \times 10^{21} \text{ m}^{-2}\text{s}^{-1}$. Pulsed reactors are proving useful for determining elements which activate to products with half-lives less than 60 s, such as ^{19}F(n,γ)^{20}F(12s), ^{76}Se(n,γ)$^{77\text{m}}$Se(17s), ^{206}Pb(n,γ)$^{207\text{m}}$Pb(0.8s) and ^{16}O(n,p)^{16}N(7.1s). In such cases the sensitivity is greatly enhanced with respect to conventional activation [7, 33].

7.4.4 Sample Rotation and Recovery

In large reactors, where the flux gradient is small ($< 1\% \text{ cm}^{-1}$), it is not necessary to rotate sample cans to ensure that all the contents receive the same flux. Many sample holders now have provision for sample rotation, especially in reactors which have small cores and hence high flux gradients.

When short-lived nuclides are to be studied, the rate of recovery of the sample from the reactor core may be critical. In such cases, the sample can, made of plastic, is blown into and out of the reactor core using compressed air to drive it through aluminium or plastic tubing. The shortest time of recovery is usually not less than 1s, so that it is difficult to study activation products with half-lives of the order of milliseconds. The plastic can is colloquially known as a rabbit, and automatic devices can be used to pro-gramme a series of rabbit irradiations and counts. If these are not employed, the hazards of radiation exposure when manually opening rabbits should be measured and kept to a minimum. Rabbits should always be received and opened in a fume cupboard to avoid inhaling ^{16}N$_2$ and ^{41}Ar.

7.4.5 Sample Decomposition during Irradiation

The decomposition of samples under neutron and gamma bombardment has already been mentioned. The work of Taylor and Rogers [39] has shown that ammonium salts, hydrates of inorganic salts and organic compounds may generate gas pressures on irradiation.

Since all nuclei gain recoil energy on capturing a neutron, they frequently break their chemical bonds as a result. This is the Szilard-Chalmers effect. Brune [40] has suggested using this effect to recover short-lived nuclides such as ^{27}Mg from activated biological material, without the need for ashing.

7.4.6 Reactor Activation Using Recoil Protons, etc.

Any sample containing hydrogen will behave as a source of fast protons during neutron activation. The ^{1}H nucleus readily accepts energy from fast neutrons, and the energetic protons produced are called recoil protons. They can induce proton reactions, such as $^{18}O(p,n)^{18}F$ in water, which may interfere in other determinations. The techniques have seldom been used in actual analyses [41]. Recoil tritons from lithium have also been suggested, but rarely used, for analysis.

7.4.7 Charged Particle Activation

Although charged particles have been used for a long time for activation analysis, they have never achieved the popularity that neutrons have for this purpose. The fluxes of charged particles available, their cost, and the cross-sections of their reactions with nuclei, are of the same order of magnitude as those for neutrons. The main disadvantages of charged particles are poor penetration and the existence of threshold energies for activation. The number of interfering nuclear reactions is usually higher for charged particle activation than it is for neutron activation.

Typical penetration depths for aluminium are given in Table 7.4. This implies that samples for charged particle work must be extremely thin. In addition they must be carefully cooled during activation. A 1 mA beam of 1 MeV protons would dissipate 1 kW in the 0.01 mm surface layer of an aluminium target.

Threshold energies and interfering reactions are illustrated in Table 7.5.

It can be seen that interferences in determining boron may be controlled by working at a suitable energy and that a variable energy cyclotron or accelerator is desirable for this work. The potential sensitivity for boron and nitrogen is good.

Charged particle activation analyses have been mostly used to determine light elements, especially first row elements, which are not readily activated by neutrons. Notable applications include determinations of boron and nitrogen in silica [42], determinations of carbon and oxygen in metals [43] and of oxygen by the $(^{3}He,p)$ reaction [44]. Applications usually involve protons or deuterons, though tritons and ^{3}He ions have interesting possibilities. Alpha particle activation is exciting little current interest, but has a sensitivity of 1 pg in determining calcium by the $^{40}Ca\,(\alpha,p)^{43}Sc$ reaction [45].

TABLE 7.4. *Penetration of Protons, Deuterons and Helions into Aluminium (mm)*

Energy (MeV)	p^{+}	d^{+}	$^{3}He^{++}$
1	0.012	0.007	0.006
5	0.19	0.11	0.03
10	0.62	0.38	0.07
20	2.05	1.25	0.24

TABLE 7.5. *Threshold Energies and Induced Activities when Producing 20 Minute* ^{11}C

Nuclear reaction	Threshold energy (MeV)	Activity (disintegrations s^{-1} μA^{-1} μg^{-1})	
		5 MeV	20 MeV
^{10}B (p,γ) ^{11}C	0.4 ⎫	630	20000
^{11}B (p,n) ^{11}C	2.76 ⎭		
^{12}C (p,n) ^{11}C	~ 19	0	400
^{14}N (p,α) ^{11}C	~ 3.5	6	60000
^{10}B (d,n) ^{11}C	2	380	~ 2000
^{11}B $(d,2n)$ ^{11}C	5.5	0	~ 2000
^{12}C (d,t) ^{11}C	~ 14.5	0	300

Note: 1 μA beam current = 6 $\times$ 10^{12} particles s^{-1}.

7.4.8   Photon Activation [46]

(γ,γ) photon activation with low energy photons (< 5 MeV) can be used to activate certain nuclides to their excited states. The cross sections for such reactions are small (10^{-34} m^2), and the reactions have not been used in analysis [47, 48].

All nuclides react with photons above the threshold energy, which varies from about 20 MeV for light nuclei to 7.5 MeV for heavy nuclei. The typical reaction is a (γ,n) reaction leading to a proton-rich isotope of the element activated. At higher energies other reactions compete, so that the (γ,n) cross section has a maximum energy which is about 20–25 MeV for light nuclei and about 15 MeV for heavy nuclei. Cross sections for (γ,n) reactions lie between 0.1 and 10 fm^2, and are usually higher for heavy nuclides [49]. Penetration is excellent and there are no problems due to photon absorption by the sample, but samples have to be small because of flux inhomogeneities.

Applications of photon activation seem to have been limited by the availability of photon sources with adequate fluxes. The main requirement is a linear accelerator or similar machine producing a monoenergetic beam of variable energy up to at least 25 MeV. The beam current is usually of the order of 10^{13} –10^{14} electrons s^{-1}, and this can be converted into bremsstrahlung photons by stopping the electrons in a cooled target of a heavy element such as platinum or tungsten. The efficiency of conversion may be as high as 30% but the photons produced are not monoenergetic and only a fraction of them have energies approaching that of the incident electron beam. Few workers seem to have measured the absolute fluxes of photons produced in their experiments.

(γ,n) photon activation has the disadvantage that nearly all the nuclides produced are positron emitters and give rise to a 0.51 MeV peak in their gamma spectrum. However, it has been used to determine light elements such as carbon, nitrogen and oxygen in organic materials, metals, ceramics etc. [50]. These elements give rise to 20 minute ^{11}C, 10 minute ^{13}N and 2 minute ^{15}O, whose chemical separation or physical resolution present problems which have largely been solved. (γ,n) activation offers distinct advantages over neutron activation for the determination of Ti and Zr, and perhaps also for Nb, Ni and Rh [51]. Photon activation with 70 MeV photons has been used to determine lanthanide elements [52].

7.4.9 Prompt Radiation Studies

Prompt radiation studies differ from true activation analysis in that they involve measurements of the radiation emitted during the activation process itself. Thus the count rate is directly proportional to the amount of parent nuclide present, the particle flux ϕ and the cross section for the reaction σ. In many cases the nuclear reaction does not produce a radioactive product. Thus hydrogen can be determined from the prompt gammas emitted during the ^{1}H(n,γ)^{2}H reaction. Analyses using these reactions have the advantages of speed and high inherent sensitivity, often higher than for neutron activation analysis [53]. Disadvantages include the difficulties of observing and resolving the prompt radiation spectrum.

Prompt gammas from (n,γ) reactions have recently attracted attention [54]. Isenhour and Morrison [53] used a thermal neutron beam issuing from a reactor in trial studies. This beam had a low flux ($2 \times 10^{10} \text{m}^{-2}\text{s}^{-1}$) and was mechanically modulated at 25–600 Hz so that the gamma spectra with and without the neutron beam on could be compared. The delayed gamma spectrum could then be subtracted from the total spectrum to obtain the prompt spectrum. This work showed that prompt gamma spectra are often complex, and that the low flux employed limited the useful applications of the technique to a few elements such as Cd, Gd, Fe, Nd and Sm [55]. Prompt gamma work has been much used for studying boreholes in situ. If higher fluxes become available, the technique could have many laboratory applications, especially for microprobe work. More specialised applications include the determination of fissionable nuclides by the (n,f) reaction, and of boron and lithium by the ^{10}B(n,α)^{7}Li and ^{6}Li(n,α)^{3}H reactions. The fission tracks or alpha tracks are measured by autoradiography in fine-grained nuclear emulsions [56], or in polycarbonate detectors [57, 58].

Pierce and Peck [59] have investigated prompt gamma emission following pulsed irradiation by charged particles such as protons, deuterons and helium-3 ions. The technique of subtraction of the delayed gamma spectrum from the total spectrum was employed. The main technical difficulties were poor penetration and surface contamination of samples, even in high vacuum. The method looks promising for determining light elements, where sensitivities of 1 μg were obtained for oxygen using the reaction ^{16}O(d,p)^{17}O + γ (0.87 MeV).

Early work on prompt neutron emission by ^{2}H and ^{9}Be, which have unusually low threshold energies for the (γ,n) reaction, has not been followed up to any great extent [60, 61]. The most promising development is the introduction of neutron time of flight spectrometry as a means of measuring the energy spectrum of the prompt neutrons. Thus Peisach and Pretorius [62], using 3 MeV deuterons to activate various gases, were able to resolve the prompt neutrons produced by (d,n) reactions from ^{2}H, ^{12}C, ^{14}N and ^{16}O, and could measure the deuterium content of natural hydrogen gas from its prompt neutron spectrum after deuteron bombardment.

7.5 RADIOCHEMICAL SEPARATIONS [63]

7.5.1 Sample Dissolution
Different samples require different agents to bring them into solution [64]. In general, metals can be dissolved in strong hydrochloric and/or nitric acids. Minerals are best fused with about ten times their weight of sodium peroxide in a nickel or zirconium crucible. Biological materials may be dissolved in sulphuric acid + hydrogen peroxide or in 1:1 perchloric acid + nitric acid, or in 1:1 molten sodium nitrate + potassium nitrate (20 X weight of sample) [65]. All these ashing techniques can be carried out in a few minutes.

7.5.2 Chemical Yields
When radiochemical steps have been proved to be quantitative, yield determinations are not needed. Where losses are expected, known masses of inactive carriers of the elements to be determined are added. These elements are converted to a suitable weighing form at the end of the separation and any losses corrected for. Alternatively a known amount of long-lived active tracer is added prior to separation, and its percentage loss determined by counting at the end of the analysis. Another method of determining chemical yields is to re-activate the carrier after separation and counting; this avoids the need to precipitate the separated radioelement in a form suitable for weighing. The yield of carrier is occasionally determined by spectrophotometry or other conventional techniques.

7.5.3 Techniques of Chemical Separation
Where single elements are required in a radiochemically pure state, there is usually a wide choice of separation techniques available. Where speed of separation is important, distillation or solvent extraction are the most suitable techniques, as these can be carried out in a few seconds or minutes at most. On the other hand when selectivity is the limiting factor, ion exchange techniques are usually superior, though they may take several hours.

When a trace element has to be determined in a matrix, the main problem is often to remove the activities due to major elements in the matrix. For example, sodium-24 often constitutes 99% or more of the activity in activated biological material. This can be removed either by dissolving the sample in 7M nitric acid and passing it down a column of hydrated antimony pentoxide [66], or by dissolving the sample in perchloric acid, adding 9 times the volume of acetone, and passing the solution down a

TABLE 7.6. *Some Multi-element Separation Schemes*

Sample dissolved in	Number of elements	Number of fractions	Separation technique	Automatic or manual	Reference
HCl	41	41	Precipitation + ion exchange	Manual	71
H_2SO_4 + HF	45	6	Distillation, absorption, ion exchange & solvent extraction	Manual	10
HF + H_2O_2	23	18	Ion exchange	Manual	72
HNO_3 + HCl	12	7	Ion exchange	Semi-automatic	73
HCl	32	14	Ion exchange	Manual	74
H_2SO_4 + H_2O_2	40	11	Fast ion exchange	Automatic	9
HNO_3	22	6	Solvent extraction	Manual	75
$HClO_4$ + HF	25	6	Solvent extraction	Automatic	76

column of finely ground sodium chloride [67, 68]. Similar separations are available for other interfering radionuclides [69, 70].

7.5.4 Multi-element Separations
A number of multi-element separation schemes are now available and information on them is summarized in Table 7.6.

Note that few of the separations aim at doing more than fractionating the radionuclides present into groups, which can then be determined by gamma spectrometry. Interesting trends include increased speed (Samsahl's latest scheme is complete in five minutes [9]); avoidance of carriers; and a versatile programming device for unit operations [76]. The phase-separating centrifuge designed by Sutton and Vallis [77] is suitable for automated radiochemistry. No automated gas chromatography separations have yet been put forward [78], but there are exciting possibilities of combining activation with mass spectrography for rapid separations [79].

7.6   DETERMINATION OF RADIOACTIVITY

7.6.1   Beta Counting and Čerenkov Counting
These techniques have been discussed in Chapters 1 and 3. Thin anthracene crystals have largely replaced Geiger-Müller counters for counting betas from solid samples. This is a convenient, cheap and efficient way of counting many pure radionuclides. Low energy beta emitters or alpha

emitters are rarely used in activation analysis but are best counted by liquid scintillation techniques. Ross [80] has shown that liquid scintillation counters can readily be adapted to Čerenkov counting for samples emitting betas with energies exceeding 0.2 MeV. The sample can be counted in aqueous solutions, and the counting efficiency increases with beta energy.

7.6.2 Gamma Spectrometry

This has been fully discussed in Chapter 2. At present there is still competition between the relatively cheap sodium iodide/Tl detectors and the more expensive Ge/Li detectors with their much improved resolution. The semiconductor detectors need to be kept at liquid nitrogen temperatures. When sodium iodide detectors are used, there is no need to use a multichannel analyser with more than 256 channels, and standard techniques of peak area determination [81] or spectrum stripping may be applied. On the other hand a germanium detector with a resolution of 2 keV would need about 4000 channels to record a 0–2 MeV spectrum, and this can lead to problems. The larger multichannel analyser is more expensive, and its readout time may be considerable. However the spectrum is usually so well resolved that spectrum stripping is unnecessary [82] (Fig. 7.3).

Since even a single spectrum from a germanium detector may involve 4000 units of data, some workers prefer to store these directly in the memory banks of a small computer. This saves time and simplifies manipulation of the data to obtain actual analyses, but greatly adds to the cost of the equipment. Thus the smallest computers used cost about £16000 ($40000) [83].

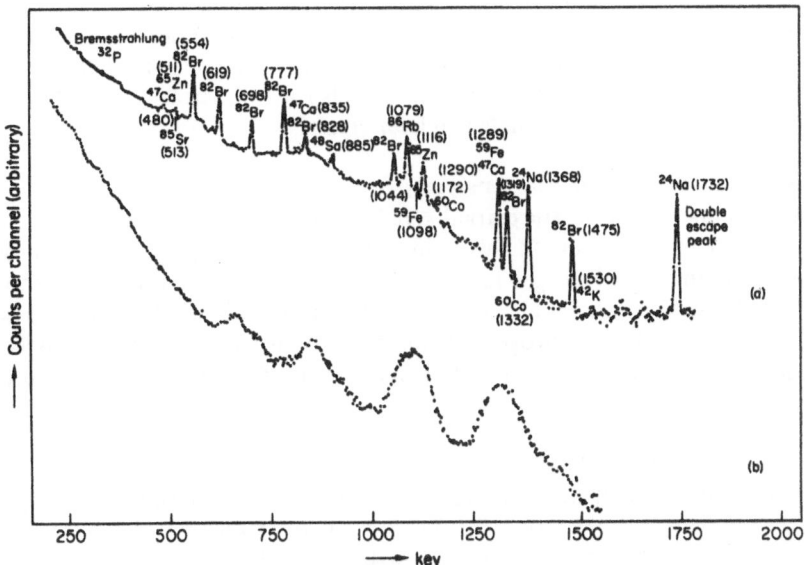

Fig. 7.3. Gamma ray spectra from standard kale, after activation for 8 days at 5_3 × 10^{16} neutrons m^{-2}s^{-1} and 9 days decay. A; 11 cm^3 Ge/Li detector. B; 7.5 cm × 7.5 cm NaI/Tl detector. From Girardi et al., *Anal. Chem.*, 37, 1085 (1967).

7.7   APPLICATIONS OF ACTIVATION ANALYSIS

It is impossible to do more than mention some of the more important applications in a review of this length. Early applications are described by Bowen and Gibbons [84].

7.7.1   Pure Chemistry

There is a good potential for inorganic microanalysis, i.e. analysis of major constituents in samples weighing only a few μg. Examples include the determination of the Ba/Fe ratio in synthetic ferrites [85] and the Cs/Rh/I ratios in $Cs_4Rh_2I_{10}$ [86]. Photoneutron activation, with its capacity of determining oxygen, could be useful here, and also in organic microanalysis (C, O and N determinations). The precision of the method may be limiting, and should be tested.

The use of activation to analyse paper chromatograms has recently been reviewed [87]. The technique has been most useful for organic phosphates, as the determination of many other elements is hampered by the high, inhomogeneous blank. A synthetic cellulose or similar material, containing negligible amounts of metals and halogens, would be very useful for this work. Most commercial papers crumble after exposure to about 10^{21} neutrons m^{-2}.

7.7.2   Pure Elements and Compounds

Activation analysis has been valuable in determining impurity levels in very pure metals (e.g. Al, Be, Fe, Ti and Zr) and in semiconductors (e.g. GaAs, Ge, Si) as well as in diamond and graphite. It is fair to say that much semiconductor technology could not have advanced so fast if activation analysis had not been available to determine concentrations of impurities, mostly at the nanogram level. There is still great interest in determining trace elements in all these materials and also in refractory oxides (TiO_2, ZrO_2), ferrites, yttrium garnets and laser crystals.

7.3.3   'On-Line' Industrial Analyses

Few 'on-line' analyses are carried out by activation, but the potential for using neutron generators in this field is high. Suitable applications might be the determination of oxygen in steel, strip, or the determination of common elements such as Al, F, Mn, or Si in ores or coal during their passage on conveyor belts. Recent progress in the field has been summarised by Bakes et al. [30] and by Ashe et al. [88] and a method for determining lead in flowing petrol has been published [89].

7.7.4   Geochemical Applications

An immense amount of work has been carried out in analysing the standard rocks G-1 and W-1 [14–16, 90]. Among the 74 elements which have been measured in these rocks only 11 (B, Be, Bi, C, Ca, Li, Mg, Ni, Pb, Rh and S) have not yet been determined by activation. Seven elements (Au, Ir, O, Os,

Re, Ru and Se) have been determined only by activation analysis, as no other methods have proved adequate. It is largely as a result of these studies that the accuracy and precision of the technique have been thoroughly tested. Now that stocks of G-1 and W-1 are largely exhausted, data are accumulating on new standard rocks [13].

These and other data have been extensively used by geochemists interested in the abundances of the elements [91], magmatic differentiation [92], and fractionation during sedimentation [93].

Equally extensive data have been accumulating over the last two decades for meteorites. At least 60 elements have been determined in both iron and stony meteorites by neutron activation, and a recent paper describes the non-destructive determination of 14 elements in chondrites [94]. No less than 41 of the 63 elements determined in the first lunar samples were determined by neutron activation [95]. These were Al, As, Ba, Br, Ca, Ce, Co, Cr, Cs, Cu, Eu, Fe, Ga, Gd, Hf, Ho, K, La, Lu, Mg, Mn, Mo, Na, Nd, Ni, Rb, Sb, Sc, Sm, Sr, Ta, Tb, Th, Ti, Tm, U, V, W, Yb, Zn, Zr. In geochemical prospecting, activation analysis has been less widely used, but it is a valuable technique for determining scarce elements such as iridium, rhenium and gold. It has been much employed to determine natural levels of two elements with unique nuclear properties, namely beryllium by the (γ,n)

TABLE 7.7. *Elements Determined by Neutron Activation in Sea Water*

Element	Concentration (ng l^{-1})	Reference	Element	Concentration (ng l^{-1})	Reference
Ag	290	98	Nd	2.8	101
As	1700–3650	99	Ni	100–3200	98
Au	11	98	Pr	0.64	101
Ba	13000	100	Rb	125	100
Ce	1.2	101	Re	8.4	106
Co	10–46	102	Ru	0.74	107
Cr	480	103	Sb	330	98
Cs	300	100	Sc	0.63	103
Cu	40–12000	104	Se	90	98
Dy	0.91	101	Sm	0.45	101
Er	0.87	101	Sr	8300000	108
Eu	0.13	101	Ta	$\leqslant$20	109
F	1320000	105	Tb	0.14	101
Fe	23000	103	Tl	19	110
Gd	0.70	101	Tm	0.17	101
Ho	0.22	101	U	3310	111
In	0.10–0.31	112	Y	13.3	101
La	3.4	101	Yb	0.82	101
Lu	0.15	101	Zn	210–17600	104
Mn	40–19000	104			

reaction [60], and uranium either by delayed neutron counting [96] or by fission-track counting [58].

7.7.5 Natural Waters

Activation analysis is a good method of measuring toxic and trace elements in fresh waters [97]. It is less satisfactory for sea water, because of the large amounts of sodium and chlorine present, but is nevertheless widely used for determining rare elements in this medium. Current trends involve great care in sampling and preconcentration techniques on board ship in order to avoid storage problems. Table 7.7. illustrates the great sensitivity of neutron activation for rare elements. Note that the concentrations of many rare elements in sea water are not constant, but vary with depth, geographical position etc.

7.7.6 Botany and Agriculture

This field has recently been reviewed by Bowen [113], who emphasizes the potential of the technique in studying the composition of soil solutions, toxic residues, and subcellular particles.

The 20 elements currently believed to be essential for plants can be divided into three groups with respect to neutron activation as follows:

n.a.a. competitive with other techniques: Co, Cu, Mn, Mo, N, Na, O, P, Se, Si, Zn

n.a.a. possible, other techniques simpler: Ca, Cl, Fe, K, Mg, S

n.a.a. not possible: B, C, H

The standard Kale powder prepared by Bowen [12] has been analysed in numerous laboratories. 47 of the 55 elements determined have been analysed by activation analysis, and no other techniques have proved adequate for Au, Br, Ce, Cs, Dy, Ga, Hf, In, Ir, La, O, Pd, Re, Ru, Sb, Sc, Sm, Ta, Th, U, W and Zr in this material.

7.7.7 Biochemistry and Medicine

This field has recently been reviewed by Comar [114]. There have been numerous applications in determining rare elements in blood, soft tissues, bones and teeth, which have emphasized the need for standardised samples of these materials.

Several clinical studies make use of activation. Comar has devised an automated analysis of iodine which is currently used for studying thyroid metabolism, and the effects of antimony in drugs used to treat bilharzia have been investigated [115]. A number of workers have tested the potential of the method in diagnosing cystic fibrosis, which increases sodium and potassium in sweat, hair and nails. Activation is obviously suitable for analysing biopsy samples, which usually weigh less than 10 mg, and it may soon be used in a survey of the causes of cardiovascular disease and some forms of cancer. The technique has proved useful in metabolic balance experiments using the occupancy principle [116], and can be used to determine stable tracers in metabolic tests on patients who might be susceptible to radiation. *In vivo* activation analysis, involving exposure of patients to low neutron doses, is remarkable in that it can determine several elements (Ca, Cl, N, Na) in living subjects, but has no clinical applications as yet.

7.7.8 Pollution

Inorganic air and water pollutants can readily be measured by neutron activation, though air filters often yield a high blank. Zoller and Gordon and Dams *et al.* [117, 118] have determined 24 and 33 elements respectively in particulates from city air by this method, and other work on similar topics may be found in the report of the Conference 'Modern Trends in Activation Analysis' (1969).

7.7.9 Archaeology

This field has been reviewed by Ashworth and Abeles [119]. Non-destructive activation has been used to determine sodium, manganese and other elements in ancient pottery and glass [120], and to discover the trace elements in metal objects, especially coins. Destructive analysis has also been performed on such objects as paint flakes from old pictures and drillings from silverware carried out with a <1 mm drill. The results have been of considerable value both in relating objects to their place of origin and in detecting forgeries.

7.7.10 Forensic Science

This topic has been reviewed by Coleman [121] and by Jervis [122]. Activation analysis has proved itself an important and useful tool in three main fields.

It has been used to detect inorganic elements such as As, Hg, P, Sb, Se and Tl in food and tissues in cases of poisoning. This has been especially important for arsenic, where administration of the element to a subject can now be detected in 1 mm lengths of individual hairs.

It has also been used to detect traces of gunpowder on the hands of subjects who have recently fired revolvers. In the U.S.A., powders contain sufficient barium and antimony to leave μg amounts of these elements on a hand after a single firing.

Finally, it has been used to compare samples to establish a common origin, which requires much background information. The technique has been used for drugs, glass [123], hair and paint flakes, but is unsatisfactory for blood samples.

REFERENCES

1. J. CHADWICK, *Proc. Roy. Soc. A*, **136**, 692 (1932)
2. G. HEVESY and H. LEVI, *Kgl. Danske Vidensk. Selsk. (Math.)* **14**, No. 5 (1936).
3. G. T. SEABORG and J. J. LIVINGOOD, *J. Amer. Chem. Soc.*, **60**, 1784 (1938).

4. L. D. P. KING and W. J. HENDERSON, *Phys. Rev.*, **56**, 1169 (1939).
5. W. E. KUYKENDALL and R. E. WAINDERDI, Rept. TEES-2565-1, College Station, Texas, (1960).
6. K. SAMSAHL, Repts. AE-54 and AE-56, *Aktiebolaget Atomenergi*, Stockholm, (1961).
7. H. R. LUKENS, H. P. YULE and V. P. GUINN, *Nucl. Instrum. Meth.*, **33**, 273 (1963).
8. S. S. MARKOWITZ and J. D. MAHONY, Proc. I.A.E.A. Symp. Radiochemical Methods of Analysis. 1, p. 419. Vienna, (1964).
9. K. SAMSAHL, Rept. AE-389, *Aktiebolaget Atomenergi*, Stockholm, (1970).
10. G. H. MORRISON, J. T. GERARD, A. TRAVESI, R. L. CURRIE, S. F. PETERSON and N. M. POTTER, *Anal. Chem.*, **41**, 1633 (1969).
11. G. B. COOK, M. B. A. CRESPI and J. MINCZEWSKI, *Talanta, 10*, 917 (1963).
12. H. J. M. BOWEN, Advances in Activation Analysis. 1, (J. M. A. Lenihan and S. J. Thomson, eds.) p. 101. Academic Press, London, (1969).
13. F. J. FLANAGAN, *Geochim. Cosmochim. Acta*, **33**, 81 (1969).
14. M. FLEISCHER and R. E. STEVENS, *Geochim. Cosmochim. Acta*, **26**, 525 (1962).
15. M. FLEISCHER, *Geochim. Cosmochim. Acta*, **29**, 1263 (1965).
16. M. FLEISCHER, *Geochim. Cosmochim. Acta*, **33**, 65 (1969).
17. A. P. SEYFANG and A. A. SMALES, *Analyst*, **78**, 394 (1953).
18. J. CHOLAK, L. J. SCHAFER and R. F. HOFFER, *Archs. Ind. Hyg.*, **2**, 443 (1950).
19. D. BRUNE, *Radiochim. Acta*, **5**, 14 (1966).
20. D. H. FREEMAN and R. A. PAULSON, *Nature*, **218**, 563 (1968).
21. D. H. FREEMAN, L. A. CURRIE, E. C. KUEHNER, H. D. DIXON and R. A. PAULSON, *Anal. Chem.*, **42**, 203 (1970).
22. F. GIRARDI, G. GUZZI and J. PAULY, *Anal. Chem.*, **37**, 1085 (1965).
23. G. LELIAERT, J. HOSTE and Z. EECKHAUT, *Nature*, **182**, 600 (1958).
24. W. S. LYON, Guide to Activation Analysis. Van Nostrand, New York, (1964).
25. E. RICCI and T. H. HANDLEY, *Anal. Chem.*, **42**, 378 (1970).
26. E. A. BURRILL and M. H. MACGREGOR, *Nucleonics*, **18**(12) 64 (1960).
27. J. PERDIJON, *Anal. Chem.*, **39**, 448 (1967).
28. I. FUJII, T. INOUYI, H. MUTO, K. ONODERA and A. TANI., *Analyst*, **94**, 189 (1969).
29. J. T. GILMORE and D. E. HULL, *Anal. Chem.*, **34**, 187 (1962).
30. J. M. BAKES, P. G. JEFFERY, D. W. DOWNTON and J. D. L. H. WOOD, Proc. SAC Conference, Nottingham, p. 277. Heffer, Cambridge, (1965).
31. D. E. FISHER, *Nature*, **222**, 866 (1969).
32. A. TANI, Y. MATSUDA, Y. YUASA and N. KAWAI, *Radiochem. Radioan. Lett.*, **1**, 155 (1969).
33. V. P. GUINN, Advances in Activation Analysis. 1, (J. M. A. Lenihan and S. J. Thomson, eds.) p. 37. Academic Press, London, (1969).
34. D. BRUNE and O. LANGSTROM, *Radiochim. Acta*, **5**, 1 (1966).

35. D. BRUNE and H. WENZEL, *Anal. Chem.*, **42**, 511 (1970).
36. D. C. BORG, R. E. SEGEL, P. KIENLE and L. CAMPBELL, *Internat. J. Appl. Radiation Isotopes*, **11**, 10 (1961).
37. D. BRUNE and K. JIRLOW, *Nukleonik* **6**, 242 (1963).
38. D. BRUNE, *Anal. Chim. Acta*, **46**, 17 (1969).
39. K. J. TAYLOR and G. T. ROGERS, UKAEA Report AERE-R 3409 Harwell, England, (1962).
40. D. BRUNE, *Anal. Chim. Acta*, **34**, 447 (1966).
41. D. C. AUMANN and H. J. BORN, Proc. Int. Conf. Modern Trends in Activation Analysis, p. 265. Texas A and M University, College Station, (1965).
42. C. ENGELMANN and G. CABANE, Proc. Int. Conf. Modern Trends in Activation Analysis, p. 332. Texas A and M University, College Station, (1965).
43. J. L. DEBRUN and P. ALBERT, *Bull. Soc. Chim. Fr.* **3**, 1020 (1969).
44. S. S. MARKOWWITZ and J. D. MAHONY, *Anal. Chem.*, **34**, 329 (1962).
45. R. PRETORIUS and E. A. SCHWEIKERT, *J. Radioan. Chem.*, **7**, 319 (1971).
46. G. J. LUTZ, *Anal. Chem.*, **43**, 93 (1971).
47. H. R. LUKENS, J. W. OTVOS and C. D. WAGNER, *Internat. J. Appl. Radiation Isotopes*, **11**, 30 (1961).
48. C. M. GORDON and R. E. LARSEN, *Radiochem. Radioan. Lett.*, **5**, 369 (1970).
49. G. J. LUTZ, *Anal. Chem.*, **41**, 424 (1969).
50. P. ALBERT, C. ENGELMANN, S. MAY and J. PETIT, *C. R. Acad. Sci. Paris*, **254**, 119 (1962).
51. E. SCHWEIKERT and P. ALBERT, Proc. I.A.E.A. Symp. Radiochemical Methods of Analysis. **1**, 323. Vienna, (1964).
52. T. KUTO and A. F. VOIGT, *J. Radioanal. Chem.*, **4**, 325 (1970).
53. T. L. ISENHOUR and G. H. MORRISON, *Anal. Chem.*, **38**, 162 (1966).
54. D. L. ALLAN, B. H. ARMITAGE and W. SPENCER, *Anal. Chim. Acta*, **53**, 401 (1971).
55. S. M. LOMBARD and T. L. ISENHOUR, *Anal. Chem.*, **41**, 1113 (1969).
56. A. W. ROGERS, Techniques of Autoradiography. Elsevier, Amsterdam, (1967).
57. R. L. FLEISCHER and D. B. LOVETT, *Geochim. Cosmochim. Acta*, **32**, 1126 (1968).
58. B. S. CARPENTER and C. H. CHEEK, *Anal. Chem.*, **42**, 121 (1970).
59. T. B. PIERCE and P. F. PECK, Proc. SAC Conference, Nottingham, p. 159. Heffer, Cambridge, (1965).
60. G. GOLDSTEIN, *Anal. Chem.*, **35**, 1620 (1963).
61. K. D. GEORGE and H. H. KRAMER, *Trans. Amer. Nucl. Soc.*, **11**, 474 (1968).
62. M. PEISACH and R. PRETORIUS, *Anal. Chem.*, **39**, 650 (1967).
63. F. GIRARDI, Proc. Modern Trends in Activation Analysis, NBS Spec. Publication 312 p. 577 (1969).
64. J. DOLEZAL, P. POVONDRA and Z. SULCEK, Decomposition Techniques in Inorganic Analysis. Iliffe, London, (1968).
65. H. J. M. BOWEN, *Anal. Chem.*, **40**, 969 (1968).

66. F. GIRARDI and E. SABBIONI, *J. Radioan. Chem.*, **1**, 169 (1968).
67. C-W. TANG and C. J. MALETSKOS, *Science*, **167**, 52 (1970).
68. H. J. M. BOWEN and J. A. COOK, *Radiochem. Radioan. Lett.*, **5**, 103 (1970).
69. D. L. MASSART, *J. Radioan. Chem.*, **4**, 265 (1970).
70. E. SABBIONI, R. PIETRA and F. GIRARDI, *J. Radioan. Chem.*, **4**, 289 (1970).
71. P. ALBERT, *Pure Appl. Chem.*, **1**, 111 (1960).
72. J. C. RICQ, *J. Radioan. Chem.*, **1**, 443 (1968).
73. F. GIRARDI and M. MERLINI, L'Analyse par radioactivation et ses applications aux sciences biologiques. Presses Universitaires de France, Paris, (1964).
74. R. E. JERVIS and K. Y. WONG, Proc. I.A.E.A. Symp. Nuclear Activation Techniques in the Life Sciences, p. 137. Vienna, (1967).
75. R. A. CHALMERS and D. M. DICK, *Anal. Chim. Acta*, **31**, 520 (1964).
76. G. C. GOODE, C. W. BAKER and N. M. BROOKE, *Analyst*, **94**, 728 (1969).
77. D. W. SUTTON and D. G. VALLIS, *J. Radioan. Chem.*, **2**, 377 (1969).
78. J. TADMOR, *Chromatog. Reviews*, **5**, 223 (1962).
79. E. A. SCHWEIKERT, *Talanta*, **15**, 883 (1968).
80. H. H. ROSS, *Anal. Chem.*, **41**, 1260 (1969).
81. S. STERLINSKI, *Anal. Chem.*, **42**, 151 (1970).
82. F. GIRARDI, G. GUZZI and J. PAULY, *Radiochim. Acta*, **7**, 202 (1967).
83. J. A. DOOLEY, J. H. GORRELL, J. M. THOMPSON and E. HOFFMAN, Proc. Modern Trends in Activation Analysis, NBS Spec. Publication 312 p. 1148 (1969).
84. H. J. M. BOWEN and D. GIBBONS, Radioactivation Analysis. OUP, (1963).
85. M. CHIBA, *J. Radioan. Chem.*, **2**, 415 (1969).
86. R. STELLA and M. Di CASA, *Radiochem. Radioan. Lett.*, **8**, 137 (1971).
87. A. Z. BUDZYNSKI and J. Z. BEER, *Anal. Chim. Acta*, **46**, 281 (1969).
88. J. B. ASHE, P. F. BERRY and J. R. RHODES, Proc. Modern Trends in Activation Analysis, NBS Spec. Publication 312 p. 913 (1969).
89. C. C. HAYWARD, G. OLDHAM and A. R. WARE, *Radiochem. Radioan. Lett.*, **6**, 381 (1971).
90. R. E. STEVENS and W. W. NILES, *U.S. Geol. Surv. Bull.*, No. 1113 p. 3 (1960).
91. H. C. UREY, *Quart. J. Astron. Soc.*, **8**, 23 (1967).
92. L. A. HASKIN and M. A. HASKIN, *Geochem. Cosmochim. Acta*, **32**, 433 (1968).
93. O. LANDSTRÖM, K. SAMSAHL and C. G. WENNER, Rept. A. E.-296 *Aktiebolaget Atomenergi*, Stockholm, (1967).
94. J. F. EMERY, J. E. STRAIN, G. D. O'KELLEY and W. S. LYON, *Radiochem. Radioan. Lett.*, **1**, 137 (1969).
95. G. H. MORRISON, *Science*, **167**, 1449 (1970).
96. S. AMIEL and M. PEISACH, *Israel J. Chem.*, **1**, 306 (1963).

97. D. P. KHARKAR, K. K. TUREKIAN and K. K. BERTINE, *Geochim. Cosmochim. Acta,* **32,** 285 (1968).
98. D. F. SCHUTZ and K. K. TUREKIAN, *Geochim. Cosmochim. Acta,* **29,** 259 (1965).
99. A. A. SMALES and B. D. PATE, *Analyst,* **77,** 188 (1952).
100. E. BOLTER, K. K. TUREKIAN and D. F. SCHUTZ, *Geochim. Cosmochim. Acta,* **28,** 1459 (1964).
101. O. T. HOGDAHL, S. MELSOM and V. T. BOWEN, Trace Inorganics in Water. (R. A. Baker, ed.), p. 308. Amer. Chem. Soc. Advances in Chemistry, **73,** (1968).
102. D. E. ROBERTSON, *Geochim. Cosmochim. Acta,* **34,** 553 (1970).
103. D. Z. PIPER and G. G. COLES, *Anal. Chim. Acta,* **47,** 560 (1969).
104. J. F. SLOWEY and D. W. HOOD, *Geochim. Cosmochim. Acta,* **35,** 121 (1971).
105. P. E. WILKNESS and V. J. LINNEBOM, *Limnol. Oceanog.,* **13,** 530 (1968).
106. E. M. SCADDEN, *Geochim. Cosmochim. Acta,* **33,** 633 (1969).
107. B. W. DIXON, J. F. SLOWEY and D. W. HOOD, USAEC Rept. TID-23295 (1966).
108. H. J. M. BOWEN, *J. Mar. Biol. Assocn. U.K.,* **35,** 451 (1956).
109. H. HAMAGUCHI, R. KURODA, K. HOSOHARA and T. SHIMIZU, *Nippon Gensh. Gakk.* **5,** 662 (1963).
110. A. D. MATTHEWS and J. P. RILEY, *Anal. Chim. Acta,* **48,** 25 (1969).
111. K. K. BERTINE, L. H. CHAN and K. K. TUREKIAN, *Geochim. Cosmochim. Acta,* **34,** 641 (1970).
112. A. D. MATTHEWS and J. P. RILEY, *Nature,* *225,* 1242 (1970).
113. H. J. M. BOWEN, Proc. I.A.E.A. Symp. Nuclear Activation Techniques in the Life Sciences, p. 287. Vienna, (1967).
114. D. COMAR, Advances in Activation Analysis. 1, (J. M. A. Lenihan and S. J. Thomson, eds.), p. 163. Academic Press, London, (1969).
115. M. M. MOLOKHIA and H. SMITH, *J. Trop. Med. Hyg.,* **72,** 222 (1969).
116. F. C. GILLESPIE, J. SHIMMINS and J. M. A. LENIHAN, *Radiochem. Radioan. Lett.,* **4,** 861 (1970).
117. W. H. ZOLLER and G. E. GORDON, *Anal. Chem.,* **42,** 257 (1970).
118. R. DAMS, J. A. ROBBINS, V. H. RAHN and J. W. WINCHESTER, *Anal. Chem.,* **42,** 861 (1970).
119. M. J. ASHWORTH and T. P. ABELES, *Nature,* **210,** 9 (1966).
120. A. ASPINALL, D. N. SLATER and P. MAYNES, *Nature,* **217,** 388 (1968).
121. R. F. COLEMAN, *J. Forensic Sci. Soc.,* **6,** 19 (1966).
122. R. E. JERVIS, Proc. I.A.E.A. Symp. Nuclear Activation Techniques in the Life Sciences, p. 645. Vienna, (1967).
123. L. T. ATALLA and F. W. LIMA, *Radiochem. Radioan. Lett.,* **3,** 13 (1970).

Chapter 8

The Use of Tracers in Inorganic Analysis

J. W. McMillan

Applied Chemistry Division, AERE Harwell, Didcot, Oxfordshire

8.1 THE APPLICATION OF TRACERS IN THE INVESTIGATION OF ANALYTICAL METHODS

8.1.1 Introduction

Tracer methods make an important contribution to the investigation of analytical procedures. An analyst seeking a new method for the determination of a particular element or ion in a new matrix may attempt to adapt an established technique or in extreme circumstances institute an entirely new method based on a new reagent or a novel principle. In either case, the investigation of the new procedure can be speeded and clarified by the use of radioactive tracer methods.

A typical analysis may be broken down into a number of stages:

1. Preliminary treatment; sample collection, concentration, ashing and solubilisation.
2. Separation; removal of interfering species by distillation, solvent extraction, ion exchange, chromatography, precipitation, isotope exchange, adsorption and electrodeposition.
3. Measurement; gravimetry, titrimetry, colorimetry, absorptiometry, emission spectroscopy, mass spectrometry, polarography, coulometry and radiometry.

Throughout the analytical procedure the main aims are quantitative recovery of the analysed species and efficient removal of interferences,

frequently through separation methods. Radioactive forms of the element
or ion being determined and of those interfering, when added to the sample,
trace the behaviour of the inactive species throughout the analysis. Measurement of the initial tracer activity and its intensity at each stage of the
analysis allows the occurrence of losses, the efficiency of individual operations and the overall recovery to be assessed rapidly and accurately. When a
particular step is suspect, radioactive tracers can be used with advantage for
its study in depth. For instance, in solvent extraction, many parameters may
require study and the number of experiments can escalate to a formidable
level. Measurement of the distribution of the investigated species between
the two phases after each equilibration can be simplified enormously by
tracer measurements, as this only entails withdrawal of specimens of each
phase and comparison of their radioactivity in a simple counting apparatus.

Tracer methods can be used successfully not only for the investigation of
methods for the measurement of a stable element or ion, but also when the
recovery of induced activity is required, as in activation analysis or radiochemical analysis. In this case separation from other radioactive species with
a high degree of decontamination is the main objective. Decontamination
factors in excess of 10^6 are frequently required, much higher than those
usually needed in conventional analysis. In these analyses quantitative
recovery is sacrificed for the sake of efficient separation, and the recovery
of the active species is often ascertained by the addition of a known amount
of inactive carrier, the technique being commonly referred to as reverse
isotope dilution (see p. 328).

The principles and techniques involved in the application of tracer
methods can be demonstrated by examples drawn from various areas of
inorganic analysis. However, a number of basic principles common to all
tracer methods are considered first.

8.1.2 Basic Considerations

When planning a radioactive tracer study of an analytical method, or one of
its stages, a number of factors must be considered. These include the availability of a suitable isotope, its chemical form, its behaviour in the system
studied, the amount of activity required, the form in which it should be
counted, and, not least, the health hazards involved. Many of these factors
interact and mutually contribute to the overall design of the experiment.

Today, a comprehensive range of radioisotopes is available commercially
[1]. Unfortunately there are a number of elements for which convenient
radioactive isotopic tracers do not exist [2, 3], e.g. elements with only
isotopes of short half-life such as oxygen (^{15}O, $t_{1/2}$ = 2.03 min) and boron
(^{12}B, $t_{1/2} < 1$ sec). For such elements, tracing with enriched inactive isotopes
is a possible alternative, but this lacks the convenience of the radioactive
isotope method because of the difficulty of following the progress of the
inactive tracer.

Perhaps the most important property of the tracer is its half-life. The
reduction in the activity is large if several half-lives elapse during the course
of the experiment so it is preferable to select an isotope with a half-life that

is long compared to the duration of the experiment. By doing so the problem of handling unduly large levels of activity at the beginning of the experiment and that of having to make large decay corrections are avoided. Decay corrections can be avoided by measuring both the activity of the recovered tracer and that of the master solution at the end of the experiment. A further disadvantage of using short-lived isotopes is that the experiments must be carried out near to a source of isotopes; a nuclear reactor or particle accelerator. This problem can be avoided if the isotope of interest is the daughter of a long-lived parent when the daughter can be 'milked' from the parent immediately before use [4], e.g. ^{90}Y from ^{90}Sr, or ^{44}Sc from ^{44}Ti. One advantage of using short-lived tracers is that their rapid decay eases the problem of ultimate disposal. If a short-lived tracer must be used, the analytical method under study may have to be broken down into short steps each compatible with the half-life of the tracer.

The purity of the radioactive tracer is of considerable importance. Two principle types of impurity exist. First, radionuclidic impurity, the presence of radioactive nuclides other than that desired. Second, radiochemical impurity, the presence of the nuclide of interest but in a chemical form differing from the one specifically needed. The first type of impurity, if non-isotopic, will lead to errors because its chemical behaviour will probably differ from that of the main tracer. If the impurity is isotopic, it may lead to erroneous decay corrections. The second type of impurity, e.g. the presence of ^{32}P-labelled pyrophosphate in orthophosphate tracer, could lead to difficulties in orthophosphate tracing studies. Radiochemical purity is normally less important in inorganic than in organic tracer studies where the presence of the tracer in the correct chemical form is absolutely essential.

The technique of tracer analysis is very dependent on the identical behaviour of the tracer and the inactive element or ion. In order to ensure this, a vital objective is the earliest possible mixing of the tracer and stable species. Naturally, chemical identity of the tracer and stable species is essential and the chemical conditions may need to be adjusted in order to achieve exchange. However, care is needed in those instances where ions rather than elements are being traced in order to avoid unwanted exchange.

An obvious difficulty arises when the tracer must be added to a solid material, perhaps to study the effects of ashing or dissolution when the results must be interpreted with great caution. The mixing of tracers with liquids and gases poses few problems. A further cause of the non-identical behaviour of radioactive tracers and the stable species is attributable to the difference in mass of the active and inactive isotopes; the isotope effect. This effect is greatest when the ratio of the isotopic masses is greatest, i.e. for hydrogen (^{1}H) compared to tritium (^{3}H), and it decreases with increasing atomic number and by incorporation of the isotope into an ion or molecule. Fortunately the effect is small for the separation processes encountered in inorganic analysis, seldom exceeding 1% per stage. While correction for the isotope effect may be necessary for the lightest elements, it can usually be ignored.

Finally, radiation effects must be mentioned. The chemical nature of a high specific activity tracer may be altered by nuclear decay. For instance, a doubly labelled organic reagent could, through the decay of one or other of the active nuclides in the molecule, produce one of two singly labelled species both of which are radiochemical impurities. It may be advisable to purify such reagents immediately before use. The problem is fortunately rare in inorganic analysis as the majority of tracer experiments are solely concerned with elemental behaviour. The storage of highly radioactive tracer solutions may lead to radiolytic gas production and such solutions must be stored in vented containers. While radiation absorption can produce other chemical changes [5], they are rarely met when using the relatively small amounts of activity needed in inorganic tracer studies.

Isotopes selected for tracer work should be easily measured, preferably with little or no special preparation before counting. Because of the penetrating nature of gamma radiation, in most circumstances, gamma emitting nuclides are to be preferred. Sodium iodide well detectors simplify the measurement of gamma emitters, the geometrical configuration allowing precipitates in centrifuge tubes, small but differing volumes of liquids and many small objects, to be counted with virtually identical efficiency. Gamma emitting radionuclidic impurities can be identified by gamma spectrometry employing either sodium iodide or lithium drifted germanium detectors [6]. The latter are to be preferred because of their superior resolution but are much more expensive than the equipment usually required in a laboratory carrying out tracer studies. Beta radiation is much less penetrating than gamma radiation and beta emitting nuclides normally require separation and some form of source preparation prior to Geiger, proportional or scintillation counting [7]. Liquid scintillation counting is particularly useful for weak beta emitters because of its high efficiency and the relative ease of source preparation [8]. Because of radiation absorption effects only strong beta emitters are usually detectable in solutions, on chromatographic columns, or adhering to the walls of vessels. However, for techniques such as paper chromatography and thin-layer chromatography both weak and strong beta emitters are detectable by scanning techniques or autoradiography [9, 10]. Due to the continuous nature of beta radiation the detection of radionuclidic impurities by beta counting is difficult though some energy discrimination can be obtained by means of beta spectrometry or the use of absorption foils. An alternative method is half-life discrimination. The absorption of alpha radiation is so marked that the use of alpha emitting tracers should be avoided unless no alternative exists.

A compromise may need to be made in the selection of a tracer for a particular experiment. Reference to a compendium of nuclear data [1, 2] will indicate the nuclear properties of radioactive isotopes and allow a reasoned choice for a particular element. The data may show that although a gamma emitting isotope is available, its half-life is too short, and a pure beta emitting isotope with a more favourable half-life may have to be chosen, e.g. ^{37}S is a gamma emitter but the half-life is only 5 minutes, while ^{35}S is a pure beta emitter with a half-life of 87 days.

After selecting a particular radioactive tracer, the amount of activity needed for an experiment must be calculated. This will be governed by a number of factors such as the accuracy and precision required, the loss of activity during the tracing study and the efficiency of detection.

The precision of the measurements is intimately associated with the counting statistics. The standard deviation of the number of counts recorded, C, is equal to $\sqrt{C}$. The precision of the tracer count is also dependent on the background count. If T_r = tracer count rate, S = tracer + background count in time t_s and B = background count in time t_b, then

$$T_r = \frac{S}{t_s} - \frac{B}{t_b}$$

and its precision is given by

$$\delta T_r = \sqrt{\frac{S}{t_s^2} + \frac{B}{t_b^2}}$$

If the background is insignificant a relative precision of 1% is obtained by accumulating 10^4 counts but if the background counting rate is equal to that of the tracer the same precision would require a tracer + background count of 6×10^4 counts and a background count of 3×10^4.

Experimental technique can aid the production of good counting statistics. For example, in measuring losses and distribution coefficients, one fraction may contain a relatively small activity compared to the other and the choice may be between measuring a small change in the high activity fraction or a relatively large change in the low activity fraction. The latter procedure is obviously the better approach, providing the background is insignificant, and this can be controlled by starting the experiment with sufficient activity. In tracer studies of radiochemical and activation methods high order decontamination factors are needed. While the precision of measurement required is probably low, limiting values only may be obtainable unless large initial activities are used. Another reason for having a tracer count well above background level is that it makes characterisation of the activity easier, so ensuring that the activity measured is not due to a nuclidic impurity.

The major reduction of radioactivity during an experiment is usually caused by deliberate or fortuitous separations of the type described above. Decay losses have already been discussed and can be held at a moderate level by good experimental design. The only remaining factor causing the reduction of activity during a tracer study is dilution, which is usually well defined.

The final parameter governing the amount of tracer activity required for an experiment is the efficiency of the counting system. In general, the number of counts (C) recorded by a counting system per nuclear disintegration (D) can be defined by an expression of the type

$$C = f_p D E_p G$$

where f_p is the fractional emission of particles, of the type being measured per disintegration, e.g. ^{51}Cr emits only 0.09 gammas of 0.320 MeV per disintegration. E_p is the detector efficiency. G is a factor which takes into account the detector/source geometry, activity reduction attributable to absorption of the radiation in the source and detector housing, backscattering effects etc.

In most practical situations the relationship between count rate and disintegration rate is best determined experimentally by counting a standard in a physical form and with a counter/source geometry identical to that used for processed sources; correction for absorption, scattering effects and geometry can thus be avoided.

After assessing the activity reduction, the efficiency of the counting system and the experimental precision required, the amount of activity needed for a tracer study can be readily calculated. It is usually small, <1 μCi and rarely >1 mCi. The health hazards associated with such levels of activity are seldom great; precautions needed are no more stringent than those for the handling of poisonous and corrosive chemicals. Guidance in the handling and disposal of tracer levels of activity can be obtained from standard works [11], legislative documents [12, 13] and official codes of practice [14, 15].

8.1.3   Applications

Radioactive tracers have been used in the investigation of a large number of analytical methods [16] best illustrated by examples. The majority of applications are found in the investigation of separation techniques which are dealt with in detail. In addition, the preliminary treatment of samples is covered, particularly storage, ashing, solubilisation and preconcentration. Finally a few examples are given of the study of complete methods.

Ashing

The danger of losing some elements by dry ashing of biological materials has been simply demonstrated by several tracer studies [16–18]. Two of these were concerned with normal high temperature ashing [16, 17]. The results were in general agreement, showing that significant amounts of Pb, Hg, As, Ag, Zn, Cu and Au can be lost even at temperatures of 400–500°C. Increase of temperature enhances these losses and extends the range of elements to include Fe, Mn, Mo, V, Co and Cr. Gorsuch [16] investigated the extent of these losses and also examined their cause. For instance, he was able to show that, while Hg was lost entirely by volatilisation, Pb, Cu, Zn and Ag were largely retained by the silica crucible. As might be expected, low temperature ashing with excited oxygen leads to the loss of fewer elements though Au, I, Ag and Hg were all lost to a significant degree.

A fundamental problem in ashing studies is to ensure identical behaviour of the radioactive tracer and the stable element. Simple addition of tracer solution to the biological material in a crucible, drying and ashing, is open to severe criticism for solid samples, as little or no exchange can occur. Labelled biological material can be produced in two ways. The first method is to activate the stable element *in situ* by irradiation. Gorsuch [16] produced ^{65}Zn-labelled hair in this way. Labelling by irradiation is, however, a non-specific process and a wide range of other activities will also be produced. This causes difficulties in the measurement of the wanted nuclide particularly if, due to its low concentration or unfavourable nuclear characteristics, its probability of production is low compared to the others. There may also be exchange problems if the active nuclide is in an altered chemical form due to nuclear de-excitation processes. High resolution gamma spectrometry may be required for identification and measurement. The second, and better, method of labelling is to grow the specimen on a medium containing the tracer. This technique has been used to produce ^{75}Se-labelled alfalfa [18]. Specimens labelled by either method must be measured intact and gamma emitting tracers only should be used.

Solubilisation

Solubilisation of materials also involves the possible loss of elements, particularly by volatilisation. Volatilisation can be used for the separation of many elements [19]. However, its occurrence during dissolution must be controlled. The simplest processes can lead to loss by volatilisation; treatment with dilute acid can release carbon as carbon dioxide, elements forming volatile fluorides, e.g. Ge, Si and As, can be lost during HF dissolutions, osmium and ruthenium are volatilised as their tetroxides by oxidative solubilisation methods. Many elements form chlorides volatile at high temperatures. Vigorous solubilisation conditions must be used with intractible materials despite the risk of such losses. The extent of that risk can be readily assessed with the aid of radioactive tracers.

Radioactive tracer studies have shown that the destruction of organic matter by wet exidation methods leads to the loss of fewer elements than dry ashing [16, 17]. Elements volatilised by acid mixtures included Hg, As, Sb, Au, Fe, Se, Os and Ru, although their loss could be easily prevented by trapping. As with dry ashing, volatility is not the only loss mechanism. For instance, traces of lead can be lost by coprecipitation with $CaSO_4$ during the wet oxidation with sulphuric acid mixtures of materials containing appreciable amounts of calcium. Another method for the destructive solubilisation of organic specimens, investigated using radioactive tracers, used equimolecular mixtures of $NaNO_3/KNO_3$ at 390°C [20]. The process only required a few minutes and some 24 elements were recovered to an extent of $>98\%$, mercury was completely volatilised while a few % of As, Br, Cl, I and S were also volatilised.

Solubilisation methods for inorganic materials closely parallel those for organic matter and similar losses of certain elements occur. A wealth of information is now available on the volatilisation of elements when treated with common dissolution reagents [16, 17, 19, 21–23], allowing confident prediction of losses that will ensue from the choice of a particular solvent. Frequently the volatilisation of one or more elements during the solubilisation of a sample may be turned to advantage to obtain simultaneous separation and solubilisation. Morrison [24] adopted this procedure for the dissolution of geological materials, obtaining simultaneous separation of As and Br. Samsahl [25] separated Br, As, Hg, Sb and Se by volatilisation and trapping during the dissolution of biological materials. Hoste *et al* introduced similar methods in the dissolution of Se [26, 27]. All these methods were based on volatility data obtained by radioactive tracer studies. While seeking a suitable method for dissolution of silicon carbide, Lowe [28] discovered, using tracers, that fusion in a sodium carbonate/nitrate mixture for two hours at 900°C led to high losses of Sb and In. Another example was that of Smales *et al.* who used radioactive tracers to investigate the dissolution of silicon in sodium hydroxide/hydrogen peroxide solutions [29]. The method was shown to be unsatisfactory because of the loss of Sb and As, probably as hydrides.

Preconcentration
The levels of the minor constituents in some samples are so low that concentration prior to measurement is absolutely essential. This situation is typified by sea water. While atomic absorption and neutron activation analysis have been used for the direct measurement of minor constituents, preconcentration is normally used, a recovery of 98–99% being required for concentrations down to 0.1 mg/litre [30]. The methods adopted included co-precipitation and co-crystallisation, solvent extraction and ion exchange; radioactive tracers provided an excellent means of evaluating the efficiency of these processes. Lai and Weiss [31] used the tracer method to investigate the preconcentration of minor elements in sea water by co-crystallisation with thionalide. They were able to find conditions which ensured > 90% recovery of some 16 elements. The main parameter studied was that of pH. Riley [32] has investigated the use of chelating ion exchange resins of the iminodiacetic acid type for the concentration of trace elements from sea water. Radioactive tracers were used to develop the method which was then applied to the determination of Cu, Ni, Zn, Cd and Co by atomic absorption spectrophotometry. While radioactive tracers aid investigations of this type, slight doubts frequently arise concerning the validity of the results because of the problem of ensuring exchange. Sea water is an extremely complex system; trace elements exist in many forms, i.e. simple ions, complex anions and undissociated compounds. In addition the readily hydrolysable elements exist as suspended solids or colloids and trace elements may be adsorbed on suspended material. Adjustment of the conditions, perhaps by acidification, may be necessary to obtain exchange and give reliable results.

Storage and Adsorption Losses
The adsorption of elements and compounds, particularly by container walls, can lead to losses during sample storage and processing. The growth of micro-organisms may also prove troublesome. Both glass and plastics have surfaces that possess adsorption sites. Glass can act both as an anion and cation exchanger. Surface degradation of plastics can lead to the formation of carboxyl and carbonyl groups that act as sites for ion exchange. On the whole losses are only of significance at low concentrations. Radioactive tracer techniques are ideal for studying this type of problem as the tracer

can be added at high specific activity, thereby avoiding significant alteration in the concentration of the inactive species, so that the variation of the activity remaining in solution with time can be determined radiometrically on a small aliquot. One example of this was the study of the adsorption of phosphate on glass using ^{32}P tracer [33]. The slight adsorption occurring on pyrex could be reduced by treating the surface with HF thereby reducing the number of hydroxyl groups. Another example was the adsorption of gold on polythene [34]. The concentration of gold in sea-water decreased in polythene containers though not in silica containers. Weiss and Lai [35] found that gold was not adsorbed on polythene if the pH was adjusted to 1.0 with hydrochloric acid.

Precipitation

Precipitation is one of the oldest and most frequently used separation methods. In most circumstances a precipitant should be both specific and quantitative. Exceptions are the broad-band precipitants used for preconcentration of trace constituents and for scavenging interfering nuclides in radiochemical separations.

Radioactive tracers can be used to assess the performance of precipitants; primarily their efficiency and specificity. The technique normally adopted is the addition of high specific activity tracer, exchange, adjustment of the precipitation conditions, precipitation, and separation and radiometric determination of the tracer in the precipitate and filtrate. When the precipitation is highly efficient, measurement of the filtrate is essential because of counting statistics. The technique can be readily extended to study washing losses and the solubility of the precipitate in various solvents.

The work of Sunderman and Meinke [36] provides excellent examples of the use of tracers for the study of precipitations. In order to establish an efficient radiochemical separation method for barium, several barium precipitations were investigated, as was the co-precipitation of many other elements. Their extensive data established that precipitation of $BaCl_2$ from ether/hydrochloric acid gave good decontamination from all the elements investigated. In an extension of this work, an investigation was made of the relative effectiveness of $La(OH)_3$ and $Fe(OH)_3$ as scavengers. They had previously been used interchangeably without any explicit knowledge of their relative performance. While some elements were scavenged to the same degree by both precipitates, others, e.g. Se, Ru, Co, and Sr, exhibited differing behaviour. Considering the different pH's required for the precipitation of $Fe(OH)_3$ and $La(OH)_3$, and the more flocculent nature of the former, the difference revealed might have been expected but had surprisingly been disregarded.

An interesting study of a conventional gravimetric method was that of the determination of sodium with zinc uranyl acetate, using ^{24}Na and ^{22}Na tracers [37]. Variation in the precipitation conditions was investigated, the recovery of the sodium zinc uranyl acetate being determined radiometrically and gravimetrically in order to observe any variation in the precipitate composition. It was found that 100% recovery occurs only after long standing, often in excess of 24 hours. Using the tracer to determine the yield the determination could be completed in a much shorter period but only when the reagent was added at 75°C. At 25°C the precipitate was found to be non-stoichiometric.

Volatilisation

Volatilisation provides a rapid and often selective method of separation for a wide range of elements. Table 8.1 gives an indication of elements that have been separated by volatilisation and the form in which they have been separated for analytical purposes. The importance of volatilisation was indicated in the section on Solubilisation. While many of the volatile species are commonly encountered and a large proportion can be produced from aqueous solutions, a significant number are rarely met. The volatilisation of highly reactive materials and those with high boiling points are only used in special circumstances, e.g. for very rapid separations [38]. Many other volatile compounds have been used to separate the elements, including sulphides, carbonyls, stable organic complexes such as oxyacetates for Be and Zn, and fluorinated β-diketones for the lanthanides.

Separation, in methods depending on volatilisation, is achieved by differentiation during the volatilisation process, fractionation during transfer, and selective collection. Gaseous evolution can be controlled by making use of differences in vapour pressure with temperature, adjustment of the oxidation state of an element in solution or by alteration of the matrix, in order to change the chemical combination of the element. Once gaseous, additional separation is possible and physical processes can be adopted such as gas chromatography, zone refining, fractional distillation, electrostatic precipitation, filtration of condensed phases and low temperature trapping. Chemical methods used are mainly based on the selective trapping of interfering substances by solid or liquid reagents. The methods of preferential collection of the species sought are similar to those used in the transfer stage. The study of the effectiveness of these processes using radioactive tracers is illustrated by a number of examples.

The separation of manganese activity from irradiated iron cyclotron targets by distillation as permanganic acid was studied by Pijck and Hoste [39]. The method is essentially simple, comprising the use of a heated mixture of sulphuric and nitric acids and potassium iodate to volatilise the manganese and water as the collecting medium. However, evidence was needed of the variation of the manganese yield with carrier concentration, target weight and distillation time, as well as the behaviour of a number of elements that could possibly interfere. These factors were readily determinable using ^{54}Mn, ^{59}Fe, ^{60}Co and ^{51}Cr tracers and gamma detection. Optimum conditions were found and the decontamination against iron and cobalt was found to be good and that against chromium acceptable.

A number of metals can be separated by distillation. The system is usually an irradiated target or radioactive source containing a more volatile daughter or irradiation product. Examples are, ^{103}Rh in ruthenium [40],

TABLE 8.1. *Elements Separable by Volatilisation as Certain Species*

	1	2	3	4	5	6	7	8	9	10	11	12	13	14	15	16	17	18
1.	H abcd																	He a
2.	Li a	Be											B bc*d	C bcd	N abcd	O abcd	F abcd	Ne a
3.	Na a	Mg											Al d	Si bd	P abcd	S abcd	Cl abcd	A a
4.	K a	Ca	Sc	Ti d	V d	Cr d*	Mn c*	Fe d	Co	Ni	Cu	Zn	Ga bd	Ge bd	As abcd	Se bcd	Br abd	Kr ad
5.	Rb a	Sr d	Y	Zr d	Nb d	Mo d	Tc cd	Ru cd	Rh a	Pd	Ag a	Cd a	In a	Sm bd	Sb bd	Te bcd	I abd	Xe ad
6.	Cs a	Ba a	La*	Hf d	Ta d	W d	Re cd	Os cd	Ir d	Pt	Au a	Hg ad	Tl a	Pb	Bi ab	Po ad	At ab	Rn ad
7.	Fr a	Ra	Ac**															

Ce*	Pr	Nd	Pm	Sm	Eu	Gd	Tb	Dy	Ho	Er	Tm	Yb	Lu
Th**	Pa d	U d	Np d	Pu d	Am	Cm	Bk	Cf	Es	Fm	Mv	No	

Form in which element volatilised

(a) Element
(b) Hydride
(c) Oxide
(c*) Permanganic acid
(c+) Boric acid
(d) Halides
(d*) Chromyl chloride

[109m]Ag in palladium [41], polonium in bismuth or lead [42], [106] Ag in rhodium [43] and many others [44]. The apparatus used for such separations by DeVoe and Meinke [45] is typical, consisting of an inductively heated carbon furnace with a vapour guide and a liquid nitrogen cooled collector, all contained in an evacuated vessel.

The determination of delayed neutron precursors such as [87]Br, [88]Br, [134] Sb, [137] I, [138] I and [87]Se, all with half-lives of less than one minute, rely on rapid separation methods [46]. The efficiency and speed of these methods can be tested using longer lived isotopes. For instance antimony and arsenic can be separated as hydrides by flash electrolysis [47]. Using [76]As and [124] Sb as tracers the method was shown to achieve a 45% recovery of arsenic and antimony in 10 seconds. Decontamination against other fission products which might generate gaseous species was also tested with tracers.

Solvent Extraction

The wide use of solvent extraction in analysis can be seen from several reviews [48–51]. The enormous amount of experiment needed to assess comprehensively the behaviour of even a single extraction system may not be so readily appreciated. Distribution ratios are influenced by many parameters: the nature of the solvent, the composition of the aqueous phase, the pH, the presence of complexing agents, the concentrations of the extracted ion and the extractant, and the temperature and duration of extraction. By using tracer methods the time required to undertake such studies can be greatly reduced.

Peppard [52] has discussed the use of tracer techniques for the determination of distribution ratios. The general procedure adopted is to equilibrate two immiscible phases with the radioactive labelled ion under study, to separate the phases and to count samples of each phase. The distribution ratio is then calculated from the counting rate of the tracer in each phase. Errors in the distribution ratio so determined can be attributed to the normal tracer problems, i.e. the presence of radionuclidic and radiochemical impurities and lack of exchange, plus factors primarily associated with the solvent extraction process. The latter include the presence of unsuspected extractants, the degradation of the extractant, the presence of the extractant in more than one form, e.g. enolic and ketonic, or monomeric and dimeric forms, partial miscibility of the phases, hydrolysis, and the entrainment of the opposing phases. Such errors must be avoided or allowed for.

An example of the use of radioactive tracers in distribution studies is that of the extraction of quaternary ammonium complexes of some 57 elements [53]. The distribution of these elements into methyl isobutyl ketone as their tetrapropyl-, tetrabutyl- and tetrahexyl-ammonium complexes from mineral acids and sodium hydroxide was determined using simple batch equilibration, aliquots of each phase being counted. The results revealed that the extractants could be used for a number of analytical separations, and general extraction trends could be seen, e.g. increase of degree of extraction with molecular weight of the ammonium salt.

Ion Exchange Chromatography

The most commonly used ion exchange materials in analytical chemistry are the organic resin exchangers but an ever increasing number of inorganic ion exchange materials such as zirconium phosphate and hydrated antimony pentoxide, cellulosic materials and liquid ion exchangers on inert supports are coming into use. The large number of applications can be seen by reference to standard works [54, 55] and various reviews [56–59].

Ion exchange separations are normally carried out by preferential elution of adsorbed ions from columns of ion exchangers. In order to be able to predict the conditions necessary for desired separations, comprehensive distribution studies are required. These can be undertaken with the aid of several other analytical techniques but the radioactive tracer method has the advantage that it can be used for measurements at very low concentrations, while it is quick and convenient, particularly for tracers emitting penetrating radiation. Four tracer methods are normally used.

1. Column effluent analysis. This should only be applied when adsorption is not too high. The procedure is to transfer the ion and tracer to the column in a small volume of solution and to elute until the concentration of the tracer in the effluent reaches the maximum. This can be determined by counting successive fractions of effluent, or by continuous measurement of the activity as the effluent leaves the column, using a flow through detector, rate meter and chart recorder (see Section 6.4.2).

The number of geometric column volumes required to reach the maximum tracer concentration is equal to

$$\frac{v}{LA} = \frac{1}{E}$$

where v = volume of effluent that has passed through, L = distance travelled by the maximum: the length of the column in this case, A is the cross section of the column and E is, by definition, the elution constant.

D_v, the volume distribution coefficient can be evaluated from

$$D_v = \frac{1}{E} - i$$

[see equation (5.7)] where i is the fractional interstitial space in the column.

2. The second method used is column scanning. The labelled ion is transferred to the column as in method 1. The column is eluted and the movement of the band of radioactivity measured by a column scanning method, or by autoradiography, for a known elution volume. E is again obtained and hence D_v. This method can only be used for tracers emitting penetrating radiation. It is best used when ions are strongly adsorbed by the exchanger.

3. A more convenient method for strongly adsorbed ions is batch equilibration. A known weight of dried resin is agitated with a known volume of solution of the desired composition and containing a measured amount of tracer. The resin and solution are contacted for several days, to ensure equilibrium, and the decrease in concentration of the tracer in solution

assayed radiometrically. The weight distribution coefficient, D_g, is obtained (see Section 5.3.2). For large values of D_g care is needed to ensure that the activity remaining in solution is not a radionuclidic impurity.

4. The final method is recommended particularly for high values of D_g, e.g. in excess of 10^4. The ion and tracer concerned are adsorbed uniformly onto the exchanger which is then equilibrated with the solution of interest. After equilibration the tracer transferred to the solution is measured. An advantage claimed is that equilibrium is reached faster than in the previous method. It is also less prone to errors caused by radionuclidic impurities because poorly adsorbed impurities will be eliminated in the initial adsorption stage.

An excellent example of a distribution study is the work of Kraus and Nelson [60] on the adsorption of 65 elements by a strong anion exchange resin from HCl solutions. In all, 55 of the elements were examined using the radiotracer method, other analytical methods only being used when no suitable tracer was available.

Even when distribution data indicate that a particular separation is possible it is good practice to test the method under normal working conditions to ensure that the degree of separation expected is achieved and that the method is quantitative. Samsahl [61] devised an automatic group separation system for the simultaneous determination of 40 elements in biological material after neutron irradiation. In all, 14 ion exchange steps were involved plus distillations to give 16 groups of elements. The recovery and reproducibility of the system were tested using known amounts of active tracers in the presence of the anticipated inactive amount of each element.

Inorganic ion exchange materials are widely used for chromatographic separations [62, 63]. All of the investigational techniques described so far can be used to study their behaviour. This applies equally to well known materials, such as alumina and silica, and newer substances such as zirconium phosphate and hydrated oxides of antimony, tin and zinc. Girardi and co-workers have been prominent investigators and exploiters of inorganic ion exchange for analytical separations [64–67]. Radioactive tracers were used extensively in their studies of the ion exchange properties of manganese dioxide, antimony pentoxide, tin oxide, cadmium oxide, zirconium phosphate, cupric sulphide, cuprous chloride and cerous oxalate. The size of the task of comprehensively assessing the properties of such materials can be seen from the number of experiments undertaken in one such study [67]. The elution behaviour of 61 elements in various acids from 9 inorganic ion exchangers plus an anionic and a cationic ion exchange resin was examined. The number of measurements approached 8,500. The speed and convenience of radiometric assay eased the analytical burden, particularly in the determination of the element retained by the column. The fundamental data obtained in these studies has been exploited extensively for analytical separations, especially in activation analysis [65, 66].

Two further separation techniques that employ ion exchange materials are thin-layer chromatography and paper chromatography. The former frequently makes use of layers of inorganic ion exchangers in thin layers on glass plates while the latter uses papers chemically modified to enhance their ion exchange characteristics, or papers loaded with either inorganic or resin ion exchangers. The fundamental difference between these sheet separation methods and column separation methods is that in the former the components to be measured always remain on the sheet after separation and are not eluted. The separation index for a component in these methods is the R_f value, its movement along the sheet relative to that of the solvent front. While the probable performance of a sheet separation system can be broadly forecast by batch equilibration or column experiments, the actual sheet separation itself is the ultimate test.

Radioactive tracers can be used to aid the establishment of R_f values, to determine whether or not components move quantitatively and to highlight certain phenomena, e.g. tailing or the formation of more than one ionic species during the separation. The components to be studied are each labelled with a tracer of identical ionic form and are spotted onto the sheet singly or as mixtures. The sheets are then developed in the normal manner and the positions of the tracer activities found by autoradiography, intact scanning, segmentation followed by counting, or by a combination of these [9, 68]. Autoradiography, described in detail in Chapter 6, has great sensitivity and gives excellent spacial resolution particularly for weak beta emitters such as ^{14}C and ^{35}S. In scanning, the sheet is moved at constant speed and is viewed by a detector through a collimator. A typical scanner is decribed by Pocchiari and Rossi [69] and many others are available commercially. Segmentation followed by counting is rather tedious but quantitative measurement can be carried out efficiently with less possibility of interference from adjacent spots.

Although radioactivity is normally added prior to separation, post separation labelling can be used, either by addition of a radioactive reagent which reacts with the spots [68, 70] or by activation of the spots by irradiation with neutrons or other bombarding species [68, 71].

Radioactive tracer methods were used by Tustanowski to demonstrate that the ion exchange separation of chloride, bromide and iodide on alumina was similar whether the exchanger was in the form of columns, as thin-layers or was loaded into paper [72]. Effluent analysis was used for the column method and autoradiography for the sheet separations. A further example of the use of radioactive tracers for the investigation of thin-layer chromatography is the work of König and Demel [73] who studied the separation of 19 ions on zirconium hypophosphate.

Electrophoretic modifications of sheet separations have found some applications in inorganic analysis and the tracer methods and detection techniques are the same as for ordinary sheet separations. Moghissi used tracer methods to study the thin-layer electrophoretic separation of sulphate and phosphate, and of calcium and barium [74]. Another related technique is electrophoretic focussing [75]. Separations by this method are rapid; the rare earths have been separated in less than 20 minutes [76]. The separation of fission products, transuranic elements and natural radionuclides were all shown to be possible.

Liquid Partition Chromatography

Many analytical separations have been carried out by liquid partition chromatography or extraction chromatography [77—80]. These separations are, effectively, multiple solvent extractions in which one liquid phase is held on an inert support while the other acts as the eluent. Normally, the supported phase is aqueous and the moving phase non-aqueous. When the normal disposition of the liquid phases is reversed the technique is called reversed phase partition chromatography (see Section 5.6.8). Because the method is allied both to liquid—liquid extraction and chromatography the investigational techniques using radioactive tracers are similar to those described for those techniques. A few selected examples are given.

A number of workers have separated the rare earths by reversed phase column partition chromatography [81—83]. The static phase used was di(2-ethyl hexyl) orthophosphoric acid supported on 'Celite', 'Kel-F' or 'Corvic', and elution was carried out, usually at elevated temperatures, with HCl and $HClO_4$ solutions as eluents. Using tracers satisfactory separation conditions were established.

Synergism in mixed complexant systems has been studied using radioactive tracers [84]. Reversed phase paper chromatographic separation of La, Ce and Am using either 2-thenoyl tri-fluoracetone or trioctyl phosphate as the static phase and hydrochloric acid as eluent was unsuccessful as the metal ions were only weakly retained. However, an equimolar mixture of the two as static phase was more strongly complexing and a good separation was obtained.

Fission products may be separated by paper chromatography. Using the ascending method and one or two dimensional development, Getoff *et al.* [85] determined R_f values for ^{90}Sr, ^{90}Y, ^{137}Cs, ^{140}Ba, ^{140}La and ^{144}Ce in a variety of solvent systems. Spot positions were located by autoradiography and quantitative measurement of separated activities was made after cutting out individual spots.

Gas Chromatography

With a few exceptions little use has been made of radioactive tracers in the investigation of gas chromatographic methods. On the other hand, gas chromatography has been used to separate and identify mixtures of radioactive elements and compounds produced by a variety of chemical and nuclear processes [86, 87]. For instance, fission products have been separated, tritiated products of the Wilsbach gaseous exposure method identified, Szilard-Chalmers processes studied and radioactive species produced by tracers in reaction studies identified. Several types of detector are available [86, 88, 89, 90]. The subject of detectors is dealt with in detail in Chapter 6.

One example of the use of radioactive tracers to investigate a gas chromatographic separation is the work of Koch and Grundy on Kr and Xe [91]. The relatively long-lived noble gas isotopes, ^{85}Kr ($t_{1/2}$ = 10.3 y) and ^{133}Xe ($t_{1/2}$ = 5.27 d) were used to study column operating conditions, the elution of the tracer gases being followed continuously by an on-line monitor. The results enabled conditions to be selected for the separation and determination of short-lived fission product gases. A very rapid gas chromatographic separation method for Kr and Xe was developed by Ockenden and Tomlinson [92]. Using a column of fine activated charcoal dispersed in glass wool Kr was eluted after 4 seconds and Xe after 27 seconds.

Radioactive tracers have also been used to study the gas chromatographic separation of rare earth chlorides [93].

Electrodeposition

Separation by electrodeposition is limited partly because it is relatively slow and partly because it seldom leads to complete separation of one element from many others. Two radiometric techniques can be used for the measurement of separation efficiency and yield; the comparison of the initial activity in the solution with the residual activity, or with the deposited activity. Direct comparison of the deposited activity with the solution activity is not possible because of geometry and attenuation problems. Consequently manipulation of one or the other is necessary prior to counting. The progress of deposition can be observed continuously, for instance, by pumping solution through an external detector. In some circumstances if a thin electrode is built into the cell wall, the progress of deposition can be followed by measuring the radiation penetrating the electrode. Dean and Reynolds [94] used tracers to determine the completeness of separation in developing an electrodeposition method for the successive separation and determination of Bi, Sb and Sn. When determining silver in cadmium by neutron activation analysis Lux [95] used preseparation of silver to avoid nuclear interference from the ^{110}Cd (n, p) ^{110m}Ag reaction. With the aid of ^{110m}Ag tracer he was able to find electrolysis conditions giving quantitative separation of as little as 10^{-8} g of silver from 1 g of cadmium. Experiments by DeVoe and Meinke [96] on electrolytic separations showed that decontamination factors of 10 or more were the exception suggesting that the method is unlikely to give the type of broad band separation needed in radiochemical methods.

An important application of electrodeposition is the preparation of counting sources [97–99]; This subject is dealt with in Section 5.8.2.

Isotope Exchange

Isotope exchange is a technique confined to radiochemical separations so radioactive tracers have to be used in its study. The subject is reviewed by Jordan [100]. The mechanism of isotope exchange separations can be described by the equation

$$MX + M^*Y \rightleftharpoons M^*X + MY$$

where, M is the stable element and M^* its radioactive form. MX and MY are combined forms of the element present in different but easily separated phases. If MX is initially present in an amount greatly in excess of M^*Y, virtually all of the M^* will combine with X and an equivalent amount of M will combine with Y, i.e. isotope exchange will take place. A few examples will illustrate the development, power and range of application of this technique.

Silver ions will exchange rapidly with solid silver chloride. Sunderman and Meinke [36] showed that about 10 mg of silver chloride deposited on a gauze would exchange with over 99% of ^{110m}Ag in solution in less than 15 minutes. Only tracer levels of ^{203}Hg and ^{210}Bi were carried in excess of 1%, some 20 other tracer ions being carried from 0.1–0.00005%. The quantitative exchange of silver tracer was found to be relatively unaffected by the presence of molar concentrations of the common mineral acids and of several salts. The speed of exchange was found to increase with temperature, making it possible to use the method for separation of short-lived silver isotopes. The method was applied to the determination of short-lived silver activity in deuteron irradiated cadmium and ^{111}Ag in irradiated palladium.

The elimination of ^{24}Na, and to a lesser extent, of ^{42}K, interferences is of paramount importance in the neutron activation analysis of biological samples. Isotope exchange of ^{24}Na on sodium chloride [101, 102] or perchlorate [102], and ^{42}K on potassium chloride [101], from aqueous organic solutions of the sample has been used successfully for that purpose. The principal tracer technique used was simply measurement of the intial solution, the effluent and the dissolved column in a standard geometry.

Amalgams have been used for isotope exchange separations, a notable example being the separation of bismuth [103]. Exchange can be achieved by shaking 10 μg of bismuth in 0.5 N sulphuric acid with 1.4% bismuth amalgam. It is essential to purge the system with nitrogen, prior to shaking, or yields will be low. An extensive study was made of the amount of amalgam required to give high yields, the effects of acids and salts at various strengths and the decontaminations obtained from a wide range of active nuclides. Almost without exception elements that exchanged had more positive reduction potentials than bismuth, i.e. Ru, Se, Ir, Hg and Ag. Fortunately, back extraction of the bismuth with cupric ion was found to eliminate these interferences.

Isotope exchange can be applied with considerable success to gas phase separations. Gamma photon activation analysis for small amounts of oxygen in materials containing excess carbon required the rapid release of ^{15}O ($t_{1/2}$ = 2.05 min), and its efficient separation from ^{11}C ($t_{1/2}$ = 20 min), prior to counting through its positron decay. Often the induced oxygen activity is released as carbon monoxide by inert gas fusion, active species being ^{12}C^{15}O, ^{11}C^{16}O and ^{11}C^{15}O. Hislop and Williams [104] found that passage of these species over Hopcalite, MnO_2 87%, CuO 13%, AgO 0.1%, at 650°C led to complete separation of the ^{11}C and ^{15}O, the former being released as carbon dioxide and the latter being retained by the Hopcalite. There is little doubt that the ^{15}O exchanged with ^{16}O during the oxidation of the carbon monoxide. The separation procedure enabled oxygen to be determined in steel and molybdenum specimens with high C:O ratios, 15:1 and 1000:1 respectively.

Studies of Complete Procedures

To complete this discussion of the use of tracers in the investigation of analytical methods two examples are given of their use as an ultimate check of the efficiency of newly developed methods. In doing so some of the important features of radioactive tracer studies will be re-emphasised.

The determination of Be in pathological samples is important because of its toxicity. When faced with the determination of submicrogram amounts of Be in urine, bone and soft tissue, Toribara and Sherman [105] realised that, although the morin fluorimetric method for Be was sensitive enough, the recovery and purification of the Be before measurement was a formidable task. Steps in the sample treatment were identified; destruction and solubilisation; concentration and separation. Recovery at each step was investigated with the aid of ^{7}Be tracer and a simple dipping Geiger-Müller detector. Wet and dry ashing, possible loss by volatilisation, precipitation and fixation by reaction vessels were studied. Purification and concentration by a variety of possible methods were tested, including precipitation, co-precipitation, ion exchange, chromatography, electrolysis at a mercury electrode and acetonyl acetone extraction. Finally, after selection of methods for each class of sample, the overall recovery of Be was tested with tracer and found in each instance to be virtually complete. Many of the experiments described were carried out by simply adding tracer solution to samples. However, to prove that this method was adequate, some experiments were performed with labelled bone, urine and soft tissue, obtained by injecting rabbits with ^{7}Be tracer.

Mercury contamination of the environment is recognised as a serious health hazard. Neutron activation analysis may be used for the determination of mercury. However, for many samples, separation of the mercury after irradiation is necessary before measurement and rapid separations are attractive when large numbers of samples must be handled. Workers at the National Bureau of Standards [106] have devised an activation analysis method that involves a fast empty tube combustion to volatilise and separate the mercury from interfering activities, other than ^{82}Br. A number of tracer experiments were performed to ensure that mercury recovery was complete. While $99.52 \pm 1.68\%$ recovery was obtained for ^{203}Hg tracer added to orchard leaves, the test did not prove that mercury could be recovered when incorporated in an organism. To check this, gold-fish were labelled with ^{203}Hg by introducing it into the fish tank. The fish survived for up to 3 weeks, and were then used in recovery experiments. Advantage was taken of the gamma emitting properties of the ^{203}Hg to measure its amount in whole fish, using a sodium iodide detector and a relatively low geometry. The ^{203}Hg recovered from each fish was transferred in solution to a vial approximately the same length and diameter as the original fish, thus enabling it to be counted in the 'same' geometry. In fact, incorporating the tracer into the fish had little effect on recovery, $98.9 \pm 1.0\%$. However, without the evidence of carefully conducted tracer experiments, considerable doubt about the recovery of mercury, a readily lost element, would have existed.

8.2   QUANTITATIVE RADIOACTIVE TRACER METHODS

8.2.1   Introduction
Quantitative analysis through the use of radioactive tracers is often regarded

as being of secondary importance in inorganic analysis compared to activation analysis. While this may be so for trace element analysis, because activation analysis has the advantage of freedom from blanks, there are many circumstances in which radioactive tracer methods are superior. For those without ready access to a nuclear reactor or a particle accelerator, activation analysis immediately loses much of its appeal, particularly if the only conveniently induced activities are short-lived. Rapid analysis by activation methods may be impossible if the induced activity is of long half-life. Generally, the counting equipment desirable for activation analysis is far more complex than that required for analyses based on radioactive tracers. Furthermore, the amounts of radioactivity handled in radioactive tracer analysis are usually smaller than those encountered in activation analysis where the induced matrix activities may be so high that remote handling techniques are essential. In some circumstances, radioactive tracer methods are more sensitive than activation methods, i.e. where cross sections and product half-lives lead to poor sensitivity in the latter, while suitable high specific activity isotopic tracers may be available through alternative production processes such as fission for use in the former. Alternatively the use of non-isotopic tracers may occasionally lead to a method with improved sensitivity. A radioactive tracer method might also be substituted with advantage when activation analysis is impossible because of interfering nuclear reactions.

Compared with conventional analytical methods, radiotracer techniques possess certain attractive features. Their prime advantage is that quantitative recovery of an added tracer is seldom necessary. Methods based on radiotracers may be applied over a wide concentration range, from the per cent level down to as low as one part in 10^{12}. They can also be extremely simple and rapid and are amenable to automation in many instances.

The basic considerations discussed in Section 8.1 apply equally to quantitative analysis by radioactive tracers. The properties of the tracer, both nuclear and chemical, will influence its usefulness and must be carefully examined before use. Radiochemical and radionuclidic purity of the tracer are of great importance, as indeed is the purity of reagents, particularly for analysis at low concentrations. Counting equipment can often be simple and usually gamma counting is preferred to alpha and beta counting though there can be exceptions to this rule.

For convenience quantitative radioactive tracer methods, sometimes referred to as radiometric methods, are classified into two broad groups: isotopic methods and non-isotopic methods. The former include isotopic dilution, substoichiometric isotope dilution, isotopic exchange and concentration dependent distribution methods, while the latter encompasses isotope-derivative, derivative dilution, isotope displacement and radiorelease methods. Finally, radiometric titration is an exception to this classification as it can basically depend on either isotopic or non-isotopic tracers. Other methods of classification are discussed in detail by Tölgyessy, Braun and Kyrš [109].

8.2.2 Isotopic Methods

All isotopic methods depend on the inverse proportionality of specific activity and inactive isotope concentration and the division of methods within this group depends solely on how this inverse proportionality is utilised.

Isotope Dilution

The principle of isotope dilution analysis is simple and well known. It was first used by Hevesy and Hobbie [107] for the determination of lead in minerals.

When a radioactive tracer is mixed with its inactive isotopes the overall specific activity will be reduced. If the specific activity of the tracer is S_o then

$$S_o = \frac{A_o}{W_o} \tag{8.1}$$

where A_o is the activity of a weight of W_o of tracer plus carrier. Then if activity A_o is added to a material containing a weight W_x of inactive isotopes of the tracer the specific activity will become S_x where

$$S_x = \frac{A_o}{W_o + W_x} \tag{8.2}$$

S_x can be measured by recovering a pure fraction W_2 of $(W_o + W_x)$ and determining its activity, A_2.

W_x is given by solution of equations (8.1) and (8.2):

$$W_x = W_o \left(\frac{S_o}{S_x} - 1 \right) \tag{8.3}$$

If W_o is small compared to W_x then

$$W_x = W_2 \frac{A_o}{A_2} \tag{8.4}$$

The essential step in this method of analysis is the recovery of the pure fraction of the element or compound being analysed and the measurement of its weight, W_2, and its activity, A_2. That the recovery of this fraction need not be quantitative is a great advantage as very efficient, though non-quantitative, separations may be adopted. The value of W_2 may be determined by many methods, such as gravimetry, colorimetry, polarography, titrimetry and coulometry, the method employed being dependent on the species and on its amount. The activities A_2 and A_o are normally measured under identical conditions. If equation (8.4) applies, because $W_x \gg W_o$, S_o need not be known precisely. However, if equation (8.3) applies, S_o must be known accurately or must be determined.

Of many applications of isotope dilution analysis published [108, 109], two have been selected to illustrate the general techniques involved. Salyer and Sweet [110, 111] have developed an isotope dilution method for the determination of cobalt in steels using anodic electrodeposition. Immediately the steel samples have dissolved the cobalt (^{60}Co) tracer solution is added so that isotopic exchange takes place as soon as possible. The amount of cobalt in the tracer is chosen to be approximately equal to that in the sample. If the weight of cobalt in the sample is small compared to that in the tracer solution, i.e. Wo $\gg$ Wx, the change in specific activity will be small and the accuracy low. The bulk of the iron is removed by ether extraction and the cobalt is then isolated efficiently, but non-quantitatively, by precipitation as potassium cobaltinitrite. The cobalt is finally electrodeposited onto a platinum disk anode as the oxide, the amount of oxide measured by weighing and the activity of the ^{60}Co tracer measured by gamma counting in a fixed geometry. Calibration measurements are made by following through the procedure with a known amount of cobalt.

Triphosphate and pyrophosphate can be determined in mixtures, containing in addition other phosphate ions, by a method based on separation by fractional crystallisation [112]. Isotope dilution enables fractional crystallisation to be used despite its being non-quantitative. Because complex ions are being determined, ^{32}P-labelled triphosphate and pyrophosphate tracers must be prepared radiochemically pure. It is also essential to prove that isotopic exchange only occurs between chemically identical phosphate ions. Experiments showed that, by controlling the pH and temperature, unwanted exchange could be avoided and a successful method was developed.

The accuracy and precision of isotope dilution analysis are dealt with in detail in specialised publications [109, 113]. Errors can arise from failure to achieve isotopic exchange (or undesired exchange in the case of complex ions or compounds), counting problems, lack of radionuclidic or radiochemical purity of the tracer or impurity of the recovered fraction. W_x should be considerably greater than W_o as mentioned above. The accuracy also depends on the measurement of W_o and W_2. W_o need not be known if it is much smaller than W_x [equation (8.4)], but W_2 must always be measured accurately. W_2 may be only a small fraction of $(W_o + W_x)$ but this is unimportant providing its weight can be determined accurately and its associated activity, A_2, is high enough to give good counting statistics. It is good practice to be able to accumulate at least 40,000 counts in no more than about 15 minutes. The ultimate sensitivity of isotope dilution analysis is limited by the specific activity of the available tracers, the blank due to reagents, and the sensitivity of the methods available to determine the isolated fraction, W_2. The last factor has limited conventional isotope dilution analysis to the microgram level where, for instance, colorimetry can be used [114]. The introduction of substoichiometric isotope dilution analysis has, however, provided a solution to this problem [115, 116].

Substoichiometric Isotope Dilution Analysis
In substoichiometric dilution analysis equal amounts of the element or species being determined are removed from the tracer solution and from the solution formed by isotope dilution. The activities in these amounts, a_o and

a_x, will be directly proportional to the specific activities of the tracer solution and that formed by isotope dilution respectively. Therefore equation (8.3) can be written as

$$W_x = W_o \left(\frac{a_o}{a_x} - 1 \right) \tag{8.5}$$

The essential prerequisite for this type of analysis is that a method must be found for the removal of equal amounts of the species being determined from two solutions, possibly of different volumes, and containing different total amounts of that species. Substoichiometric separation methods allow the attainment of this end through the reaction of the analysed species with a substoichiometric amount of a reagent and separation of the product.

Substoichiometric separations have been established using electrolysis, solvent extraction of metal chelates or of ion-association complexes, ion exchange combined with the formation of water soluble metal chelates, displacement and precipitation. The theoretical bases for these various methods have been worked out [115, 116].

Referring to equation (5.10) (page 190) it is seen that the extraction constant, K_{ex}, for the solvent extraction of a metal chelate complex is given by

$$K_{ex} = \frac{[ML_n]_o}{[M^{n+}]_w} \frac{[H^+]_w^n}{[HL]_o^n} \tag{8.6}$$

For the substoichiometric principle to apply, virtually all ($\geqslant 99\%$) of the HL must react with the excess metal ion M^{n+} and the product ML_n must be separable by solvent extraction. If the pH of the solution is too low the chances of the reaction occurring are decreased. It can be shown [116] that the threshold pH above which 99% of the metal chelate is formed is given by the equation

$$pH \geqslant -0.01 \log C_{HL} - \frac{1}{n} \log K_{ex} \tag{8.7}$$

where C_{HL} is the original concentration of the chelating agent in the organic phase. However, if the pH is too high the chelating agent will dissociate and pass into the aqueous phase, preventing formation of the metal chelate. The upper pH limit can be determined from the expression

$$pH \leqslant pK_{HL} + \log D_{HL} + \log \frac{V_{org.}}{V_{aq.}} \tag{8.8}$$

where pK_{HL} is the negative logarithm of the dissociation constant of the chelating agent, D_{HL} is its distribution coefficient and $V_{org.}$ and $V_{aq.}$ are the volumes of the organic and aqueous phases respectively. A further limitation on the upper value of pH is the formation of insoluble hydrated oxides by many metals in alkaline solution.

A relatively wide range of chelating agents form metal chelates with extraction constants high enough to allow their use for substoichiometric extraction of milligram quantities of metals. However, for isotope dilution analysis at the microgram level and below the choice is limited to reagents forming metal chelates with very high extraction constants. In addition the reagents must be stable at high dilution. Dithizone and cupferron meet these requirements. Selectivity in substoichiometric extraction depends principally on the values of the equilibrium constants of the metal complexes formed by the chelating agent. The metal forming the complex with the highest equilibrium constant can be determined in the presence of any other metal with little chance of interference, e.g. mercury with dithizone [117] and iron with cupferron [118]. The determination of metals with lower equilibrium constants depends on the absence of appreciable amounts of metals with higher equilibrium constants, their prior removal, e.g. Hg, Ag and Cu by dithizone from Cd [119], or their masking with a suitable masking agent [117]. The selectivity of separations by substoichiometry, where there is no excess of reagent, is greater than with methods in which an excess of reagent is employed, as in the latter, the excess reagent will form extractable chelates with other metals unless prevented from doing so by pH adjustment or by some other device.

The procedures adopted in the development and in the use of methods based on substoichiometric extraction of metal chelates are generally similar. The conditions for substoichiometric extraction are first established. These include assessing the influence of pH, the time and reagent concentration, on the extraction of a constant amount of the chelate and, where necessary, the need for masking or pre-extraction of interfering elements. An actual analysis commences with sample preparation and the addition of the tracer at an early stage. Pre-separation may then be carried out, followed by chelation and substoichiometric extraction, possibly in the presence of masking agents. After phase separation the activity of a fixed volume of organic phase is counted, probably in a well detector, if the tracer is a gamma emitter. An identical volume of tracer solution is also subjected to substoichiometric extraction and the activity of the extracted chelate determined. It is essential to ensure that an excess of metal is present in both tracer solution and sample solution before extraction. Provided this limitation is observed non-quantitative pre-separation methods may be used in sample preparation.

Many substoichiometric methods of this type have been developed [109, 116] including some with excellent sensitivities, e.g. zinc [120], mercury [117] and iron [121] down to 10^{-9} g and copper [122] down to about 10^{-10} g.

Displacement (or replacement) substoichiometry is similar in concept to the method based on the extraction of metal chelates. The basic reaction is

$$nML_m + mN \rightleftharpoons mNL_n + nM$$

For the determination of metal M a substoichiometric amount of metal ion N is added, and for the determination of metal N a substoichiometric amount of ML_m is added, the essential common factor being the greater stability of NL_n. The theory of the method has been developed [147] and model experiments carried out [148]. However, applications have been restricted to activation analysis, where the reverse isotope dilution principle is employed [149, 150]. In activation analysis the method can be modified to allow the recovery of groups of metals [151, 152].

In the solvent extraction of ion association compounds an anion, Q^{n-}, reacts with a large organic cation, P^+, to form an extractable complex, P_nQ,

$$nP^+ + Q^{n-} \rightleftharpoons (P_nQ)_{org.}$$

The equilibrium constant for the reaction is

$$K = \frac{[P_nQ]_{org.}}{[P^+]^n [Q^{n-}]} \tag{8.9}$$

For $\geqslant 99\%$ of a cation to react with a monovalent anion, it can be shown that

$$K \geqslant \frac{C_P}{(0.01 \, C_P)(C_Q - C_P)} \frac{V_{aq.}}{V_{org.}} \tag{8.10}$$

where C_P and C_Q are the original concentrations of the cation and anion respectively and $V_{aq.}$ and $V_{org.}$ are the volumes of the aqueous and organic phases [116]. In an analogous manner large organic anions can be used for the extraction of cations.

A number of reagents exist that form ion association complexes with equilibrium constants high enough to allow their use for substoichiometric analyses at the milligram level and above. However, these have only been applied in the recovery of active nuclides produced by neutron irradiation by reverse isotope dilution, e.g. tetraphenylarsonium chloride for manganese [123], rhenium [124], gold ($AuCl_4^-$) [124] and tetraphenylboron [124] and dipicrylamine [125] for caesium.

Substoichiometric extraction of the ion association complex formed between antimony chloride ($SbCl_6^-$) and methyl violet may be used for the isotope dilution analysis of antimony in amounts as low as a few micrograms [126] and the method was successfully applied to the determination of antimony in lead. A method for fluoride is based on the formation of its tetraphenylstibonium complex [146].

Selectivity in substoichiometric isotope dilution analysis by the extraction of ion association complexes, is achieved by the same means as for the extraction of metal chelates and the procedures for methods development and use are also similar.

Substoichiometric separations may also be achieved through the formation of water soluble metal chelates, MY, and the removal of the excess metal cations, M, by ion exchange. In order to form the water soluble chelate in $\geqslant 99.9\%$ yield the stability constant, β_{MY}, must exceed a limiting value:

$$\beta_{MY} \geqslant \frac{C_{H_nY}}{C_M - C_{H_nY}} \frac{\sum\limits_{n=0}^{} \frac{[H]^n}{k_0, \ldots, k_n}}{0.001 \, C_{H_nY}} \tag{8.11}$$

where C_{H_nY} and C_M are the original concentrations of the chelating agent and metal ions, [H] is the hydrogen ion concentration and $k_o, \ldots, k_n$ are the dissociation constants of the chelate H_nY [116]. The stability constants of many metal ion complexes with nitrilotriacetic acid, ethylenediamine tetraacetic acid and 1,2-diaminocyclohexane tetraacetic acid are high enough for substoichiometric isotope dilution analysis down to at least the microgram level. Metal complexes with these reagents are anionic so the removal of uncomplexed metal ions with cationic exchangers is very convenient. However, other separation methods such as extraction [127], electrophoresis [128] and paper chromatography are possible alternatives.

Selectivity depends on the relative values of the stability constants of the metal complexes. The one with the highest stability constant may be determined without interference from any other metals while specificity for those with lower stability is achieved through pre-separation or masking. Methods development involves the investigation of the range of concentration over which substoichiometric reaction occurs, the effect of pH, the time required for the reaction to reach equilibrium and the influence of interfering ions.

Examples of the use of this method are the determination of iron in sodium iodide [129], in which ethylenediamine tetraacetic acid was the substoichiometric complexing agent and as little as 10^{-9} g of iron could be determined, and the determination of indium [130] with a sensitivity of 10^{-11} g.

Precipitation reactions are only suitable for substoichiometric separations at the milligram level, being too insensitive for smaller amounts.

Electrolysis as a means of substoichiometric separation was suggested by Růžička [131], who with Beneš [132] developed a method for the determination of silver. The technique adopted was to place two electrolysis cells in series, one containing tracer and the other tracer plus sample, and to deposit equal amounts of silver in each cell by passing a substoichiometric quantity of current. While simple in principle the problem of competitive deposition of interfering ions has to be overcome in practice. Iodide has also been determined by an electrolytic substoichiometric method [133].

Substoichiometric isotope dilution analysis may be automated and the Technicon Auto Analyser has been used [134]. An automatic method for the determination of mercury has been devised which uses its substoichiometric extraction with zinc dithizonate [135].

The application of substoichiometric separation methods to the recovery, by reverse isotope dilution, of radioactivity induced in samples is in some respects much simpler than their application in isotope dilution. In the former, equal amounts of inactive carrier are added to the sample and to the radioactive standard, and choosing an ideal substoichiometric amount of reagent is simple. In isotope dilution, however, the choice of the amount of reagent is more difficult as the concentration of the determined species in the sample and tracer solutions may differ widely. While this could be avoided by keeping the concentration of the tracer solution high the danger exists, that on addition of the sample, the change of specific activity could be negligible. Use of a very small amount of reagent is undesirable as the activity recovered from the solutions could be low, particularly from the sample solution. While it is apparent that the amount of the determined species in the tracer solution and in the sample solution, before addition of tracer, should be similar and that the maximum substoichiometric fraction should be recovered, it is not easy to achieve this ideal. In addition, the isotope dilution method suffers more from blank problems and the presence of interfering ions.

A modified method, due to Grashchenko and Sobotovich [136] tends to overcome some of the objections described above. If equal amounts, W_0, of a tracer with activity, A_0, are added to the sample containing W_x of the species being determined, and to a standard solution containing W_s of that species, and $W_0 \ll W_x$ and W_s, then the specific activities of the solutions are

$$S_x = \frac{A_0}{W_x}$$
(8.12)

and

$$S_s = \frac{A_0}{W_s}$$
(8.13)

If equal amounts, W_T, of the species are removed substoichiometrically from each solution, then their respective activities are

$$a_x = S_x W_T = \frac{A_0}{W_x} \cdot W_T$$
(8.14)

and

$$a_s = S_s W_T = \frac{A_0}{W_s} \cdot W_T$$
(8.15)

Solution of these equations gives

$$W_x = W_s \frac{a_s}{a_x}$$
(8.16)

For this method a high specific activity tracer is needed. The method is most accurate when $W_x = W_s$, as the activities can be recovered in high yield and will be equal.

Two other variants of substoichiometric IDA are of interest; double substoichiometric isotope dilution [137, 138] and multiple radioactive isotope dilution [139].

In double substoichiometric isotope dilution, identical amounts, A_0, of a tracer are added to equal aliquots of sample solution each containing an amount, W_x, of the element being determined. To one is added a known amount of inactive isotopes, W_1, and to the other a second amount, W_2.

The weight of the element contributed by the tracer, W_o, must be negligible compared with W_x, W_1 and W_2.

The specific activities of the two isotope dilution solutions are

$$S_1 = \frac{A_o}{W_1 + W_x} \tag{8.17}$$

$$S_2 = \frac{A_o}{W_2 + W_x} \tag{8.18}$$

If equal amounts, W_q, are removed substoichiometrically their activities will be

$$a_1 = S_1 W_q = \frac{A_o}{W_1 + W_x} \cdot W_q \tag{8.19}$$

and

$$a_2 = S_2 W_q = \frac{A_o}{W_2 + W_x} \cdot W_q \tag{8.20}$$

Solution of these equations gives

$$W_x = \frac{a_2 W_2 - a_1 W_1}{a_1 - a_2} \tag{8.21}$$

Optimum conditions occur when $W_1 = W_x$ and $W_2 \gg W_x$.

This method has been used for the determination of silver by dithizone extraction and of iron by the complexone ion exchange technique [138]. A successful attempt has been made to use this principle combined with differential constant-voltage coulometry [138, 140]. The developed method was applied to the determination of cadmium in zinc [141].

In multiple radioactive isotope dilution [139], different activities, A_1 and A_2, of a tracer of known specific activity, S_o, are added to equal aliquots of the sample solution each containing an amount, W_x, of the element being determined. The specific activities of the two solutions after isotope dilution are

$$S_1 = \frac{A_1}{W_1 + W_x} = \frac{S_o W_1}{W_1 + W_x} \tag{8.22}$$

and

$$S_2 = \frac{A_2}{W_2 + W_x} = \frac{S_o W_2}{W_2 + W_x} \tag{8.23}$$

If equal amounts, W_p, are substoichiometrically removed from each solution, their respective activities are

$$a_1 = S_1 W_p = \frac{S_o W_1}{W_1 + W_x} \cdot W_p \tag{8.24}$$

and

$$a_2 = S_2 W_p = \frac{S_0 W_2}{W_2 + W_x} \cdot W_p \tag{8.25}$$

Solution of these equations gives

$$W_x = \frac{W_1 W_2 (a_2 - a_1)}{W_2 a_1 - W_1 a_2} \tag{8.26}$$

Optimum conditions occur when $W_1 = W_x$ and $W_2 \gg W_1$ [142]. The method has been used for the determination of antimony in lead [139]. Under optimum conditions equation (8.26) simplifies to

$$W_x = W_1 \left(\frac{a_2}{a_1} - 1 \right) \tag{8.27}$$

as $W_2 a_1 \gg W_1 a_2$. This equation is essentially identical to equation (8.5), i.e. that applicable to ordinary substoichiometric isotope dilution, the only difference being that in multiple radioactive isotope dilution the activity of the tracer solution is derived from a solution containing a small proportion of sample. A disadvantage of choosing $W_2 \gg W_1$ is that the substoichiometric separation method must be applicable to solutions containing widely different amounts of the determined species.

Double substoichiometric isotope dilution also suffers from the latter disadvantage. In addition, it is less sensitive than ordinary substoichiometric analysis, as W_0 must be negligible compared with W_1 and W_2. An advantage of the method is that the specific activity of the tracer need not be known accurately.

The final modification is that devised by Suzuki and Kudo [143]. This involves irradiation of the sample, usually with neutrons, dissolution, division of the solution into two halves, addition of a weight, W_s, of the element to one of the solutions, followed by purification and substoichiometric separation of equal amounts of the element from both solutions. If the activities of the recovered material are A_x and A_s and the initial weight of the element in each solution is W_x, then

$$W_x = W_s \frac{A_s}{A_x - A_s} \tag{8.28}$$

Zinc, copper and silver have been determined by this technique [143–145]. The method suffers, however, from the disadvantages of both tracer methods and activation analysis, i.e. the blank problems of the former and the need to separate from high levels of extraneous activity in the latter.

Isotope-Exchange Methods

Isotope-exchange reactions of the following type are utilised in this technique:

$$M^*X + MY \rightleftharpoons (M, M^*)X + (M, M^*)Y$$

where (*) indicates radioactive labelling. The method depends on the ease of separating MX and MY, e.g. by solvent extraction, and the most

strongly complexing ligand, be it X or Y, must be associated stoichiometrically with one portion of M before exchange. If the amount of M associated
with Y, $[M]_y$, is unknown, and that with X, $[M]_x$, is known, then

$$[M]_y = [M]_x \frac{A_y}{A_x} \tag{8.29}$$

where A_y and A_x are the activities associated with ligands Y and X respectively, after equilibration and separation.

The amount $[M]_y$ can be determined in other ways. Thus the change in
activity of the phase originally labelled may be measured. If the original
activity associated with the ligand X was A_o and on exchange it becomes A_x

$$[M]_y = [M]_x \frac{(A_o - A_x)}{A_x} \tag{8.30}$$

A further alternative is to measure the activity, A_y, of the unlabelled phase
after equilibration and refer to an activity calibration graph obtained by
equilibrating standards. If the original amount of the element associated
with ligand X was $[M]_x$ and its activity A_o then

$$[M]_y = [M]_x \frac{A_y}{A_o - A_y} \tag{8.31}$$

Pierce has pointed out [153] that if $[M]_x \gg [M]_y$ then $A_o \gg A_y$ and
equation (8.31) simplifies to

$$[M]_y = [M]_x \frac{A_y}{A_o} \tag{8.32}$$

He showed that silver in aqueous solution could be determined by isotope
exchange with ^{110m}Ag labelled silver dithizonate in an organic solvent. The
activity in the aqueous phase after exchange was directly proportional to
the original silver content within the range 0–1 μg. However, in the range 1
and 12 μg the relationship became non-linear as $[M]_x$ approached $[M]_y$. In
this case equation (8.31) can be rearranged [154] to

$$\frac{1}{A_y} = \frac{[M]_x}{A_o} \frac{1}{[M]_y} + \frac{1}{A_o} \tag{8.33}$$

A plot of $1/A_y$ against $1/[M]_y$ gives a straight line of slope $[M]_x/A_o$ and an
intercept on the ordinate of $1/A_o$. This relationship has been shown to fit
Pierce's data in the range 0–12 μg of silver.

An isotope exchange method for mercury has been described by Handley
[155]. A number of metals form extractable complexes with di-*n*-butyl
phosphorothioate. The distribution coefficient of the mercury complex into
water immiscible organic solvents is $> 10^3$. Exchange of aqueous mercury
with ^{203}Hg-labelled mercury di-*n*-butyl phosphorothioate has permitted the
measurement of mercury down to 0.1 μg. The most accurate results were
claimed to be obtained when the activities in the two phases after equilibra-

tion were equal. An alternative approach was to label the aqueous phase. This had the advantage that tracer of high specific activity could be added, prior to preliminary separation, the percentage recovery measured radiometrically and the weight recovered determined by isotopic exchange. Interference by metals forming more stable complexes with the reagent than mercury make it essential to eliminate this by preliminary separation or by masking.

Richter [156] devised a method for the determination of iodine in natural waters that involved radiometric yield determination after preliminary separation, followed by the measurement of the weight of iodine recovered by isotope exchange. The method was used to measure down to 1 μg of iodine per litre of water. Oxby and Dawson have developed a method for iodine based on the same exchange reaction [157].

An example of a method relying on the measurement of the decrease in activity of the originally labelled phase is one for the determination of silver [158]. Silver in aqueous solution is exchanged with [110m]Ag labelled silver diethyldithiocarbamate in carbon tetrachloride. The decrease in activity of the organic phase is a measure of the silver in the aqueous solution. The method has been used for the determination of silver in lead at the 2–3 ppm level. Mercury and cyanide interfere.

Isotope-exchange is similar to substoichiometric isotope dilution analysis and the selectivity, sensitivity and accuracy are controlled by the same factors which have been discussed in the previous section.

Concentration-Dependent Distribution Methods
Concentration-dependent distribution (CDD) or substoichiometric radiometric methods involve the use of calibration graphs relating the initial concentration of the determined species either to its distribution in a two phase system or to that of a substance reacting with it [159, 160].

The theoretical basis of CDD methods has been thoroughly expounded. Kyrš has discussed four variants: heterogeneous sorption conforming to the Langmuir or Freundlich isotherms, extraction with a constant amount of an extracting agent and sorption or extraction in the presence of a constant amount of a chelating agent [159]. A theoretical treatment has also been given by De Voe [140].

Although CDD can be highly sensitive and may be used when substoichiometric isotope dilution is inapplicable because of the limiting effect of stability constants, it has not proved popular, and a limited number of applications have been reported. Adsorption methods have been developed for caesium using Prussian Blue [161], and for Co, Ni, Zn, Tl, Ag, Cd and Fe using manganese dioxide [138]. In both cases the colloidal adsorbent was dispersed on filter paper. Cobalt may be determined by its adsorption on a cation exchanger in the presence of a substoichiometric quantity of EDTA. Graphical plots reveal that the cobalt remaining in solution is related to the initial cobalt concentration well below that at which the substoichiometric isotope dilution principle is applicable, demonstrating the higher sensitivity of CDD.

Any isotopic method involving the use of calibration curves could be classified as a CDD method. An isotope dilution method based on ring-oven precipitation is a good example [162]. In this method an active tracer, containing a weight, W_0, of the element to be determined, is added to material containing an unknown amount of that element, W_x. A constant amount, W_1, of $(W_0 + W_x)$ is precipitated and separated from the remainder by the ring-oven method. The specific activity, S, of the two fractions is identical, so theoretically

$$S = \frac{A_x}{W_0 + W_x - W_1} = \frac{A_1}{W_1} \qquad (8.34)$$

where A_x and A_1 are the activities of the portions containing $(W_0 + W_x - W_1)$ and W_1 respectively. Solving for W_x gives

$$W_x = W_1 \left(\frac{A_x}{A_1} + 1 \right) - W_0 \qquad (8.35)$$

A calibration graph can be obtained relating A_x/A_1 to W_x, for constant values of W_0 and W_1. Theoretically this should give a straight line relationship. However, deviations occur at low values of W_x. These can be explained on the basis of a decrease in W_1 as W_x becomes small, and is possibly a function of the solubility product of the precipitate. Calcium, phosphate and iron have been determined by this method at the microgram level.

From the definition of CDD methods it can be seen that non-isotopic procedures could be devised using the principle. Some isotope-derivative methods could be regarded as non-isotopic CDD methods.

Reverse Isotope Dilution

Reverse isotope dilution is usually considered to be the opposite of normal isotope dilution, namely, the addition of a known amount of inactive carrier to an unknown amount of radioactive entity, chemical exchange, recovery of a pure fraction of the substance, and determination of the chemical yield and radioactivity of the recovered fraction. Frequently the total amount of a radionuclide requires determination and the carrier and radionuclide are simply brought to the same chemical state to ensure exchange. If, however, the fraction of a radionuclide in a particular chemical form is required, carrier of identical chemical form must be added and the isolation process must avoid exchange with the radionuclide combined in other chemical forms. Inorganic applications usually involve the determination of the total amount of a radionuclide and carrier methods for activation analysis are designed to achieve that end. However, some studies, such as the investigation of the chemical form of nuclides after their formation by nuclear processes obviously must employ carriers of specific chemical forms.

Inactive chemical carrier is normally added in considerable excess of the amount initially associated with the radionuclide being determined. Consequently, the initial activity, A_I, is related to the recovered activity, A_R, by

$$A_I = A_R \frac{W_I}{W_R} \tag{8.36}$$

where, W_I and W_R are the initial and recovered weights of chemical carrier.

The advantage of such a method for the recovery of a specific radionuclide from an excess of other radionuclides, as in activation analysis, is readily apparent. Non-quantitative separations with high selectivity can be used and losses corrected through the measurement of the chemical yield. Many examples of the use of this technique can be found in standard works on the measurement of radionuclides [163] and activation analysis [164].

Substoichiometric recovery of carrier can be employed in reverse isotope dilution [116]. The large decontamination factors frequently needed normally require the use of preliminary separations. However, the ultimate use of a substoichiometric separation eliminates the measurement of the chemical yield, but involves the identical treatment of a radioactive standard. The weight of carrier used in reverse isotope dilution frequently allows the use of reagents unsuitable for substoichiometric isotope dilution at the microgram level as was discussed on page 322.

Equation (8.36) requires modification if the weight of inactive material initially associated with the radionuclide, W_P, is appreciable compared with the added carrier. In this case

$$A_I = A_R \frac{(W_I + W_P)}{W_R} \tag{8.37}$$

Consequently W_P must be determined if it is not known.

In the determination of ^{89}Sr and ^{90}Sr in samples of bone the inactive strontium present in the bone may contribute significantly to the total strontium after the addition of carrier, so it must be determined, e.g. by spectrographic analysis, in order to correct the chemical yield [165, 166]. Alternatively a known amount of ^{85}Sr tracer can be added with the inactive carrier and the yield found radiometrically [167]. The radiometric method depends on the measurement of the ^{85}Sr, via its gamma emission, on there being no interference in the beta counting of ^{89}Sr and ^{90}Sr (^{85}Sr emits no beta radiation) and the absence of ^{85}Sr in the sample.

The use of a radiometric yield determination is most important for elements with no stable isotopes, e.g. protactinium, where the alpha active ^{231}Pa has been used as a tracer in the determination of the beta active ^{233}Pa [168].

In activation analysis the situation described by equation (8.37) rarely occurs. When it does, the effect of W_P on the yield can be readily assessed, because the specific activity of the nuclide of interest, S_I, is known from the measurement of the irradiation standard. In fact, W_P is the quantity normally required in activation analysis.

Using

$$S_I = \frac{A_I}{W_P} \text{ (from irradiation standard)}$$

and

$$S_R = \frac{A_R}{W_R} \text{ (specific activity of the source recovered from sample)}$$

and substituting $S_I W_P$ for A_I and S_R for A_R/W_R in equation (8.37)

$$W_P = \frac{W_I S_R}{S_I - S_R} \tag{8.38}$$

The technique has been applied successfully in the determination of nickel in irradiated stainless steel oxides through re-irradiation, separation and measurement of ^{65}Ni [169].

Substoichiometric reverse isotope dilution can be used for the determination of the carrier in preparations of radioisotopes. The procedure adopted is simply substoichiometric separation of identical amounts of the carrier from the original solution containing an unknown amount, W_P, and from an equal volume containing an added additional known weight, W_I. The activities of the two fractions will be a_I and a_R respectively and W_P will be given by

$$W_P = \frac{W_I a_R}{a_I - a_R} \tag{8.39}$$

which is identical in form to equation (8.38). The method has been applied to the determination of zinc in high specific activity tracer solutions using solvent extraction [170] and of yttrium and europium in radioactive preparations using EDTA complexation with ion exchange separation [171].

8.2.3   Non-isotopic Methods

The principle of non-isotopic methods is that the amount of radioactivity derived from a non-isotopically labelled reagent, when it reacts with the species being determined, is quantitatively related to the amount of that species. The derived activity may be measured directly, or may need to be recovered chemically in which case the chemical recovery is often measured by reverse isotope dilution. An alternative name for this group of methods is radioreagent techniques [108]. It includes isotope-derivative, derivative-dilution, radio-release and isotope displacement methods.

Isotope-Derivative and Derivative-Dilution Methods

Isotope-derivative and derivative-dilution methods are basically identical. The species being determined is made to react with a radioreagent of known specific activity to form a radioactive compound. This compound is separated from excess radioreagent and its activity measured. Assuming quantitative recovery, the amount of the species being determined, W_x, is calculated from

$$W_x = \frac{A_x}{S_r} E \tag{8.40}$$

where A_x is the activity of the compound formed; S_r the specific activity of the radioreagent and E is the ratio of the equivalent weights of the species being determined and the radioreagent.

Certain conditions are essential for the application of the method. Firstly, the reaction must be made quantitative by correct choice of reagent and reaction conditions; a selective reagent may aid subsequent radiochemical purification of the product. Secondly, the radiochemically pure reaction product must be recovered quantitatively, or in known yield. The recovered compound does not, however, need to be chemically pure as its weight is not required in equation (8.40). Yield correction is possible through addition of macro amounts of inactive product and measurement of the chemical yield, or alternatively by addition of reaction product labelled with a second radioisotope and radiometric determination of the yield.

Many species that can be determined by direct quantitative precipitation may also be determined by the isotope derivative technique as long as the reagent can be radioactively labelled [108]. Thus phosphate labelled with ^{32}P may be used for elements forming sparingly soluble phosphates, silver labelled with [110m]Ag for the determination of chloride, chromate, molybdate etc., and sulphate labelled with ^{35}S for lead and barium. Weighing of the precipitates is, of course, unnecessary but the advantage of these methods over classical precipitation is debatable. Classical precipitation methods are frequently very precise while for a radiometric method 10^6 counts are needed to give a counting precision of 0.1% and geometry and absorption problems must be carefully controlled to give a good overall precision.

Precipitation methods may be used to good advantage in association with paper chromatography. Lead can be determined down to 0.1 μg in presence of a hundredfold excess of several common metals [172]. After separation lead is determined by sparying the chromatogram with ^{32}P-labelled phosphate, eluting excess reagent and measuring the ^{32}P associated with the lead spot. Sulphate labelled with ^{35}S has been used for the determination of as little as 2 ng Ca, 5 ng Sr, 8 ng Ba and 10 ng Pb [173]. Labelled phosphate has been applied to the measurement of beryllium and zirconium at the 0.1 ng level [173].

Modifications of isotope-derivative methods have been used to avoid the problem of ensuring that the reaction between the determined species and the radioreagent is quantitative. Reaction conditions have been rigidly controlled and calibration curves used, constructed from the measurement of standards [140, 174]. Alternatively the determined species has been isotopically labelled making possible the determination of the reaction yield and also the overall yield when preliminary separations are needed [175].

A variety of paper chromatographic methods relying on calibrated radioreagent finishes have been reported by de Voe [140]. Separated iron and copper were determined by reaction with ^{32}P-labelled phosphate; as little as 0.1 μg of each could be measured. Ni, Cu, Co, Cd and Mn were all determined with a sensitivity of at least 1 μg by first precipitating their hexacyanoferrates and subsequent exchange with [59]Fe.

$$3M_2[Fe(CN)_6] + 4{}^*Fe^{3+} \rightarrow {}^*Fe_4[Fe(CN)_6]_3 + 6M^{2+}$$

(* indicates radioisotopic labelling). The method favoured for manganese was precipitation of MnO_2 in situ and equilibration of the paper with a buffer solution of ^{60}Co tracer. The adsorbed cobalt activity was proportional to the amounted of manganese present. In all these methods calibration curves were produced by measuring known amounts of each element.

A method relying on isotopic labelling for yield correction has been used for the determination of cobalt by extraction with ^{35}S-labelled zinc diethyldithiocarbamate [175]. After extraction, the excess reagent is removed and the cobalt determined from the ^{35}S activity associated with it. However, as the replacement of zinc by cobalt is not quantitative, its extent is determined by isotopic labelling with ^{60}Co.

Radio-Release and Isotope-Displacement Methods
In radio-release and isotope-displacement methods the reactions involved are such that a radioactive tracer is displaced or liberated from its initial combined form by the species being determined, the amount of tracer released being proportional to the determined species. Methods of this type have been reviewed by Tölgyessy [176]. The numerous variations of this method are illustrated by the following examples.

A number of methods involve the release of a gaseous radioactive tracer and several of these are based on the release of ^{85}Kr from krypton clathrate compounds [177, 178]. Levels down to 1 part of ozone in 10^{10} of air can be measured by the release of ^{85}Kr from its quinoline clathrate [178]. Water vapour interferes, if not removed, but SO_2 and H_2S do not. The same reagent may be used for the indirect determination of SO_2 [177]. The SO_2 reacts with sodium chlorite and the resulting chlorine dioxide oxidises the clathrate and releases ^{85}Kr. The method is sensitive to 0.05 ppm SO_2. It does however suffer from appreciable interference from nitrogen dioxide and ozone.

An alternative method of incorporating ^{85}K in solids is by bombardment with ionised krypton or by diffusion of krypton into the surface at high temperatures and pressures [179]. The krypton is confined to a surface layer 10^2 to 10^3 nm deep. It can be released from the material by chemical attack. Kryptonated pyrolytic graphite and copper have been examined as materials for the determination of oxygen [179]. Copper is the more satisfactory material, the reaction rate being dependent on the reciprocal of the absolute temperature for incompletely oxidised surfaces. By adjusting the temperature, oxygen levels between 10^5 and 10^{-5} ppm can be measured. The converse of this reaction, the reduction of kryptonated platinum oxide has been used for the determination of hydrogen [180]. To determine hydrogen in air it is necessary to dilute the platinum oxide with alumina, when 10 ppm of hydrogen can be measured and the normal constituents of air do not interfere. Kryptonated silica [181] and silver iodide [182] have been used for the determination of hydrogen fluoride in air. The calibration curves are linear and there is little or no interference from HCl or HBr at 150 ppm.

Sulphur dioxide can be measured in air at concentrations as low as 10^{-4} ppm by its reaction with ^{131}I-labelled potassium iodate [183]. The reaction is

$$5SO_2 + 2K^{131}IO_3 + 4H_2O \rightarrow K_2SO_4 + 4H_2SO_4 + {}^{131}I_2$$

The iodine released may be extracted into chloroform and counted. Hydrogen sulphide interferes significantly, if present.

A number of methods have been used for determining species in solution. Dissolved oxygen in water can be determined by its reaction with thallium metal labelled with [204]Tl [184].

$$4Tl + O_2 + 2H_2O \rightarrow 4Tl^+ + 4OH^-$$

The ionic thallium released is determined radiometrically. The sensitivity attained in practice was 0.2 ppm. Theoretically it should be possible to determine 1 part in 10^{12}. Nitrate interferes and must be absent.

Both vanadate and dichromate are used as non-active species for tracing water. In acid conditions silver labelled with [110m]Ag reacts with vanadates [185] or dichromates [186] to release silver ions. The reaction is stoichiometric and the released activity readily measured. The quoted sensitivities are 0.1 ppm and 0.01 ppm for vanadate and dichromate respectively. Iron and chloride, normally present in natural waters, interfere but this can be eliminated by masking.

A number of isotopically labelled oxalates have been examined as possible reagents for the determination of calcium by displacement [187]. Manganese was displaced stoichiometrically by calcium on passing a solution through a short column of manganese oxalate. The [54]Mn released can be readily measured. A sensitivity of 16 μg per ml is obtainable and the measurement completed in less than 5 minutes. Magnesium interferes to a small extent.

Cyanide in solution may be determined by passing the solution through a column of silver iodide labelled with [110m]Ag and measuring the activity in the eluent [188]. Two moles of cyanide are needed to dissolve one of silver. The method is rapid and sensitive; down to 0.5 μg cyanide. Mineral acids, ammonia, chlorides and thiocyanate do not interfere, but thiosulphate does.

An interesting variant of the column technique is one in which the solubility of the column material is suppressed by the species being determined. A method for chloride is based on the decrease in solubility of silver chloride with increasing chloride concentration [189]. By using silver chloride labelled with [110m]Ag the silver activity released can be correlated with chloride concentration. Chloride levels of 1–10 μg per ml may be determined.

Displacement of one metal from its complex with an organic reagent by another may be used as the basis of a radiometric method of analysis. Thus an unknown amount of a metal forming a complex of high stability will release a proportionate amount of activity from a labelled metal complex of lower stability. If the released and complexed activities can be separated, e.g. by solvent extraction, and measured, they can be used to determine the amount of the displacing metal. This principle is used in the determination of mercury by displacing silver labelled with ^{110m}Ag from silver dibutyl phosphorothioate [155]. The released and complexed silver are separated by solvent extraction. The method allows the determination of as little as 0.1 μg of mercury. Lead may be determined similarly by displacing thallium labelled with ^{204}Tl from its diethyl dithiocarbamate complex [190].

Many radiorelease methods are sensitive, simple and rapid. In consequence they are attractive and may often be readily adapted for automatic operation.

8.2.4   Radiometric Titrations

Radiometric titrations are those in which the course of the titration is followed by the change of radioactivity within the system. In order to detect the end-point in a radiometric titration it is essential to follow the chemical state of the tracer activity, a process that inevitably involves a separation. Radiometric titrations could be described as separation titrations. This is made clearer by a hypothetical example. A metal labelled with an isotopic tracer, M*$^+$, when titrated with an organic reagent, HL, forms a product M*L. The total radioactivity within the titration vessel remains constant irrespective of the amount of organic reagent added. However, the chemically combined form of the tracer changes and at the end point it is completely associated with organic ligand. By separating and measuring the tracer activity present in the ionic or organically bound form, the progress of the titration may be determined. Repetitive or continuous measurements of this type during the addition of titrant allow the construction of a titration curve and the determination of the equivalence point.

The type of titration curve obtained will depend on whether the activity is introduced into the system as a labelled titrant, titrand or indicator, or perhaps a combination. It will be influenced by the choice of the measured component, and may be complicated by the successive titration of several species and the use of direct or back titration. A number of typical but idealised titration curves are shown in Fig. 8.1. The curves are largely self explanatory, the inflection points indicating the equivalence points. From Fig. 8.1(c) it is obvious that the use of both labelled titrant and titrand is unnecessary. The curves in Figs. 8.1(d) and 8.1(e) are virtually identical, but, in the latter case it is essential to continue the titration beyond the second inflection point to determine the second titrand.

Similar curves are produced irrespective of whether the titration involves precipitation reactions, complex formation or redox reactions. What differ, however, are the methods of separation required. In precipitation titrations phase separation occurs naturally, while in the other systems auxiliary methods of separation must be employed such as solvent extraction, ion exchange, paper chromatography, or amalgamation.

The main advantage of the method is the sensitivity of end-point detection, which is only limited, in theory, by the specific activity of the radioactive tracers. A further advantage is that the method will also function in coloured, heterogeneous, non-aqueous and other unusual media, because the

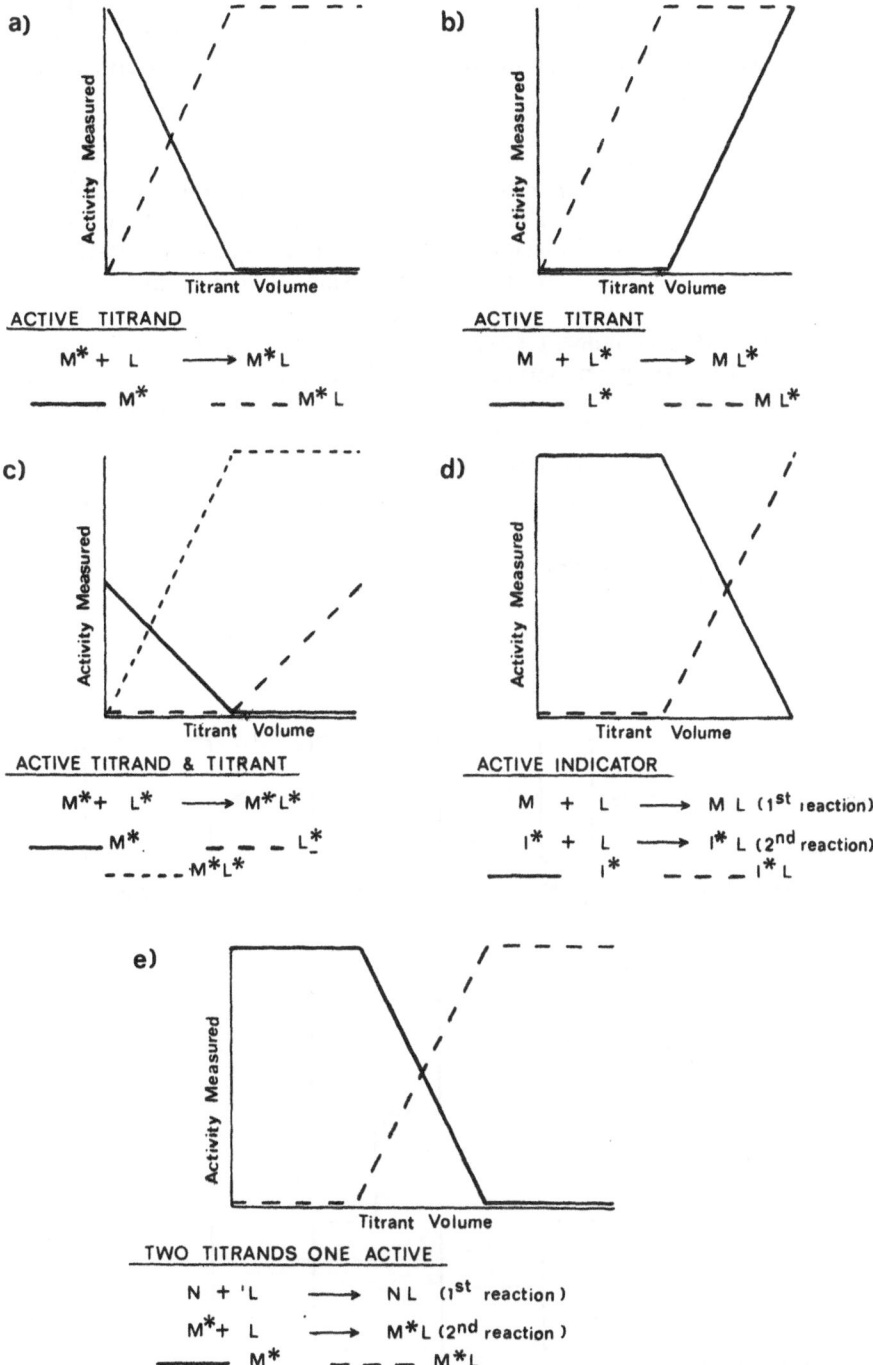

Fig. 8.1. Ideal radiometric titration curves.

radioactivity measurement is not affected. Radiometric titration has been reviewed [191, 192] and is the subject of a monograph [193]. The range of techniques employed in radiometric titration, and various applications, are indicated by examples.

Precipitation Titrations

Historically the first radiometric titrations developed were those involving precipitation reactions. In 1940, Langer described the principle of radiometric titration based on precipitation and filtration equipment [194], ^{32}P being used as indicator in titrations involving phosphate precipitation [195].

Indeed, the predominant separation method employed in radiometric titrations based on precipitation has been filtration. Equipment has been designed for both discontinuous [193–197] and continuous titration [193, 198]. The essential features of both are shown in Fig. 8.2. During discontinuous titration, the titrant is added in batches and the solution pumped through a filter into a counter. The solution is then returned to the titration

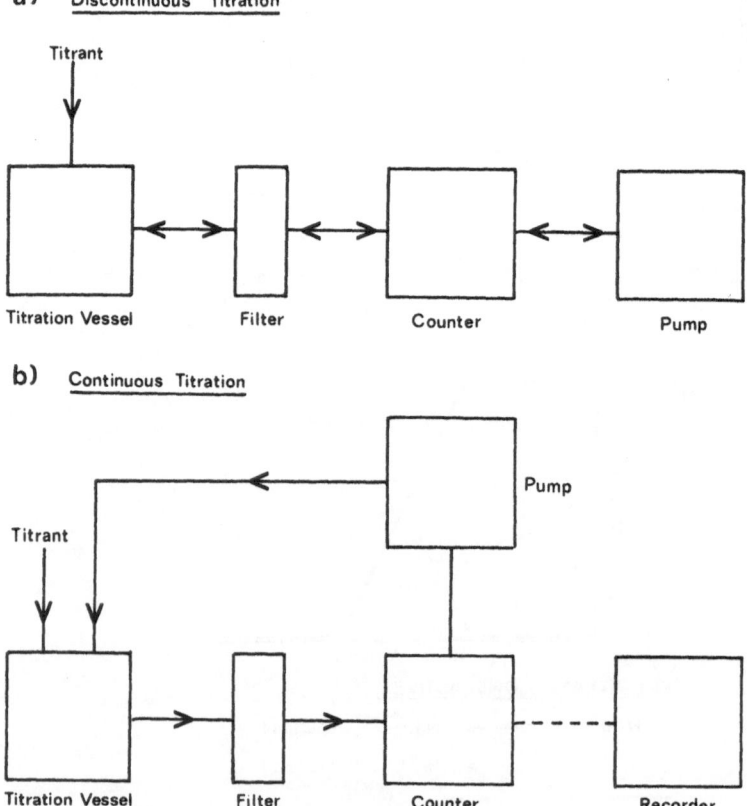

Fig. 8.2. Techniques for precipitation titrations using separation by flitration.

vessel before adding the next batch of titrant. In continuous titration, titrant is continuously added while solution is constantly withdrawn through a filter and circulated through the counter before return to the titration vessel. The counter output is recorded and may be used to control the rate of addition of titrant near the end-point. It is notable that the normal practice is to measure the filtrate rather than the precipitate. The counter used may be a scintillation detector or a Geiger counter, and is usually shielded. The activity used in a titration must be sufficient to give good counting statistics; it is therefore desirable to measure about 10,000 counts per minute.

An alternative to filtration is centrifugation [199]. Using this technique the method of titration is modified. A constant volume of titrand is added to a series of centrifuge tubes, different amounts of titrant are added and the volume made constant. After centrifuging the precipitate, the activity of each supernate is measured either by withdrawing an aliquot or by using a submersible beta-counter [200, 201]. The titration curve may then be plotted and the equivalence point determined.

Flotation has also been used to separate precipitates in radiometric titrations [202–204]. On shaking the aqueous phase and certain metal/organic precipitates with an immiscible organic solvent, the precipitate collects at the solvent interface. If the solvent has a specific gravity heavier than water, an aliquot of the aqueous phase may be readily removed for counting. Using carbon tetrachloride as flotation agent, cobalt labelled with ^{60}Co has been determined by precipitation as the pyridine thiocyanate complex, employing potassium thiocyanate as titrant [202]. Separation by flotation is claimed to be easier, faster and more complete than centrifugation.

The determination of silver by radiometric titration based on precipitation illustrates the use of an active titrand or titrant. Silver labelled with ^{110m}Ag may be titrated in nitric acid media with a dilute solution of chloride [205]. Measurement of the aqueous phase produces a titration curve similar to that in Fig. 8.1(a) M*. Alternatively, use of iodide labelled with ^{131}I produces a titration curve like that in Fig. 8.1(b) L* [206]. Indicators are normally applied when there is no suitable active label for the titrand or titrant. Copper may be titrated with potassium hexacyanoferrate using ^{65}Zn as indicator because copper hexacyanoferrate precipitates before the zinc complex [199]. A titration curve identical with that in Fig. 8.1(d) I* is obtained by measurement of the aqueous phase. The same method may also be used for the determination of mixtures of copper and zinc when the titrarion curve is similar to that in Fig. 8.1(e) M*.

Normally radiometric titration based on precipitation is limited to the determination of milligram quantities by the solubility product of the precipitate and the difficulty in handling small amounts of precipitate. However, special techniques such as titration in the presence of ion exchange resins and the use of collectors have improved the sensitivity of the method to microgram levels [191]. For example, iodate bound to anion exchange resin can be titrated with 10^{-4} M silver solutions labelled with ^{110m}Ag (191). Also, microgram amounts of zirconium may be titrated with ^{32}P-labelled phosphate if the zirconium phosphate is collected on freshly precipitated silver iodide [191].

Complexometric Titrations

Radiometric titration based on complex formation is frequently more sensitive than that based on precipitation, but suffers the disadvantage of requiring auxiliary separation of the reacting species or product. Separation methods described include solvent extraction, ion exchange and paper chromatography. A further modification is that based on solid indicators.

In methods based on solvent extraction one of the reaction components, usually the reaction product, can be extracted into an immiscible organic solvent; e.g. a metal ion may react with an organic ligand to produce a readily extractable metal-organic complex. Normally both the aqueous and organic phases are measured and a set of titration curves similar to that in Fig. 8.1 obtained, depending on the active labelling used. Several titration techniques may be applied. A procedure similar to that described for the centrifugation mode of the precipitation method may be used. Equal aliquots of an aqueous solution of the determined species are added to a series of separating funnels or centrifuge tubes, and different amounts of the complexing agent and a constant volume of solvent added. After mixing and separation of the phases, aliquots are removed for measurement, and a titration curve constructed. The alternative method is to carry out the titration discontinuously in one vessel, removing, measuring, and returning aliquots of one or both phases between each addition of complexing agent and semi-automatic apparatus for doing this has been described [198, 207, 208].

Examples of titrations employing labelled titrands are the determination of cobalt [209], mercury [209], silver [208] and zinc [208] by titration with dithizone. These titrations are so sensitive that less than a microgram of each metal may be determined. Non-isotopic titration of a metal ion can be carried out providing its extraction constant is higher than that of the indicator. For example, zinc has been determined by titration with dithizone using ^{60}Co as indicator, and the method can also be used for the successive determination of mixtures of zinc and cobalt [209]. In certain circumstances back titration can be used with advantage. The palladium dithizone complex only forms quickly in the presence of excess reagent. Palladium may be determined by measuring the excess reagent with silver [208]. The method depends on palladium dithizonate having a higher stability constant than the silver complex.

Ion exchange is used in radiometric titrations when complexing agents such as E.D.T.A. are employed. The type of reaction involved is

$$M^{2+} + H_2Y^{2-} \rightleftharpoons MY^{2-} + 2H^+$$

Cation exchangers can be used to separate uncomplexed metal ions, and anion exchangers the complexed metal. A procedure often adopted is to add increasing amounts of complexing agent to a series of test solutions of equal volume, and to pass each mixture through a short column of ion exchange resin. The tracer activity in each eluent is then determined and a titration curve plotted. By this means ^{114m}In tracer has been used for the determination of indium isotopically and cobalt non-isotopically at the microgram level [210]. The latter determination depends on the stability of the cobalt complex of E.D.T.A. being much higher than that of indium. Ion exchange columns may be replaced by ion exchange membranes that can be immersed in each test solution, allowed to equilibrate, and then counted after removal and rinsing [211]. The preferred technique is to add both anion and cation resin membranes to each solution, as this lessens the problem of physical entrainment of the non-adsorbed ion. The method has been used for the determination of nanogram amounts of cobalt and zinc, using isotopic labelling, and of zinc using non-isotopic labelling with ^{60}Co. A further modification of the method is to suspend ion-exchange resin in solution, and after the addition of each portion of titrant, to pump filtered solution to a counter [212–214]. Titration of zinc with E.D.T.A. produces a titration curve like that in Fig. 8.1(a) M* when using ^{65}Zn tracer. The method may be applied even using $10^{-6}M$ solutions.

Both paper chromatography and focussing electrophoresis can be used to separate ions from their E.D.T.A. complexes. The activity of the isotopically labelled and separated species are both measured. Providing an excess of ion is present, the variation of the ratio of activities will allow the determination of the amount of the ion. The paper chromatographic procedure has been used for the determination of traces of cobalt and terbium [215], and focussing electrophoresis for the determination of yttrium [216].

The use of solid indicators in radiometric titration, based on complexation, bears some resemblance to certain radiorelease methods. In the latter the radioactivity released from a solid is proportional to the amount of the determined species, while in the former the solid indicator releases activity at the end of the titration, the activity released frequently being proportional to the excess titrant. Titration curves of the type shown in Fig. 8.1(d) are usually obtained. Two types of indicator have been used: those releasing complexed tracer in solution [217, 218], and kryptonated materials that release gaseous ^{85}Kr at the end-point [179, 192, 219].

Titration techniques and apparatus employed for radiometric titration based on precipitation are equally applicable to solid indicator methods in which the indicator activity is released in solution. For titrations based on kryptonated indicators, two main groups of techniques have been used [180, 192]: those that measure the released ^{85}Kr, and those that measure the kryptonated indicator either directly in the titration vessel, or after removal to a counter.

Silver iodate labelled with ^{110m}Ag has been used as a solid indicator for the determination of calcium, strontium and magnesium by titration with E.D.T.A. [218]. The reactions involved are

$$M^{2+} + H_2Y^{2-} \rightleftharpoons MY^{2-} + 2H^+ \text{ (Initial reaction)}$$

$$*AgIO_3 + H_2Y^{2-} \rightleftharpoons *AgY^{3-} + IO_3^- + 2H^+ \text{ (indicator reaction)}$$
(solid)

The technique depends on the stability of the MY^{2-} complex being greater than that of the $*AgY^{3-}$ complex which in turn must be more stable than $*AgIO_3$. Iron may be determined using yttrium oxalate labelled with ^{91}Y as solid indicator [220]. Back titration of aluminium has been carried out using silver iodate as indicator [221]. Aluminium is reacted with excess E.D.T.A. in the presence of silver iodate which in turn releases silver activity. Back titration with calcium first removes excess E.D.T.A., then reduces the silver activity by reprecipitating silver iodate. The calcium does not react with the aluminium E.D.T.A. complex. A titration curve similar to that in Fig. 8.1(e) M* is obtained, the end-point being the second inflection.

Kryptonated silver iodate may be substituted for isotopically labelled silver iodate, in the titration of calcium, magnesium and strontium, using the reactions described above, ^{85}Kr being released at the end point [192]. Kryptonated yttrium oxalate may also be used as indicator in the determination of iron by E.D.T.A. titration [192]. Nickel may be determined using kryptonated silver as indicator and potassium cyanide as titrant. Back titrations are not possible with kryptonated indicators as the ^{85}Kr is lost irreversibly from the system.

Redox Titrations

Solid indicators are also used in radiometric titrations based on redox reactions. The reactions involved are usually as follows:

$$A_{Ox} + B_{Red} \rightleftharpoons A_{Red} + B_{Ox} \quad \text{(Initial reaction)}$$

$$*IL + B_{Red} \rightleftharpoons *IB_{Red} + L \quad \text{(Indicator reaction)}$$

For kryptonated indicators the second reaction differs somewhat:

$$IL(*Kr) + B_{Red} \rightleftharpoons IB_{Red} + L + *Kr \quad \text{(Indicator reaction)}$$
$$\text{(gaseous)}$$

The method can only be applied if all of species A_{Ox} is reduced before the reductant B_{Red} complexes I and releases the active tracer, so that IB_{Red} must be more stable than IL. The titration techniques and apparatus described for radiometric titrations based on complex formation and utilising solid indicators are also used for the equivalent redox titrations.

Silver thiocyanate labelled with ^{110m}Ag has been employed for the determination of iodine by thiosulphate titration [207]. The identical determination may be carried out substituting kryptonated silver thiocyanate as indicator [192].

Phase separation by solvent extraction has been used in radiometric titrations based on redox reactions. Thiosulphate can be titrated with iodine solutions labelled with ^{131}I [207]. Extraction of iodine with carbon tetrachloride and measurement of the aqueous or solvent phases allows the course of the titration to be followed. Titration curves similar to those in Fig. 8.1(b) are obtained. The reverse titration has also been carried out successfully [207].

Labelled amalgams have been effectively employed for radiometric end-point determination in redox titrations [202, 207]. The reactions are of the form

$$A_{Red} + B_{Ox} \rightleftharpoons A_{Ox} + B_{Red} \quad \text{(Initial reaction)}$$

$$*I(Hg) + B_{Ox} \rightleftharpoons *I + Hg + B_{Red} \quad \text{(Indicator reaction)}$$

The oxidation of the amalgamated species by B_{Ox} must not occur prior to the complete reaction of A_{Red}. At the end-point activity is released into the aqueous phase in proportion to the excess titrant. The titration curve produced by measurement of the aqueous phase is similar to that in Fig. 8.1(d) I*L.

Zinc amalgam labelled with ^{65}Zn has been used as indicator in the determination of ascorbic acid with ferric ions [207]. The same oxidant may be used for the titration of hydroquinone in the presence of cadmium amalgam labelled with ^{115}Cd [207].

Titrations Employing Radioisotope Sources

The absorption and backscattering of beta particles [191, 222, 223] and the absorption of neutrons [192] from radioisotope sources, have been used to follow the course of titrations. Such methods may be broadly classified as radiometric titrations.

REFERENCES

1. 'Radiochemicals 1971'. The Radiochemical Centre, Amersham, Bucks. U.K. (1970).
2. C. M. LEDERER, J. M. HOLLANDER and I. PERLMAN, Table of Isotopes. 6th Edition, John Wiley, New York, (1967).
3. W. SEELMANN-EGGEBERT, G. PFENNIG and H. MÜNZEL, Chart of the Nuclides. 3rd Edition, Institut für Radiochemie, Karlsruhe (1968).
4. M. W. GREEN and M. HILLMAN, *Intern. J. Appl. Radiation Isotopes*, 18, 540, (1967).
5. R. J. BAYLY and E. A. EVANS, Storage and Stability of Compounds Labelled with Radioisotopes. Review 7, The Radiochemical Centre, Amersham, Bucks. U.K. (1968).
6. C. E. CROUTHAMEL, F. ADAMS and R. DAMS, Applied Gamma-Ray Spectrometry. 2nd. Edition, Pergamon, Oxford, (1970).
7. Nuclear Instruments and Their Uses. (A. H. Snell, ed.) Vol 1. John Wiley, New York, (1962).
8. J. B. BIRKS, The Theory and Practice of Scintillation Counting. Pergamon Press, Oxford (1964).
9. Thin-Layer Chromatography. A Laboratory Handbook. (E. Stahl, ed.), 2nd Edition, Allen and Unwin, London, (1969).
10. R. J. BLOCK, E. L. DURRUM and G. ZWEIG, Paper Chromatography and Paper Electrophoresis. Academic Press, New York, (1958).
11. D. A. LAMBIE, Techniques for the Use of Radioisotopes in Analysis. A Laboratory Manual. E. and F. N. Spon, London, (1964).
12. Radioactive Substances Act, 1960 H.M.S.O. London (1960).

13. The control of Radioactive Wastes. H.M.S.O. London (1959).
14. Legislation and Codes of Practice Governing the Use of Radioactive Materials. Radioisotopes Review Sheet E2. Isotopes Division A.E.R.E., Harwell (1965).
15. Code of Practice for the Protection of Persons Exposed to Ionising Radiations in Research and Teaching. H.M.S.O. London (1964).
16. T. T. GORSUCH, *Analyst*, **84**, 135 (1958).
17. J. PIJCK, J. GILLIS and J. HOSTE, *Intern. J. Appl. Radiation Isotopes*, **10**, 149, (1961).
18. C. E. GLEIT and W. D. HOLLAND, *Anal. Chem.*, **34**, 1454 (1962).
19. T. S. WEST, *Anal. Chim. Acta*, **25**, 405 (1961).
20. H. J. M. BOWEN, *Anal. Chem.*, **40**, 969 (1968).
21. J. I. HOFFMANN and G. E. F. LUNDELL, *J. Research Natl. Bur. Standards*, **22**, 465 (1939).
22. G. HERRMANN and H. O. DENSCHLAG, *Ann. Revs. Nucl. Sci.*, **19**, 1 (1969).
23. K. SAMSAHL, Aktiebologet Atomenergi Report AE-82 Stockholm (1962).
24. G. H. MORRISON, J. T. GERARD, A. TRAVESI, R. L. CURRIE, S. F. PETERSON and N. M. POTTER, *Anal. Chem.*, **41**, 1633 (1969).
25. K. SAMSAHL, *Anal. Chem.*, **39**, 1480 (1967).
26. C. BALLAUX, R. DAMS and J. HOSTE, *Anal. Chim. Acta*, **45**, 337 (1969).
27. C. BALLAUX, R. DAMS and J. HOSTE, *Anal. Chim. Acta*, **47**, 397 (1969).
28. L. F. LOWE, H. D. THOMPSON and J. D. CALI, *Anal. Chem.*, **31**, 1951 (1959).
29. A. A. SMALES, D. MAPPER and A. J. WOOD, U.K. Atomic Energy Authority Report C/R-2254 (1957).
30. Chemical Oceanography. (J. P. Riley and G. Skirrow, eds.) Vol 2. Academic Press, New York and London (1965).
31. M. G. LAI and H. V. WEISS, *Anal. Chem.*, **34**, 1012 (1962).
32. J. P. RILEY and D. TAYLOR, *Anal. Chim. Acta*, **40**, 479 (1968).
33. W. HASSENTEUFEL, R. JAGETSCH and F. F. KOCZY, *Limnol. Oceanogr.*, **8**, 152 (1963).
34. R. W. HUMMEL, *Analyst*, **82**, 483 (1957).
35. H. V. WEISS and M. G. LAI, *Anal. Chim. Acta*, **28**, 242 (1963).
36. D. N. SUNDERMAN and W. W. MEINKE, *Anal. Chem.*, **29**, 1578 (1957).
37. W. J. ROSS, *Anal. Chem.*, **37**, 168 (1965).
38. S. AMIEL, Nuclear Chemistry. (L. Yaffe, ed.) Vol. II p. 251. Academic Press, New York and London, (1968).
39. J. PIJCK and J. HOSTE, *Anal. Chim. Acta*, **26**, 501 (1962).
40. W. C. PARKER and Y. GRUNDITZ, *Nucl. Instr. Meth.*, **14**, 71 (1961).
41. J. R. DeVOE, U.S. At. Energy Comm., Rept. AECU-4610 (1959).
42. K. W. BAGNALL, R. W. M. D'EYE and J. H. FREEMAN, *J. Chem. Soc.*, 2320 (1955).
43. W. G. SMITH, *J. Inorg. Nucl. Chem.*, **17**, 382 (1961).
44. J. R. DeVOE, The Application of Distillation Techniques to Radiochemical Separations. Nuclear Science Series NAS-NS-3108, U.S. National Academy of Sciences (1962).
45. J. R. DeVOE and W. W. MEINKE, *Anal. Chem.*, **35**, 2 (1963).

46. L. TOMLINSON, Proceedings of I.A.E.A. Panel on Delayed Fission Neutrons, p. 61. Vienna (1968).
47. L. TOMLINSON, *Anal. Chim. Acta*, 31, 545 (1964); 32, 157 (1965).
48. G. H. MORRISON and H. FREISER, *Anal. Chem.*, 36, 93R (1964).
49. J. STARÝ, *Talanta*, 13, 421 (1966).
50. D. F. PEPPARD, Advances in Inorganic Chemistry and Radiochemistry. (H. J. Emeleus and A. G. Sharp, eds.) Vol. 9, p. 25. Academic Press, New York, (1966).
51. H. FREISER, *Anal. Chem.*, 40, 522R (1968).
52. D. F. PEPPARD, *Ann. Revs. Nucl. Science*, 21, 365 (1971).
53. W. J. MAECK, G. L. BOOMAN, M. E. KUSSY and J. E. REIN, *Anal. Chem.*, 33, 1775 (1961).
54. O. SAMUELSON, Ion Exchange Separations in Analytical Chemistry. John Wiley, New York, (1963).
55. J. INCZEDY, Analytical Applications of Ion Exchangers. (English Translation) Pergamon Press, Oxford, (1966).
56. R. KUNIN and F. X. McGAVREY, *Anal. Chem.*, 36, 142R (1964).
57. H. F. WALTON, *Anal. Chem.*, 38, 79R (1966).
58. H. F. WALTON, *Anal. Chem.*, 40, 51R (1968).
59. H. F. WALTON, *Anal. Chem.*, 42, 86R (1970).
60. A. KRAUS and F. NELSON, Proc. 1st Intern. Conf. Peaceful Uses At. Energy, 7, 113 (1956).
61. K. SAMSAHL, P. O. WESTER and O. LANDSTRÖM, *Anal. Chem.*, 40, 181 (1968).
62. C. B. AMPHLETT, Inorganic Ion Exchangers. Elsevier, Amsterdam, (1964).
63. M. J. FULLER, *Chromatog. Revs.*, 14(1), 45, (1971).
64. F. GIRARDI, R. PIETRA and E. SABBIONI, Euratom Report, No. EUR 4287e (1969).
65. C. BIGLIOCCA, F. GIRARDI, J. PAULY, E. SABBIONI, S. MELONI and A. PROVASOLI, *Anal. Chem.*, 39, 1634 (1967).
66. F. GIRARDI and E. SABBIONI, *J. Radioan. Chem.*, 1, 169 (1968).
67. F. GIRARDI, R. PIETRA and E. SABBIONI, *J. Radioan. Chem.*, 5, 141 (1970).
68. K. RANDERATH, Thin-layer Chromatography. (D. D. Libman Translator) Academic Press, New York and London, (1966).
69. F. POCCHIARI and C. ROSSI, *J. Chromatog.*, 5, 377 (1961).
70. P. C. Van ERKELENS, *Nature*, 172, 357 (1953).
71. J. SHERMA and H. H. STRAIN, *Anal. Chem.*, 34, 76 (1962).
72. S. TUSTANOWSKI, *J. Chromatog.*, 31, 270 (1967).
73. K-H. KÖNIG and K. DEMEL, *J. Chromatog.*, 39, 101 (1969).
74. A. MOGHISSI, *Anal. Chim. Acta*, 30, 91 (1964).
75. M. PAUWELS, R. GIJBELS and J. HOSTE, *Anal. Chim. Acta*, 36, 210 (1966).
76. K. AITZETMÜLLER, K. BUCHTELA and F. GRASS, *Anal. Chim. Acta*, 38, 249 (1967).
77. E. HEFTMANN, *Anal. Chem.*, 36, 14R (1964).
78. E. HEFTMANN, *Anal. Chem.*, 38, 31R (1966).
79. G. ZWEIG, *Anal. Chem.*, 40, 490R (1968).
80. G. ZWEIG and R. B. MOORE, *Anal. Chem.*, 42, 349R (1970).
81. E. CERRAI and C. TESTA, *J. Inorg. Nucl. Chem.*, 25, 1045 (1963).
82. T. B. PIERCE and R. S. HOBBS, *J. Chromatog.*, 12, 74 (1963).

83. J. W. WINCHESTER, *J. Chromatog.*, **10**, 502 (1963).
84. N. CVJETICANIN, *J. Chromatog.*, **34**, 520 (1968).
85. N. GETOFF, H. BILDSTEIN, V. PFEIFER, H. SORANTIN and H. TITZE, *Atompraxis*, **12**, 563 (1963).
86. F. CACACE, *Nucleonics*, **19**, (5), 45 (1961).
87. J. TADMOR, *Chromatog. Revs.*, **5**, 223 (1963).
88. D. C. NELSON, R. C. HAWES, D. PAULL and P. C. RESSLER (JR), *Develop. Appl. Specr.*, **4**, 323 (1965).
89. J. P. ADLOFF, *Chromatog. Revs.*, **4**, 19 (1962).
90. S. P. CRAM and J. L. BROWNLEE, *J. Gas Chromatog.*, **6**, 305 (1968).
91. R. C. KOCH and G. L. GRUNDY, *Anal. Chem.*, **33**, 43 (1961).
92. D. W. OCKENDEN and R. H. TOMLINSON, *Can. J. Chem.*, **40**, 1594 (1962).
93. T. S. ZVAROVA and I. ZVARA, *J. Chromatog.*, **44**, 604 (1969).
94. J. A. DEAN and S. A. REYNOLDS, *Anal. Chim. Acta.*, **11**, 390 (1954).
95. F. LUX, *Radiochim. Acta*, **1**, 20 (1962).
96. J. R. DeVOE and W. W. MEINKE, *Anal. Chem.*, **35**, 3 (1963).
97. R. F. MITCHELL, *Anal. Chem.*, **32**, 326 (1960).
98. M. Y. DONNAN and E. K. DUKES, *Anal. Chem.*, **36**, 393 (1964).
99. T. H. HANDLEY and J. H. COOPER, *Anal. Chem.*, **41**, 381 (1969).
100. P. JORDAN, *Chimia*, **21**, 148 (1967).
101. C-W. TANG and C. J. MALETSKOS, *Science*, **167**, 52 (1970).
102. H. J. M. BOWEN and J. A. COOK, *Radiochem. Radioan. Lett.*, **5**, 103 (1970).
103. F. E. ORBE, I. H. QURESHI and W. W. MEINKE, *Anal. Chem.*, **35**, 1436 (1963).
104. J. S. HISLOP and D. R. WILLIAMS, *Radiochem. Radioan. Lett.*, **7**, 129 (1971).
105. T. Y. TORIBARA and R. E. SHERMAN, *Anal. Chem.*, **25**, 1594 (1953).
106. H. L. ROOK, T. E. GILLS and P. D. LaFLEUR, *Anal. Chem.*, **44**, 1114 (1972).
107. G. HEVESY and R. HOBBIE, *Z. Analyt. Chem.*, **88**, 1 (1932).
108. S. A. REYNOLDS and G. W. LEDDICOTTE, *Nucleonics.*, **21**, 128 (1963).
109. J. TÖLGYESSY, J. BRAUN and T. KYRŠ, Isotope Dilution Analysis. Pergamon Press, Oxford, (1972).
110. D. SALYER and T. R. SWEET, *Anal. Chem.*, **28**, 61 (1956).
111. D. SALYER and T. R. SWEET, *Anal. Chem.*, **29**, 2 (1957).
112. O. T. QUIMBY, A. J. MABIS and H. W. LAMPE, *Anal. Chem.*, **26**, 661 (1954).
113. H. WEILER, *Intern. J. Appl. Radiation Isotopes*, **12**, 49 (1961).
114. W. D. RALPH, T. R. SWEET and I. MENCIS, *Anal. Chem.*, **34**, 92 (1962).
115. J. RŮŽIČKA and J. STARÝ, *Atom. Energy Revs.*, **2**, 3 (1964).
116. J. RŮŽIČKA and J. STARÝ, Substoichiometry in Radiochemical Analysis. Pergamon Press, Oxford, (1968).
117. J. RŮŽIČKA and J. STARÝ, *Talanta*, **8**, 535 (1961).
118. J. STARÝ, J. RŮŽIČKA and M. SALAMON, *Talanta*, **10**, 375 (1963).
119. I. KAŠPAREC, Substoichiometric determination of zinc and cadmium by isotope-dilution analysis. Thesis, Faculty of Technical Nuclear Physics, Prague, (1966).

120. J. STARÝ and J. RŮŽIČKA, *Talanta*, **8**, 296 (1961).
121. J. STARÝ and J. SMIŽANSKÁ, *Anal. Chim. Acta*, **29**, 545 (1964).
122. J. RŮŽIČKA and J. STARÝ, *Talanta*, **9**, 617 (1962).
123. A. ZEMAN, J. PRASILOVA and J. RŮŽIČKA, *Talanta*, **13**, 457 (1966).
124. I. P. ALIMARIN and G. A. PEREZHOGIN, *Talanta*, **14**, 109 (1967).
125. J. KRTIL and M. BEŽDEK, *Talanta*, **15**, 1423 (1968).
126. I. E. ZIMAKOV and G. S. ROZHAVSKII, *Trudy. Kom. Analit. Khim. Izd. Akad. Nauk SSSR*, **9**, (12), 231 (1958).
127. G. B. BRISCOE and A. DODSON, *Talanta*, **14**, 1051 (1967).
128. V. KNOBLOCH and J. STARÝ, *Zh. Analit. Khim.*, **20**, 1160 (1965).
129. J. STARÝ and J. RŮŽIČKA, *Talanta*, **8**, 775 (1961).
130. J. STARÝ and J. RŮŽIČKA, *Talanta*, **11**, 691 (1964).
131. J. RŮŽIČKA, *Coll. Czech Chem. Comm.*, **25**, 199 (1960).
132. J. RŮŽIČKA and P. BENEŠ, *Coll. Czech Chem. Comm.*, **26**, 1784 (1961).
133. P. BERONIUS, *Acta Chem. Scand.*, **15**, 1151 (1961).
134. J. RŮŽIČKA and M. WILLIAMS, *Talanta*, **12**, 967 (1965).
135. G. B. BRISCOE, B. COOKSEY, J. RŮŽIČKA and M. WILLIAMS, *Talanta*, **14**, 1457 (1967).
136. S. M. GRASHCHENKO and E. V. SOBOTOVICH, Radiokhim. metody opredeleniya mikroelementov. p. 73 Izd. Nauka, Moscow (1965).
137. National Bureau of Standards Technical Note 248. (J. R. DeVoe, ed.) p. 10, U.S. Dept. of Commerce, Washington, D.C. (1964).
138. National Bureau of Standards Technical Note 276. (J. R. DeVoe, ed.) p. 111, U.S. Dept. of Commerce, Washington, D.C. (1966).
139. I. E. ZIMAKOV and G. S. ROZHAVSKII, *Zavod. Lab.*, **24**, 922 (1958); *Ind. Lab.*, **24**, 1030 (1958).
140. National Bureau of Standards Technical Note 404. (J. R. DeVoe, ed.) p. 141 U.S. Dept. of Commerce, Washington, D.C. (1966).
141. P. W. PELLA, A. R. LANDGREBE, J. R. DeVOE and W. C. PURDY, *Anal. Chem.*, **39**, 1781 (1967).
142. I. P. ALIMARIN and G. N. BILIMOVICH, *Intern. J. Appl. Radiation Isotopes*, **7**, 169 (1960).
143. N. SUZUKI and K. KUDO, *Anal. Chim. Acta*, **32**, 456 (1965).
144. K. KUDO, *Radioisotopes* (Tokyo), **15**, 209 (1966).
145. K. KUDO, *Radioisotopes* (Tokyo), **16**, 199 (1967).
146. I. A. CARMICHAEL and J. E. WHITLEY, *Analyst*, **95**, 393 (1970).
147. I. OBRUSNÍK and A. ADAMEK, *Talanta*, **15**, 433 (1968).
148. T. BRAUN, L. LADÁNYI, M. MAROTHY and I. OSGYANI, *J. Radioan. Chem.* **2**, 263 (1969).
149. A. ADAMEK, I. OBRUSNIK, F. KUKULA and M. KŘIVÁNEK, Proc. I.A.E.A. Conf. on Nuclear Activation Techniques in the Life Sciences, p. 189. Vienna, (1967).
150. I. P. ALIMARIN, YU V. YAKOVLEV and O. V. STEPANETS, *J. Radioan. Chem.*, **11**, 209 (1972).
151. F. KUKULA and M. ŠIMKOVÁ, *J. Radioan. Chem.*, **4**, 271 (1970).
152. A. ELEK, J. BOGÁNCS and M. SZABÓ, *J. Radioan. Chem.*, **4**, 281 (1970).
153. T. B. PIERCE, *Analyst*, **85**, 166 (1960).
154. J. R. VAN SMIT, *Analyst*, **90**, 366 (1965).

155. T. H. HANDLEY, *Anal. Chem.*, **36**, 153 (1964).
156. H. G. RICHTER, *Anal. Chem.*, **38**, 772 (1966).
157. C. B. OXBY and J. B. DAWSON, Proc. I.A.E.A. Symp. on Radiochemical Methods of Analysis, p. 229. Vienna, (1965).
158. S. HIRANO, A. MIZUIKE and E. NAKAI, *Radioisotopes* (Tokyo), **13**, 118 (1964).
159. M. KYRŠ, *Anal. Chim. Acta*, **33**, 245 (1965).
160. A. R. LANDGREBE, L. T. McCLENDON and J. R. DeVOE, Proc. I.A.E.A. Symp. on Radiochemical Methods of Analysis, p. 321. Vienna, (1965).
161. M. KYRŠ and L. KADLECOVÁ, *Anal. Chim. Acta*, **33**, 481 (1965).
162. H. WEISZ and D. KLOCKOW, *Mikrochim. Acta*, 1082 (1963).
163. Monographs on Radiochemistry. National Academy of Sciences, National Research Council, Nuclear Science Series. U.S. National Academy of Sciences.
164. H. J. M. BOWEN and D. GIBBONS, Radioactivation Analysis. Clarendon Press, Oxford (1963).
165. F. J. BRYANT, A. MORGAN and G. S. SPICER, U.K. Atomic Energy Authority Report R 3030, Harwell, (1959).
166. A. PARKER, E. H. HENDERSON and G. S. SPICER, U.K. Atomic Energy Authority Report AM 101 (1965).
167. D. H. KEEFER, L. F. EDMONDSON and R. E. ISAACKS, *Health Physics*, **11**, 193 (1965).
168. W. JENKINS, J. W. McMILLAN and T. B. REES, U.K. Atomic Energy Authority Report AM 105 (1967).
169. J. W. McMILLAN, M. PERKINS and T. W. SANDERS, unpublished work.
170. P. BENEŠ and J. RŮŽIČKA, *Jaderna energie*, **12**, 179 (1966).
171. V. KNOBLOCH and J. STARÝ, *Zh. Analit. Khim.*, **20**, 1160 (1965).
172. P. C. VAN ERKELENS, *Anal. Chim. Acta*, **25**, 570 (1961).
173. G. A. WELFORD, E. L. CHIOTIS and R. S. MORSE, *Anal. Chem.*, **36**, 2350 (1964).
174. J. TADMOR, *Anal. Chem.*, **36**, 1565, (1964).
175. P. C. VAN ERKELENS, *Anal. Chim. Acta*, **26**, 46 (1962).
176. J. TÖLGYESSY, *Jaderna energie*, **14**, 169 (1968).
177. C. O. HOMMEL, R. L. BERSIN, A. M. FILIPOV, F. J. BROUSAIDES and Y. TOKIWA, U.S. At. Energy Comm., Report NYO-2767, Tracerlab. Inc. (1962).
178. Isotopes Radiation Technol., U.S. At. Energy Comm., **1**, 53 (1963).
179. D. J. CHLECK, R. MAEHL, O. CUCCHIARA and E. CARNEVALE, *Intern. J. Appl. Radiation Isotopes*, **14**, 581; 593; 599 (1963).
180. D. J. CHLECK, Proc. I.A.E.A. Symp. on Radiochemical Methods of Analysis, Vol. II p. 273 (1965).
181. V. JESENAK, M. M. NAOUM and J. TÖLGYESSY, *Radiochem. Radioan. Lett.*, **13**, 199 (1973).
182. O. CUCCHIARA and P. GOODMAN, U.S. At. Energy Comm., Report NYO-2757-6, Parametrics Inc. (1967).
183. H. H. ROOS and W. S. LYON, Proc. I.A.E.A. Symp. on Radiochemical Methods of Analysis, Vol. II p. 285. Vienna, (1965).
184. H. G. RICHTER and A. S. GILLESPIE, *Anal. Chem.*, **34**, 1116 (1962).
185. A. S. GILLESPIE and H. G. RICHTER, *Anal. Chem.*, **36**, 2473 (1964).

186. H. G. RICHTER and A. S. GILLESPIE, *Anal. Chem.*, **37**, 1146 (1965).
187. H. J. M. BOWEN, *Analyst*, **96**, 220 (1971).
188. H. J. M. BOWEN, *Analyst*, **97**, 728 (1972).
189. K. BURGER, *Acta Chim. Acad. Sci. Hung.*, **40**, 17 (1964).
190. P. C. VAN ERKELENS, *Anal. Chim. Acta*, **26**, 32 (1962).
191. T. BRAUN and J. TÖLGYESSY, *Talanta*, **11**, 1277 (1964).
192. J. TÖLGYESSY, *Isotopenpraxis*, **7**, 208 (1971).
193. T. BRAUN and J. TÖLGYESSY, Radiometric Titrations. Pergamon Press, Oxford (1967).
194. A. LANGER, Radiometric Titration Method. U.S. Patent 2,367,949 (1940).
195. A. LANGER, *J. Phys. Chem.*, **45**, 639 (1941).
196. N. I. POLEVAYA, N. N. CHERNOVA and S. L. MIRKINA, Vsesoyuz. Nauch. Issledovatel. Geol. Inst. Sbornik, No. 1, 119 (1955).
197. I. A. SIROTINA and I. P. ALIMARIN, *Zh. Analit. Khim.*, **12**, 367 (1957).
198. J. F. DUNCAN and F. G. THOMAS, Proc. Australian Atomic Energy Symposium, Radioisotopes Physical Sciences, p. 637 (1958).
199. I. M. KORENMAN, F. R. SEYANOVA, E. A. DEMINA and M. I. SAPOSHNIKOVA, *Zavod. Lab.*, **22**, 1143 (1956).
200. F. B. SCOTT and W. F. DRISCOLL, Proc. I.A.E.A. Conf. on Radioisotopes in the Physical Sciences and Industry, Vol. II p. 339 Vienna (1962).
201. D. S. BERRY and B. F. SCOTT, U.S. At. Energy Comm., Report TID-7655 p. 291 (1963).
202. T. BRAUN, *Chimie Analytique*, **46**, 61 (1964).
203. T. BRAUN, I. GALATEANU and I. MAXIM, *Nuclear Technic.* (Bucharest), **7**, 20 (1959).
204. I. MAXIM, T. BRAUN and I. GALATEANU, *J. Inorg. Nucl. Chem.*, **10**, 166 (1959).
205. V. I. PLOTNIKOV, *Zavod. Lab.*, **24**, 927 (1958).
206. J. TOLGYESSY and P. SCHILLER, *Magy. Kém. Folyóirat*, **63**, 269 (1957).
207. T. BRAUN and E. KÖRÖS, Proc. I.A.E.A. Symp. on Radiochemical Methods of Analysis, Vol. II p. 213, Vienna (1965).
208. H. SPITZY, *Mikrochim. Acta*, 789 (1960).
209. J. F. DUNCAN and F. G. THOMAS, *J. Inorg. Nucl. Chem.*, **4**, 376 (1957).
210. J. STARÝ, J. RŮŽIČKA and A. ZEMAN, *Talanta*, **11**, 481 (1964).
211. J. TÖLGYESSY, J. KONEČNÝ and T. BRAUN, *Nuclear Applications*, **3**, 383 (1967).
212. A. HEIJINK and H. L. POLAK, *J. Radioan. Chem.*, **2**, 235 (1969).
213. A. HEIJINK and H. L. POLAK, *J. Radioan. Chem.*, **3**, 359 (1969).
214. A. HEIJINK and H. L. POLAK, *J. Radioan. Chem.*, **4**, 63 (1970).
215. E. SCHUMACHER and W. FRIEDLI, *Helv. Chim. Acta*, **43**, 1013 (1960).
216. E. SCHUMACHER and H. J. STREIFF, *Helv. Chim. Acta*, **41**, 1771 (1958).
217. T. BRAUN, I. MAXIM and I. GALATEANU, *Nature*, **182**, 936 (1958).
218. T. BRAUN, I. MAXIM and I. GALATEANU, *Zh. Analit. Khim.*, **14**, 542 (1959).

219. D. J. CHLECK, R. MAEHL and O. CUCCHIARA, U.S. At. Energy Comm., Report, NYO-2757-3, Parametrics Inc. (1963).
220. R. LIEBERMAN, C. W. TOWNLEY, C. T. BROWN, J. E. HOWES, R. A. EWING and D. N. SUNDERMAN, U.S. At. Energy Comm. Report, BMI-15 08 (1961).
221. T. BRAUN, *Acta Chim. Acad. Sci. Hung.*, 41, 199 (1964).

Chapter 9

Tracer Methods in Organic and Biochemical Analysis

R. P. Ekins

Institute of Nuclear Medicine, The Middlesex Hospital Medical School
Nassau Street, London W1N 7RL

9.1 INTRODUCTION

In the 30 years since artificial radioisotopes first became generally available, they have found increasing use as tools in analytical chemistry. Indeed, this area of application of radioisotopic techniques has witnessed an explosive growth in the last decade following the development and increasingly widespread exploitation of radioimmunoassay and other saturation assay methods (discussed in detail later in this section) in clinical and research biochemistry. The reasons for this growth, particularly in these fields, are not difficult to identify. Because of the sensitivity and simplicity of isotopic measurement techniques, the reactions in an assay system of exceedingly low concentrations or amounts of a compound under test may be readily monitored. The limits of sensitivity of radioisotopic analytical methods can therefore frequently be extended to levels lower by orders of magnitude than those attainable by many conventional physico-chemical methods, and thereby encompass the exceedingly minute concentrations at which many compounds of biological importance exert their effects in living matter. In consequence, large areas of biomedical science (such as endocrinology) which have hitherto remained shrouded in obscurity for want of analytical techniques of sufficient sensitivity have recently been opened up for the first time to accurate quantitative investigation.

It is principally for this reason that despite some initial reluctance (stemming partially from unfamiliarity, partially from the cost of appropriate facilities and instrumentation) displayed by clinical chemists and other biochemical analysts, radioisotopic analytical techniques are being urgently adopted as diagnostic and research tools in medical and university institutions at the present time. This development has fortunately been assisted by the instrument manufacturing and pharmaceutical industries both of which, in the past few years, have devoted increasing resources to this field. In particular, the introduction and increasing availability of commercial assay kits is contributing significantly to the rapid movement of radioisotopic analytical methods from specialist research groups to many service analytical laboratories.

Nevertheless the exponential growth of interest in these techniques, and the removal of some of the more daunting barriers to their widespread application, is not without its dangers and disadvantages. Not all commercially available assay kits are above reproach: not infrequently they yield inaccurate and misleading results arising either from inadequacies in the methodology on which they are based, or to a particular vulnerability which some display to lack of experimental expertise on the part of the analyst. Indeed, the sometimes deceptive simplicity of many such analytical procedures, whether or not available in kit form, brings them within easy reach of groups of users who, because of an almost complete ignorance of certain of the basic principles governing radioisotope methodology, or of the more subtle problems which affect many current assay techniques, may be led to derive, and to publish, totally erroneous conclusions based on their use. For this reason there is a particular need for vigilance in the reading of the extensive literature now emerging in many diverse areas of medicine and biochemistry, which depends on the exploitation of radioisotopic analytical methods, and a special responsibility on the part of biochemists, clinicians and others dependent upon these methodologies to develop a deeper understanding of their limitations and the pitfalls which attend their use.

Because of their enormous range of application, and the widespread interest they have engendered in consequence, the radioimmunoassay (RIA) and other saturation assay (or 'radioligand' assay methods) will form the principal topic of this chapter. Nevertheless because there are common elements in the principles underlying all the radioisotopic analytical techniques, and because some of the older approaches are finding new forms of expression at the present time, it is necessary to examine, albeit briefly, the diverse ways in which isotopic methods have hitherto been used in the analysis of organic materials.

The principal radioanalytical methods used in biology and medicine may be grouped under 4 principal headings:

(i) isotope dilution analysis
(ii) activation analysis
(iii) labelled reagent methods (including derivative analysis, 'immuno-radiometric assay')
(iv) saturation analysis or radioligand assay (including radioimmunoassay)

There is, it must be emphasised, a large measure of arbitrariness in this subdivision and, as indicated earlier, some of the fundamental principles on which they rest are shared between these different groups. There are legitimate reasons, for example, for regarding many of the radioligand assays, including radioimmunoassay, merely as a form of 'dilution analysis', and it is not unknown (though uncommon) to find such techniques referred to in this way by authors who object to a proliferation of vogue terminology. Nevertheless, whilst it must be recognised that the above subdivision may not be entirely justified on conceptual grounds, these terms have traditionally been used to denote certain well-defined groups of analytical techniques, each with characteristic experimental problems and requirements, and therefore form a convenient basis for subsequent discussion in this chapter.

9.2   DILUTION ANALYSIS

9.2.1   General Method
The fundamental principle of the isotope dilution method is shown in Fig. 9.1. In this illustration, the compound of interest, designated as P, is assumed to occur as a component of a complex mixture. If a mass M^* of the identical radioactively labelled compound (P^*) is added to the mixture then, if the mass of P initially present is given by M, and the added activity by R, the resulting specific activity of the compound, assuming complete mixing or 'equilibration' of the labelled and unlabelled moieties, is given by

$$S = \frac{R}{M + M^*}$$

whence

$$M = \frac{R}{S} - M^* \tag{9.1}$$

Thus assuming both R and M^* are known (or alternatively, that M^* is so small that it can be legitimately ignored), the mass M can be derived by isolating P from other components of the mixture and estimating the specific activity (S) of the purified product. Quantitative recovery of P through the extraction and purification steps is not of primary importance,

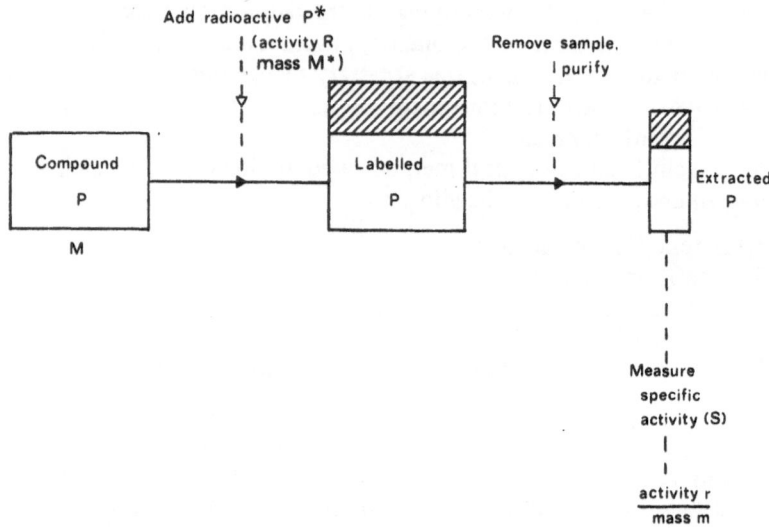

Fig. 9.1. Basic principle of dilution analysis. The shaded area represents the labelled material which, in practice, must be permitted to equilibrate throughout the pool of compound P prior to isolation of the purified sample.

since the specific activity of the isolated sample, however small, can normally be assumed to be identical to that of the total pool of P from which it is derived. It is clearly necessary, however, that sufficient of the purified compound should remain at the end of the procedure on which to perform an adequate specific activity determination.

The principle of isotope dilution analysis can, however, be viewed in another light. Let us assume that the specific activity of the isolated material is estimated by measurement of the activity (r) and the weight (m) of the purified sample, then $S = r/m$ and (neglecting M^*)

$$M = \frac{R \times m}{r} \text{ [from equation (9.1)]} \tag{9.2}$$

i.e. the weight of compound P present in the original mixture is given by the weight of the isolated sample divided by a recovery factor (r/R) representing the fraction of the added radioactivity recovered. In short, the function of the radioactive compound can be viewed as simply to provide a measure of the recovery of the original compound through the isolation procedure.

Thus any method of analysis in which a radioactive tracer is used to monitor the recovery from a mixture of the corresponding inactive material prior to its gravimetric (or photometric) measurement, can legitimately be regarded as a dilution technique. One of the first examples to be described was the determination of individual amino acids in a mixture [1]; a known quantity of the isotopically labelled acid was added to the mixture, the acid subsequently isolated in pure form, and the recovered activity in the weighed sample determined.

Clearly, there are many circumstances in which this approach is desirable and is adopted. Typically they arise when the substance to be assayed is present in a milieu in which it cannot be directly measured and from which it cannot be quantitatively isolated, usually because of large and variable losses through the purification steps. However, the usefulness of the technique can sometimes extend beyond merely providing a recovery estimate *per se*. Occasionally situations arise in which the purity of the isolated substance may be open to doubt and must be tested; a useful indication of purity may be provided by the demonstration of the attainment of a constant (apparent) specific activity through successive purification stages. It is clear, of course, that such evidence relies on the assumption of initial purity of the *labelled* material, and of complete identity of behaviour of the labelled and unlabelled compounds in any chemical or physical separation steps to which they are jointly subjected.

9.2.2 Reverse Isotope Dilution Analysis

The principle of isotope dilution analysis is also encountered in situations in which the substance to be measured is itself initially radioactive, and is present in the original mixture either at a high dilution, or in conjunction with other similarly labelled contaminating compounds. In either of these circumstances an extraction and purification step prior to radioassay may be necessary. Customarily the identical *inactive* compound is added to the system in known amount, the purification steps performed (to the point where constant specific activity of the isolated material is achieved) and the recovery of the labelled compound estimated from a measurement of the *weight* of recovered material. This approach, usually referred to as 'reverse isotope dilution analysis', finds particular application as a step in some forms of activation and derivative analysis as discussed in Sections 9.3 and 9.4; it is also useful in studies involving the administration of labelled compounds to experimental animals and the subsequent determination of labelled metabolic degradation products (e.g., [2, 3, 4]).

A further refinement of this approach, employed in essentially similar circumstances, comprises the addition of the desired compound labelled with a second isotope. We may, for example, visualise a metabolic study involving the administration of a tritiated compound to an experimental animal resulting in the occurrence of a range of tritium labelled metabolites in the tissues. Suspected metabolic products, labelled with carbon-14, may subsequently be added to the homogenised tissues, re-extracted and purified to a constant $^3H/^{14}C$ ratio. The recovery of the ^{14}C compound in the purified product reflects that of the tritiated metabolite, and hence enables an estimate of its total activity in the tissue to be made. The special advantage of such use of a second label is the elimination of the gravimetric measurements entailed in conventional reverse isotope dilution measurements, enabling those powerful purification techniques, such as paper chromatography, electrophoresis, etc., which are particularly applicable to small amounts of material, to be employed. This approach also finds application in 'double isotope derivative assays' as subsequently described in Section 9.4.

9.2.3   Isotope Dilution Analysis Using a Specific Binding Reagent

Conventional techniques of isotope dilution analysis, as emphasised earlier in this section, typically enable losses incurred in a purification procedure to be corrected for prior to gravimetric measurement; however, they do not in themselves provide a means for carrying out this final measurement. Hence the sensitivity of the overall analysis, which is essentially governed by the sensitivity of the final gravimetric determination, will not normally be improved by the dilution technique (and, indeed, may suffer if losses incurred in purification procedures are high). Clearly, therefore, isotope dilution analysis *per se* does not immediately provide a solution to the analytical problems posed by the extremely minute concentrations at which many compounds exert their effects in biological systems.

Nevertheless, approaches have been suggested which circumvent the need to make a direct measurement of the mass of the purified sample. Most of these essentially depend upon permitting the isolated material to react with a complexing or 'binding' reagent (designated subsequently in this discussion as Q) in such relative proportions that the isolated compound (P) is in excess of, and saturates, Q (see Fig. 9.2). Following separation of the reaction products, the activity in the bound (P$\overline{Q}$) moiety is estimated. Meanwhile the system can be 'calibrated' by allowing a standard sample of compound P, of known specific activity (S_s) likewise to react with an iden-

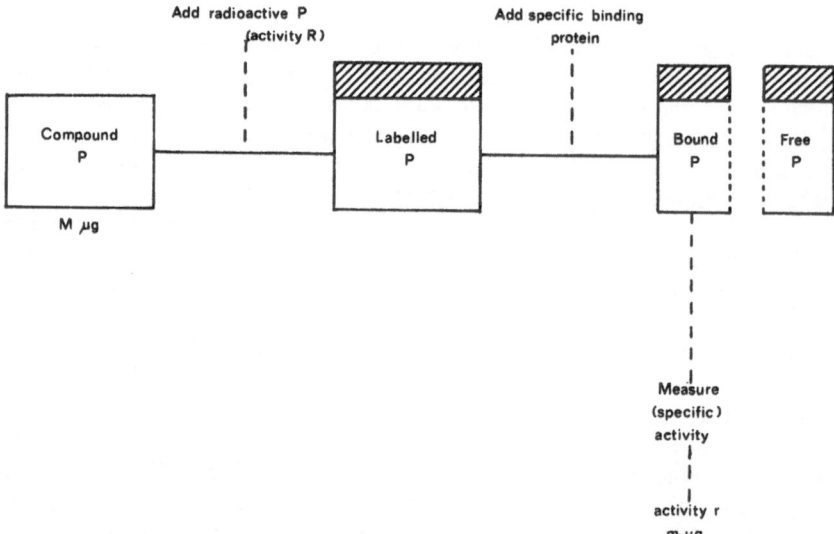

Fig. 9.2. Dilution analysis using a specific binding reagent (e.g. specific binding protein). 'Purification' of the isolated sample is dependent upon the specificity of the reagent the binding capacity of which defines the magnitude of m.

tical amount or concentration of Q. The products of reaction are separated in the same manner, and the activity in the bound fraction similarly estimated.

If the masses of both the unknown and standard preparations of P bound to Q are assumed identical, the *activities* appearing in the bound fractions are proportional to the specific activities of the unknown and standard samples. i.e. if S_u is the specific activity of the unknown

$$S_u = \frac{r_u}{r_s} S_s \qquad (9.3)$$

where r_u, r_s are the activities recorded respectively in the bound fractions corresponding to the unknown and standard samples.

Since S_s is known, S_u can be readily calculated, and hence the total mass of compound P originally present estimated from equation (9.1). However in such a system it is clear that

$$S_u = \frac{r_u}{m} \qquad (9.4)$$

where m = the mass of compound P bound to Q. Hence, from equation (9.1)

$$M = \frac{R \times m}{r_u} - M^* \qquad (9.5)$$

$$= \frac{const.}{r_u} - M^* \qquad (9.5a)$$

if R (i.e. M^*) is held constant.

Equation (9.5) thus implies that, if we add a constant amount (M^*) of labelled compound (with activity R) to variable known amounts of inactive compound P, the relationship between the mass of inactive compound and the reciprocal of the fraction of activity (r_u/R) appearing in the bound (PQ) moiety will be linear, the resulting straight line having a slope equal to $1/m$ (where m is the binding capacity of reagent Q, assumed to be constant) (see Fig. 9.3). An unknown amount of compound P can hence be estimated, assuming the same experimental steps are followed, by interpolation of the observed value for the response 'metameter' (1/fraction bound) onto the standard response curve.

There is no essential difference in principle between the latter procedure, entailing the use of several standards, and the first approach to estimating the unknown, in which only a single standard is relied upon. The plot of the response curve obviates the (trivial) calculations demanded by equations (9.1) and (9.3); moreover the use of multiple standards increases the statistical confidence that can be placed in the measurement of the unknowns.

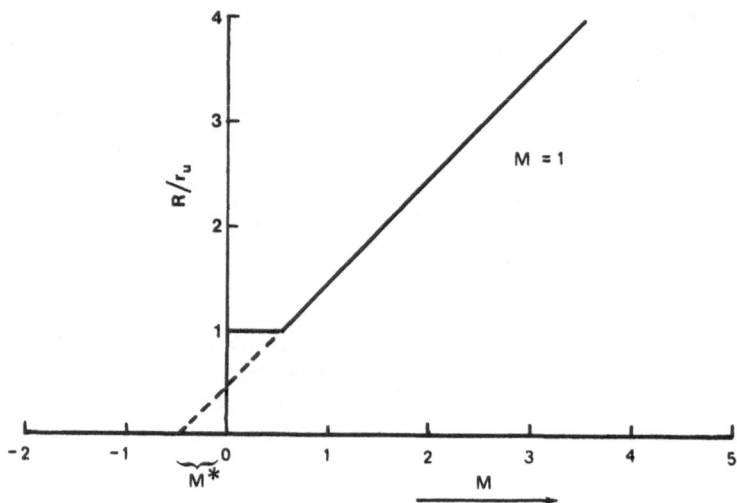

Fig. 9.3. Dilution analysis using a specific binding reagent. Relationship between reciprocal of fraction of activity in the bound fraction as a function of the mass M of inactive compound P. Note that R/r_u cannot fall below unity.

More importantly, however, this approach largely eliminates the assumptions implicit in the calculations based on equations (9.1) and (9.3), i.e. that the binding capacity of Q (i.e. m) is constant for all values of M, and, secondly, that labelled and unlabelled forms of compound P display identity of chemical behaviour in the system.

The form of isotope dilution analysis described in this section has relied, in practice, upon a variety of classes of specific binding reagents. In the biological sciences, specific antibodies, specific naturally occurring serum or tissue binding proteins, and specific enzymes represent the principle groups that have been so used. However, although assays based on such reagents can often legitimately be regarded as particular examples of dilution analysis, they have come to be known by a variety of more specialised terms (e.g. radioimmunoassay, protein binding assay, radioenzymatic assay, radio receptor assay, etc.). Moreover as indicated above, some of the fundamental assumptions which are customarily associated with conventional isotope dilution analysis frequently do not hold in these protein binding assay techniques, and their theoretical basis is consequently more complicated. They will therefore be considered separately and more fully in Section 9.5.

9.3   ACTIVATION ANALYSIS

This form of analysis has been discussed at length in Chapter 7, and its further inclusion here is intended primarily for the sake of completeness, and as an introduction to the derivative assay techniques discussed in Section 9.4.

The basic principle of activation analysis is set out in Fig. 9.4. The method, essentially depends upon the radioactivation of the sample by some

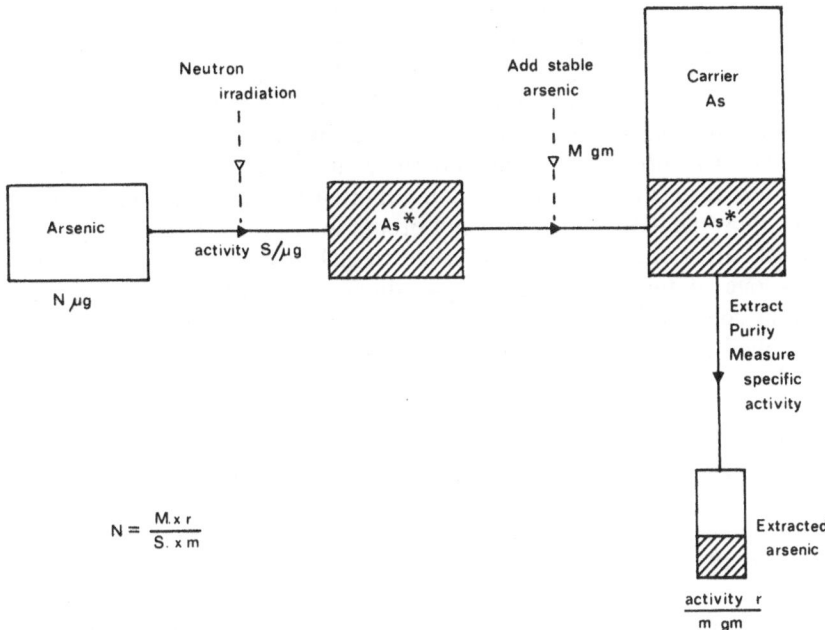

Fig. 9.4. Activation analysis of arsenic (schematic). The induced specific activity (S) is estimated by simultaneous irradiation of an adjacent standard of pure arsenic. Addition of a known amount of inactive 'carrier' arsenic following irradiation and prior to isolation of a purified radioactive sample represents an example of 'reverse' isotope dilution analysis.

form of nuclear particle or photon bombardment, usually by slow neutrons from an atomic reactor. Subsequently the amounts of the elements of interest can be deduced from a measurement of the corresponding radio-isotopic activities thereby induced. Standardisation of the activation process is usually effected by irradiating a standard sample of the element of interest of known weight in close proximity to the unknown, ensuring that both samples are exposed to the same radiation flux.

The method is essentially applicable only to the measurement of elements; however it is occasionally possible to determine by this means the amount of a particular compound present in a biological sample, provided that a suitable element is present solely as a constituent of the compound of interest or that the latter can be isolated prior to the activation step. In this way, estimates have been made of the concentration of organic iodine compounds such as mono- and di-iodotyrosines in blood [5]; another interesting application has been the estimation of ATP derived from house-flies, based on the activation of the phosphorus present to ^{32}P [6]. In such studies, the usual procedure has been to isolate the required compound by paper chromatography and subsequently to expose the paper chromatogram to particle bombardment. Naturally it is imperative that the chromatographic paper used in such studies should be relatively free of the elements concerned.

Such applications are nevertheless uncommon, and the method has principally been employed in the biological sciences for the determination of trace elements in blood, tissues and other biological materials.

A particular difficulty which often arises in such contexts is the highly complex composition of the material to be analysed, and the wide variety of trace elements present. It is therefore frequently necessary to adopt extreme measures to distinguish the radiations emitted by the desired radioisotope from those emitted by other radioisotopes in the sample. Occasionally, in simpler situations, it may be possible to isolate the radiation of interest using one of the physical techniques discussed elsewhere in this volume, such as conventional gamma-ray pulse height analysis. Such techniques have been rendered considerably more powerful in the last few years following the development of high resolution semi-conductor crystal detectors; nevertheless the current limitations on the volume of such detectors restricts their use to situations in which high sensitivity is not of principal importance.

In most circumstances chemical separation offers the most practical method whereby the amount of a particular radioisotope present in the irradiated sample may be determined. In these situations, recourse is usually made to the reverse isotope dilution method discussed in Section 9.2.2 as illustrated in Fig. 9.4. In the example depicted in this Figure, inactive arsenic in known amount is added to the biological sample subsequent to irradiation, followed by chemical separation of the pure material [7]. Nevertheless in such procedures it is not unusual to find traces of radioactive contaminants persisting through the extraction steps, despite exhaustive purification, and it is frequently necessary to validate the efficiency and completeness of the separation by gamma-ray spectrometry at the final stage. Indeed it is often more rewarding not to attempt an exhaustive chemical isolation of individual elements, but to restrict chemical procedures to simple group separation, and to rely on a subsequent gamma-ray analysis to identify the individual radioisotopes of interest. An example of this approach is provided by Samsahl, Brune and Wester [8] who have assayed up to 30 trace elements simultaneously in a variety of tissue samples.

9.4   LABELLED REAGENT ASSAYS

9.4.1   Derivative Analysis
It is perhaps helpful to visualise labelled derivative analysis as representing essentially an activation analysis technique wherein the element or compound to be measured is activated, not by particle or photon bombardment, but by combination with a radioactive reagent of known specific activity. Thereafter the chemical procedures, the difficulties encountered and the calculations involved are closely analogous to those associated with conventional activation analysis.

Some of the earliest derivative assay methods relied on naturally occurring radioisotopes. Solutions of chromate, for example, were analysed [9] by treatment with an excess of a radiolead (thorium B) solution of known concentration and activity; the activity in the precipitated lead chromate was subsequently determined, and the amount of chromate present calculated using the formula

$$M = \frac{X}{S}$$

where M = mass of chromate in sample (Eq)

    X = activity of precipitate (cpm)

    S = specific activity of radiolead used (cpm/Eq)

The method was first applied to the assay of organic materials by Keston, Udenfriend & Cannan in 1946, who exploited the radioactive reagent [131] I p-iodosulphonyl (pipsyl) chloride in the assay of mixtures of amino acids [10]. A number of other reagents have subsequently been employed. They include ^{3}H-acetic anhydride [11], used in assays for a variety of steroid hormones, thyroid hormones and other amino acids, 4-methoxy-3-[36]Cl-benzoic acid [12], p-[131] I-benzoyl chloride [13], ^{14}C-fluorodinitrobenzene [14] and ^{35}S-thiosemicarbazide [15]. More recently the area of application and practical usefulness of labelled reagent methods has been significantly extended by the exploitation of [131] I-labelled or [125] I-labelled antibodies, specific in their reaction with individual proteins, polypeptide hormones, the thyroid and steroid hormones, viral antigens and other similar groups. These methods, collectively termed 'immunoradiometric assay' by Miles and Hales [16] are characterised by a much greater intrinsic specificity than is readily achieveable using the relatively simple chemical reagents listed above, which usually necessitate extensive and time consuming purification steps to isolate the desired derivative from other labelled reaction products.

In their original method, Keston et al. added, to the resulting mixture of [131] I-labelled pipsyl derivatives, a large excess of the unlabelled pipsyl derivatives of the amino acid of interest. The derivative was then reisolated, purified, and its specific activity finally determined. (These steps represent, in a manner closely analogous to the procedures adopted in conventional activation analysis, a further example of the 'reverse isotope dilution' technique). However, in its simple and original form the method required the isolation of a weighable amount of derivative, and essentially precluded the use of the more powerful micro-methods of isolation and purification now available. This problem was overcome by the addition either of known quantities of the ^{14}C-labelled amino acid before, or the corresponding ^{35}S-labelled derivative after, preparation of the [131] I-labelled pipsyl derivative [17]. Purification of the derivative was followed by a measurement, in the isolated sample either of the [131] I/^{35}S, or [131] I/^{14}C ratio. These alternative procedures, both usually referred to as 'double isotope derivative analysis' are illustrated in Figs. 9.5a and 9.5b. The mass of the amino acid of interest can subsequently be calculated from the expression

$$M = \frac{X}{fS} - M^* \qquad (9.6)$$

where M = mass of amino acid in the original sample (mEq)

    X = [131] I activity recovered in isolated derivative (cpm)

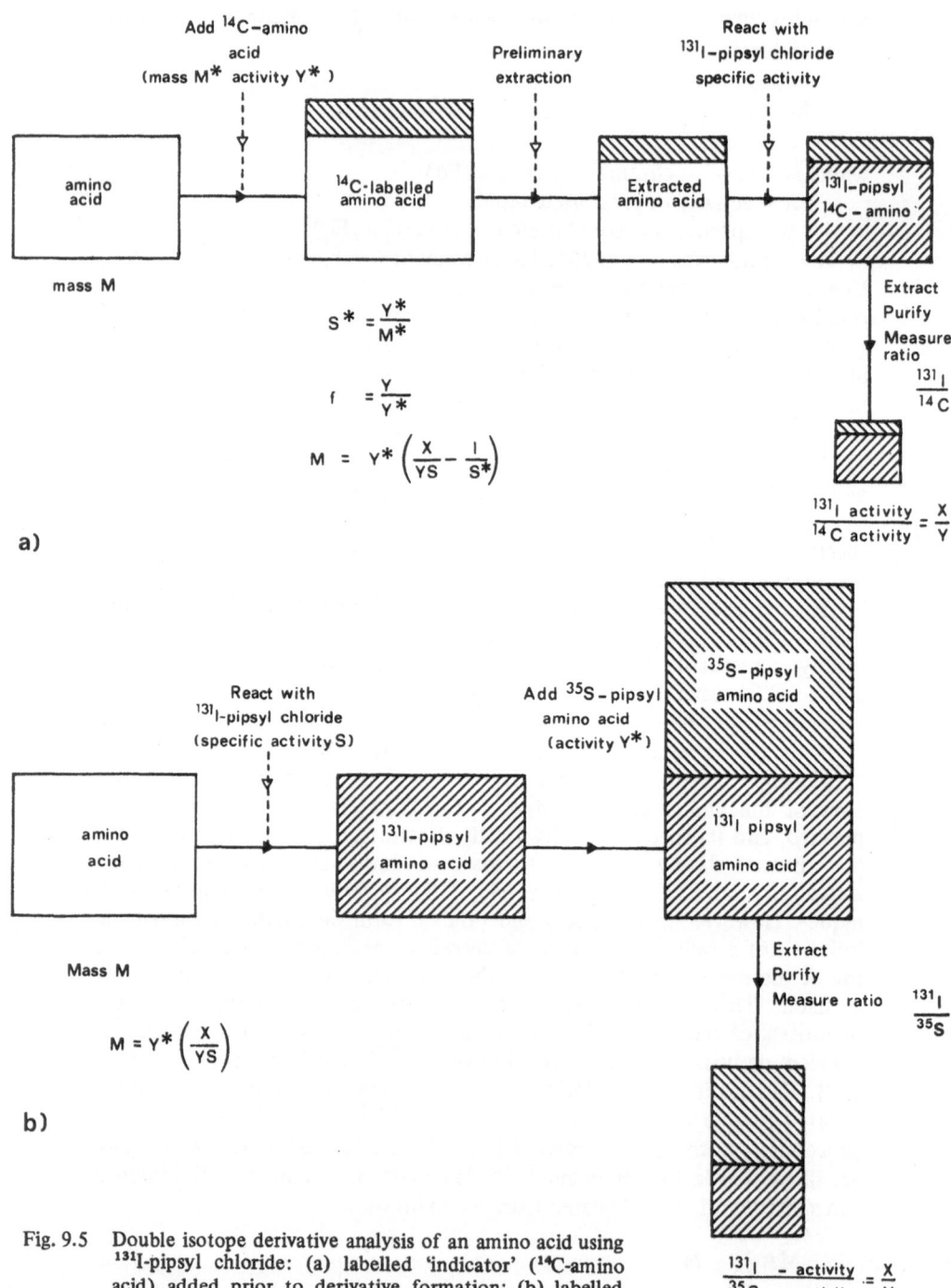

Fig. 9.5   Double isotope derivative analysis of an amino acid using
^{131}I-pipsyl chloride: (a) labelled 'indicator' (^{14}C-amino
acid) added prior to derivative formation; (b) labelled
'indicator' (^{35}S-pipsyl amino acid) added after derivative
formation.

f = fraction of added ^{14}C or ^{35}S activity recovered
S = specific activity of the ^{131}I-pipsyl reagent (cpm/mEq)
M* = mass of ^{14}C-labelled amino acid (if added) (mEq)

Equation (9.6) can readily be rearranged in the form

$$M = Y^* \left(\frac{X}{YS} - \frac{1}{S^*} \right) \tag{9.7}$$

where Y = ^{14}C or ^{35}S activity recovered (cpm)
Y* = ^{14}C or ^{35}S activity initially added (cpm)
S* = specific activity of ^{14}C-labelled amino acid (if added) (cpm/mEq)

Equation (9.7) underlines the dependence of the assay solely on the final determination, in the purified sample, of the ratio of the two activities (X/Y) since S, S* and Y* are presumed known.

The two alternative procedures adopted to monitor the recovery of either the original amino acid, or the derivative, through the isolation procedure both clearly represent isotope dilution techniques. Nevertheless they differ in the advantages they respectively offer. The addition of ^{14}C-labelled amino acid prior to derivative formation (Fig. 9.5a) enables a preliminary extraction to be carried out, thereby significantly reducing the quantity and cost of ^{131}I-labelled reagent required, and the difficulty with which unwanted labelled derivatives or excess reagent are subsequently removed. Moreover the reaction leading to derivative formation need not be carried to completion, any losses of the amino acid arising from incomplete reaction (or during any preliminary extraction) being revealed by the recovery of ^{14}C-amino acid in the final product. Nevertheless this approach must be used with caution, since the addition of a large mass of labelled compound before derivative formation may swamp the endogenous compound present, and thereby greatly reduce the precision of its measurement. This problem clearly necessitates the use of high specific activity 'indicators' (exemplified by ^{14}C-amino acid in the above example) so as to minimise the amount of labelled material required to yield acceptable count rates in the final purified derivative sample.

In contrast, the alternative procedure shown in Fig. 9.5b is largely unaffected by the specific activity of the labelled derivative used as indicator since, within limits, the mass of ^{35}S derivative introduced into the system at the later stage is unimportant. However, it suffers from the disadvantage that the initial derivative formation step must, in this case, be carried to completion; any amino acid not undergoing reaction, or any other losses occuring before addition of the labelled derivative indicator cannot subsequently be corrected for.

The specific activity of the reagent used, whichever of the approaches shown in Fig. 9.5 is adopted, is clearly an important determinant of the sensitivity of labelled reagent assays. However, probably the greatest practical problem which these techniques present, and one which frequently prevents attainment of the sensitivity which the specific activities of available labelled reagents and indicators would otherwise permit, is the difficulty of complete isolation of the desired derivative from unused reagent and from other contaminating labelled reaction products. Very careful separation techniques, typically including repeated chromatography on columns, paper or silica gel, may be required before radiochemical purity of the isolated derivative falls within acceptable limits.

A commonly applied criterion of purity of the derivative is the demonstration of constancy in the ratio of the two radioisotopes present at successive stages of the isolation procedure (or of constancy of specific activity if a second isotope is not used for monitoring recovery through these steps). Another accepted index of purity is a constant radioisotope ratio across the chromatogram peak representing the isolated derivative (i.e. in successive sectors of the peak from the leading to trailing boundaries). However, Laragh, Sealey and Klein [18] in measurements of aldosterone in urine using ^{14}C-acetic anhydride, have observed the presence of an isotope effect on chromatographic mobility resulting in relative displacement of ^{3}H-aldosterone acetate with respect to both ^{14}C-labelled and unlabelled derivatives. Although such effects are not large, they can potentially lead to a relative enrichment of one radioisotope in the final derivative and thus to an incorrect estimate of the substance under assay, particularly if a number of successive chromatographic steps are required in the purification procedure. To obviate errors stemming from this source, it is essential that the zone transferred from any one chromatogram to the next should be sufficiently large to encompass entirely the peak comprising the required derivative labelled with each of the different isotopes used.

Despite careful checks on radiochemical purity, blank values (i.e. estimates obtained by assaying samples containing none of the compound of interest) may nevertheless be both high and variable in relation to the amount of the compound in test samples. It is therefore imperative that the magnitude and variation of the blank be studied experimentally, both to establish the limit of sensitivity of the method (cf. discussion of sensitivity in Section 9.5), and the confidence that can be placed on low values. The problems surrounding the attainment of high sensitivity in double isotope derivative techniques have been discussed in detail by Brodie and Tait [19] who, in an examination of the application of the method to the steroid hormone assays, have extensively investigated the factors affecting the optimal choice of labelled reagent, indicator and other similar assay parameters.

9.4.2 Assay of Specific 'Binding' Proteins Using Labelled Ligands
Although derivative assay methods have long been potentially applicable to a wide range of compounds of biological interest, the technical problems referred to earlier have in general restricted their use to research laboratories equipped with sophisticated analytical equipment and to situations in which a very low throughput of samples has been acceptable. A group of labelled reagent assay methods, hitherto small, which has escaped these limitations has been that in which one of the reactants, usually a protein, has displayed

particularly high chemical specificity, thereby reducing the technical problems otherwise incurred in isolation of the labelled derivative. Illustrative of such applications are the measurement of serum thyroxine-binding globulin using labelled thyroxine [20], of serum iron binding protein (transferrin) with ^{59}Fe [21] and the determination of the vitamin B_{12} binding proteins (transcobalamins), present in serum or gastric juice ('intrinsic factor') using ^{58}Co- or ^{60}Co-labelled vitamin B_{12} [22, 23]. Such applications have been characterised by the use of very simple techniques for the separation of the reaction products, e.g. the adsorption by activated charcoal or an ion exchange resin of residual unreacted labelled material [24]. A more sophisticated example of such techniques is provided by Pearlman and Crépy [25] who employed Sephadex gel to separate bound and free labelled testosterone in incubation mixtures containing a specific testosterone binding protein present in serum. The Sephadex particles, acting as an essentially inert molecular sieve and thereby permitting entry only of the unbound steroid, enable the reaction between steroid and binding protein to be examined at different concentrations of steroid in the system without disturbance of the underlying chemical equilibrium. Such measurements allow estimation both of the amount of binding protein present, and of the equilibrium constant of the reaction. Similar techniques have subsequently been extended to the analysis of more complex protein mixtures comprising more than one binding protein, although the resulting calculations are laborious, and necessitate the use of computer curve-fitting techniques in the estimation of the required parameters.

9.4.3 Immunoradiometric Assay Techniques

Labelled reagent techniques have, within the past 5 years, entered a new era following the introduction, by Miles and Hales in 1968 [16], of the immunoradiometric assay technique relying on ^{125}I-labelled antibodies specific in their reaction with the compound of interest. Originally applied to the measurement of insulin, this technique is currently being extended to the assay of non-protein compounds such as the thyroid and steroid hormones, and promises ultimately to cover a very wide area of application in biochemistry and biology. The fundamental basis for this belief lies in the current development of techniques (described more fully in Section 9.5) for the raising, in laboratory animals, of highly specific antisera directed against compounds of progressively decreasing molecular size. It therefore becomes increasingly realistic to regard virtually any compound of biological interest as a potential antigen, and to predict the ultimate development of an immunoassay system by which it can be measured.

The immunoradiometric assay methods combine the very high specificity of antibody/antigen reactions with the delicacy of radioactive detection methods, attributes which they share with the radioimmunoassay techniques discussed in Section 9.5. Although the two methods differ in principle, it is clear therefore that their ultimate areas of application are likely to be essentially identical, and that the two approaches are, to a large extent, competitive. Although Miles and Hales originally claimed, on semi-intuitive grounds, that the immunoradiometric assays were capable of greater sensitivity, this contention has not been unequivocally confirmed in a more rigorous theoretical analysis [26] nor has it been borne out in experimental practice. The relative merits of the two approaches are therefore likely to reside in the experimental and logistic advantages which they respectively display.

The steps involved in a typical immunoradiometric assay are shown in Fig. 9.6. Because of the specificity of the antibody reaction, the problems associated with the isolation of the desired reaction product are clearly reduced as compared with conventional derivative assay techniques, and only removal of residual excess labelled antibody from the antigen/antibody complex is required. The means whereby this is achieved, and the technique employed for the preparation of the specific labelled antibody, both rely on the use of an 'immunoadsorbent', specific in its reaction with the desired antibody, and it is the preparation and use of this material which constitutes the most critical and important feature of immunoradiometric methods.

Preparation of Specific Labelled Antibodies
The techniques adopted for the raising of antisera against substances (such as insulin) which are intrinsically immunogenic, or compounds such as cyclic AMP which are not, are dealt with in detail in Section 9.5. In general the material used for the immunisation of laboratory animals is inevitably impure, and results in a wide spectrum of antibodies in the immune serum produced. The isolation of the desired species of antibody using an immunoadsorbent — a material prepared by coupling the corresponding antigenic compound to a solid matrix such as powdered cellulose — demands therefore that the coupled antigen be particularly highly purified. Moreover, the matrix material should itself display little intrinsic affinity for protein of any kind present in the immune serum.

The immunoadsorbents which have proved most useful in practice have been prepared by coupling the antigen to diazonium derivatives of powdered cellulose [27], although cyanogen-bromide activated cellulose [28] or Sepharose [29] have also been used. Addition of a small quantity of immunoadsorbent to undiluted antiserum results in uptake by the adsorbent of specific antibody. The resulting complex is freed from residual serum proteins by washing and the antibody then radioactively labelled (whilst still bound to the solid matrix) by iodination with ^{125}I [30]. Following further washing to remove residual radioiodine and damaged antibody, the specific labelled antibody may be decoupled from the matrix by elution with an acidic buffer, usually in the region of pH 2. The purified labelled antibody preparation is collected in buffer of pH 8; if not used immediately it may be stored for periods of some months in a highly stable form following recombination with further immunoadsorbent.

Assay Procedure
This consists essentially of the incubation, for periods varying from 1 hour up to several days, of an appropriately diluted solution of labelled anti-

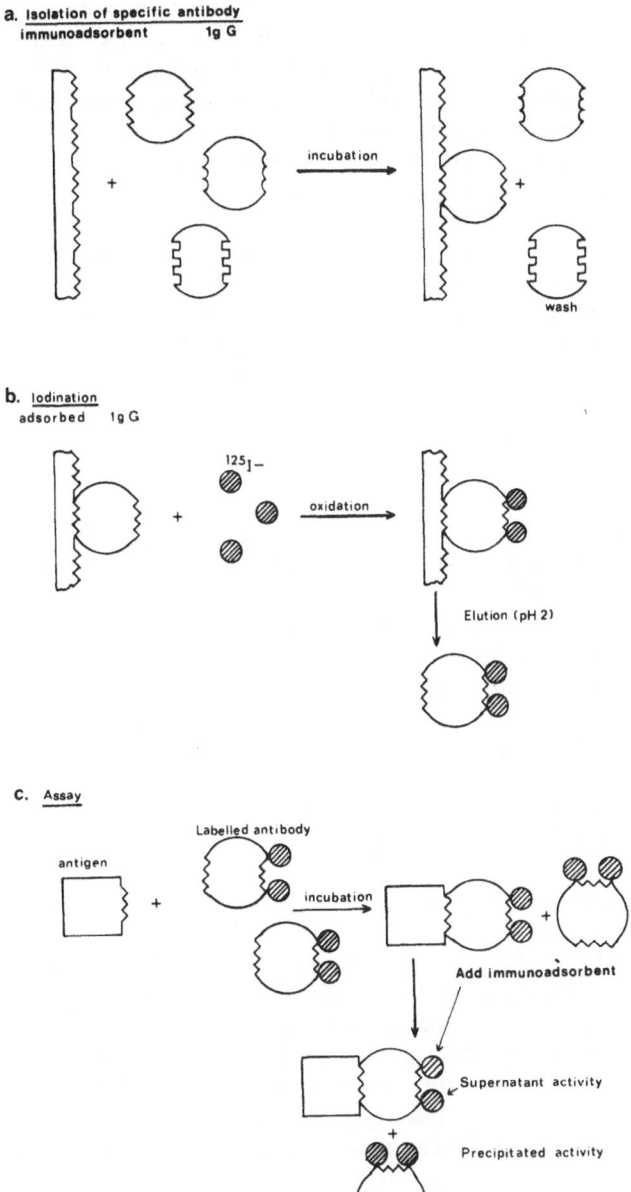

a. Isolation of specific antibody
immunoadsorbent   1g G

incubation

wash

b. Iodination
adsorbed   1g G

$^{125}I-$

oxidation

Elution (pH 2)

c. Assay

Labelled antibody

antigen

incubation

Add immunoadsorbent

Supernatant activity

Precipitated activity

Fig. 9.6. Steps in immunoradiometric assay: (a) separation of specific antibody from other immunoglobulins and serum proteins in antiserum using immunoadsorbent; (b) iodination of specific antibodies and subsequent elution from immunoadsorbent; (c) reaction of labelled antibodies with antigen and removal of excess antibody.

body, either with the solution under test or with a set of standard solutions made up to resemble, as closely as possible, the composition of the test solution (thus reducing the possibility of non-specific effects on the antibody—antigen reaction). Following incubation, excess antibody may conveniently be removed by addition of immunoadsorbent to the mixture, thus precipitating the unused antibody and permitting an estimate to be made of the amount of antibody—antigen complex by radioassay of the supernatant.

Because of the reversibility of antibody/antigen reactions, the addition of large amounts of immunoadsorbent to the system results in dissociation of the initial antigen/antibody complex, and would indeed ultimately lead to the total removal of supernatant activity. For this reason, antibodies charac- terised by high dissociation rate constants cannot be employed in assays of this type. In general, the amounts of reagents and periods of incubation used (both in the initial reaction, and in the excess antibody separation step) are matters of fine judgement necessitating considerable experimenta- tion.

In this simple form the immunoradiometric method has been applied to a wide range of proteins, including insulin [16], human growth hormone [31], follicle-stimulating hormone and luteinising hormone [32], parathy- roid hormone [33], IgG [34], porcine and human calcitonins [35], angio- tensin I [36], and ferritin [37]. Although the reactions between lower molecular weight compounds such as the steroid and thyroid hormones, and the antibodies directed against them are frequently associated with some- what higher dissociation rate constants than are characteristic of protein/ antibody reactions, it is becoming evident that this is not as likely to impede the development of immunoradiometric assays for such compounds as was originally feared, and that techniques for these will in due course be reported [38].

Amongst advantages claimed for the immunoradiometric technique *vis-à- vis* radioimmunoassay is the relative ease with which specific antibodies can be iodine-labelled (by virtue of the number of tyrosine residues in their structure) as compared with the difficulties which occasionally arise when the antigen (i.e. the substance to be assayed) must be so labelled, as is implicit in RIA. Moreover, labelled antibodies once prepared, are charac- terised by long-term stability, both during storage and during subsequent incubation in the assay milieu (e.g. in diluted serum), in contrast to the tendency of certain labelled antigens to deteriorate rapidly and catastrophic- ally in either or both of these situations.

Some of these practical advantages may be offset, however, by the addi- tional complexity entailed in the preparation of the required reagents (albeit some of the necessary starting materials are now commercially available) and by the rather heavier demands the technique makes upon purified antigens and specific antibodies which may be in relatively short supply.

9.4.4 Two-Site or 'Sandwich' Type Immunoradiometric Assay
A further refinement of immunoradiometric assays has subsequently been

developed by Addison and Hales [39] which further increases the sensitivity and specificity implicit in the original techniques. This extension relies on prior extraction of the compound under test from a biological fluid using an unlabelled antibody pre-coupled to an insoluble matrix such as cellulose paper or powder. Provided that, as in the case of all but the smallest poly-peptide molecules, the antigen possesses more than one immunologically active determinant, the amount of antigen adhering to the coupled antibody can subsequently be determined by incubation with a second, labelled, anti-body, prepared in the manner described in the preceding section. A typical response curve obtained in an assay for human growth hormone, using this method, is shown in Fig. 9.7.

In practice the pre-coupled antibody can be a relatively impure, wide spectrum, preparation comprising the immunoglobulin fraction of a suitable immune serum derived by precipitation with sodium sulphate. However, improved specificity is achieved if the antibodies coupled to the solid matrix

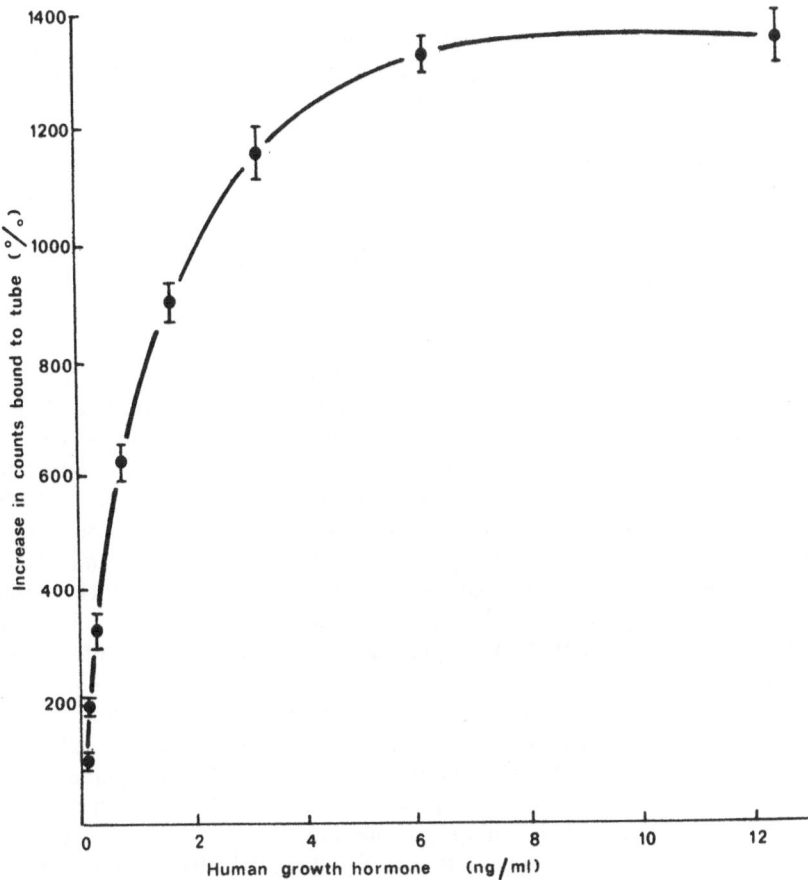

Fig. 9.7. Typical response curve observed using two-site assay of human growth hormone (Woodhead *et al.* [29]).

are selected to differ from the second, labelled, antibody with respect to the portion of the antigen molecule against which they are directed. Thus if the coupled antibody is directed against, for example, the carboxy-, or C-, terminal end of a polypeptide molecule, whilst the labelled antibody is raised against the amino, or N-, terminal portion, then the assay system will respond solely to molecules comprising the intact polypeptide sequence and will be unaffected by degradation products consisting of molecular fragments, albeit some of these will react with one or other of the antibodies present in the system. Exemplifying this approach, O'Riordan et al. [40] have developed a 'sandwich' type immunoradiometric assay for native bovine parathyroid hormone which is totally unresponsive either to amino-terminal or carboxy-terminal fragments. The necessity for specificity inherent in these analytical systems has been underlined by observation of the presence, in serum, of large amounts of polypeptide fragments following administration of, for example, ACTH [41] and more generally, by the studies of peripheral metabolism of polypeptide hormones described by Habener et al. [42].

A further example of the exploitation of the specificity characteristics of double antibody sandwich assay systems is provided by the development of techniques for the measurement and sub-typing of Australia (or hepatitis B) antigen, a material which is characteristically present in the form of virus-like particles in the blood of patients displaying an acute form of hepatitis B. Techniques for the detection of the antigen are of considerable social importance. Fatalities can occur as a result of transfusion of blood from donors who may be healthy carriers of the disease. Although conventional radioimmunoassay techniques may be employed for detection of the virus, they are time consuming and technically laborious, and not suitable for the large scale screening of transfusion blood.

The labelled antibody sandwich method originally described by Ling and Overby [43] is currently regarded as the most practical approach, and now forms the basis of commercial assay kits which are in routine use in the US and many other countries. The technique relies on the prior adsorption, onto the interior walls of polypropylene incubation tubes, of immune guinea pig anti-Australia immunoglobulins; the test specimen is subsequently incubated in the tube overnight (N.B. assays necessitating significantly shorter incubation periods are currently under development). After thorough washing, a second labelled antibody is added to the tube and, following a further incubation of approximately 1 hour, activity adhering to the tube wall is measured. The situation is rendered somewhat more complex by the occurrence of subtypes of hepatitis B antigen whose epidemiological significance as yet remains unknown. These subtypes contain a common antigenic determinant designated 'a', together with mutually exclusive determinants 'd' and 'y'. In the technique described by Cameron and Dane [44], polypropylene tubes coated respectively with specific anti-d or anti-y antibodies are used, the labelled antibody being predominantly of anti-a specificity. Coating antibodies are initially prepared from mixed anti-ay or anti-ad antisera from which the anti-a components are removed by passage through affinity columns prepared from ad or ay antigen coupled to Sepharose.

The use of antibody coated tubes in this context clearly reduces the problems associated with cellulose or Sepharose coupled antibody systems wherein the repeated washings required to remove excess labelled antibody are tedious and may lead to undue dissociation of the labelled antibody/antigen complex system from the solid matrix.

Aside from the increase in specificity implicit in the use of a double antibody recognition system, prior removal of the antigen from the (biological) fluid under test may substantially increase overall assay sensitivity whilst decreasing vulnerability of the assay to non-specific effects. Thus relatively large volumes of the test fluid may be extracted (provided the adsorbed antibody is not thereby overloaded) resulting in concentration of the desired antigen. Meanwhile, subsequent incubation with labelled antibody can be carried out in standard buffer solutions thus eliminating the problems which occasionally arise when labelled proteins are incubated for long periods in serum and other enzymatically active fluids.

9.4.5 'Double Antibody' Immunoradiometric Assay

A further elegant refinement of the immunoradiometric principle has recently been described by Miles [45] and Woodhead, Addison and Hales [46] which relies on the use of a universal labelled reagent comprising ^{125}I-labelled anti-IgG. The labelled antibody, prepared by immunisation of laboratory animals with rabbit or guinea pig IgG, and subsequent labelling by the techniques previously described in this chapter, is capable of reacting with the specific antibodies conventionally employed in either direct or two-site immunoradiometric assay methods. By judicious selection of appropriate antibodies, it is possible to extract a desired antigen using a solid matrix-coupled (rabbit) antibody, and to allow the antigen to react with a second unlabelled (guinea pig) antibody; the Ab-Ag-Ab complex is itself subsequently incubated with the anti-guinea pig IgG-labelled antibody.

The advantage of this and similar variations clearly resides in the possibility of preparing and storing only a single labelled reagent for use in all immunoradiometric assays, an attribute which offers considerable practical advantages whilst minimising the relatively heavy consumption of sometimes scarce antisera which is implicit in the conventional immunoradiometric approach.

In conclusion of this section, it should be emphasised that, although immunoradiometric methods are in their infancy and they have not as yet gained the popularity of the now conventional radioimmunoassay techniques, they share with these a near universality of potential application and therefore must merit careful consideration when the setting up of any new assay system is contemplated. In certain circumstances the relative advantages which this approach displays and to which attention has here been drawn, may render the immunoradiometric method the procedure of choice.

9.5   RADIOIMMUNOASSAY AND SATURATION ANALYSIS

9.5.1   General Principle

As has already been indicated in Section 9.2, radioimmunoassay (RIA) and other saturation assay methods may be regarded as a subset of dilution analysis techniques, and it is evident that, in certain situations the simple theory applicable to conventional dilution analysis can be legitimately applied to them. Nevertheless because of the complexity of the binding protein systems on which they rest, and because they are frequently employed in situations in which the reactants are present in exceedingly high dilution, there are advantages in considering these methods from a different conceptual viewpoint.

In general, the distinguishing principle of saturation analysis methods is that, by allowing the substance to be measured (compound P) to react with a specific receptor reagent (Q) of limited capacity, substance P — by saturating, or tending to saturate Q — is partitioned into two moieties, such that the ratio of the two varies as a function of the total amount of P present. An unknown amount of P may thus be quantitated by comparing its distribution with distributions yielded by a set of standards — comprising known amounts of P — which are permitted to react with Q in identical circumstances. A convenient method of observing the distribution, particularly when the concentrations of the reactants are low, involves the preliminary addition of a trace amount of radioactive P to the unlabelled compound, and this approach has predominantly been used in the past in the application of such methods. Nevertheless it should be emphasised that the basic principle of the method does not depend on the use of a radioactive indicator; the availability of any physical, chemical or biological means of quantitating the distribution of compound P between the two compartments enables the assay principle to be exploited.

The principle is more rigorously expressed in Fig. 9.8. In this diagram the specific saturable reagent Q is shown as typically comprising a specific binding protein or antibody and these constitute the major groups of specific 'receptor' reagents used in biochemistry, albeit other classes of molecules are occasionally employed (see Table 9.1). As implied in Fig. 9.8, it is sometimes necessary to extract and isolate compound P from its biological milieu, in which circumstances labelled P may be employed in a dual role: that of indicating recovery of inactive P through the extraction procedure and subsequently of revealing the distribution of P between the reacted (or bound) and unreacted (or free) moieties. Chemical identity of labelled P with inactive P is necessary in the first of these roles, since the labelled compound must exactly mimic the unlabelled material through all steps of the extraction procedure. However in those situations in which no extraction of P is required it is unnecessary for the radioactive and unlabelled compounds to be chemically identical, provided that the two compete for reaction sites associated with Q, and hence the distribution of the label reflects the distribution and amount of unlabelled P in the system.

It is evident on intuitive grounds that the specific activity of the labelled

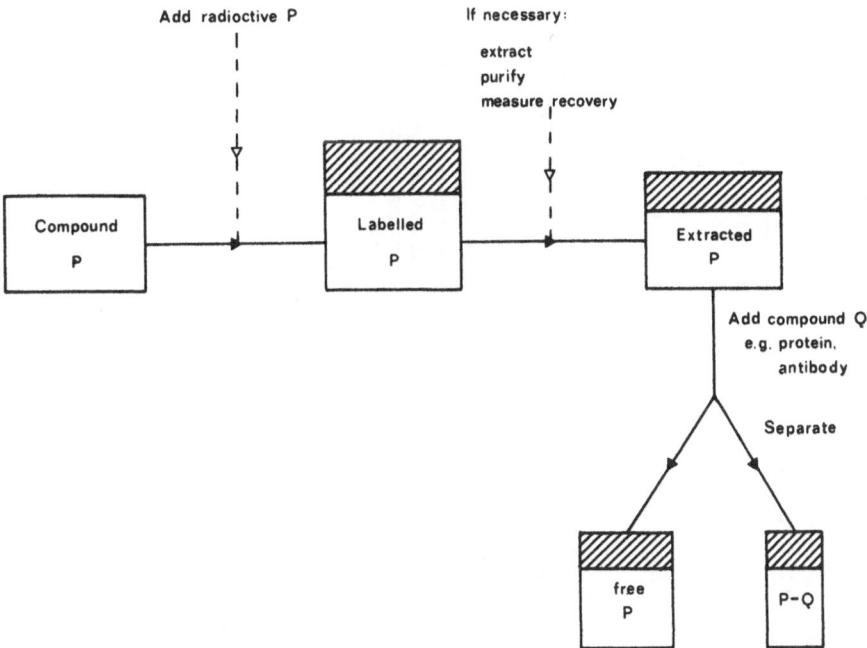

Fig. 9.8. Principle of saturation analysis (schematic).

material used must be sufficiently high to ensure that adequate count rates are achieved using amounts of the label which are commensurate with the amount of P in the system. Likewise the amount of Q used must be of the same order as that of P, for a workable assay. In short, there exists an optimum choice of reagent concentrations (i.e. of labelled P and of Q) which yield best precision of measurement of any specified concentration of inactive P.

In addition to these requirements, application of the principle requires a convenient separation method to sequestrate bound P from residual free material prior to radioassay, since both moieties will normally remain in solution following reaction. A very wide variety of separation techniques has been devised, the particular choice of which normally depends upon the nature of the substance to be assayed, and on the specific reagent Q used in the system. None, however, has emerged as a general technique applicable to all assays of this type.

The final step in these procedures consists in the measurement of distribution of radioactivity in the moieties derived from each sample and the plotting of the 'standard' or 'response' curve against which unknowns are to be compared. Isotopic distributions can be evaluated by counting both free and bound activities, or if the total activity in each incubation tube is known (as is usually the case), in either the free or the bound moiety alone. The statistical implications of each of these approaches is different however and the concentrations of reagents should, in principle, be adjusted depending upon which procedure is adopted. (Though there is no theoretical advan-

TABLE 9.1 *General Designations: Saturation Analysis, Displacement Analysis, Radioligand Assay, Radiostereo Assay, Competitive Radioassay, Competitive Protein-Binding Assay*

Class of specific reagent	Common names(s)	Some classes of compounds to which applied	Examples
Antibodies	Radioimmunoassay	Polypeptide hormones	Insulin, HGH, ACTH, LH, FSH, etc.
		Proteins	Albumin, thyroglobulin, IgE, IgG, α-fetoprotein
		Steroid hormones	Aldosterone, testosterone, oestradiol, progesterone etc.
		Thyroid hormones	Thyroxine, triiodothyronine
		Cyclic nucleotides	cAMP, cGMP, cIMP, cUMP
		Tumor antigens	Carcinoembryonic antigen (CEA)
		Viral antigens	Australia antigen
		Prostaglandins	
		Drugs	Morphine, digoxin, digitoxin
		Enzymes	Dopamine β-hydroxylase, fructose-1,6-diphosphatase
		Carrier proteins	TBG, intrinsic factor
		Vitamins	Folic acid
Serum intracellular-binding proteins, also other specific binders (probably proteins) found in biologic fluids (e.g., milk, gastric secretions, etc.)	Protein-binding assay; competitive protein-binding assay	Thyroid hormones	T_4, T_3
		Steroid hormones	Cortisol, progesterone, oestradiol, etc.
		Vitamins	Vitamin B_{12}, vitamin D, folic acid
		Trace elements	Iron
		Cyclic nucleotides	cAMP, cGMP
Cell membrane receptors	Radioreceptor assay	Polypeptide hormones	ACTH, LH, TSH
Enzymes	Radioenzymatic assay	Vitamins	Folic acid
		Cyclic nucleotides	cAMP, cGMP
Microorganism	Radiomicrobiologic assay	Vitamins	Folic acid
Inorganic reagents	Substoichiometric assay	Metals	Mercury, iron

tage associated with the particular choice of the fractions counted, it is probably better practice, at least initially, to count activities in both free and bound moieties as a check for any inadvertent error which may occur in the introduction of radioactivity into each incubation tube.)

Standard curves are conventionally plotted in a wide variety of ways, though none has any particular theoretical advantage. Amongst the more popular response 'metameters' are the bound to free ratio (B/F), the free to bound ratio (F/B), the percentage or fraction of total activity bound (b), the percentage or fraction free (f), b/b_o (where b_o represents the fraction or percentage bound at the zero 'dose' point on the response curve), $1/b$, and more recently logit b/b_o (i.e. log $\frac{b/b_o}{1-b/b_o}$). The ligand* concentration or 'dose' is usually plotted on either arithmetic or logarithmic scales (see Fig. 9.9). Probably the major practical advantage associated with certain of the currently popular coordinate systems is that they yield response curves which are essentially linear over a large part of the ligand concentration range. This observation is of particular assistance in the manual drawing of response curves, and the calculation and statistical evaluation of assay results, though of diminishing importance when computer methods are used in the processing of assay data. Rodbard and co-workers have drawn particular attention to the advantages of the logit/log plot which linearises the response curve in many assay systems, though not in all. This transform has served as the basis of a number of computer programs published by Rodbard and his collaborators relating to saturation assay data processing [47].

9.5.2 Areas of Application and Nomenclature

The saturation assay method was first applied successfully in 1960 to the measurement of serum hormones by Yalow and Berson [48], who employed a specific antibody, and independently by Ekins [49] using a specific naturally occurring binding protein to measure insulin and thyroxine respectively. Subsequently the method was used for the assay of growth hormone [50], vitamin B_{12} [51], glucagon [52] and ultimately many other hormones, vitamins and trace compounds found in blood and urine. The use of enzymes as saturable reagents was introduced by Rothenberg in 1965 [53] and more recently intracellular receptor proteins [54, 55] and cellular membrane receptors [56] have been employed in the same role. Probably the most important development in the last few years has been the extension of antibody methods to a number of classes of compounds (e.g. nucleotides [57], thyroid [58], steroid hormones [59] and prostaglandins [60]) which, because of their small molecular size, are not intrinsically immunogenic. This development has enormously widened the area of exploitation of the assay principle which now potentially embraces almost all compounds of biochemical interest.

---

*The compound to be measured is frequently referred to as a 'ligand', since the reaction normally involves a stable binding of the test substance.

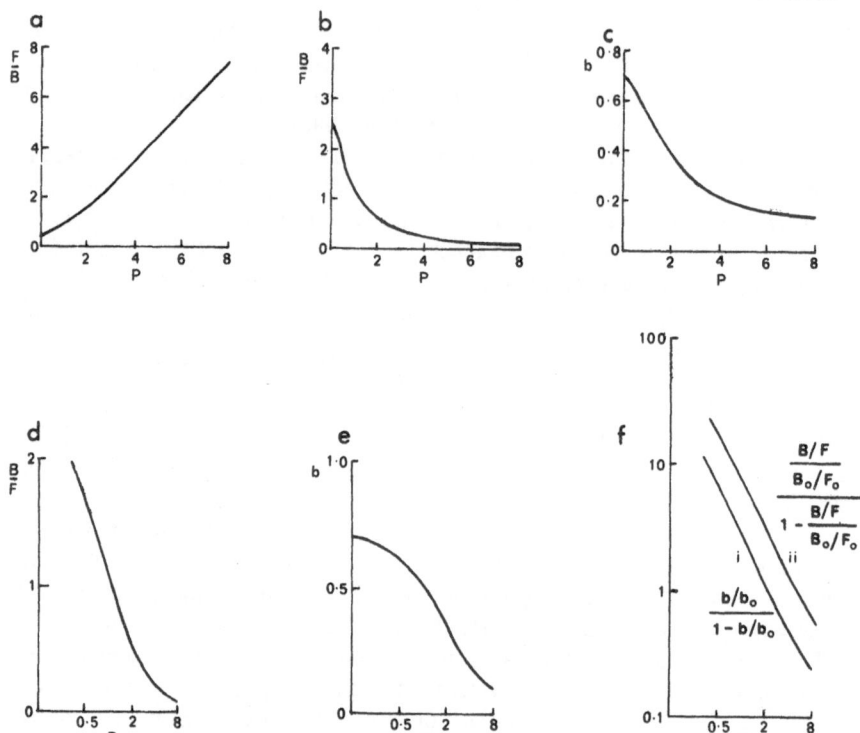

Fig. 9.9. Typical methods of plotting saturation assay response curves. The same basic
data are represented in each of the plots shown:
(a) F/B against ligand concentration
(b) B/F against ligand concentration
(c) fraction of total activity, bound (b) against ligand concentration
(d) B/F against log ligand concentration
(e) fraction of total activity bound (b) against log ligand concentration
(f) curve i: $\log \dfrac{b/b_0}{1 - b/b_0}$ (i.e. logit b/b_0) against log ligand concentration

curve ii: $\log \dfrac{\dfrac{B/F}{B_0/F_0}}{1 - \dfrac{B/F}{B_0 F_0}}$ (i.e. logit $\dfrac{B/F}{B_0/F_0}$) against log ligand concentration.

Note that the region of steepest slope of the response curve depends upon
the co-ordinate systems used (cf. discussion on 'sensitivity' in text).

Table 9.1 indicates some typical areas of application which currently
include more than 200 compounds of biological importance though this
number expands almost daily. Meanwhile the nomenclature which has
grown in the field is diverse, and not universally agreed. A number of terms
such as 'saturation assay', 'radioligand assay', 'competitive radioassay', have
been suggested which are not related to a particular class of binding reagent,
though none is entirely satisfactory or has gained general acceptance. In

particular, objection has been raised (especially by biochemists) to the use of the term 'competitive' in this context since in many assays (though not in all) the labelled and unlabelled ligands are chemically identical, and the reactions involved are not competitive in the accepted chemical sense. The expression 'saturation analysis' has likewise provoked dissent, on the grounds that the binding reagent employed is not necessarily saturated [61]. Meanwhile the recent development of methods relying on identical analytical principles, though not on the use of radioactive tracers, underlines the limitations of certain of the suggested expressions.

9.5.3 Assay Design

The minute concentrations at which many active compounds are encountered in biological systems frequently present a formidable analytical challenge. Thus, despite the inherent delicacy of saturation assay techniques, it is often vital to exploit fully the sensitivity of which the methods are potentially capable. For this reason, particular attention has been paid to optimal assay design, both in the development of the underlying theory of the method, and in the largely empirical studies involved in setting up assays of acceptable sensitivity in the analytical laboratory.

Considerable controversy [62, 63] and wasted experimental effort has nevertheless arisen in this area of endeavour in the past, largely in consequence of a fundamental lack of understanding of the basic concepts of precision and sensitivity displayed by many investigators in the field. The notable lack of agreement regarding the significance of these terms has led directly to fundamental conflict regarding the theoretical principles governing the setting up of optimal assay systems [64, 65]. For a proper understanding of the diversity of experimental approaches to assay design which are currently adopted in this area, it is therefore necessary to examine in detail the concepts of assay sensitivity and precision, and the differing interpretations that are commonly placed upon them.

'Sensitivity' is frequently regarded (and has been authoritatively defined [66]) as the 'slope of the assay dose response curve'; hence an assay is usually considered to be more 'sensitive' if the slope of the response curve is steeper. Guided by this concept, the majority of investigators, in the development of assay techniques, strive to increase response curve slope in order to enhance the sensitivity (and precision) of their methods.

Paradoxically the fallaciousness of this simplistic approach has been made especially apparent by saturation assay methods, since the variety of response 'metameters', each of equal validity, that have been employed in plotting assay data is unusually extensive, and has revealed the anomalies to which this definition of sensitivity can give rise. Thus, assuming that both the bound and free activities in each incubation tube are counted, it is clearly equally legitimate, when plotting dose response curves, to express the distribution of activity either as the free to bound (F/B) or bound to free (B/F) ratio (see Fig. 9.9). However, the effect of a particular experimental stratagem, such as a reduction in the amount of binding reagent Q in each incubation tube, will often be contradictory in each of these two co-ordinate systems, increasing the slope of the response curve in the first, whilst decreasing the slope in the second. Moreover, the region of steepest slope of the B/F curve is observed at low ligand concentration, which corresponds to the region of *least* slope in the reciprocal F/B plot.

Nevertheless many investigators continue to describe regions of high response curve slope as being of 'high sensitivity', implicitly disregarding the dependance of their conclusion solely on the characteristics of the co-ordinate system which they are using. In particular, Berson and Yalow, in a number of theoretical expositions relating to radioimmunoassay design, (e.g., [67, 68]) have concluded that, because the slope of the B/F response curve increases as the ligand concentration is reduced, a minimal amount of tracer ligand must be used to maximise assay 'sensitivity'. They have also shown that the slope of the curve relating the fraction of tracer bound to inactive ligand concentration is maximal when one-third of the tracer is bound [68]. They have therefore concluded that a concentration of reagent capable of binding 33% of the minimal usable amount of tracer should be employed in the assay system [68] *. Under these conditions it is readily demonstrable that the slope of the response curve is proportional to the equilibrium constant (K) of the ligand/binding reagent reaction, (correctly) implying that high assay sensitivity is dependent upon a high energy of the fundamental reaction.

Such precepts possess a certain intuitive appeal, and notwithstanding the controversial nature of the arguments on which they are based, give rise to highly sensitive assays. They have therefore largely governed the approach to the setting up of sensitive techniques adopted by the majority of workers in the field. However, the factors controlling the ability of an assay system to quantitate very small amounts of ligand are considerably more complex than is apparent in these oversimplified and sometimes misleading concepts.†

The key to an understanding of this situation lies in a recognition that the term 'sensitivity' can, in analytical work of *any* nature, only meaningfully be used in a sense synonymous with the assay 'detection limit' or 'lower limit of detection', despite explicit recommendations to the contrary [70]. This view of sensitivity implies that an assay is rendered more 'sensitive' if, and only if, the assay detection limit is reduced.

Fig. 9.10 illustrates in general terms the meaning of this concept and its essential dependence upon both the slope of the response curve, and the error‡ (ΔR_o) incurred in the measurement of the response at zero dose.

---

*In other publications, Berson and Yalow, relying on similar arguments concluded that a concentration of binding reagent capable of binding 50% of the tracer should be used (69).

†In consequence assay systems which depart markedly in their overall design from that which is currently conventional can be demonstrated to yield greater assay sensitivity and precision (in the sense in which these terms are subsequently defined).

‡There is no general agreement, in the relevant scientific literature, on the definition of 'error' in this context. Usually the standard deviation of replicates, or some multiple of this parameter, is used, although some authors take a more empirical view [71]. Accordingly there is no universally accepted quantitative definition of the 'minimum detectable amount', or 'lower detection limit' [72].

More generally, Fig. 9.10 demonstrates that the lower limit of detection (ΔH_o) represents a special case of the more general concept of precision. Thus ΔH, representing the precision of measurement of a dose (or ligand concentration) H, is similarly dependant both on the slope and the error in the response (ΔR_H) at the corresponding point on the response curve. In short, the 'sensitivity' of an assay system can be defined as the precision of measurement of zero dose. Neither the detection limit nor, more generally, the precision of measurement of any ligand concentration H, are dependant on the co-ordinate system used, and translation of assay data to any other co-ordinate frame will not affect the magnitude of either of these parameters.

Following from these definitions, it is evident that only in relatively unusual circumstances is it a principal objective of assay design to achieve maximal assay sensitivity; the aim of the analyst is more correctly represented as the measurement of a finite (albeit usually small) concentration, or amount, of the substance under test with maximum precision. This in turn implies that the optimal assay system is that which, by use of optimal reagent concentrations, and optimal deployment of available assay time, minimises the magnitude of ΔH_T, the error in the measurement of H_T (where H_T denotes the anticipated, or target, concentration of the test substance in the unknown sample).

The attainment of this objective, in saturation assays relying on radioactive markers, constitutes a difficult theoretical and practical problem. The magnitude of ΔH_T is dependant upon three principal parameters – the slope of the response curve at the point corresponding to the dose H_T, and two independent sources of error – the counting errors arising in the radio-

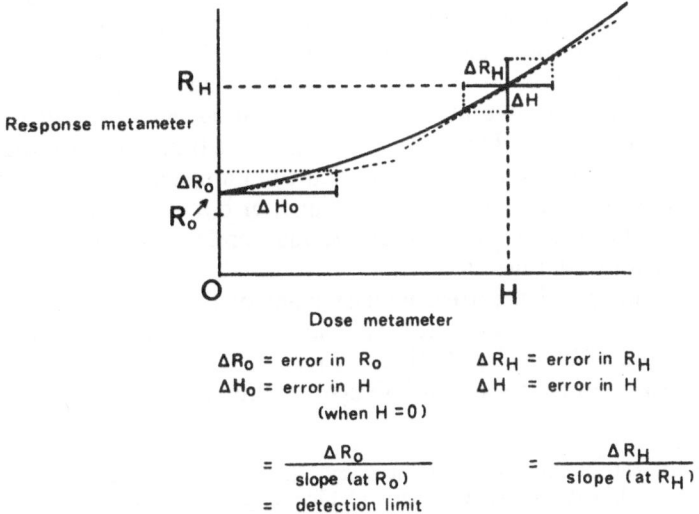

ΔR_o = error in R_o ΔR_H = error in R_H
ΔH_o = error in H ΔH = error in H

(when H = 0)

$$= \frac{\Delta R_o}{\text{slope (at } R_o)} \qquad = \frac{\Delta R_H}{\text{slope (at } R_H)}$$

= detection limit

Fig. 9.10. Generalised concepts of sensitivity and precision. Note that the detection limit (ΔH_o) is defined by the precision of measurement of zero 'dose'.

assay, and the experimental errors arising from pipetting and other manipulations — which together make up the error ΔR_H in the response measurement. Each of these parameters is influenced by a range of factors, some of which are within the control of the analyst (e.g. reagent concentrations, times of incubation, counting, etc.) whilst others are characteristic of his personal expertise, the experimental techniques and equipment used, etc. and are therefore largely outside his control. In this highly complex situation, it has proved impossible to lay down simple theoretical guidelines specifying optimal assay design. Nevertheless, by setting up mathematical models of the assay system, and by the use of computer optimisation techniques, it is now possible to gain more meaningful insights into optimal usage of assay reagents [73].

One of the most important conclusions to emerge from this approach is the overriding influence of experimental errors (i.e. those errors arising from pipetting, the separation of free and bound moieties, adsorption of reagents on to glassware, etc.) on optimal assay design. This may be illustrated by consideration of a hypothetical 'disequilibrium'* assay system in which the bound moieties only are radioassayed, and in which the *relative* error in the estimate of the bound fraction is observed to be constant. Under these circumstances, it is demonstrable that highest assay sensitivity will be achieved by delaying addition of labelled ligand until shortly before the termination of the reaction. The concentration of radioligand used should be large (relative to the concentration of binding reagent and unlabelled ligand in the system), whilst the fraction of label bound will fall to a low level (typically less than 5%).

Such a conclusion differs markedly from the traditional concept of an optimal assay system, in which the fraction of labelled ligand bound (in the absence of unlabelled ligand) is adjusted to around 50%. The optimal assay system existing under the conditions postulated above also differs considerably from that which is defined by computer analyses on the assumption that the *absolute* error in the estimate of the bound fraction is constant. The optimal reagent concentrations and incubation times predicted in the latter case are shown in Figs. 9.11 and 9.12; the setting up of an assay for maximum sensitivity (target concentration 0) in a manner defined in these figures would yield a higher fraction bound, or b_0, value than in the first example, although it would nevertheless fall considerably below the 50% value conventionally called for.

The advantages of a correct understanding of the principles underlying optimal assay design are of more than academic importance. As exemplified by the studies of Wide *et al* [74], assay techniques which correspond more closely than normal to optimal design conditions may yield results of

---

*The traditional method of setting up assays involves the simultaneous mixing of unlabelled and labelled ligands together with binding reagent at the commencement of the reaction period, which is terminated when equilibrium is reached. An apparent increase in sensitivity is demonstrated to occur [75], when addition of the labelled ligand is delayed. Such assays are frequently termed 'disequilibrium' techniques.

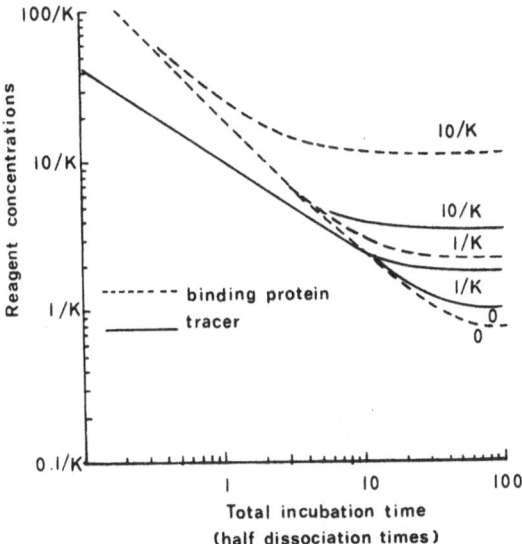

Fig. 9.11. Optimal concentrations of reagents as a function of total incubation time for target ligand concentrations of zero, 1/K and 10/K in a disequilibrium assay system in which the experimental error in the response (i.e. σR) is assumed constant at all points along the response curve. In order to effect a measure of generalisation, all concentrations are expressed in terms of 1/K, and incubation times of the half time of dissociation of the ligand complex. Particular assumptions have been made in the computation regarding the magnitude of σR, the specific activity of the tracer and other relevant assay parameters.

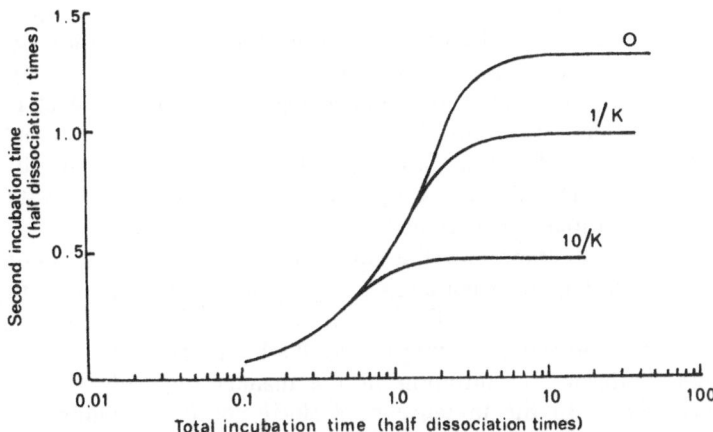

Fig. 9.12 Optimal 'second' incubation time (after addition of tracer ligand) as a function of total incubation time and target ligand concentration (see Fig. 9.11). The 'first' incubation time, preceeding the addition of tracer ligand is given by the difference between the total and second incubation times.

acceptable precision in a much shorter overall assay time, and thereby contribute greatly to their value, particularly in clinical situations where the rapid availability of an assay result may be crucial to its usefulness. In addition, the improvement in assay sensitivity consequent upon an optimal design is occasionally vital in enabling lower concentrations of the test compound to be assayed with acceptable precision. However, perhaps the most important benefit to emerge from the use of mathematical modelling and computer techniques is the guidance they yield on experimental strategy in setting up new assay methods, and the light they shed upon the nature of the parameters which essentially define the accuracy of assay results. These parameters and the methods whereby they are evaluated are discussed below.

Equilibrium Constant (K) and Antibody Titre
In those assays in which reagents are simultaneously mixed and are subsequently allowed to attain equilibrium, the equilibrium constant governing the reaction is a fundamental determinant of assay sensitivity [64, 76] *. Meanwhile, the 'titre' of an antiserum (a term which largely reflects the concentration of specific antibody that it contains) is relatively unimportant in this context, albeit it governs the number of assays that a given amount of antiserum will yield. Antibody titre is usually assessed by incubating together the appropriate labelled antigen with varying concentrations of antiserum. The activity bound to antibody is plotted as a function of antiserum dilution, typically yielding a sigmoid antigen binding curve (see Fig. 9.13). To define roughly the working titre, a dilution of antiserum is customarily selected yielding approximately 50% binding of the labelled antigen; it should be noted, however, that this dilution of antiserum is dependant upon the concentration of antigen used.

The energy of the antibody binding sites (as reflected by the equilibrium constant) is not revealed by the shape or position of the dilution curve *per se*; however, as indicated in Fig. 9.13 dilution curves yielded by progressive dilutions of antigen (or other ligand in analogous systems) approach coincidence at an antigen concentration which is dependant upon the antibody equilibrium constant. Some indication of the energy of the reaction can therefore be gained by setting up antibody dilutions with two different (low) concentrations of the labelled antigen.

A more exact estimate of the equilibrium constant can be derived by setting up reaction tubes containing varying quantities of antigen together with a fixed quantity of antibody as previously defined by the antiserum dilution curve and subsequently plotting the bound/free ratio as a function of bound antigen concentration in the manner originally described by Scatchard [77]. In the presence of a single species of binding site the resulting plot is a straight line with a (negative) slope equal to the equilibrium constant. However, because of heterogeneity of binding sites, it is not

---

*The maximum sensitivity attainable in a saturation assay is of the order of ϵ/K where ϵ = the relative experimental error in the free/bound ratio (76).

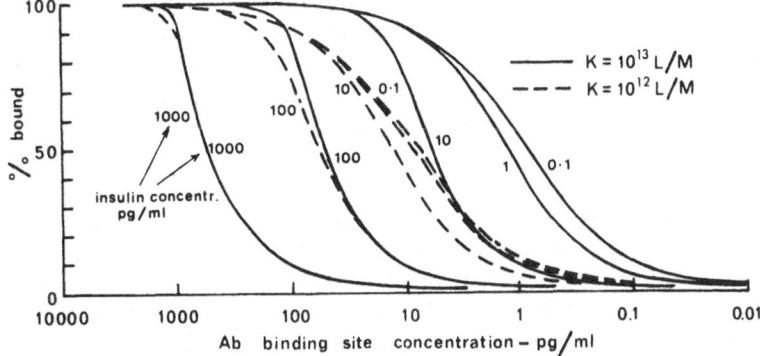

Fig. 9.13. Theoretical antibody dilutions curves for two antisera (insulin) each
assumed to contain the same concentration of antibody, but with equilibrium constants differing by a factor of 10. Note that when incubated with
a high concentration of labelled insulin, the two antisera yield essentially
superimposable dilution curves. With decreasing concentrations of labelled
antigen, the respective dilution curves diverge, such that, at the lowest
concentration of antigen, the 'titres' of the two antisera differ by a factor
of 10. In practice, the antibody of highest equilibrium constant is revealed
by the greatest disparity between the curves yielded by the two lowest
concentrations of antigen used.

uncommon to observe curvature of the plot [78] although other causes
(such as non-identity of labelled and unlabelled antigens) can give rise to
similar effects. In all such circumstances, the 'apparent' or 'effective' equilibrium constant changes with ligand concentration [76].

Knowledge of the equilibrium constant of the binding reaction is
essential for computer optimization of saturation assays, and for simplified
nomographic methods based on similar principles [65, 76]. For investigators who rely on the more conventional semi-intuitive approach to assay
design, a formal determination of the equilibrium constant is not necessary,
although an estimate of its magnitude gives a rapid indication of the
potential sensitivity attainable in the system.

Association and Dissociation Rate Constants (k_a and k_d)
An estimate of the constants governing the kinetic characteristics of binding
reactions are only obligatory for investigators using computer design
methods. A knowledge of the dissociation rate is, nevertheless, useful in a
consideration of the practical techniques to be used in the separation of free
and bound moieties following the binding reaction. A rapidly dissociating
complex renders more hazardous the use of — for example — adsorption
techniques for the isolation of the free fraction, since small differences in
times of exposure to adsorbent may lead to significant variation in dissociation of complex between successively processed incubation tubes, and hence
to errors in assay results.

The usual procedure for the measurement of the dissociation rate
constant relies on a prior incubation of labelled ligand with binding reagent;

a large excess of unlabelled ligand is then introduced into the mixture, and the activity remaining in the bound fraction is assessed at different time intervals [78]. Many protein/antibody reactions are characterised by exceedingly slow dissociation and little decrease in bound activity may be seen over several days. Other interactions, such as between many steroids and normally occuring serum binding proteins, may, in contrast, show rapid dissociation effects.

The association rate constant is less readily estimated from observations of the forward reaction rate since the calculation demands a knowledge of the concentrations of ligand and of binding sites in the system, neither of which may be accurately known. However, since $k_a/k_d = K$, an estimate of k_a may be made based on measurements of K and k_d as described above.

Specific Activity of Labelled Ligand

In those saturation assay systems in which an identical amount of labelled ligand is added to all incubation tubes (i.e. to standards and unknowns) it is not essential that the specific activity of the labelled ligand should be accurately known. Nevertheless a rough estimate is always desirable insofar as the analyst must ensure that the weight of labelled ligand introduced into each incubation tube is acceptable in relation to the 'target' concentration of unlabelled ligand. For the more exact assay design techniques, the specific activity is a required parameter, and a relatively precise estimate must be made.

The specific activity of a radioiodine-labelled protein can be estimated by careful measurements of the recoveries of radioiodine and protein through the iodination and subsequent purification procedures, assuming that the starting amounts of both are known [30]. Alternatively the specific activity of the product, and indeed of any labelled ligand, can be estimated by treating dilutions of labelled ligand as unknowns in the assay procedure [79]. Neither of these approaches is completely satisfactory: the first involves the assumption, inter alia, that all proteins, including impurities, in an initial preparation are labelled to the same specific activity, whilst the second presumes identity of reaction energy of labelled and unlabelled ligands. The latter assumption is likely to be valid in the case of labelled reagents which are chemically identical to the corresponding unlabelled material, but is more open to question when chemical modification is entailed, such as is implicit in the iodination of proteins.

The quoted specific activities of compounds supplied by radiopharmaceutical manufacturers must normally be regarded as suspect, and cannot be relied on in this context.

It should be emphasised that in those techniques involving an extraction step, and in which the radioactive indicator is used to monitor recovery (as indicated in Fig. 9.8), it is essential that the weight of the labelled material initially added should be known [cf. Section 9.2.1, and equation (9.1)].

Although it is evident that high specific activity radioligands are advantageous insofar as their use results in a reduction of the weight of material that must be added to the unlabelled ligand in the assay system to maintain acceptable count rates, the emphasis on high specific activity has often been misplaced in the past. Increase in radioligand specific activity becomes progressively less rewarding in terms of its effect on assay sensitivity [80], although such increase is always accompanied by a concomitant decrease in the counting times necessary to achieve acceptable precision in the radio-assay of free and/or bound fractions. However, attenuation of counting times must be weighed against the technical difficulties and economic penalties which often accompany the preparation and use of high specific activity tracers, and in many circumstances it is more fruitful to extend counting times and rely on lower specific activity material.

Error Function

The more rigorous assay design techniques discussed earlier in this section have drawn particular attention to the importance of the relationship between the experimental error in the response metameter and the magnitude of the response as a fundamental determinant of optimal assay design. For the application of computer methods this relationship must be explicitly defined.

Rodbard *et al.* [81] have attempted, by using modelling techniques, to predict theoretically the experimental error function from a knowledge of the errors arising in each component operation in the assay system. However, it is doubtful whether any practical advantages stem from this type of analysis, and it is customary practice, in the author's laboratory, to estimate the magnitude and variation of the error along the response curve experimentally [82]. This may be readily achieved by setting up a relatively large number of replicates for each of a number of points along the curve.

At the present time investigators using less sophisticated assay design techniques have tended to disregard the significance of experimental errors, with a detrimental effect on their experimental strategy when establishing new assay methods.

Counting Time and Effective Incubation Volume

These parameters influence assay design insofar as an increase in either the counting time allocated to each incubation tube, or the volume of incubate available for a radio-assay, implicitly reduces the concentration of radio-ligand required in the assay system to yield acceptable counting statistics [83]. Nevertheless, as is the case with increase of specific activity, increase in either counting time or effective incubation volume becomes progressively less rewarding in terms of increase in assay sensitivity and precision. Although many investigators adhere to a standard practice of counting all samples to not less than 10,000 counts, it is readily demonstrable that this procedure is seldom justified, and that other experimental errors in the assay system often play the dominant role in restricting the precision of the final ligand measurement.

9.5.4 Assay Specificity

The most obvious determinant of the specificity of a saturation assay is the intrinsic specificity of the protein binding sites implicated in the reaction. Clearly if a second ligand, R, present in test samples cross-reacts with the specific reagent Q, and hence 'competes' with labelled P in the reaction, then it will simulate P in the assay and lead to false results. The extent to which this occurs is expressed in the concept of 'relative potency', which represents the relative concentrations of the ligands P and R which change the response variable to an identical extent. However, the relative potency of two compounds in a saturation assay system is not constant but is dependant on the distribution of radioactive ligand in the system [47, 65], i.e.

$$\text{relative potency} = \left(f \times \frac{K_P}{K_R} \right) + b$$

where f =   fraction of total radioactivity free
      b =   fraction of total radioactivity bound
      K_P, K_R =the equilibrium constants characterising the reactions of P
                 and R respectively [84]

(It is assumed that radioactive P is used as the indicator in the system). This equation implies that cross-reacting compounds, provided they display different reaction energies, will react with a changing potency ratio throughout the concentration range; this indicates in turn that response curves yielded by different ligands (plotted in terms of log dose) will not be exactly parallel. Thus, a commonly applied criterion of lack of identity between the standard preparation of the ligand P, and the reactive compound in a test sample, is lack of parallelism of the response, or dilution, curves yielded by each. Nevertheless there are circumstances in which such curves will not significantly diverge [65]. This is readily illustrated by numerical example. Let us assume that K_R is 100-fold less than K_P. Assuming that 50% of the labelled ligand is bound in the absence of unlabelled material, the relative potency of P and R at this point on the response curve will be 50.5. However, in the presence of large amounts of the unlabelled ligand, such that the fraction of labelled ligand bound approaches zero, the relative potency of the two compounds will approximate to 100. Thus the two response curves will diverge only to an extent represented by a factor of 2 on the log dose ordinate. Such lack of parallelism is, in practice, often overlooked by analysts, particularly when the dilution curve yielded by the unknown sample unavoidably extends over a restricted range of ligand concentrations, and the divergence is correspondingly less apparent. However, a much greater degree of non-parallelism will be manifested in situations in which the competitor compound R displays the higher equilibrium constant.

These simple theoretical conclusions are valid only when a single species of binding site is implicated in the binding reaction. However, antibody/ antigen reactions in particular are generally of greater complexity, especially those in which the antigenic compound is a polypeptide of large molecular size. Antiserum raised against such material usually comprises a number of species of antibody each directed against a different antigenic determinant in the polypeptide molecule. A typical radioimmunoassay can therefore be viewed as the sum of a number of component systems, each of which may be individually susceptible to cross-reaction with polypeptide fragments, polymerised or precursor polypeptides, or even biologically unrelated polypeptides, which include, within their structure, the molecular configuration against which a particular species of antibody in the immune serum is directed. The heterogeneity of typical antisera clearly increases their vulnerability to cross-reacting compounds, nevertheless it is relatively improbable that a cross-reacting antigen will compete in each of the individual antibody/labelled antigen subsystems, and marked divergence of dilution curves yielded by dissimilar antigens is correspondingly more likely. Nonetheless, parallelism of dilution curves yielded by standards and unknowns is a notoriously unreliable, though necessary, test of their immunological or 'ligandic' identity [64], and full reliance can only be legitimately placed on assay results if they are confirmed in a variety of immunoassay systems relying on different antisera.

Non-specific effects can invalidate assay results in a number of other ways. The biological sample under test may contain enzymes, binding proteins, or endogenous antibodies which, by reacting with the ligand P, and thereby effectively reducing its concentration, vitiate the measurement. Such effects can be overcome by the inclusion of enzyme inhibitors such as aprotinin (Trasylol) [85], blockers of protein binding such as ANS (8-anilino naphthalene sulphonic acid [86], or by prior extraction of the ligand from its biological milieu.

Substances present in the test sample may otherwise inhibit or potentiate the binding reaction between ligand and binding reagent, either by alteration of the equilibrium constant or by blocking or exposing binding sites. The presence of salts, urea and high concentrations of protein are amongst the influences known to affect binding protein and antibody reactions, rendering the assay of biological fluids, such as urine, particularly vulnerable to non-specific disturbance [87].

A further important source of non-specificity arises from differential effects, in standard and unknown incubation mixtures on the separation system used to isolate free and bound fractions. Many such methods rely on adsorption of free ligand by solid adsorbents such as activated charcoal, many of which are dependant on the composition of the solutions in which they are used. Pre-coating of charcoal with Dextran [88] is often employed to reduce such effects; however, this stratagem is only of limited value [89, 90] and the only reasonably reliable method of avoiding errors from this source is to ensure that, as far as is possible, standard and unknown incubation mixtures are as similar as possible, particularly in the concentrations and spectrum of proteins they contain. To this end it is now increasingly customary to include hormone-free plasma or serum as a constituent of standard incubation mixtures when assaying unknown serum or plasma samples. Likewise in studies on the release of cyclic AMP from cells subjected to a variety of stimuli, it has proved essential that standard tubes should contain all relevant reagents present in the test samples [91].

9.5.5 Assay Standards

Assays in which the compound of interest is of relatively simple and well-defined structure, such as the thyroid or steroid hormones, or a nucleotide such as cyclic AMP, present relatively few problems regarding the provision of assay standards of acceptable purity. Considerable practical and conceptual problems arise, however, when the substance to be assayed cannot be completely defined by chemical or physical means, as exemplified by many of the protein hormones of large molecular size. In such situations, there are many factors which may result in differences between standard and the substance assayed in the test sample. Some of these factors have been touched on briefly in the previous section; these are related particularly to the possible presence, in the test sample, of a wide variety of cross-reacting compounds (such as metabolic degradation products) which can simulate the test compound in the assay, albeit they may do so with differing potency. In addition, difficulties may arise because the standard itself may comprise a relatively impure mixture of compounds, some of which are reactive within the assay system.

Impurity in the standard preparation may arise in many ways. For example, the glycoprotein hormones [such as thyrotrophin (TSH), follicle-stimulating hormone (FSH), luteinising hormone (LH)], which are each prepared by extraction from pituitary glands, are especially susceptible to contamination with other hormones, and their degradation products. They are also liable to desialylation and to other chemical or structural alterations in consequence of exposure to low pH, to freeze drying, etc. and it is therefore particularly difficult to isolate a homogeneous and uncontaminated preparation of any of these substances. Moreover, it is currently believed that each of these hormones contains, within its structure, a major common subunit (the α-subunit [92]), implying that antibodies directed against this moiety will react with each hormone if present in the assay system.

Further complexity is evident when it is recalled that the preparations used as immunogens for the raising of antisera will likewise represent heterogeneous mixtures of the different hormones. The antisera thus raised will contain a miscellany of antibodies present in concentrations not necessarily directly related to the concentrations of contaminating hormones in the immunogen. Moreover, labelled hormone used in the assay is equally unlikely to comprise a single molecular species.

The consequence of the combination of these factors is that it is virtually impossible to calibrate any standard glycoprotein hormone preparation against another, and to obtain a relative potency estimate for the two which is likely to be reproducible in another assay system using different reagents [93]. Indeed, even using identical reagents, two different standard preparations are likely to yield different relative potencies depending on, for example, the concentration of antibody employed (and hence the fractional binding of labelled hormone observed, as discussed in the previous section) or on the particular protocol of first or second incubation times if a disequilibrium assay system is employed.

Fortunately the total confusion which might be expected to ensue from a situation of this complexity has, in the past, been partially mitigated by the comparative scarcity of even partially purified glycoprotein hormones and other similar substances. Hence investigators throughout the world have largely and necessarily relied on identical, or a limited range of, reference hormone preparations and antibodies, distributed by such agencies as the World Health Organization, the National Pituitary Agency in the United States, and the Medical Research Council in the United Kingdom. Nevertheless this partial constraint has not, for the reasons indicated above, ensured comparability of assay results emanating from different laboratories.

More detailed discussion of the vexing problems which surround the provision of suitable standard preparations, particularly for the protein hormones, are to be found in reviews by Bangham and Cotes [94, 96], Cotes [95], and Ross [97]. The particular difficulties associated with the standardisation of the glycoprotein hormones have been discussed by Bangham and Borth [98], Bangham et al. [99] and Jacobs and Lawton [93].

9.5.6 Radioactive Labelling Techniques

It is not appropriate, in this section, to attempt a comprehensive review of the labelling procedures employed for the synthesis of the wide range of radioactive compounds potentially exploitable in saturation assay methods. Many such compounds, particularly those labelled with ^{14}C and ^{3}H, are commercially available, whilst their synthesis lies beyond the resources of the ordinary user. Moreover, in practice, assays of protein hormones, and increasingly of non-protein compounds, have relied on the use of radio-iodinated derivatives differing chemically from the test substance, but nevertheless retaining the capacity to compete with it in a reaction with the specific binding reagent (though not invariably with an identical potency).

Although ^{131}I was the isotope originally employed for these purposes, it displays a number of disadvantages compared to a second iodine isotope, ^{125}I, by which it has almost totally been supplanted in the past few years. Although ^{131}I possesses the shorter half-life (8 days as compared with 60 days) and would therefore be expected to yield iodinated products with a more than 7-fold greater specificity activity, ^{131}I solutions supplied by radio-pharmaceutical manufacturers are invariably contaminated with relatively large amounts of stable iodine. The low isotopic abundance of these preparations largely nullifies the theoretical superiority of ^{131}I over ^{125}I. Moreover ^{125}I is easily counted in scintillation counters with an efficiency roughly twice that of ^{131}I. In consequence of these factors, the effective specific activities of ^{125}I preparations are considerably greater than those of ^{131}I, even at the time of their initial preparation, and this superiority increases with time due to the higher decay rate of the latter. Other advantages characterising ^{125}I are the lower energy of its γ-emission, leading to minimisation of the radiation hazards that it presents, and (as a result of its lack of β-emission) a reduction of radiation breakdown of compounds in which it is incorporated. The 'softness' of its emission also permits the use of γ-scintillation detectors of smaller crystal volume, which are both cheaper and less susceptible to radiation background, and hence potentially facilitates the construction of sample counters of considerably reduced bulk and cost. Unfortunately only relatively few manufacturers are currently marketing counting equipment specifically designed for the radioassay of ^{125}I.

Radioiodine can readily be incorporated into proteins and polypeptides by iodination of the tyrosine residues in the molecule when these are present. For reaction to take place, the isotope (which is usually supplied as sodium iodide) must first be oxidised to elemental iodine, following which iodination of protein is extremely rapid. The most popular oxidation method at the present time relies on the bleaching agent, chloramine-T, which is mixed together with protein and the radioactive solution, usually in a phosphate buffer of pH 7.4, under which conditions the reaction proceeds most rapidly [30, 100]. Almost immediately after bringing the reagents into contact sodium metabisulphite is added to the mixture, quenching the reaction, and reducing unused iodine back to iodide. The reaction products are then usually immediately applied to a small Sephadex G-50 column, previously equilibrated with a neutral protein to prevent adsorption of the labelled material and the iodinated protein and residual iodide eluants separated. However, for some of the less homogeneous or stable proteins, larger columns are often employed in order to isolate a more highly purified protein moiety from those fractions (containing degraded or aggregated protein, or other protein impurities) which, on subsequent incubation with an excess of specific antibody, are shown to be of attenuated immuno reactivity. Other purification methods rely on adsorption to, and subsequent elution from, small columns of cellulose, silicates and other similar materials [101].

Despite the apparent ease of the chloramine-T iodination method, it is not entirely reliable in practice for reasons stemming, at least in part, from the composition of the radioiodine preparations supplied by radiopharmaceutical manufacturers. Some suspicion also hangs over the use of a relatively powerful oxidant, such as chloramine-T in circumstances in which it is brought into direct contact with proteins of uncertain stability. Moreover, the mere handling of certain proteins in pure solution in the absence of other protective 'carrier' proteins can lead to their breakdown, and, may also contribute to the variable success that normally attends the use of this procedure. Thus, although the method is very widely used, other techniques are currently under evaluation which attempt to avoid some of its more suspect features. One such technique relies on chloramine-T to oxidise Cl⁻ (present as NaCl) in a small gas-tight, double compartmented, vessel containing, in one of the two compartments, the mixture of protein and radioiodine. Chlorine, diffusing from the oxidation mixture contained in the second, liberates nascent iodine and results in relatively efficient protein iodination [102]. A similar technique, devised and used by the author prior to the advent of the original chloramine-T method, relied on oxidation of radioiodide in the outer compartment of a Conway micro-diffusion cell using sulphuric acid and potassium dichromate as oxidants. Diffusion of iodine into the central protein-containing compartment resulted in successful iodination [103]. Although slightly less convenient than the chloramine-T method which almost immediately displaced it, this technique presents the unique advantage of isolating the protein from all other reactants, including the noxious contaminants suspected of being present in commercial radioiodide preparations. More recently it has been observed that similar results can be achieved using chloramine-T by separating the oxidation reaction from the protein solution, relying on gaseous diffusion of iodine from the oxidation mixture.

Other approaches have relied on electrolytic oxidation of iodine [104, 105], on the use of iodine monochloride [106] and more recently on enzymatic oxidation [107, 108]. In the latter approach, lactoperoxidase catalyses oxidation of iodide and its subsequent incorporation into protein. Avoiding the use both of possibly destructive oxidants and of reducing agents (since the enzyme is relatively short-lived), this method has been claimed by Thorell and Johansson [109] to yield iodinated proteins showing a lesser degree of 'damage' than is encountered with other techniques.

A recent method described by Bolton and Hunter [110] relies on prior iodination, using the conventional chloramine-T procedure, of the active acylated agent 3-(p-hydroxyphenyl) propionic acid N-hydroxysuccinimide ester. If protein is subsequently added to a dried residue of the product, a condensation reaction occurs with any free amino group (N-terminal or ε-lycine), the iodinated acyl group being attached by a new peptide bond. Claimed by its originators to yield labelled proteins with a higher degree of immunoradioactivity than those directly labelled with chloramine-T, it offers the considerable advantage of being applicable to those proteins, such as ovalbumin, secretin and porcine parathyroid hormone, which contain no tyrosine residues, and which cannot therefore be directly iodinated.

A similar concept has been applied to the iodination of other compounds, such as the steroid hormones, which cannot (successfully) be directly iodine-labelled. Thus Midgley et al. [111] and others [112, 113] have labelled steroids by iodination of a tyrosine or histamine residue either pre-coupled to the steroid molecule, or (to avoid damage to the steroid hormone) conjugated following the iodination step. Tyrosine conjugated steroids appear to be difficult to prepare and to degrade relatively rapidly, but the histamine derivatives retain their immunoreactive integrity for several months [114].

In addition to the practical advantages, e.g. gamma counting, displayed by iodinated rather than ^{3}H- or ^{14}C-labelled ligands in this context, the fact that antisera are prepared against ligand protein conjugates (as described in the next section) and therefore tend to react more avidly with labelled conjugate rather than with the ligand alone, can, in certain circumstances, lead to an increase in assay sensitivity [65, 115]. Nevertheless if the avidity of the antibody towards the labelled conjugated material in comparison with the unconjugated compound is too great, then a reduction of assay sensitivity is the likely consequence [65]. It is largely for this reason that many investigators have experienced difficulties in measurements of compounds (such as certain drugs) for which tritium or ^{14}C labels have not been commercially available, and for which they have had necessarily to rely on iodine-labelled conjugates prepared essentially as described above.

It should be noted in this context, moreover, that although conjugated and unconjugated compounds frequently cross-react with relative potencies which permits the assay of the unconjugated material, the chemical modifications implicit in conjugation invariably obliterate the ability of the resulting compound to react with naturally occurring binding proteins. Thus although iodinated tyrosyl derivatives of cyclic AMP are successfully used in the radioimmunoassay of this compound [116], analogous assays relying on naturally occurring binding proteins can only be established using chemically unaltered [3]H- or ^{32}P-labelled cyclic AMP.

Incorporation of iodine *per se* can occasionally so reduce the reactivity of iodinated proteins that a marked reduction in assay sensitivity results. Different antisera are, not unexpectedly, affected to a varying extent in their reaction with proteins following iodination: some will detect the presence of a single iodine atom in the protein molecule whilst others are indifferent to this degree of chemical alteration, and will be affected only by a greater iodine content [117]. For this reason, it is conventional to attempt to restrict the average extent of iodination to one atom of iodine per molecule of protein or less.

A second important reason for avoiding a greater radioiodine content is the phenomenon of 'radiation catastrophe' described by Yalow and Berson [118]. This possibility arises when a ligand molecule is doubly labelled with radioactive iodine. Disintegration of the first active atom results in chemical dislocation of the host molecule, and the production of labelled fragments whose chemical and immunological behaviour in the system are grossly impaired or destroyed. The radioimmunoassay is conducted, in such circumstances, against an increasing 'background' of unreactive labelled material, and although some degree of radioactive impurity is usually tolerable in saturation assay systems, the loss of assay precision consequent on a long incubation period may prove unacceptable. For these reasons, it is occasionally desirable to isolate the mono-iodinated ligand [119], a procedure which is possible with compounds of small molecular size, though usually difficult with larger molecular weight proteins. Nevertheless, it must be borne in mind that even in a situation in which the entire protein molecule population is doubly labelled, then the generation of labelled damaged protein fragments arising as a direct result of 'decay catastrophe' will be little greater than 1% per day (when using carrier-free ^{125}I). Provided the labelled protein is freshly purified before use, the damage arising even in a quite prolonged incubation would be unlikely to vitiate the assay.

9.5.7 Preparation of Antibodies and Other Binding Reagents
Antibodies have provided the largest single class of specific binding reagents employed in saturation assay techniques. The raising of antisera against compounds, such as proteins of molecular weight greater than 5000 which

are naturally immunogenic, is of course a well established art. The appropriate immunogen is injected into a laboratory animal accompanied by an 'adjuvant' — for example, 'complete' Freund's adjuvant, comprising a suspension of mycobacteria in mineral oil and emulsifier — the function of which is to stimulate non-specifically the production of B-lymphocytes derived from bone marrow [120] which are the essential precursor of the plasma cells responsible for antibody formation [121].

For compounds of smaller molecular size, it is normally necessary initially to conjugate them to larger carrier molecules which are themselves intrinsically immunogenic, and whose function appears to be to simulate thymus-derived 'T'-lymphocytes [122] and ultimately B-lymphocyte formation. Such conjugation has usually been effected with common proteins such as bovine serum albumin (BSA) or thyroglobulin, or with synthetic polymers such as polylycine [123] albeit experience with the latter has proved it to be a poor immunogenic carrier for the steroid [124] and thyroid hormones.

The chemical methods involved in the preparation of steroid derivatives suitable for subsequent conjugation have been described by Erlanger *et al.* [125]; they typically involve the formation of an oxime at the position 3 of the steroid A ring, or hemisuccinate in the side chain at the 17 position in ring D. Conjugation is commonly effected using the mixed anhydride [126] or carbodi-imide reactions, although the latter results in inter- and intra-protein cross linkage and the first is therefore usually preferred. Other conjugation reactions, used particularly for the linking of drug molecules in proteins have been reviewed by Marks *et al.* [127]; they include the reaction of -OH groups with phosgene to form chlorocarbonates, which are subsequently coupled to protein using the Schotten-Baumann reaction.

In general, it has proved advantageous to attach haptenic* compounds at a point chosen to expose the distinctive features of the molecule, since it is the position furthest from the point of linkage to the carrier protein which is chiefly implicated in antibody recognition, and which largely determines antibody specificity. Thus, steroid molecules coupled through position 3 show particular specificity with respect to variations in the D ring. Conversely, coupling through, or adjacent to, the 17 position leads to antibodies which are unable to distinguish between steroids, such as testosterone and progesterone, which only differ at this position [128] but are sensitive to alterations in the A ring. For these reasons techniques for the conjugation of steroids at positions remote from biologically reactive groups in the A and D rings have been developed, typically through the 6-position as exemplified by the oestradiol-6 conjugate described by Exley *et al.* [129] and others [130, 131].

Although a hapten protein molecular ratio of not less than 15:1 has been reported as necessary to provoke a good immunological response [132],

---

*The term 'hapten' is used to describe compounds which are not intrinsically immunogenic, but against which, by the procedures described in this section, antibodies can be raised and with which they will react.

significantly lower ratios (8:1 or 4:1) have been successfully employed in the author's studies in the raising of antisera against the two thyroid hormones, thyroxine (T4) and triiodothyronine (T3). Indeed others have claimed [133] that immunogenicity of the conjugate may be *increased* when the number of haptenic determinants is reduced.

Most antisera for RIA purposes have been raised in small laboratory animals, such as guineapigs or rabbits, and although the volumes of antisera that can be derived from these species are small, the high working dilutions at which a good antiserum is customarily employed (i.e. usually between $1:10^4$ to $1:10^6$) renders even a restricted supply sufficient for most local purposes. Nevertheless with the increasing need for well-characterised antisera which can be used for routine purposes on an international scale, attention is currently being devoted to the raising of antisera in larger mammals such as sheep, horses and goats. There is no clearly established indication that the species of animal used is of overriding importance in the successful raising of antisera although some investigators express species preferances and provide supporting evidence for their beliefs, albeit this is usually anecdotal in character. Most workers in this field prefer to use random rather than in-bred animals in order to widen the gene pool under challenge, and thus increase the chances of raising a usable antiserum [134].

The search for optimal routes of immunization has likewise engendered beliefs more mythological than scientific in basis. Traditionally, relatively large amounts of immunogen (of the order of 1–5 mg) were injected in the oily adjuvant emulsion either subcutaneously or intramuscularly, often at regular fortnightly or monthly intervals. In the late 1960's the injection of the immunogen directly into a number of lymph nodes at open operation became fashionable, following the initial success of Boyd and Peart [135] in using this technique. A more prolonged study of its efficacy failed to establish unequivocally its superiority over established methods, and it has subsequently been abandoned by the majority of workers in the field. More recently, the multiple intradermal procedure [136, 137] has been extensively adopted, in which very small amounts of immunogen (of the order of 1–5 μg/site) are injected into some 40 or more sites spread over the body surface of the recipient animal. This procedure appears to yield near maximal response within a few weeks of initial immunisation, and subsequent boost injections have relatively little effect. Since the ultimate yields, and energy and specificity characteristics, of antisera raised by this technique are generally not inferior to those obtained using orthodox methods, its relative speed and economy of sometimes scarce immunogens have clearly commended themselves to the majority of investigators currently engaged in the raising of specific antisera for RIA purposes.

The screening and characterisation of antisera can frequently prove more laborious than the initial work involved in their production. As previously indicated in this chapter, the two essential criteria by which a particular antiserum is judged are its energy of reaction, and its specificity characteristics. The titre, or working dilution, of the antiserum is clearly only of logistic importance; nevertheless most investigators would normally feel uncomfortable using an antiserum so deficient in antibody content that its final working dilution fell below 1:1000, and would strive by booster injection to raise its titre.

The specificity of an antisera is normally assessed in two ways: first by exposing it to reaction with a number of compounds in which there exist a priori reasons for believing that cross-reaction with the test compound might occur; secondly, by applying customary tests of specificity within the assay system itself as previously discussed in Section 9.5.4. Neither of these checks alone, or in combination, is a complete guarantee of antibody specificity, nor of the validity of the final assay result.

The first approach clearly depends on a prior knowledge of the potentially cross-reacting compounds likely to be found in test samples, and the availability of these compounds in pure form. Occasionally, because of the absence of the latter, tests of cross-reaction may suggest poor antibody specificity when the converse may subsequently be shown to be the case. For example, Jacobs and Lawton [93] have observed apparent cross-reaction of a relatively highly purified preparation of thyrotrophin (TSH) and labelled luteinising hormone (LH) when incubated with an antiserum commonly used for LH assay. However no effect on the binding of labelled LH was observed using serum from a subject to whom thyrotrophin-releasing hormone (TRH) had previously been administered, and in which high levels of TSH were both anticipated and subsequently confirmed. Conversely, an antibody may survive specificity checks with a range of potential cross-reactants, but may nevertheless be subsequently demonstrated to respond to substances in test samples whose existance was previously unsuspected. Such cross-reaction may be suggested either by non-parallelism of standard and unknown dilution curves, or disagreement of assay results using different antisera as previously described.

It is largely because of the problems encountered in the raising and characterisation of specific antisera, particularly for polypeptides of large molecular size, that attention has been devoted in the past 2 or 3 years to the development of specific plasma membrane receptor assays [138]. Target cell receptors might be anticipated to display a greater, biologically-related, specificity than do many antisera, and to possess energy characteristics commensurate with the concentrations at which the matching protein is encountered in biological systems. 'Radio-receptor' assays, based on the saturation assay principle, have now been developed for corticotrophin (ACTH) [139], prolactin [140, 141], LH and HCG [142].

Receptor preparations for the assay of ACTH were originally prepared from homogenates of an ACTH-responsive mouse adrenal tumour. They were isolated by initial homogenisation of the tissue in a Dounce homogenizer, followed by differential centrifugation of the homogenate. The selected particle fraction was subsequently solubilised in an emulsion of phosphatidylethanolamine in Tris-HCl buffer, the suspension being repeatedly passed through a French press before centrifugation to remove residual insoluble particulate matter. Finally, the clear supernatant was dialysed and employed as binding reagent for ACTH measurement. More recently, preparations of adrenal receptors for assay purposes from freshly excised glands of normal laboratory animals [143] and from slaughterhouse bovine adrenals [144] have been described.

Analogous procedures for the isolation of gonadotrophin receptors in the particulate subcellular fraction of rat testes have likewise been described, and provide the basis for assays of LH and HCG.

Intracellular receptors involved in the biological action of the steroid hormones and the cyclic nucleotides have likewise been utilised for assay purposes. Their preparation is generally of a simple nature, involving homo-genisation of the target tissues, centrifugation, and removal of the super-natant [145], in which the receptor protein is located. Such techniques have led to the development of highly sensitive assay systems for oestrogens [146] and for cyclic AMP [147].

9.5.8 Separation Methods

The phrase 'separation methods' is here used in relation to the methods whereby free and bound ligand are isolated prior to the radioassay of either or both fractions, which measurement comprises the end-point of those saturation assay methods which rely on radioactive tracers*. This step almost invariably necessitates a physico-chemical processing of the sample, since ligand-receptor reagent complexes do not spontaneously precipitate from the reaction mixture under the concentration conditions prevailing in most assay systems.

A large number of separation techniques have been described and are commonly employed. The choice of the most suitable technique facing the investigator when setting up a new assay procedure is governed by several criteria:

(i) the separation method should ideally separate the two fractions totally and instantaneously, halting association and dissociation reac-tions in the incubation mixture, whilst not disturbing the chemical equilibrium established during the preceding incubation period,

(ii) the reproducibility of replicate separations should be high,

(iii) the method should be specific, i.e. its efficiency should not be affected by the composition of the incubation mixture,

(iv) the method should possess sufficient latitude to permit some degree of imprecision in the laboratory procedures on which it rests.

In addition to these primary requirements, the separation method should be cheap and simple to perform; it should also rely on readily available reagents, and lend itself to simple laboratory mechanisation.

Failure to meet criterion (i) will result in activity properly belonging to the bound fraction appearing in the free fraction, and vice versa, or both, a phenomenon which has been termed 'misclassification error' by Rodbard

---

*Non-radioactive methods are currently being developed for the measurement of the ligand distribution between free and bound moieties (see Refs. [148, 149 and 150]). Amongst the advantages claimed for such methods, the elimination of a separation step, obligatory in the radioactive methods, is perhaps the most important.

and Cooper [81]. Misclassification error may be important for two reasons. First, it will result in an apparent loss in the energy of the reaction as evidenced by a reduction in the equilibrium constant displayed by the receptor reagent. (Note: In a system in which a single receptor site is implicated and therefore characterised by a single equilibrium constant, misclassification of free and bound activities will frequently cause a curvature of the Scatchard plot, incorrectly suggesting the presence of heterogeneous binding sites characterised by a range of equilibrium constants). Reduction in the effective energy of the binding reaction will result in a loss in assay sensitivity, and, more generally, in a loss in precision of all ligand measurements although the effect will be of less consequence at higher target concentrations [151]. The second possible consequence of misclassification error arises from the variability in the magnitude of the effect arising, for example, from a difference in the composition of assay incubation mixtures or in the regime to which they are exposed. Such variability is frequently displayed in assay systems dependant on specific reagents characterised by a high dissociation rate. Addition of a solid adsorbent to the system to sequestrate free ligand will (unless careful precautions are observed) result in exposure of the ligand complex to adsorbent material for slightly differing time intervals, and variability in the fraction of ligand, dissociated from the receptor reagent, appearing in the free fraction. Effects of this kind clearly lead to an overall loss in assay precision. Of these two consequences of misclassification of free and bound moieties, the apparent loss in reaction energy is normally of minor importance, and only significantly affects the precision of assay results when the magnitude of the effect is large.

In general, assay precision (and hence sensitivity) are more profoundly affected by changes in the efficiency of the separation procedure, and variation in the magnitude of the misclassification error between incubation tubes. In practice, all separation techniques will introduce errors into the assay system arising from this source, the magnitude and variation of which will be characteristic of the particular technique adopted, and the care with which it is employed. Though infrequently discussed in the literature, the error characteristics of the separation system used often define the overall error observed in the measurement of the response metameter and therefore play a fundamental role in determining the optimal design of the assay system and the precision which it will yield. For this reason, the optimal concentration of reagents, and the fraction of labelled ligand bound to the specific reagent to yield maximal assay sensitivity, are largely defined by the nature of the separation system, as illustrated by the experimental studies of Wide using antibody-coupled solid phase, techniques [74, 151].

Susceptibility to the composition of incubation mixtures is now a well-recognised characteristic displayed by many separation methods, a phenomenon which can give rise to very large assay inaccuracies unless appropriate precautions are taken to minimise the effect. Such susceptibility is shown, for example, by many solid adsorbents used to remove free ligand from reaction mixtures, whose efficiency may be significantly affected by the presence of proteins in the incubation milieu. Generally such effects may be mitigated by ensuring that all incubation mixtures are similar in composition, particularly with respect to the nature of the proteins that they contain.

Currently the most commonly employed separation techniques fall under six main headings:

(i) chromatographic and electrophoretic methods depending on differential mobility of free and bound moieties in or on a solid matrix material [49, 152, 153, 154, 155, 156],

(ii) techniques involving adsorption of free or bound fractions by an adsorbent such as activated charcoal, silica or ion exchange resin [79, 157, 158, 159, 160],

(iii) chemical methods relying on organic solvents, salts or acids to precipitate bound activity [161, 162, 163, 164, 165, 166, 167],

(iv) immunologic methods involving reaction between the specific binding reagent (e.g. the 'first' antibody) and an antibody directed against it (the 'second' antibody) resulting in precipitation of the ligand/reagent/antibody complex [168, 169, 170],

(v) solid phase techniques in which the binding reagent is adsorbed to a solid support, such as polystyrene tubes or discs, or covalently linked to activated Sepharose or cellulose particles [171, 172, 173],

(vi) solid phase, second antibody, methods, combining an initial reaction between ligand and 'first antibody', in the liquid phase with a subsequent reaction with second antibody pre-coupled to a solid matrix material [174].

The chromatographic and electrophoretic separation techniques have, in recent years, fallen into disuse largely because they are too cumbersome for application to large numbers of samples. Moreover, the sample volume which they can handle is often restricted, and the counting techniques entailed are time consuming and inconvenient.

Adsorption methods have gained widespread popularity as a result of their ease of use and applicability to large numbers of samples. In contrast with some of the chromatographic techniques, the entire volume of an incubation mixture can be treated, with an implicit reduction of the specific activity of the tracer required to attain acceptable counting statistics in the final radioassay. One of the most widely used adsorbant materials is activated charcoal, originally employed in assays for vitamin B_{12} [79] and insulin [175]. In solutions containing an adequately high concentration of protein, charcoal adsorbs materials of low molecular size and weight, leaving higher molecular material, including protein bound substances, in the supernatant fraction.

When applying the charcoal technique to an assay system, care must be exercised in establishing the optimal quantity of charcoal, which, whilst adsorbing most of the free ligand, minimises adsorption of the bound material. In many assay systems, a fairly wide charcoal weight range will be found to exist over which the differential adsorption of free and bound moieties remains roughly (though never exactly) constant; beyond these limits, however, misclassification effects become increasingly and unacceptably severe. Amongst the disadvantages of the charcoal method are its vulnerability to variation in the composition of incubation mixtures (particularly in their protein content) and the perturbation of the chemical equilibrium that is generated as a result of the total removal by the charcoal adsorbent, of free ligand in the system. As discussed in Section 9.5.4, 'coating' of charcoal with a high molecular weight substance such as Dextran or albumin [88] is a currently fashionable strategem adopted to prevent adsorption of protein bound ligand, although its efficacy has frequently been questioned [89, 90, 176], and the only effective method of minimising non-specific effects on charcoal adsorption is to ensure that the protein composition of all incubation mixtures is essentially identical [82, 89].

Disturbance of the chemical equilibrium of the ligand-receptor reagent reaction — an effect which charcoal shares with the majority of solid adsorbents — is of particular significance in systems displaying high dissociation rates, when it gives rise to the phenomenon (inappropriately) known as ligand 'stripping'. The effect on the overall assay system can be largely overcome by keeping the temperatures of incubation mixtures (and hence the dissociation rate displayed by the ligand complex) low throughout the period of charcoal addition, and by ensuring that, as far as is possible, the time of exposure to charcoal is identical in all assay tubes.

Other adsorbents possess most of the disadvantages of charcoal, and are less commonly used except in special situations. In contrast Sephadex gel particles display a much lower (and often negligible) adsorption avidity towards most free ligands, and thus the use of this material as a basis for separation of free and bound fractions in a number of saturation assay techniques thus relies predominantly on exclusion of protein-bound material from the solution contained in the interstices of the particles, whilst entry of free is permitted. Because chemical equilibrium is undisturbed by this procedure and since, therefore, the time of exposure of the reaction mixture to the Sephadex particles is not critical, it is particularly valuable in systems characterised by high dissociation rates.

Many chemical precipitation methods likewise avoid the problems inherent in solid adsorbent methods arising from progressive dissociation of the bound material. Ammonium sulphate, ethanol, polyethylene glycol, and aqueous dioxane are amongst the salts and organic solvents which have been successfully employed to precipitate protein bound material. Nevertheless in certain assays a progressive increase of the free ligand appearing in the supernatant, presumably arising from disruption of the ligand complex, has been observed [47]. Moreover, experimental conditions resulting in optimal separation of free and bound material are sometimes quite critical, and minor variations in the quality of the solvents used may significantly alter their efficacy.

Immunologic separation methods have been specifically confined to radioimmunoassay techniques, and have relied upon a 'second' antibody to precipitate the soluble antigen-antibody complex as formed in the first antibody reaction. Although the precipitation step is normally prolonged to 24 hrs. or more in order to minimise non-specific effects on the second antibody reaction (which are particularly intrusive in short reaction times) the chemical equilibrium is not disturbed by the procedure [178] and to a large extent it fulfils the requirements of an ideal separation system listed earlier in this section. Unfortunately precipitating antisera, which are usually raised in large animals challenged with an IgG preparation of the animal species from which the first antibody was derived, must fulfil a number of exacting requirements in order to be usable. These include lack of cross-reaction with other IgG that may be in the assay system, a good plateau region defining the permissible relative concentrations of first and second antibody, and a high titre [179]. Although precipitating antisera are available commercially, they are relatively expensive and in limited supply.

Second antibody techniques can be employed in either of two modes, involving pre- or post-precipitation of the first antibody. The pre-precipitation techniques, introduced by Hales and Randall [170] avoids the use of carrier, non-immune globulin which is a necessary feature of the post-precipitation methods, and is therefore more economical of precipitating antiserum. Moreover the precipitation of first antibody is carried out in carefully controlled conditions prior to the addition of other components in the incubation mixture, and non-specific effects are therefore greatly reduced. Pre-precipitation of the first antibody appears to alter its affinity characteristics, however, and the sensitivity of the technique is somewhat reduced in consequence.

Pre-precipitated antibody techniques resemble in many respects other 'solid phase' assay systems which have recently attracted considerable attention. These depend upon the use of antibody precoupled to solid material, either by simple adsorption or by covalent linkage. Polystyrene and polypropylene strongly adsorb certain proteins, including immunoglobulins, and readily adsorb antibody when incubated with a suitably diluted or Rivanol-treated serum [171]. Alternatively, antibodies may be directly coupled to cyanogen-bromide activated Dextran, cellulose, or Sephadex [172], or indirectly to the same materials via an antigen link. Although such techniques offer a number of considerable advantages, doubts have been expressed in the past arising from their apparent lack of precision, and their relatively high consumption of antibody. Nevertheless, it is likely that under appropriately optimised conditions, these criticisms can be negated, and that the solid phase method will attract a large increase in popularity in the future.

Finally, the second antibody and solid phase techniques have been recently combined in a method devised by den Hollander and Schuurs [174] in which the precipitating serum is conjugated to a solid matrix (Sepharose or cellulose). Although displaying a number of advantages over conventional second antibody techniques, the cost of the reagents is high, and the use of the method is currently confined to certain commercial assay kits.

9.5.9 Conclusion

In this chapter, particular attention has been devoted to the saturation assay/radioimmunoassay methods in consequence of the prominence that these have attained in clinical and biochemistry in the last decade. Nevertheless, in a review of restricted length it is clearly impossible to discuss the immense variety of technical development which has taken place in the field, and the reader is referred to the much fuller accounts found in the bibliography of this chapter. It must be emphasised that improvements in the technique continue unabated and it is likely that much of the content of this chapter will be out-dated within a relatively short period of time.

BIBLIOGRAPHY

R. L. HAYES, F. A. GOSWITZ and B. E. P. MURPHY (Editors) Radio-isotopes in Medicine: In Vitro Studies, U.S.A.E.C., Oak Ridge, Tennessee, Conf. 671111, (1968).

J. W. McARTHUR and T. COLTON (Editors) Statistics in Endocrinology, Massachusetts Institute of Technology Press, (1970).

M. MARGOULIES (Editor) Protein and Polypeptide Hormones, Part I, II and III, Excerpta Medica Foundation, Amsterdam, 1968, (1969).

E. DICZFALUSY (Editor) Karolinska Symposia on Research Methods in Reproductive Endocrinology. 1st Symposium: Immunoassay of Gonado-trophins, Stockholm, 1969. 2nd Symposium: Steroid Assay by Protein Binding, Geneva, (1970). (Reproductive Endocrinology Research Unit, Karolinska Institute, Stockholm).

K. E. KIRKHAM and W. M. HUNTER (Editors) Radioimmunoassay Methods, Churchill Livingstone, Edinburgh, (1971).
In Vitro Procedures with Radioisotopes in Medicine, Vienna: I.A.E.A., (1970).

F. G. PERON and B. V. CALDWELL (Editors) Immunologic Methods in Steroid Determination, Appleton-Century-Crofts, New York (1970).

M. MARGOULIES and F. C. GREENWOOD (Editors) Structure-activity Relationships of Protein and Polypeptide Hormones, Excerpta Medica Foundation, Amsterdam, Int. Congr. Ser. No. 241, (1972).

W. D. ODELL and W. H. DAUGHADAY (Editors) Principles of Competitive Protein-Binding Assays, J. B. Lippincott Co., Philadelphia and Toronto, (1971).

S. A. BERSON and R. S. YALOW (Editors) Methods in Investigative and Diagnostic Endocrinology, vol. 2a: Peptide Hormones, North-Holland Publishing Co., Amsterdam and London, (1973).

P. H. SÖNKSEN (Editor) British Medical Bulletin: Radioimmunoassay and Saturation Analysis, Vol. 30 n. 1, Medical Dept., The British Council, London, (1974).

REFERENCES

1. D. SHEMIN and G. L. FOSTER, *Ann. N.Y. Acad. Sci.*, **47**, 119 (1946).
2. C. C. LEE, L. W. TREVOY, J. W. T. SPINX and L. B. JAQUES, *Proc. Soc. Exp. Biol. Med.*, **74**, 151 (1950).
3. C. HEIDELBERGER, H. I. HADLER and G. WOLF, *J. Am. Chem. Soc.*, **75**, 1303 (1953).
4. R. W. SCHAYER, *J. Biol. Chem.*, **196**, 468 (1952).

5. A. DIMITRIADOU, P. C. R. TURNER and T. R. FRASER, *Nature*, **197**, 446 (1963).
6. F. P. W. WINTERINGHAM, *Intern. J. Appl. Radiation Isotopes*, **1**, 57 (1956).
7. A. A. SMALES and B. D. PATE, *Analyst*, **77**, 188 (1952).
8. K. SAMSAHL, D. BRUNE and P. O. WESTER, *Intern. J. Appl. Radiation Isotopes*, **16**, 273 (1965)
9. R. EHRENBERG, *Biochem. Z.*, **164**, 183 (1925).
10. A. S. KESTON, S. UNDENFRIEND and R. K. CANNAN, *J. Am. Chem. Soc.*, **68**, 1390 (1946).
11. P. AVIVI, S. A. SIMPSON, J. F. TAIT and J. K. WHITEHEAD, Proceedings 2nd Radioisotope Conference (Oxford), (J. E. Johnson, ed.), p. 313, Butterworths, London, (1954).
12. P. SORENSEN, *Anal. Chem.*, **27**, 389 (1955).
13. W. M. STOKES, F. C. HICKEY and W. A. FISH, *Anal. Chem.*, **28**, 207 (1956).
14. B. L. BROWN, W. S. REITH and W. TAMPION, *Biochem. J.*, **97**, 30P (1965).
15. A. RIONDEL, J. F. TAIT, M. GUT, S. A. S. TAIT, E. JOACHIM and B. LITTLE, *J. Clin. Endocrinol. Metab.* **23**, 620 (1963).
16. L. E. M. MILES and C. N. HALES, Protein and Polypeptide Hormones. (M. Margoulies, ed.), p. 61, Excerpta Medica Foundation, Amsterdam, (1968).
17. A. S. KESTON, S. UDENFRIEND and M. LEVY, *J. Am. Chem. Soc.*, **72**, 748 (1950).
18. J. H. LARAGH, J. E. SEALEY and P. D. KLEIN, Proc. I.A.E.A. Symp. on Radiochemical Methods of Analysis, Vol. II p. 353. Vienna, (1965).
19. A. H. BRODIE and J. F. TAIT, Methods in Hormone Research. (R. I. Dorfman, ed.), Vol. I, p. 323, Academic Press, New York, (1969).
20. S. TANAKA and P. STARR, *J. Clin. Endocrinol. Metab.*, **19**, 84 (1959).
21. T. PETERS, T. J. GIOVANNIELLO, L. APT and J. F. ROSS, *J. Lab. Clin. Med.*, **48**, 274 (1956).
22. C. A. HALL and A. E. FINKLER, *Blood*, **27**, 611 (1966).
23. S. ARDEMAN and I. CHANARIN, *Brit. J. Haemat.*, **11**, 305 (1965).
24. V. HERBERT, M. FISHER, K-S. LAU, C. GOTTLIEB, N. R. GEVIRTZ and L. R. WASSERMANN, *Am. J. Clin. Nutr.*, **16**, 385 (1965).
25. W. H. PEARLMAN and O. CRÉPY, *J. Biol. Chem.*, **242**, 182 (1967).
26. D. RODBARD and G. H. WEISS, *Anal. Biochem.*, **52**, 10 (1973).
27. A. E. GUREVICH, O. B. KUZOVLEVA and A. E. TUMANOVA, *Biochemistry* (N.Y.) **26**, 803 (1962).
28. J. C. HENDRICK and P. FRANCHIMONT, *Ann. Biol. Clin.*, **30**, 113 (1972).
29. J. S. WOODHEAD, G. M. ADDISON and C. N. HALES, *Brit. Med. Bull.*, **30**, 44 (1974).
30. F. C. GREENWOOD, W. M. HUNTER and J. S. GLOVER, *Biochem. J.*, **89**, 114 (1963).
31. L. E. M. MILES and C. N. HALES, *Lancet*, **2**, 492 (1968).
32. L. E. M. MILES, *J. Clin. Endocrinol. Metab.*, **33**, 399 (1971).
33. G. M. ADDISON, C. N. HALES, J. S. WOODHEAD and J. L. H. O'RIORDAN, *J. Endocrinol.*, **49**, 521 (1971).

34. G. M. ADDISON and C. N. HALES, Radioimmunoassay Methods. (K. E. Kirkham and W. M. Hunter eds.), p. 447, Churchill Livingstone, Edinburgh, (1971).
35. E. W. D. COLT, L. E. M. MILES, K. L. BECKER and N. J. SHAH, *J. Clin. Endocrinol. Metab.*, **32**, 285 (1971).
36. A. B. KURTZ, *Horm. Metab. Res.*, **3**, 203 (1971).
37. G. M. ADDISON, M. R. BEAMISH, C. N. HALES, M. HODGKINS, A. JACOBS and P. LLEWELLIN, *J. Clin. Path.*, **25**, 326 (1972).
38. G. M. ADDISON, Personal Communication (1974).
39. G. M. ADDISON and C. N. HALES, *Horm. Metab. Res.*, **3**, 59 (1971).
40. J. L. H. O'RIORDAN, G. M. ADDISON, J. S. WOODHEAD, H. T. KEUTMANN and J. T. POTTS, Endocrinology (1971), Proceedings of the Third International Symposium, (S. Taylor, ed.), p. 386, Heinemann Medical, London, (1972).
41. G. M. BESSER, D. N. ORTH, W. E. NICHOLSON and J. WOODHAM, *J. Endocrinol,* **46**, i-ii (1969).
42. J. F. HABENER, G. V. SEGRE, D. POWELL, T. M. MURRAY and J. T. POTTS, *Nat. New Biol.*, **238**, 152 (1972).
43. C. M. CLING and L. R. OVERBY, *J. Immunol.*, **109**, 834 (1972).
44. C. H. CAMERON and D. S. DANE, *Brit. Med. Bull.*, **30**, 90 (1974).
45. L. E. M. MILES, C. P. BIEBER, L. F. ENG and D. A. LIPSCHITZ, Proc. I.A.E.A. Symp. on Radioimmunoassay and Related Procedures in Medicine, Vol. I, p. In Press.
46. J. S. WOODHEAD, G. M. ADDISON and C. N. HALES, *Brit. Med. Bull.*, **30**, 44 (1974).
47. D. RODBARD and J. E. LEWALD, *Acta Endocrinol.*, (Copenhagen), Suppl. **147**, 79 (1970).
48. R. S. YALOW and S. A. BERSON, *J. Clin. Invest.*, **39**, 1157 (1960).
49. R. P. EKINS, *Clin. Chim. Acta*, **5**, 453 (1960).
50. R. D. UTIGER, M. L. PARKER and W. H. DAUGHADAY, *J. Clin. Invest.*, **41**, 254 (1962).
51. R. M. BARAKAT and R. P. EKINS, *Lancet*, **2**, 25 (1961).
52. R. H. UNGER, A. M. EISENTRAUT, M. S. MCCALL and L. L. MADISON, *J. Clin. Invest.*, **40**, 1280 (1961).
53. S. P. ROTHENBERG, *Nature*, **206**, 1154 (1965).
54. S. G. KORENMAN, *J. Clin. Endocrinol. Metab.*, **28**, 127 (1968).
55. A. G. GILMAN, *Proc. Nat. Acad. Sci.* (Washington), **67**, 305 (1970).
56. R. J. LEFKOWITZ, J. ROTH and I. PASTAN, *Science,* **170**, 633 (1970).
57. A. L. STEINER, D. M. KIPNIS, R. UTIGER and C. W. PARKER, *Proc. Nat. Acad. Sci.* (Washington), **64**, 367 (1969).
58. B. L. BROWN, R. P. EKINS, S. M. ELLIS and E. S. WILLIAMS, Further Advances in Thyroid Research. (K. Fellinger and R. Hofer, eds.), Vol. 2, p. 1107, Wiener Medizinischen Akadamie, Vienna, (1971).
59. G. E. ABRAHAM, Proc. I.A.E.A. Symp. on Radioimmunoassay and Related Procedures in Medicine, Vol. I In Press.
60. L. LEVINE and H. VAN VUNAKIS, *Biochem. Biophys. Res. Commun.* **41**, 1171 (1970).
61. B. E. P. MURPHY, Radioisotopes in Medicine: In Vitro Studies. (R. L. Hayes, F. A. Goswitz and B. E. P. Murphy, eds.), p. 3, U.S.A.E.C., Div. Tech. Inform., Oak Ridge, Tenn., (1968).

62. R. P. EKINS, Protein and Polypeptide Hormones. Excerpta Med. Congress Series no. 161 (M. Margoulies, ed.), p. 672, Excerpta Medica Foundation Amsterdam, (1968).

63. Statistics in Endocrinology. (J. W. McArther and T. Colton eds.), p. 379, The MIT Press, Cambridge, Mass. and London, (1970).

64. R. S. YALOW and S. A. BERSON, Radioisotopes in Medicine: In Vitro Studies. (R. L. Hayes, F. S. Goswitz and B. E. P. Murphy, eds.), p. 7, U.S.A.E.C., Div. Tech. Inform., Oak Ridge, Tenn., (1968).

65. R. P. EKINS, G. B. NEWMAN and J. L. H. O'RIORDAN, Radio-isotopes in Medicine: In Vitro Studies. (R. L. Hayes, F. S. Goswitz and B. E. P. Murphy, eds.), p. 59, U.S.A.E.C., Div. Tech. Inform., Oak Ridge, Tenn., (1968).

66. L. B. MACURDY, H. K. ALBERT, A. A. BENEDETTI-PICHLER, H. CARMICHAEL, A. H. CORWIN, R. M. FOWLER, E. W. D. HUFFMAN, P. L. KIRK and T. W. LASHOF, *Anal. Chem.*, 26, 1190 (1954).

67. R. S. YALOW and S. A. BERSON, Statistics in Endocrinology. (J. W. McArther and T. Colton, eds.), p. 327, The MIT Press, Cambridge, Mass. and London, (1970).

68. R. S. YALOW and S. A. BERSON, Protein and Polypeptide Hormones. Excerpta Med. Congress Series no. 161 (M. Margoulies, ed.), p. 71, Excerpta Medica Foundation, Amsterdam, (1968).

69. S. A. BERSON and R. S. YALOW, The Hormones. (G. Pincus, K. V. Thimann and E. B. Astwood, eds.), Vol. IV, p. 557, Academic Press, New York, (1964).

70. G. H. MORRISON and R. K. SKOGERBOE, Trace Analysis: Physical Methods. (G. H. Morrison, ed.), p. 1, Interscience, New York, (1965).

71. W. M. HUNTER, Protein and Polypeptide Hormones. Excerpta Med. Congress Series no. 161 (M. Margoulies, ed.), p. 551, Excerpta Medica Foundation, Amsterdam, (1968).

72. H. KAISER and H. SPECKER, *Z. Analyt. Chem.*, 149, 46 (1965).

73. R. P. EKINS, Proc. I.A.E.A. Symp. on Radioimmunoassay and Related Procedures in Medicine, Vol. I. In Press.

74. L. WIDE, S. J. NILLIUS, C. GEMZELL and P. ROOS, *Acta Endocrinol.*, (Copenhagen), Suppl. 174, (1973).

75. E. SAMOLS and D. BILKUS, *Proc. Soc. Exp. Biol. N.Y.*, 115, 79 (1964).

76. R. P. EKINS and B. NEWMAN, *Acta Endocrinol.*, (Copenhagen), Suppl. 147, 11 (1970).

77. G. SCATCHARD, *Ann. N.Y. Acad. Sci.*, 51, 660 (1949).

78. S. A. BERSON and R. S. YALOW, *J. Clin. Invest.*, 38, 1196 (1959).

79. R. P. EKINS and A. M. SGHERZI, Proc. I.A.E.A. Symp. on Radio-chemical Methods of Analysis, Vol. I p. 239, Vienna, (1965).

80. Protein and Polypeptide Hormones. (M. Margoulies ed.), Part 3, p. 612, Excerpta Medica Foundation Amsterdam, (1969).

81. D. RODBARD and J. A. COOPER, Proc. I.A.E.A. Symp. on In Vitro Procedures with Radioisotopes in Medicine, p. 659, Vienna, (1970).

82. J. D. M. ALBANO, R. P. EKINS, G. MARTIZ and R. C. TURNER, *Acta Endocrinol.* 70, 487 (1972).

83. R. P. EKINS, G. B. NEWMAN and J. L. H. O'RIORDAN, Statistics in Endocrinology. (J. W. McArther and T. Colton eds.), p. 345, The MIT Press, Cambridge, Mass. and London, (1970).

84. R. P. EKINS, G. B. NEWMAN, R. PIYASENA, P. BANKS and J. D. H. SLATER, *J. Steroid Biochem.*, 3, 289 (1972).
85. A. M. EISENTRAUT, N. WHISSEN and R. H. UNGER, *Am. J. Med. Sci.*, 255, 137 (1968).
86. I. J. CHOPRA, *J. Clin. Endocrinol. Metab.*, 34, 938 (1972).
87. J. GIRARD and F. C. GREENWOOD, *J. Endocrinol.*, 40, 493 (1968).
88. V. HERBERT, K. S. LAU, C. W. GOTTLIEB and S. J. BLEICHER, *J. Clin. Endocrinol. Metab.*, 25, 1375 (1965).
89. R. P. EKINS, Protein and Polypeptide Hormones. Excerpta Med. Congress Series no. 161, (M. Margoulies ed.), p. 633, Excerpta Medica Foundation. Amsterdam, (1969).
90. M. A. BINOUX and W. D. ODELL, *J. Clin. Endocrinol. Metab.*, 36, 303 (1973).
91. J. D. M. ALBANO, G. D. BARNES, D. V. MAUDSLEY, B. L. BROWN and R. P. EKINS, *Anal. Biochem.*, 60, 130 (1974).
92. T. H. LIAO and J. G. PIERCE, *J. Biol. Chem.*, 245, 3275 (1970).
93. H. S. JACOBS and N. F. LAWTON, *Brit. Med. Bull.*, 30, 55 (1974).
94. D. R. BANGHAM and P. M. COTES, *Brit. Med. Bull.*, 30, 12 (1974).
95. P. M. COTES, Proc. I.A.E.A. Symp. on Radioimmunoassay and Related Procedures in Medicine, Vol. I In Press.
96. D. E. BANGHAM and P. M. COTES, Radioimmunoassay Methods. (K. E. Kirkham and W. M. Hunter, eds.), p. 345, Churchill Livingstone, Edinburgh, (1971).
97. G. T. ROSS, Principles of Competitive Protein-Binding Assays. (W. D. Odell and W. H. Daughaday, eds.), p. 325, J. B. Lippincott, Philadelphia and Toronto, (1971).
98. D. R. BANGHAM and R. BORTH, *Acta Endocrinol.* (Copenhagen), 71, 625 (1972).
99. D. R. BANGHAM *et al.*, *J. Clin. Endocrinol. Metab.*, 36, 647 (1973).
100. W. M. HUNTER, *Brit. Med. Bull.*, 30, 18 (1974).
101. W. M. HUNTER, Radioimmunoassay Methods. (K. E. Kirkham and W. M. Hunter, eds.), p. 3, Churchill Livingstone, Edinburgh, (1971).
102. W. R. BUTT, *J. Endocrinol.*, 55, 453 (1972).
103. R. N. BANERJEE and R. P. EKINS, *Nature*, 192, 746 (1961).
104. U. ROSA, F. PENNISI, R. BIANCHI, G. FEDERIGHI and L. DONATO, *Biochim. Biophys. Acta*, 133, 486 (1967).
105. P. G. MALAN, L. JAYARAM, N. J. MARSHALL and R. P. EKINS, *J. Endocrinol.*, (1974) In Press.
106. E. SAMOLS and H. S. WILLIAMS, *Nature*, 190, 1211 (1961).
107. J. J. MARCHALONIS, *Biochem. J.*, 113, 299 (1969).
108. M. MORRISON, G. S. BAYSE and R. G. WEBSTER, *Immunochemistry*, 8, 289 (1971).
109. J. I. THORELL and B. G. JOHANSSON, *Biochim. Biophys. Acta.*, 251, 363 (1971).
110. A. E. BOLTON and W. M. HUNTER, *J. Endocrinol.*, 55 xxx (abstract) (1972).
111. A. R. MIDGLEY Jr., G. D. NISWENDER, V. L. GAY and L. E. REICHERT Jr., *Recent Prog. Horm. Res.*, 27, 235 (1971).
112. E. D. GILBY, S. L. JEFFCOATE and R. EDWARDS, *J. Endocrinol.*, 58, xx (abstract) (1973).
113. P. W. NARS and W. M. HUNTER, *J. Endocrinol.*, 57, xlvii (abstract) (1973).

114. R. EDWARDS, E. D. GILBY and S. L. JEFFCOATE, Proc. I.A.E.A. Symp. on Radioimmunoassay and Related Procedures in Medicine, Vol. II Vienna, in Press.

115. A. R. MIDGLEY Jr. and G. D. NISWENDER, *Acta Endocrinol.*, (Copenhagen), Suppl. 147, 320 (1970).

116. A. L. STEINER, R. E. WEHMANN, C. W. PARKER and D. M. KIPNIS, Advances in Cyclic Nucleotide Research. (P. Greengard and G. A. Robison, eds.), Vol. 2, p. 51, Raven Press, New York, (1972).

117. A. FRIEDLANDER and R. E. CATHOU, Radioimmunoassay Methods. (K. E. Kirkham and W. M. Hunter, eds.), p. 94, Churchill Livingstone, Edinburgh, (1971).

118. R. S. YALOW and S. A. BERSON, Protein and Polypeptide Hormones. Excerpta Med. Congress Ser. no. 161, (M. Margoulies, ed.), p. 36, Excerpta Medica Foundation, Amsterdam (1968).

119. R. S. YALOW and S. A. BERSON, *Trans. N.Y. Acad. Sci.*, 28, 1033 (1966).

120. I. M. ROITT, M. F. GREAVES, G. TORRIGIANI, J. BROSTOFF and J. H. L. PLAYFAIR, *Lancet*, 2, 367 (1969).

121. H. WIGZELL and O. MÄKELÄ, *J. Exp. Med.*, 132, 110 (1970).

122. N. A. MITCHISON, *Eur. J. Immunol.*, 1, 68 (1971).

123. E. HABER, L. B. PAGE and G. A. JACOBY, *Biochemistry*, N.Y., 4, 693 (1965).

124. C. S. WALKER, S. J. CLARK and H. H. WOTIZ, *Steroids*, 21, 259 (1973).

125. B. J. ERLANGER, F. BOREK, S. M. BEISTER and S. LIEBERMAN, *J. Biol. Chem.*, 234, 1090 (1959).

126. B. J. ERLANGER, F. BOREK, S. M. BEISER and S. LIEBERMAN, *J. Biol. Chem.*, 228, 713 (1957).

127. V. MARKS, B. A. MORRIS and J. D. TEALE, *Brit. Med. Bull*, 30, 80 (1974).

128. S. L. JEFFCOATE, Radioimmunoassay Methods. (K. E. Kirkham and W. M. Hunter, eds.), p. 151, Churchill Livingstone, Edinburgh, (1971).

129. D. EXLEY, W. M. JOHNSON and P. D. G. DEAN, *Steroids*, 18, 605 (1971).

130. H. R. LINDNER, E. PEREL, A. FRIEDLANDER and A. ZEITLIN, *Steroids*, 19, 357 (1972).

131. S. L. JEFFCOATE and J. E. SEARLE, *Steroids*, 19, 181 (1972).

132. G. D. NISWENDER and A. R. MIDGLEY, Immunologic Methods in Steroid Determination. (F. G. Peron and B. V. Caldwell, eds.), p. 149, Appleton-Century-Crofts, New York, (1970).

133. R. E. TIGELAAR, R. L. RAPPORT, J. K. INMAN and H. J. KUPFERBERG, *Clin. Chim. Acta*, 43, 231 (1973).

134. B. A. L. HURN, *Brit. Med. Bull.*, 30, 26 (1974).

135. G. W. BOYD and W. S. PEART, *Lancet*, 2, 129 (1968).

136. J. VAITUKAITIS, J. B. ROBBINS, E. NIESCHLAG and G. T. ROSS, *J. Clin. Endocrinol. Metab.*, 33, 988 (1971).

137. G. T. ROSS, J. L. VAITUKAITIS and J. B. ROBBINS, Structure-activity Relationships of Protein and Polypeptide Hormones. Excerpta Med. Congress Ser. no. 241, (M. Margoulies and F. C. Greenwood, eds.), p. 153, Excerpta Medica Foundation, Amsterdam, (1972).

138. D. SCHULSTER, *Brit. Med. Bull.*, 30, 28 (1974).

139. R. J. LEFKOWITZ, J. ROTH and I. PASTAN, *Science*, 170, 633 (1970).

140. R. W. TURKINGTON, *J. Clin. Invest.*, **50**, 94a (abstract) (1971).
141. R. P. SHIU and H. G. FRIESEN, *Clin. Res.*, **20**, 932 (abstract) (1972).
142. K. J. CATT, M. L. DUFAU and T. TSURUHARA, *J. Clin. Endocrinol. Metab.*, **32**, 860 (1971).
143. A. R. WOLFSEN, H. B. McINTYRE and W. D. ODELL, *J. Clin. Endocrinol. Metab.*, **34**, 684 (1972).
144. R. A. McILHINNEY and D. SCHULSTER, *J. Endocrinol.*, **57**, xlvi (abstract) (1973).
145. B. L. BROWN, J. D. M. ALBANO, R. P. EKINS, A. M. SCHERZI and W. TAMPION, *Biochem. J.*, **121**, 561 (1971).
146. S. G. KORENMAN, *Endocrinology*, **87**, 1119 (1970).
147. C. MACKIE, M. C. RICHARDSON and D. SCHULSTER, *FEBS Lett.*, **23**, 345 (1972).
148. K. E. RUBENSTEIN, R. S. SCHNEIDER and E. F. ULLMAN, *Biochem. Biophys. Res. Commun.*, **47**, 846 (1972).
149. J. M. ANDRIEU, S. MAMAS and F. DRAY, Proc. I.A.E.A. Symp. on Radioimmunoassay and Related Procedures in Medicine, Vol. II Vienna. In Press.
150. J. LANDON, Steroid Immunoassay. (Tenovus Workshop, Cardiff, April 3–5, 1974) In Press.
151. R. P. EKINS, *Brit. Med. Bull.*, **30**, 3 (1974).
152. S. A. BERSON, R. S. YALOW, A. BAUMAN, M. A. ROTHSCHILD and K. NEWERLY, *J. Clin. Invest.*, **35**, 170 (1956).
153. W. M. HUNTER and F. C. GREENWOOD, *Biochem. J.*, **91**, 43 (1964).
154. H. ØRSOV, *Scand. J. Clin. Lab. Invest.*, **20**, 297 (1967).
155. E. HABER, L. B. PAGE and F. F. RICHARDS, *Anal. Biochem.*, **12**, 163 (1965).
156. W. M. HUNTER, Recent Research on Gonadotrophic Hormones. (E. T. Bell and J. A. Loraine, eds.), p. 92, Livingstone, Edinburgh, (1967).
157. C. W. GOTTLIEB, K. S. LAU, L. R. WASSERMAN and V. HERBERT, *Blood*, **25**, 875 (1965).
158. G. ROSSELIN, R. ASSAN, R. S. YALOW and S. A. BERSON, *Nature*, **212**, 355 (1966).
159. R. C. MEADE and H. M. KLITGAARD, *J. Nucl. Med.*, **3**, 407 (1962).
160. E. P. FRENKEL, S. KELLER and M. S. McCALL, *J. Lab. Clin. Med.*, **68**, 510 (1966).
161. W. D. ODELL, J. F. WILBER and W. E. PAUL, *J. Clin. Endocrinol. Metab.*, **25**, 1179 (1965).
162. L. G. HEDING, Labelled Proteins in Tracer Studies. Proc. Conf. held in Pisa, Jan. 1966 (L. Donato, G. Milhaud and J. Sirchis, eds.), p. 345, European Atomic Energy Community, Brussels.
163. K. THOMAS and J. FERIN, *J. Clin. Endocrinol. Metab.*, **28**, 1667 (1968).
164. B. DESBUQUOIS and G. D. AURBACH, *J. Clin. Endocrinol. Metab.*, **33**, 732 (1971).
165. G. M. GRODSKY and P. H. FORSHAM, *J. Clin. Invest.*, **39**, 1070 (1960).
166. T. CHARD, M. MARTIN and J. LANDON, Radioimmunoassay Methods. (K. E. Kirkham and W. M. Hunter, eds.), p. 283, Churchill Livingstone, Edinburgh, (1971).
167. M. L. MITCHELL and J. BYRON, *Diabetes*, **16**, 656 (1967).

168. R. D. UTIGER, M. L. PARKER and W. H. DAUGHADAY, *J. Clin. Invest.*, **41**, 254 (1962).
169. C. R. MORGAN and A. LAZAROW, *Diabetes,* **12**, 115 (1963).
170. C. N. HALES and P. J. RANDLE, *Biochem. J.*, **88**, 137 (1963).
171. K. CATT and G. W. TREGEAR, *Science,* **158**, 1570 (1967).
172. L. WIDE and J. PORATH, *Biochim. Biophys. Acta.*, **130**, 257 (1966).
173. S. DONINI and P. DONINI, *Acta Endocrinol.*, (Copenhagen), Suppl. **142**, 257 (1969).
174. F. C. DEN HOLLANDER and A. H. W. M. SCHUURS, Radioimmuno-assay Methods. (K. E. Kirkham and W. M. Hunter, eds.), p. 419, Churchill Livingstone, Edinburgh, (1971).
175. V. HERBERT, K. S. LAU, C. W. GOTTLIEB and S. J. BLEICHER, *J. Clin. Endocrinol. Metab.*, **25**, 1375 (1965).
176. G. M. A. PALMIERI, R. S. YALOW and S. A. BERSON, *Horm. Metab. Res.*, **3**, 301 (1971).
177. J. G. RATCLIFFE, *Brit. Med. Bull.*, **30**, 31 (1974).
178. R. BORTH, *Acta Endocrinol.*, (Copenhagen), **65**, 453 (1970).
179. W. M. HUNTER and P. C. GANGULI, Radioimmunoassay Methods. (K. E. Kirkham and W. M. Hunter, eds.), p. 243, Churchill Livingstone, Edinburgh, (1971).

Chapter 10

Determination of Radioactivity Present in the Environment

M. S. Baxter

Department of Chemistry, The University, Glasgow G12 8QQ

10.1 RADIOACTIVITY IN THE ENVIRONMENT

10.1.1 Introduction

The development of radioactivity monitoring equipment to sophisticated levels has permitted detection of otherwise unseen phenomena of major significance to man. In recent times refined techniques have yielded data of basic importance to the biosciences, geosciences and cosmosciences alike. Dating methods have become key contributors to the understanding of our geological and archaeological past. Analyses of biospheric materials have equipped us with some idea of the health hazards presented by a wide spectrum of radioisotopes. At this time, when man's influence on his surroundings has reached the quasi-vertical stage of exponential increase, perhaps the prime contribution of radioactivity studies lies in the definition of the delicately balanced equilibria on which a stable environment depends. Through their function as tracers of terrestrial geophysical and geochemical processes, radioisotopes can expand our awareness of this earth and thereby, perhaps, influence our future relationship to it.

The aim of this chapter is to outline methods and progress in the measurement of environmental radioactivity. No attempt has been made to set forth a comprehensive text covering all isotopes. Rather it is hoped that general methods of approach are defined, to be adapted by the reader to his own particular areas of interest.

10.1.2   Natural Radioactivity

All materials, living and dead, can be shown to contain at least traces of natural radioactivity. The relative abundances of the natural radioisotopes in the present terrestrial reservoirs are directly derived, though modified during geological time, from the isotopic composition at the time of formation of the solar system. During evolution of the earth, many of the shorter-lived activities have decayed below detectable levels so that the residual radio-activity can be subdivided into (a) the primary activity, due to isotopes of half-lives comparable to or greater than the estimated age of the earth (4.6 $\times$ 10^9 years), and (b) the secondary activity, resulting from radioactive decay of the primary nuclides. The primary component therefore includes U-238 ($t_{1/2}$ 4.5 $\times$ 10^9 y), Th-232 (1.39 $\times$ 10^{10} y) and U-235 (7.13 $\times$ 10^8 y) which are the parents of the three remaining families of natural radio-isotopes (Fig. 10.1). The daughters of these isotopes, from the parent to terminal stable lead isotopes represent the secondary element of natural radioactivity. The extinct component may be illustrated by Pu-241 ($t_{1/2}$ 14 y), probably the parent of a further decay series of which the only remain-ing isotope is Bi-209 ($t_{1/2}$ 2.7 $\times$ 10^{17} y). Besides the active species of the three decay series some shorter-lived transuranic isotopes (Pu-239, Np-237) are induced through capture of naturally occurring neutrons by uranium isotopes. Spontaneous and neutron-induced fission of U-235 does not occur in nature to a significant degree. In addition there are a number of nuclides which exist singly, i.e. without radioactive parents. These include a group (C-14, H-3, Be-7) produced continuously by the interaction of cosmic ray particles with terrestrial elements and a primordial group (K-40, Rb-87, Sm-187) of long half-life which represent residual species from the time of nucleosynthesis.

In consideration of the natural radiation flux the influence of the iso-topes mentioned must be combined with the effects of cosmic radiation. Cosmic rays originate in galactic space and, on arrival at the earth's atmo-sphere, comprise about 79% protons, 20% alpha particles and 1% nuclei of higher mass, mainly carbon, nitrogen and oxygen. The interactions of these highly energetic primary radiations with atmospheric nuclei yield a flux of electrons, gamma rays, neutrons and mesons. At sea level, mesons are by far the major component of this flux (80%) and are, of course, a principle cause of the background radiation observed in the detectors described in previous chapters.

Natural radioactivity is not distributed uniformly throughout the earth. For example, most mineral springs contain high concentrations of radium

(a) U-235 SERIES

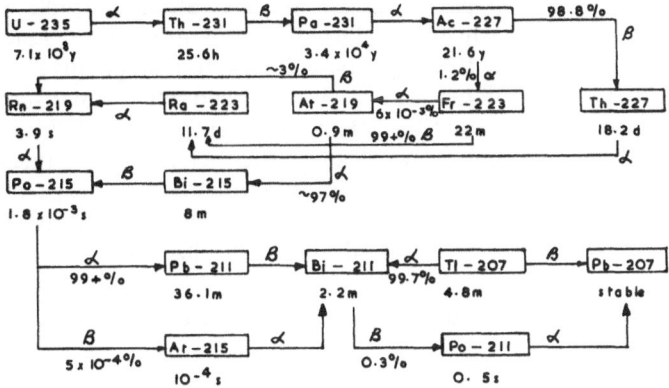

(b) U-238 SERIES

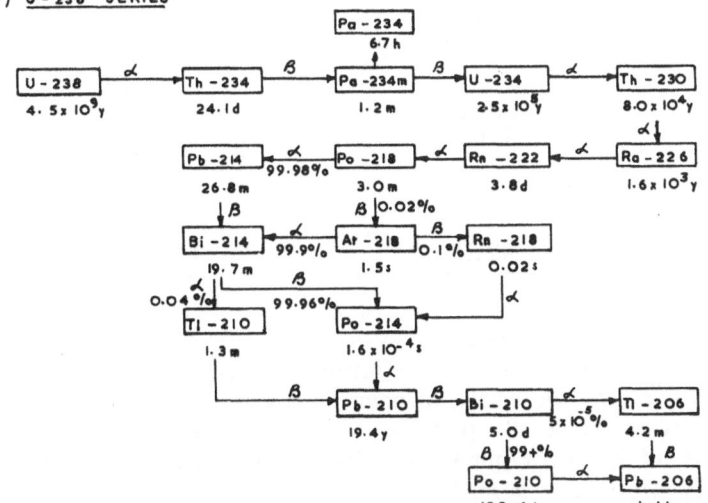

(c) Th-232 SERIES

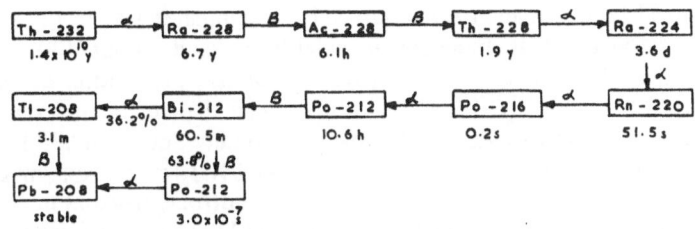

Fig. 10.1. The primary decay series, (a) U-235 (b) U-238 (c) Th-232, showing the modes and half-lives of decay (y = year, d = day, h = hour, m = minute, s = second).

and its daughter, radon, with concentrations about a million times greater than in normal water supplies. Similarly areas in Brazil and India demonstrate whole populations living on alluvial deposits of monazite sands containing 0.1% thorium and its daughters. This lack of homogeneity, however, is an inverse function of the general chemical and physical uniformity of each terrestrial system so that the effect decreases from the earth's crust to the oceans to the atmosphere, as mixing efficiencies increase. Indeed the tendency for natural isotopes to achieve a uniform distribution, particularly in the oceans, biosphere and atmosphere provides a basis for many research and dating applications to be discussed later in the chapter.

TABLE 10.1. *Fission and Fusion Yields of Nuclear Devices*

Period	Fission and fusion yield (Mton)	
	Air tests	Surface tests
1945–51	0.19	0.57
1952–54	1.0	59.0
1955–56	11.0	17.0
1957–58	57.0	28.0
1959–60	Test moratorium	
1961	120.0	
1962	217.0	
1963–68	16.3	0.2
1945–68	422.5	105.0

10.1.3 Artificial Radioactivity

During the past 25 years man has introduced vast quantities of artificial radioisotopes into the earth's dynamic reservoirs through the testing of nuclear devices. Table 10.1 presents a cumulative summary of the yields, in megatons, of the many hundreds of individual events between 1945 and 1968 [1]. Distinction is made between 'ground' and 'air' bursts since the radionuclide yield and the subsequent pattern of debris fallout are significantly different in each case. Thus 'ground' tests involve interaction of the nuclear cloud with rocks, sand and soil and generally produce more short-term local fallout than altitude bursts. For example, if 10% of the energy of a 1-megaton bomb is expended in volatilising sand, about 40000 tons of debris are added to the fireball [2]. Conversely, production of the small number of nuclides (C-14, H-3, Ar-39) which depend on the reaction of neutrons with atmospheric nuclei is approximately double in 'air' bursts. To a first approximation, an 'air' burst produces 2×10^{26} neutrons/megaton of fission and fusion as compared to 1×10^{26} neutrons/megaton in a 'surface' burst [3]. Since the fission of U-235 and Pu-239 releases a mixture of over 200 isotopes of 35 elements the fallout is 'dirty' in comparison to the debris from a fusion device based on the reaction of light species such as H-2 and H-3. It is customary, however, for primarily fusion devices to be triggered

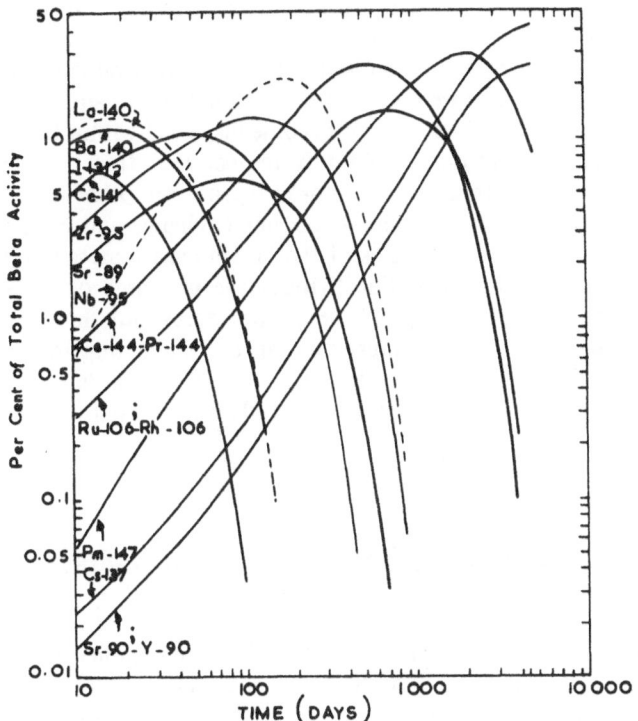

Fig. 10.2. The yields of major nuclides in fallout from megaton weapon tests (after Hallden *et al.* [4]).

and sometimes enhanced by fission reactions so that the fallout from these weapons still contains significant levels of fission products. The yields of the principle fission products from megaton tests and their temporal variations are presented in Fig. 10.2. While most product nuclides are relatively short-lived some, notably Sr-90, Y-90 and Cs-137, increase slowly as they form from radioactive decay of their parent isotopes. Besides the species induced in air, several are produced in 'ground' tests by neutron interaction with soil constituents. Of these the only long-lived nuclides of some significance are Ca-45 ($t_{1/2}$ 165 d) and Fe-55 ($t_{1/2}$ 2.6 y).

As the majority of weapon tests were carried out in upper latitudes of the northern hemisphere, the original distribution of the artificial radio-activity was markedly non-homogeneous. Thus fallout levels, in particular of the short-lived species, were notably concentrated in the latitudes of testing. Since then, however, atmospheric mixing processes have been acting towards equilibration of the longer-lived nuclides so that, by 1970, virtual homogeneity was attained for all remaining atmospheric activities. This is not yet true for the isotopes taken up by the biosphere and oceans where turnover times are appreciably longer (ca. 40 and 500 years respectively, cf. 1 to 5 years for the atmosphere). For long-term and global applications, the data in Table 10.1 can often be adequately approximated by assuming that the artificial activities were released to the atmosphere of the northern hemisphere during the time interval 1961–1962. An invaluable opportunity was, and is, therefore presented for study of the kinetics and mechanisms of transport of these isotopes as they become dispersed within the total atmosphere, oceans and biosphere. In this respect the 'tracer' treatment has enabled some understanding of important processes otherwise difficult to study, e.g. movement of air masses, oceanic exchange and biospheric turnover patterns. In addition studies of the health hazard presented to man by both natural and artificial radioactivities have increased our awareness to, and understanding of, 'acceptable' exposure doses and their effects. In a future where nuclear power must play an ever-increasing role as a world wide energy source it is fitting that the potential hazards both of reactor operation and waste disposal should be fully appreciated by scientists and public alike.

10.2   GENERAL SAMPLING METHODOLOGY

10.2.1   The Environment
Monitoring of environmental radioactivity has been applied with success to (a) the atmosphere, (b) the biosphere, soil and ground waters, (c) the oceans and their sediments, (d) the earth's crust, (e) materials of extra-terrestrial origin.

Perhaps the only areas to escape measurement have been the earth's mantle and core. In this chapter we shall review all five fields listed above and examine techniques and applications in each.

10.2.2   Atmospheric Analysis
With the exception of C-14, H-3 and the long-lived rare gas isotopes, the nuclides created in nuclear reactions are distributed in the atmosphere as particulate matter. The absorption of fission products by airborne particles occurs during cooling of the fireball and, through variable diffusion and condensation properties, fractionation of these isotopes takes place. Thus one fission product may become associated with small particles, another with large particles, so that each nuclide should in theory receive individual consideration. At every altitude of the atmosphere particle transport is governed by an interaction of various geophysical processes, gravitational settling, eddy mixing, precipitation scavenging, molecular diffusion, particle coagulation, ion migration and atmospheric mean motions. It is the net effects of the interactions of these mechanisms which can be studied in some detail by determinations of the temporal, latitudinal and altitudinal fluctuations of fallout isotopes.

To collect and monitor atmospheric particulates two common techniques are available: (1) air-filter collection, and (2) surface deposition. Air-filtration sampling, the most popular method, involves the drawing or forcing of air through a paper, cellulose, glass fibre or mineral fibre filter at a controlled and known rate for a measured time period. Paper and cellulose filters have the inherent advantage that they can readily be dissolved or ashed prior to radiochemical analysis The glass and mineral fibre samplers, on the other hand, normally have slightly higher collection efficiencies. Pumping velocity and duration is dependent on the equipment available and on the time scale of the effect being investigated although it has been shown [5] that collection efficiencies are a function of pump velocity. Typically, for ground level air sampling, flow rates of 20 to 70 ft^3/min. are employed and, used with a 5 μ pore filter, an efficiency of 100% is acheived for collection of artificial nuclide-bearing particulates. Natural radioactivities are associated with smaller particles, less than 0.5 μ, so that finer pore size filters are required along with increased pumping velocities. For upper tropospheric and stratospheric dust collection it has been shown that with few exceptions the radioisotopes are carried by particles between 0.04 μ to 0.30 μ diameter [6]. Using monitored aircraft ram pressure for sampling, 100% nuclide collection is possible. For descriptions of specific examples of filter usage the reader is referred to detailed works [5, 7–9].

Surface deposition techniques involve collection of debris particles settling down either as dry dust or associated with rainfall. Collection of dry particulates can be achieved using large area acetate films coated with waterproof adhesive and exposed to the atmosphere on horizontal frames. Dust particles remain on the screen later to be isolated by ashing the acetate film. A disadvantage of the technique is that leaching of some soluble isotopes by the rain itself can effect the observed levels of these species even though not significantly reducing the gross activities. However the most frequently applied sampling arrangement for deposited dusts entails collection of both precipitation and dry deposition using large funnels to transport the samples into flasks or tanks. For collection of large samples it is not unusual for entire roofs to be covered with plastic sheeting with run-off facilities to storage drums. The method of choice is determined by the purpose of the sampling programme and the sensitivity required. Following sample collection, filtration or ion exchange processes can be used to isolate the nuclides.

It is perhaps worthy of note here that, for sampling of radioactivities in all terrestrial reservoirs, collection locations should be selected with care to minimise sources of contamination. Research laboratories, nuclear power stations and large industrial areas, for example are all potential sources of localised perturbations which can yield non-representative data. In handling particulates, filters and sampling equipment must be scrupulously cleaned between experiments. Many laboratories coat sample storage drums and auxiliary materials with clear lacquer to minimise loss of radionuclides to metal surfaces.

Collection of particulates forms the basis of analytical techniques for the majority of radioisotopes (naturally occurring Be-7, Be-10, Pb-210, Pb-212, Pb-214, Si-32, P-32, Na-22; artificially produced Sr-90, Cs-137, Ce-144, Sb-125, Ru-106, Mn-54). Following collection, samples are either counted directly or treated, by radiochemical procedures (Chapter 5) to isolate and detect specific nuclides. Direct counting of untreated particulates is frequently employed only when γ-emitters, the most commonly analysed and biologically significant fallout isotopes, are under investigation. This technique has become particularly valuable since the introduction of Ge(Li) detectors which permit a 30 to 40-fold improvement in spectral resolution over NaI scintillators (see Chapter 2). Using a γ-ray spectrometer in conjunction with a multichannel analyser, individual nuclides are identifiable by their decay energy spectra. In some cases, for example, simultaneous analysis of the 6-component system Cs-137, Sb-125, Ru-106, Ce-144, Zr-95 and Mn-54, it is necessary to permit a suitable time-lag before measurement to permit decay of interfering shorter-lived species, e.g. Ce-141, Ru-103. For other isotopes of short half-life, Na-24 ($t_{1/2}$ 15 h), Cl-38 (37 m) and Cl-39 (55 m) for example, rapid measurement is necessary. To separate nuclides dissolved in rain water, e.g. Be-7, Si-32, radiochemical techniques are used (e.g. precipitation, with carrier, as $Be(OH)_2$ and SiO_2 respectively) [10, 11]. As β-emission is not confined to unique and identifiable energies, radiochemical procedures are commonly employed for detection of β-emitters. Thus Sr-90 is isolated as $SrCO_3$ or $SrNO_3$, Ru-106 as Ru metal, Ce-144 as Ce_2O_3, Pm-147 as metal plated on Pt discs. Tracer carriers indicate the yield of the radiochemical processes. For some purposes it is adequate to monitor a filter sample directly by straightforward Geiger-Muller counting and subsequent analysis of the decay curve [5] Several species, notably the α-emitters Rn-222 and the β-emitter Pb-210 are frequently estimated by analysis of a daughter nuclide, e.g. Pb-214 or Po-214 and Po-210 respectively [12–14], using the assumption that secular equilibrium exists between parent and daughter. In general, however, levels of α-emitters are relatively low and can be monitored successfully, without elaborate chemical pretreatment of samples, using α-spectrometry.

While air-filtration provides the basis for the majority of dust-borne nuclide analyses, methods of detection of the non-particulates cannot be so generalised. The abundant species in this group are C-14, H-3 and the rare gas isotopes, notably Ar-37, Ar-39, Kr-81, Kr-85, Rn-220 and Rn-222.

Atmospheric C-14 collection is achieved at ground level by absorption of CO_2 in KOH solution in an open dish and at higher altitudes by uptake of CO_2 on molecular sieve material (Linde molecular sieve material, type 4A) borne by aircraft. In the former case, 8N KOH solution is normally exposed for two or four weeks, shielded from rainfall and in an area removed from heavy industry (i.e. far from sources of inactive CO_2 from fossil fuel combustion). CO_2 is released from molecular sieve samples by a steam displacement technique [15] and is reabsorbed in KOH. The K_2CO_3 resulting from both techniques is subsequently acidified with H_3PO_4 to regenerate the CO_2 which may be counted after purification [16] or hydrogenated to CH_4, C_2H_6, C_2H_4 or C_6H_6 [17–19]. Of these all are counted by proportional counting incorporating anticoincidence circuitry except C_6H_6 for which a liquid scintillation detection system is used.

As H-3 exists primarily in environmental waters, rainfall is normally collected. To increase the ratio H-3/H-1 enrichment procedures, carefully standardised, are carried out by electrolysis or thermal diffusion [20, 21]. The H-3-enriched H_2O is reduced over Mg and the product H_2 either mixed with inactive CH_4 [22] or converted to CH_4, both for gas proportional counting. The rare gas isotopes Ar-37, Ar-39, Kr-81 and Kr-85 are also counted in proportional counters following isolation from air by rectification procedures [22, 23]. Mixtures of these gases with CH_4 are frequently used as the presence of the latter gas greatly improves counting characteristics. Rn isotopes represent the only significant source of α-activity in the atmosphere. Rn-222 can be monitored *per se* by α-scintillation after air filtration and adsorption on charcoal [24], or by less direct methods, as mentioned previously. Direct measurement of Rn-220, thoron, is difficult because of its short half-life (54.5 s), and it is more practical to monitor for example the α-activity of Rn-222 and Rn-220 daughters collected on air-filters. After 7 hours Rn-222 daughters decay and the Rn-220 daughters can be measured so that extrapolation to the end of the collection period permits both Rn-220 and Rn-222 concentrations to be calculated [25].

Conclusion

The technique required for analysis of a specific atmospheric nuclide depends on several critical properties:

(a)   its physical state, i.e. particulate, gaseous or water-dissolved. Air-filtration, chemical or physical absorption, and chemical precipitation respectively are the appropriate sampling techniques.

(b)   its decay mode; γ-emitters can often be monitored by spectrometry without chemical separation. β-emitters, and γ-emitters in the presence of conflicting nuclides, customarily require separation by radiochemical procedures. α-emitters may be analysed directly by scintillation counting or by detection of daughter activities. Radiochemical procedures vary for each isotope but general techniques are outlined in Chapter 5.

(c)   its half-life; many species of short half-life are estimated by monitoring more readily measurable daughter isotopes assuming conditions of secular equilibrium. In some cases it is necessary to permit conflicting nuclides to decay while in others the decay curve itself permits identification of the component species.

10.2.3   Biospheric Sampling

The biosphere, as discussed here, comprises all living organisms and includes soil, humus and related products of decay and the coexisting ground waters. The ecological systems of interaction between diverse species result in eventual distribution of radionuclides throughout the biosphere. Natural or artificial nuclides introduced into soils become incorporated into plant life via root uptake. Atmospheric C-14 is absorbed by plant photosynthesis. In addition, settling particulates are intercepted on foliar surfaces with subsequent metabolic absorption by the plants themselves or direct transfer to animals which consume the vegetation. The influence of the food chain on biospheric radioactivity investigations is exceptionally important. Decay of plant and animal life returns nuclides to the soil for recycling In addition, many living species have the ability to concentrate specific isotopes by factors as high as 10^5. This normally occurs in aquatic organisms which concentrate minerals essential to their metabolism and thereby, via food cycles, introduce their radioactive isotopes in increased burdens into the life materials of higher animals. Man's position at the end of the food chain therefore entitles him to accumulate the concentrated activities of his less elevated biospheric partners. Besides exposure to internal radioactivity man receives significant doses both from nuclides in his immediate environment, such as in buildings and streets, and from the flux of cosmic radiation which penetrates the atmosphere.

In measuring biospheric activities two basic and inter-related objectives are commonly sought: (1) to determine metabolic and transfer pathways and (2) to estimate potential radiation levels and resultant health hazards (primarily to man). By definition and by virtue of the diversity of species and physiochemical systems within its limits, the biosphere is a most complex terrestrial reservoir. Sampling and analysis of one sample type frequently yields data which bears little general relationship to another species or to the biosphere as a whole Such heterogeneity represents one major difficulty in biospheric sampling when data of a general nature is required. In theory a subdivision of the reservoir into its myriad individual sub-groups would be necessary for optimum treatment. For example turnover times of C-14 and hence of carbon, can vary from several months in grains, several years in human bone, several centuries in tree tissues, several millenia in humus, and several million years in rock carbonates, marble and coal-bearing deposits [26–28]. Estimation of a 3 year mean turnover time for C-14 in the biosphere [26] therefore demands a qualified interpretation and an understanding of the internal variations within the reservoir.

Studies aimed at radiation hazard estimation are concerned primarily with measurements within the food chain and in man himself. As gross indicators of human health hazards from both local and world-wide fallout, analyses of Sr-90, the most critical species of long half-life (27.7 y), and I-131, the short-lived counterpart (8.1d), are of major value. Although measurements of as many sample types as possible would be optimal, it has been shown that milk assay is a highly satisfactory indicator of the hazards to man [29]. Since the thyroid concentrates iodine, the dose received in that organ from I-131, introduced almost exclusively in milk, may be a limiting factor in hazard evaluation. In countries where milk is a common dietary component, about 30–50% of the absorbed Sr-90 is introduced to the body in milk. In non-milk-drinking countries green vegetables and cereals are the significant sources [29]. Thus to indicate the activity levels in thyroid and bone materials, I-131 and Sr-90 analyses of milk (and in some cases vegetables) are appropriate techniques.

While the rapid assay procedures discussed above can be used for monitoring critical contamination levels following nuclear weapon tests or reactor leaks the relatively long time elapsed since a major event has allowed studies of 'minor' radionuclide distributions and of less critical but extremely interesting biospheric processes. More complete food chain monitoring has been achieved with determinations of concentration factors and subsequent correlations with activity levels in a wide variety of animal and human tissues. Soils and grass, for example, are excellent media for trapping debris and can yield sensitive isotope data. Hydrological studies can not only define the nuclide concentrations available to the living world but also may reveal the rates of water movement in specific areas.

Techniques of biospheric sampling and analysis involve the entire range of radiochemical and counting procedures at our disposal. A highly successful technique for food analysis is γ-spectrometry using a Ge(Li) detector in anticoincidence with a plastic phosphor guard counter [30]. Many food samples present problems of unusual geometry, self-absorption and sample-detector distance. However, biological materials, from human tissues to foodstuffs, commonly contain 50–90% water, so that freeze-drying greatly reduces sample volume and permits its accommodation in a convenient container, such as an aluminium can. Liquid samples can also be shell-frozen by mixing with a dry ice-propanol mix before dehydration under vacuum. It is interesting to note that gaseous nuclides, such as xenon or krypton species, can also be monitored by this versatile technique. The sample can, sealed to prevent losses or spoilage of materials, may be reopened after γ-analysis for chemical analysis or β-analysis of beta emitters. The method is of particular value in assay of complex mixtures in which large energy peaks coexist with small adjacent ones and is therefore a major technique in studies of slow uptake of low-level activities by biospheric materials.

β-Analysis of beta emitters in solid samples from the biosphere commonly involves preliminary ashing followed by dissolution and radiochemical separation of the desired species. Besides the higher activity fission products, the low-level β-emitters C-14 and H-3 are present in the basic elements of all living systems and may therefore constitute a significant long-term radiation hazard. Experimental procedures for these two species are virtually identical to those described in the previous section except that, for C-14 analysis, the organic sample material is combusted under controlled conditions to yield the required CO_2 gas.

α-Emitters are normally identified, without elaborate chemical treatment, using α-spectroscopy. Where environmental samples are of exceptionally low activity concentration by ashing is employed. It is noteworthy that although the short range of α-particles renders them relatively harmless as external radiation, their incorporation into living tissues is of major concern since their localised ionising ability is considerable and since, in any case, many α-emitters have associated γ-emission characteristics.

Finally, it should be noted that 'total' α, β or γ monitoring by simple counting techniques often provides a useful guide to biological radiation hazards [25], particularly where rapid surveillance is necessary immediately after a major release of radioactivity.

10.2.4 Oceanic Sampling

The 1970's will hopefully be a decade of oceanographic studies on a large scale. The distribution and transport of currents, particulates and isotopes within this vast reservoir (3.6 × 10^8 km^2 area, 1.37 × 10^9 km^3 volume) is presently only faintly understood. The practical problems of systematic and large-scale sampling are obvious. However, during the 1970's, the majority of the nuclear burden released by the concentrated series of weapon tests of the early 1960's will have entered and penetrated the oceans. As these isotopes diffuse downwards to subsurface levels towards an equilibrium distribution, scientists have an invaluable, perhaps unique, opportunity to monitor these tracers and thereby clarify at least some of the uncertainties. Ocean waters represent a 'sink' for nuclides deposited or dissolved from the atmosphere, eluted by rainfall and river run-off, precipitated or dissolved from decaying biospheric materials, absorbed from the rich sediment floor, injected by seafloor volcanism and introduced more deviously by man's nuclear-waste disposal techniques. Because of this complexity of feed-in mechanisms, oceanographic sampling sites need careful selection to avoid localised effects. In addition the oceans are believed to incorporate many distinct water bodies, e.g. Antarctic Bottom Water (A.A.B.W.) and North Atlantic Deep Water (N.A.D.W.), which have defining properties of density, salinity and temperature and which have specific limits of geography and depth. Thus radioisotopes may be used as tracers of the transport and mixing characteristics of these major currents. The oceans are covered by a shallow and relatively warm surface layer, well mixed by wind action, to the depth of the thermocline (ca. 100 m) in which the rate of temperature decrease with depth reaches a maximum relative to the more gradual decreases observed in the deep ocean masses. Since it is the exchange of heat and salt at the air/sea interface which controls the surface water density and hence the movement of the deep currents, determinations of surface isotope residence and transfer times can provide basic data about the processes of water body formation and downward mixing. In addition, ocean waters provide the primary nutrients and minerals essential to planktonic, and higher, life forms. Many of these species concentrate minerals such as copper, zinc and iron so that levels of the radioisotopes of these elements, e.g. Zn-65, Fe-55, Co-60, Mn-54, in the organisms are increased by many orders of magnitude relative to the water environment. According to Eisenbud [29] concentration factors as high as 150,000, 850,000, 100,000 and 70,000 are observed in phytoplankton, algae, larvae and fish respectively. This effect of enormous activity enhancement quite obviously demands continuous oceanic survey by the ultimate species in the food chain. Besides water analyses, studies of ocean sediments have opened up a field of particular interest as these deposits not only exchange natural radio-isotopes with the overlying water but also preserve a record of the geological history of the earth-ocean system. The rate of sedimentation and the chemical and isotopic composition of the precipitated materials are sensitive to climatic and environmental changes. Where it can be assumed that ratios and abundances of the natural species such as Th-230, Pa-231, U-234, U-238 in the sediments have not been perturbed during their history other than by radioactive decay, sediment ages and accumulation rates can be estimated.

Water sampling techniques normally involve lowering an opened container which can be triggered to close efficiently by dropping a weighted 'messenger' down the connecting cable. The depth at closing is estimated either by measuring the length and angle of cable released or by sounding or, most commonly, by using a reversing thermometer [31]. All deep water samplers thus incorporate a container (drum, plastic bag, converted beer barrel [31–33]), an easily triggered spring-loaded closing device and a depth monitoring facility. For depth profile of small samples a 'Nansen' bottle series can be used. When the first bottle is closed it releases a second 'messenger' which triggers the closing device of the next bottle and so forth until samples from surface to deep ocean are collected. Surface ocean sampling can of course be achieved either by an open bucket or bottle collection system or by pumping through plastic tubing. Sediments may be sampled either by a jawed grab device lowered by cable and closed automatically or by a cylindrical corer which is piston-driven into the sediment floor, thereby preserving the chronological sequence of the deposits. The latter technique enables recovery of unconsolidated deposit cores of 10–20 metres length, representing perhaps a time span from a few hundred thousand to several million years. This period, the Pleistocene epoch, is of major interest since it includes the eras of large-scale climatic change and in addition bore witness to the evolution of man.

Due to the chemical complexity of sea water, radiochemical separation is a common preliminary step in analysis. Filtration or centrifuging separates off the particulate material. For C-14 analysis, 50–200 litres of water are treated with dilute acid and the CO_2 released is absorbed in KOH bubblers. Cs-137 is concentrated on ammonium molybdophosphate ion exchange crystals and Sr-90 coprecipitated with calcium oxalate [34]. Th-228 is precipitated with $Fe(OH)_3$ prior to α-spectrometry; Ra-228 is monitored as Ac-228 by preferential solvent extraction with TTA [35]. For sediment core dating, ionium, thorium and protactinium can be analysed by continuous sweeping of Rn, Tn and An from a solution containing their parents in HCl into an α-proportional gas counter [36]. More commonly, uranium and thorium isotopes can also be measured after separation by $Fe(OH)_3$ precipitation, ion exchange techniques and electroplating on platinum discs [37]. Thus specific methods often involve elaborate and specialised chemical procedures for each isotopic species so that generalisations cannot be made here. To study in more detail some examples of radiochemical separation techniques the reader is referred to Chapter 5.

10.2.5 Crustal Sampling – Rocks and Minerals

The phenomenon of radioactivity was first discovered when, by chance, some uranium-bearing minerals fogged a photographic plate. Since then refined techniques of monitoring the radioisotopes in rocks and minerals have led to the understanding of many geological processes, e.g. mountain building, ore formation, volcanism. Virtually all rocks exhibit radioactivity, almost exclusively through decay of K-40, U-238, U-235, Th-232 and their daughter isotopes. In most rock types radioactive elements are distributed rather heterogeneously so that the major α-activity, from uranium, thorium and their daughters, is concentrated in inclusions comprising less than one thousandth of the rock itself. Most of the potassium activity occurs in the major minerals feldspar and mica. Uranium occurs as the accessory minerals uraninite, orthite, titanite, apatite and zircon. It is known that the earth's crust contains much more radioactivity than the underlying high density silicate mantle and the internal iron-nickel core. If this were not the case the heat generated from radioactive decay would be more than sufficient to melt the entire earth! In addition the continental landmasses, essentially granitic in composition, contain over three times more natural activity than the basaltic ocean floors. It is believed, however, that radioactive heating from the underlying mantle contributes a larger proportion to the earth's heat flow than either the continental or oceanic crust material so that despite their differing activities the crustal thermal gradient is approximately constant on a worldwide scale.

Natural radioisotopes in rocks and minerals are most commonly measured for purposes of age determination or to establish the location of valuable mineral deposits. In geochronological studies the U-Th-Pb, Rb-Sr and K-Ar methods have provided the large majority of radiometric ages. In selecting samples it should be borne in mind that, owing to the variability in concentration of parent nuclide, certain minerals and rocks are best suited for particular dating methods. For example, zircon, uraninite and pitchblende can normally be dated satisfactorily by U-Th-Pb techniques, while muscovite, biotite, hornblende, glauconite, sanidine and whole volcanic rocks may be dated by the K-Ar method. The Rb-Sr method is best suited to muscovite, biotite, lepidolite, microcline, glauconite and whole metamorphic rock. The U-Th-Pb approach, which depends on the decay schemes U-238 $\xrightarrow{8\alpha,6\beta}$ Pb-206, U-235 $\xrightarrow{7\alpha,4\beta}$ Pb-207, Th-232 $\xrightarrow{6\alpha,4\beta}$ Pb-208 (Figure 10.1), encompasses several dating techniques, e.g. U-Pb, Th-Pb, Pb-Pb, U-He [38]. Uranium is isolated radiochemically by acid dissolution or treatment with fused alkalies [39]. Chemical separation is achieved by ion-exchange methods [40] or by liquid-liquid extraction [41]. Uranium abundances are measured by one of the following techniques; spectrophotometry, neutron activation analysis, direct γ-spectrometry, α-spectroscopy, fluorimetry, autoradiography, mass-spectrometry, optical spectroscopy or X-ray fluorescence [42]. Lead can be assayed by X-ray fluorescence, optical spectroscopy, isotope dilution and mass-spectrometry. The Rb-Sr method, which depends on the decay of Rb-87 to Sr-87, requires two samples from the same rock mass to establish the Sr-87/Sr-86 ratio for common strontium and the product Rb-87/Sr-87 ratio [43]. Analyses are exclusively by mass-spectrometry. K-Ar dating is possible through the β-decay of K-40 to Ar-40, the former being determined by neutron activation, flame photometry or optical spectrography, the latter by mass-spectrometry [44]. The Re-Os method (Re-187 $\xrightarrow[4.3 \times 10^{10} \text{ y}]{\beta}$ Os-187) is valuable for dating meteorites and some ore minerals using neutron activation analysis [45]. Also within the primary U-Th decay series there are a number of intermediate isotopes with half-lives in the order of $10^3 - 10^5$ years, some of which show ratios, e.g. Th-230/Pa-231, Th-230/Th-232, which can be correlated with sample age in a similar manner as for ocean sediment dating.

To detect areas of rich uranium, thorium or potassium deposits several techniques are currently employed — portable Geiger counters and scintillometers, semi-portable γ-ray spectrometers, down-hole radiometric probes, airborne scintillometers and γ-ray spectrometers [46]. In addition, X-ray fluorescence analysis and activation analysis permit detection of most valuable metal ores. There are, of course, several areas of the world where natural radioactivities greatly exceed normal levels and possibly expose the local populace to hazardous levels of radiation. The monazite sands of eastern Brazil, associated with high concentrations of Th-232 and daughters, yield an ambient γ-radiation over a hundred times higher than, for example, New York levels [47]. Areas such as these are, fortuitously, uncommon.

10.2.6   Sampling Extraterrestrial Materials

Meteorite studies are of major interest as they indirectly yield data concerning (a) the origin and internal composition of the solid earth, (b) the nature of cosmic rays, their temporal and spatial variations and (c) the origin of meteorites themselves, a topic which bears directly on nucleosynthesis, the origin of the universe and the theory of interaction of solid matter in space. The radioisotopes occurring in meteorites are produced, along with several inactive trace species, primarily by nuclear spallation reactions, the reaction of cosmic ray particles, mainly high energy protons, with target elements in the meteoritic mass. Many nuclides produced in this way have either too short half-lives or are masked by the stable abundance of a meteorite constituent, so that the only major species detectable on earth are the rare gas isotopes and the relatively rare nuclides of other elements, e.g. Ar-37, Ar-39, H-3, Cl-36. To interpret the cosmic ray-produced isotope distribution in a meteorite it is also necessary to know the spectrum of the primary cosmic ray flux and the secondary particles produced by nuclear interactions. This spectrum changes with depth in the target meteorite and can be estimated only approximately. To avoid this problem, it has been shown that the yields of neighbouring isotopes are relatively insensitive to the energy spectrum of cosmic ray particles [48]. By estimation of ratios such as Cl-36/Ar-36, Ar-38/Ar-39, K-40/K-41 the exposure ages of the meteorites, i.e. their time in space flight, following parent body disruption, can be determined. These ratios combine a radioactive nuclide and a stable cosmogenic isotope, which is, in effect, a measure of the integrated cosmic ray flux. Similarly, by comparison of two radioactive nuclide activities of different half-lives, e.g. Ar-37/Cl-36, Ar-37/Ar-39, a measure of the temporal and spatial constancy of the cosmic flux is obtained. Assuming equilibrium between spallation production and radioactive decay rates of both nuclides and knowing their production ratio, the observed concentration ratio becomes a measure of the cosmic ray intensity for different times

in the meteorite's past. In fact, many chemical features of meteorite composition suggest an origin in asteroidal or lunar-size bodies with an age of
about 4.6×10^9 years. Since the radiogenic, or gas-retention, ages date the
time of crystallisation of the parent body matter, they are always greater
than the cosmic ray exposure ages.

Meteoritic Ar-37, Ar-39, Cl-36, H-3, Kr-81, Kr-85, Xe-127, Xe-133 and
Rn-222 are normally measured by gas proportional counting in small
volume detectors after vacuum melting and gas separation using gas phase
chromatography techniques [49]. For analyses of lunar samples identical
procedures are necessary. The level of precision and internal agreement
achieved in recent years by the network of lunar analysis laboratories has
perhaps constituted one of the most impressive demonstrations of the
accuracy and efficiency of modern analytical methods. Some findings of
these studies are discussed in a later section.

10.3   APPLICATIONS OF RADIOACTIVITY MEASUREMENT

10.3.1   Isotopes as Geochemical/Geophysical Tracers

In this and future sections of this chapter some examples of radioisotope
applications are presented. It is hoped that, besides their intrinsic value,
these will both enlarge upon experimental procedures and also illustrate the
approach required for specific problems.

Recent studies of the distribution of fission products and natural radioisotopes have improved our understanding of atmospheric transport
processes. Ground level filter sampling, high altitude collection by aircraft,
rocket and balloon sampling on a systematic basis have shown that, with
respect to mixing rates, the atmosphere requires subdivision into several
reservoirs. Transport patterns appear to be extremely complex with considerable variations in exchange rates as a function of latitude, altitude,
season and year. Fig. 10.3 presents the normal schematic outline of the
atmosphere based on mean thermal characteristics. Radioisotope monitoring
has shown that above an altitude of 120 km there is significant hold-up of
fallout materials, in particular of the lighter gaseous products such as C-14
and H-3. This delay is the result of competition between upward molecular
diffusion and eddy mixing. Thermospheric transport processes, principally
by eddy diffusion, appear to require about 16 days for complete mixing of
molecular debris from 120 km to the mesopause, as determined by atomic
lithium observations [50, 51]. Mixing within the mesosphere, via large-scale
circulation, eddy diffusion and, for large particles gravitational sedimentation, occurs with a mean residence time of about 25 days. The large-scale
transport is believed to follow a meridional pattern with ascent over the
summer pole, descent over the winter pole and well-defined movement from
the summer to winter hemisphere above 50 km. Since detailed data for the
thermosphere and mesosphere are severely limited by sampling and analytical problems these interpretations are subject to future modification. In the
stratosphere and troposphere, however, more detailed sampling has been

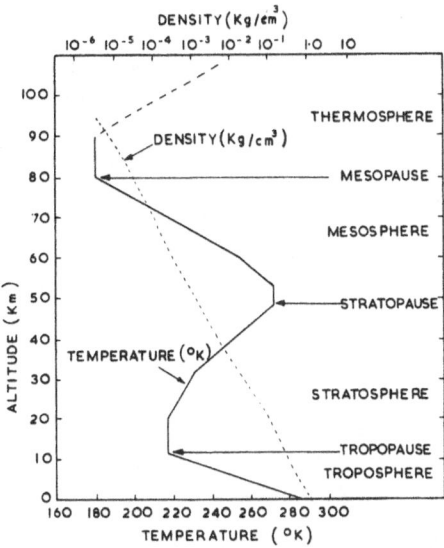

Fig. 10.3. Temperature and density profile of the atomosphere.

effected and the picture of circulation patterns, although by no means certain, appears much clearer. Perhaps the most conclusive type of study is typified by the tracer experiments of 1958 and 1959 when quantities of Rh-102 and Cd-109 were released into the lower thermosphere. Downward mixing of the nuclides into the lower stratosphere was observed to occur selectively at high latitudes of both hemispheres so that vertical mixing in the polar stratosphere appeared to be associated with the polar vortex which forms in high latitudes in the autumn, expands towards the equator in winter and breaks up in early spring [52, 53]. The peaks in stratospheric and tropospheric concentrations of these nuclides occurred some 18 and 30 months respectively after their injection in the thermosphere. The tracers thus required two periods of descent from stratopause to 20 km and exhibited a mean residence time in the atmosphere of about three years.

Fig. 10.4 shows the recent decrease in atmospheric C-14 content following the 1963 tropospheric maximum values [54]. These results reveal that upper and lower stratospheres behave as separate reservoirs with respect to CO_2 exchange. In addition the temporal and spatial fluctuations of artificial C-14 give strong support to diffusion theories of stratosphere-troposphere exchange as opposed to large-scale circulation models. Transfer of C-14, hence of CO_2 and presumably air also, between stratosphere and troposphere appears to result mainly from the seasonal displacement, folding and local rupture of the tropopause boundary and from horizontal diffusion through the tropopause gap regions. Besides the apparent lack of homogeneity within the stratosphere, the overall transfer rate from stratosphere to troposphere is dependent on the diffusion processes which transfer CO_2 from higher levels to the tropopause region, so that the concept of a mean residence time for stratospheric C-14 is a function of both the location and

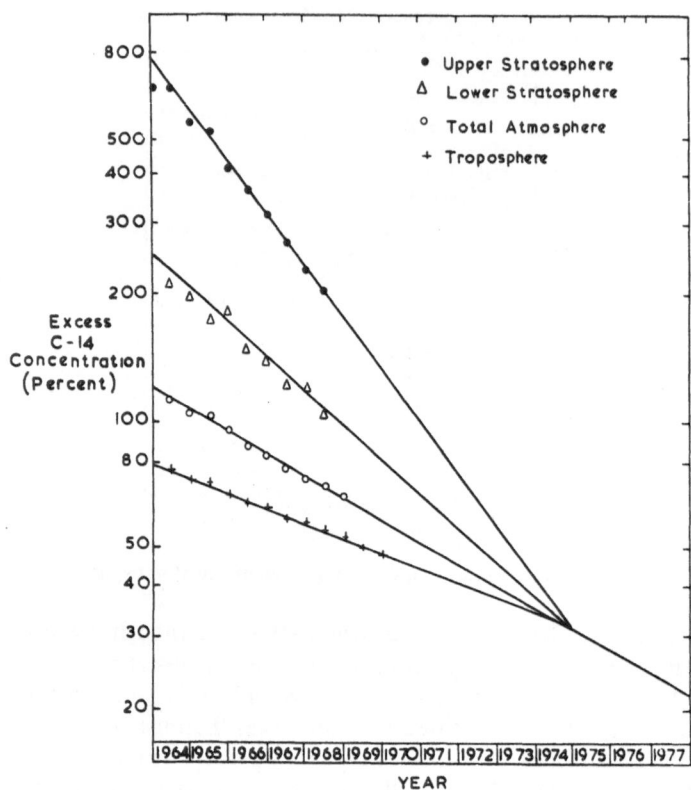

Fig. 10.4. Temporal fluctuation of excess C-14 in the atmospheric reservoirs (after Walton *et al.* [54]).

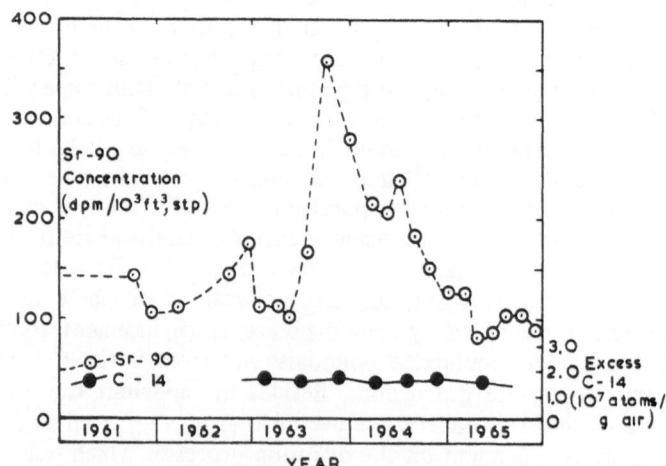

Fig. 10.5. Variations of Sr-90 and C-14 concentrations at 35°–40°S near 20 km. altitude (after Martell [56]).

season of the original nuclide injection. An additional variable has been postulated as a result of studies of pre-bomb C-14 levels in the troposphere [55] using biospheric samples as indicators of atmospheric activities. Fluctuations of over ± 3% were observed in inverse correlation with the solar cycle. Too large an effect to be explained by variation of the C-14 production rate, the fluctuation has been attributed to a solar-sensitive variation of the stratosphere-troposphere exchange processes. Thus the concept of a constant residence time is qualified further to permit yearly variations. An interesting theory [56] based in part on the Sr-90 and C-14 data presented in Fig. 10.5 suggests that a majority of the nuclear debris produced at high altitudes actually was injected by the rising thermonuclear clouds into the northern mesosphere and not, as previously thought, into the stratosphere. Artificial C-14 is produced by escape neutrons throughout a large air volume so that most of it would be left behind at lower levels as the clouds containing fission products and induced activities rose to the northern mesosphere. Assuming slow interhemispheric mixing within the stratosphere the fission products, including Sr-90, would mix rapidly within the mesosphere and then downwards into the southern stratosphere to produce the observed maxima.

Exchange of nuclides within the troposphere is believed to be relatively rapid. Dust samples collected at ground level on air filters indicate yearly maxima in fission product concentrations, e.g. Cs-137, Sr-90, Ce-144, between February and July at temperate and high latitudes. Cosmic ray-produced isotopes, e.g. Be-7 and P-32, also exhibit these seasonal variations. Originally [57], it was believed that these were due entirely to large downflows of stratospheric air of high nuclide content at preferred latitudes. However, more recent data [58, 59] have suggested that seasonal variations in vertical and meridional mixing within the troposphere itself may also be causal factors. Thus, in measurements of tropospheric Be-7/P-32 ratios, it has been demonstrated that seasonal variations of about 30% occur, whereas, if stratosphere-troposphere exchange perturbations were the sole mechanism, fluctuations of about 100% would be expected as the stratospheric source is considerably richer in these nuclides (about 500 times for Be-7).

Among the radongenic isotopes, Pb-210, a daughter of soil Rn-222, appears to be of considerable value as an atmospheric tracer. The highest concentrations are found within a layer several kilometers thick immediately above the tropopause [14]. Seasonal variations of Pb-210 concentration within this layer indicate that atmospheric transport could be controlled by an equilibrium between eddy diffusion and gravitational settling and that in the future the tracer may be used to illuminate the finer details of trans-tropopause exchange. On a smaller scale, short-lived nuclides can be used as indicators of localised atmospheric events [60]. For example, measurements of Cl-38 and Cl-39 activities in rainfall have defined the efficiency and time-scales involved in precipitation scavenging of aerosols. Thus Cl-38 ($t_{1/2}$ 37 m) and Cl-39 (55 m), measured rapidly by AgCl precipitation and γ-spectrometry, have half-lives of the same order of magnitude as the formation time of rain. On the assumption that Cl-38 and

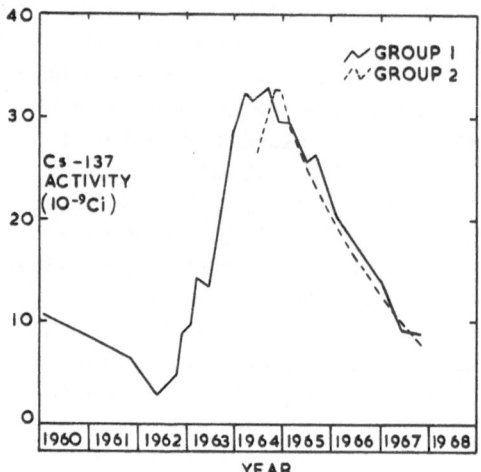

Fig. 10.6. Recent whole-body levels of Cs-137 in two groups of Scandinavians (after Lindblom [61]).

Cl-39 activities decrease with increasing drop size, the rates of nucleation may be studied.

Within the biosphere, research has been concentrated on establishing fission product levels, with particular emphasis on human activities. Fig. 10.6 presents the whole body Cs-137 burdens determined in two groups of Scandinavians [61]. Maximum levels occurred 2–3 years after the period of major weapon testing. This delay is attributable to hold-up firstly in the stratosphere and latterly in nuclide incorporation by the biosphere. The relationship

$$Q(i) = 0.80 \ [F(i) + F(i-1) + F(i-2)]$$

expresses the average body burden, $Q(i)$, in nCi for the year i as a function of the total annual deposition, $F(i)$, in mCi/km^2 (1 nCi = 10^{-6} mCi = 3.7 × 10^1 d.p.s.) [62]. The decrease in average body burden in recent years has a half-life of 1.5 years, indicative of a biological half-life of 90 days. The mean genetic Cs-137 dose at maximum in 1964 was less than 2 mrad/year compared to the average background dose of 125 mrad/year so that, if considered alone, this excess activity may be considered insignificant. However, the existence of many additional nuclides which present excess radiation hazards makes consideration of only one species rather misleading. Concentration effects within the body, for example the ability of the lung to concentrate particulates, can lead to locally active regions where doses considerably greater than the mean are experienced. Table 10.2 presents the body tissue dose due to external and internal radiation from purely natural sources. In general, worldwide fallout from nuclear weapon tests has not markedly enhanced radiation levels. However, local fallout, debris from detonation of 'surface' devices and leakages from nuclear reactors have occasionally radically increased ambient dose rates. For example, the

TABLE 10.2. *Body Tissue Dose Rates from Natural Radiation**

Radiation origin	Dose rates (mrem$^+$/year)		
	Gonad	Haversian canal	Bone marrow
(a) *External:*			
Cosmic rays (including neutrons)	50	50	50
Terrestrial (including air)	50	50	50
(b) *Internal:*			
K-40	20	15	15
Ra-226 and daughters	0.5	5.4	0.6
Ra-228 and daughters	0.8	8.6	1.0
Pb-210 and daughters	0.3	3.6	0.4
C-14	0.7	1.6	1.6
Rn-220	3	3	3
Total	125	137	122

* Data represent mean values for 'normal' regions and are based on United Nations report [63].
+ 1 mrem = 10^{-3} rem; 1 rem (roentgen equivalent man) is the dosage, in rads, multiplied by the relative biological effectiveness (RBE; for β, X and γ rays RBE is 1, for fast neutron, α and proton radiation 10). A rad is the dosage which deposits 100 ergs/gram of material.

inadvertant release of fission products at the Windscale reactor in 1957 released about 20000 Ci of I-131 alone and raised local radiation dosages to about 4 mr/hour (1 mr = 10^{-3} roentgen).

Within the oceans the movement of currents and major subsurface water masses are only now being studied with precision. Prior to the nuclear era, naturally-occurring isotopes were distributed in the oceans at near-equilibrium concentrations so that the less sensitive counting equipment of these times prevented systematic detection of small differences in isotopic distributions. As mentioned previously, the artificial nuclides are now entering the deeper water masses from the surface ocean (residence time about 10 years) so that monitoring of their rates and directions of transport should permit accurate definition of oceanographic mass transfer processes. In this respect, the isotopes C-14 and H-3 should be of considerable value. Fig. 10.7 shows some C-14 concentration depth profiles in the Pacific as a function of time [64]. Though many such profiles are needed for detailed interpretation it is clear that artificial C-14 has been absorbed in the oceans and that, at this particular sampling station several distinct water masses exist. As the relatively shallow and well-mixed surface layer begins to reach equilibrium with the atmosphere, the rate of decrease of atmospheric C-14 will be controlled primarily by the exchange rate between the surface and deep sea. Studies of oceanic C-14 distributions are, however, somewhat complicated by the apparent interference of the marine biosphere. Evidence has been

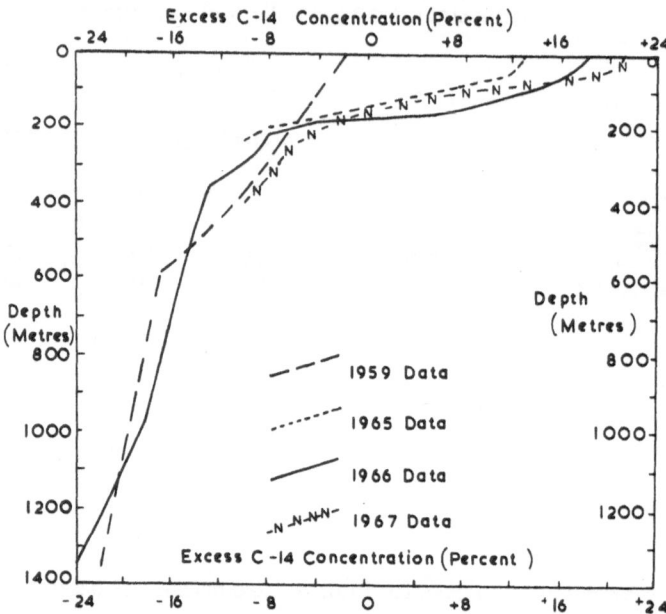

Fig. 10.7. Vertical profiles of C-14 in the North Pacific Ocean (after Fairhall and Young [64]).

found [65] suggestive that CO_2 uptake by the oceans may be influenced, at least in part, by enzymatic processes. Besides this, a general C-14 concentration minimum at mid-depths in the Pacific has been attributed to increased C-12 levels from marine particulate life [66]. Tritium, on the other hand, appears to provide a more direct method of monitoring ocean water flow although the relatively short half-life (12.7 y) is two orders of magnitude less than the deep ocean turnover time. In fact, this could be turned to advantage in cases where, for example, subsurface ocean masses of old water low in tritium receive a surface water contribution rich in the artificial nuclide. The movement of the surface artificial H-3 downwards and through the deep ocean system, with little or no natural background levels, should then be detectable.

Uranium-thorium decay series nuclides are also of value in oceanographic studies. Sample analyses have established that the ratio Th-228/Th-232 is very high, with excess levels of Th-228. Ra-228 levels are of particular interest since the 6.7 year half-life makes it applicable to tracing marine processes occurring on a 1–30 year time scale, such as monitoring mixing through the thermocline or vertical eddy diffusion rates near the ocean floor. It appears that Ra-228 is supplied to the ocean primarily by diffusion from Th-232-bearing sediments such as are found in nearshore and deep-ocean deposits [36]. These feed-in mechanisms enable mixing rates of nearshore and main ocean surface waters to be estimated since the former are originally richer in Ra-228. Similar possibilities exist in deep water studies where faster bottom water mixing produces decreased Ra-228 levels.

On the grounds of mixing rate determinations such as these, optimum dumping sites for, say, nuclear or chemical wastes can be, and have been selected to minimise mixing and resultant interaction with the marine biosphere. Experimental techniques involve coprecipitation of Ra isotopes with $BaSO_4$ from about 200 litres of water followed by conversion of the precipitate to $BaCl_2$. Counting samples are monitored either by extraction of Ac-228 with yttrium carrier or TTA extraction or by permitting daughter Th-228 to grow to partial equilibrium with Ra-228 during storage for at least four months. In both cases samples are counted on planchets using gas flow proportional counters.

10.3.2 Isotopes in Dating Techniques

It was mentioned earlier that specific dating methods are applicable to particular sample types. In addition, however, each technique has a specific age range of applicabity. Thus tritium dating in hydrology is confined to time-scales of up to 100 years, radiocarbon dating to less than 5×10^4 years, K-Ar to greater than 10^6 years and U-Th-Pb commonly to greater than 5×10^7 years. The dating method appropriate to a specific project is therefore not only a function of sample type but of sample age also. From this viewpoint of age ranges, it is significant to note the value of the uranium series methods which fill the 'gap' in Pleistocene chronology between eras covered by the C-14 and K-Ar approaches. The measurement of both Th-230/U-234 and Pa-231/U-235, for example, in coral or ocean sediments provides a stringent age determination method whereby non-concordant ages indicate deviation from the basic assumptions of a 'closed system' [67]. Thus, although complicated by migration and irregular distribution of the daughter species, in favourable cases deposition rates can be estimated. For example, the South Pacific Basin which is relatively far from land and of low biological activity (two sources of sediments) contains mainly red clay deposits with sedimentation rates below 1 mm/1000 years. At the other extreme, in the Caribbean Sea, close to land and highly active biologically, precipitation rates of the mainly carbonate sediments appear to exceed 1 cm/1000 years.

The assumption that a sample has remained a 'closed system', i.e. without addition or loss of the particular species in question, is common to virtually all nuclear dating methods and is unfortunately frequently proved invalid. In carbon-14 dating any non-contemporaneous carbonaceous materials such as twiglets or humic acids entrained during burial or exposure can result in gross inaccuracy. In a recent study of a series of British mortars of known age, for example, C-14 ages too old by an order of magnitude were observed and attributed to the presence of inactive carbonate in the original lime preparation [68]. In methods, such as U-He and K-Ar, which involve the assumption of gas retention within the minerals, leakage of the daughters during geological time has often occurred. In addition excess primordial gases may be present. Similarly in U-Th-Pb techniques original lead may be present or alternatively the isotopic distribution may be altered by weathering. There are indeed many pitfalls encountered in virtually all dating methods. Nevertheless, with care in sampling and interpretation of data much has been learned about terrestrial evolution. Through radioactive dating methods, a concentration of crustal ages at about 3.5×10^9 years is found indicating that at that 'moment' in geological history mantle/crust separation occurred. Dating of volcanic-ash falls and lava flows, deposited instantly in geological time and often interstratified with fossiliferous sediments, has provided the best reference points for defining time scales and planetary evolution. Contintental drift, or tektonic plate, theories, whereby new magmatic matter upwells from oceanic ridges to supersede the older seafloor as it moves towards the continents, have been proved in part by correlating K-Ar ages of oceanic basalts with distance from the ridges. The movement and distribution of mankind in more recent, but yet distant, times have been monitored by C-14 dating of debris material. In this way it was shown that, contrary to popular belief, man appeared in North America only about 11,000 years ago to spread rapidly through the continent. Glacial retreat and advance and vegetational development have also been defined, the latter in conjunction with pollen analysis. For very recent times, the tritium dating method enables measurement of subterranean water flow rates and of lake storage times providing the water ages are appreciable relative to the isotopic half-life. Nuclear geochronology therefore spans the complete range of geological and historical times with approaches which, although often limited, are essentially powerful and soundly based.

10.3.3   Isotopes in the Cosmosciences

Perhaps the most exciting feature of recent lunar analyses was the finding, by counting and mass-spectrometric methods, that the lunar mare surfaces appear extremely old. The K-Ar, Rb-Sr and U-Th-Pb data indicate that crystallisation of the mare basaltic rocks ended about 4×10^9 years and 3×10^9 years ago in Mare Tranquillitatis and Oceanus Procellarum respectively. Some samples of lunar soil and breccia, however, gave concordant U-Th-Pb and Rb-Sr ages of 4.6×10^9 years indicative that lunar evolution began perhaps more than half a billion years prior to crustal crystallisation. The disparity between crystalline rock ages from different mare suggests that a period of about a billion years was required for maria formation. These ages therefore indicate a highly active formation and differentiation sequence during the first two billion years of lunar history followed by a relatively static period. To learn more of the pre-mare period of lunar evolution, an interval obliterated in the terrestrial record, we must await future analyses of the older 'highland' materials. By combining evidence from radionuclide and stable isotope measurements it has been shown [49, 69] that the recent 'top' surface of specific lunar rocks was at one time at the 'bottom', i.e. that tumbling around and partial burying has occurred. In addition these studies revealed that rock erosion, in the absence of a significant atmosphere but subject to a flux of micrometeorite, solar and cosmic particles, is about 10^{-7} cm/year. We know also, from exposure age estimations, that some rocks have been on or within a few centimetres of the surface for at least ten million years and within two metres for at least five hundred million years.

On the basis of the known galactic cosmic ray flux and the target cross-sections, at least half of the H-3, Ar-37 and Ar-39 activities are produced by solar cosmic rays. Indeed much of the value of lunar studies now centres around the use of the findings to indicate solar wind, cosmic ray flux and solar abundance data.

The lunar abundances of cosmogenic nuclides such as Be-7, Na-22, Al-26, Ti-44, Sc-46, V-48, Cr-51, Mn-54, Co-56 and Co-57 are considerably different from meteoritic levels [70]. This is the result firstly of lunar production primarily by solar proton interactions and secondly of the different chemistry of the surface materials. In addition the isotopes produced by solar cosmic ray interactions in meteorites are normally lost by ablation during passage of the meteorite through the earth's atmosphere. Analyses of these isotopes by non-destructive γ-spectrometry confirmed that surface exposures of rocks and fines are long compared with the 0.74 My half-life of Al-26. Th/U ratios, approximately constant in Apollo 11 data at 3.8, indicate very little geochemical differentiation and are in good agreement with a common nucleosynthesis for lunar and terrestrial materials.

A somewhat negative but significant application of cosmic radioactivity lies in the assumption that many primary activities are now extinct. The most important examples with respect to meteoritics are I-129 ($t_{1/2}$ 1.7 $\times$ 10^7 y) and Pu-244 ($t_{1/2}$ 8.2 $\times$ 10^7 y) since both decay to form measurable excesses of stable Xe isotopes, Xe-129 by β-decay from I-129 and Xe-131 to Xe-136 by spontaneous fission of Pu-244. Through an interesting example of analogy and scientific intuition, Kuroda [71] has drawn a markedly strong comparison between the physical processes involved in the formation of particulates from nuclear weapon detonations and of the solar system from supernova explosion. By analysis of the analogy, he was able to deduce that the time interval between element synthesis in stars and formation of the solar system was $1-3 \times 10^8$ years.

Radioisotopes of cosmogenic origin are first extracted by chemical or physical means followed by low level counting in small Geiger, or, more commonly, proportional counters. Background reduction through extensive shielding and anticoincidence circuitry is required. In some cases γ-spectroscopy is valuable. Thus γ-γ coincidence counting, a non-destructive method, can be applied to all positron emitters, e.g. Na-22, Al-26 [72]. Measurement of Ar-39/Cl-36 ratios in iron meteorites are consistent with theoretical data so that average cosmic ray intensities appear to have been constant within the last 10^5 years [73]. Estimation of Al-26/Be-10, Be-10/Cl-36 and Al-26/Cl-36 ratios has indicated that the cosmic ray intensity has been constant within a factor of two during the last 10^7 years [74]. Accurate analyses of Ar-37/Ar-39 in freshly-fallen meteorites have shown that local variations in cosmic ray flux occur, with 10–20% increase per A.U. (astronomic unit) distance from the sun. Measurement of isobaric pairs, e.g. He-3/H-3, Ne-22/Na-22, Ar-36/Cl-36 provides a measure of the meteorite exposure ages. These studies reveal that most stone samples have exposure ages below 10^8 years with a concentration below 10^7 and extreme values of 2×10^4 and 2.2×10^8 years [75, 76]. For iron meteorites the mean exposure age is over ten times longer than for stones with many values around 7×10^8 years and extremes of 4×10^6 and 2.3×10^9 years [77, 78]. Radiogenic ages suggest that meteorite crystallisation occurred about $1-5 \times 10^9$ years ago [79] so that it appears that meteorites existed in large parent bodies, of asteroidal size, prior to breakup, when cosmic ray exposure began, with presumably different meteorite types resulting from breakup of different parent bodies at the times indicated by the exposure age clusters.

10.3.4   Isotopes in Health Analysis

Man has always been exposed to radiation both externally and internally. Soon after the discovery of radioactivity it was found that exposure to high concentrations of Ra-226 and Rn-222 from medical treatments and self-luminous paints resulted in early death via anemias, bone lesions and tumours. It is now generally accepted that excessive radiation exposure can be a causal factor in, amongst others, cases of erythema, epilations, trans-epidermal injury, cellular damage, cancers, tumours, kidney damage, degeneration of the central nervous system, lung fibrosis, loss of fertility, genetic mutations and early death. Accordingly, with the likelihood of future nuclear detonations and the certainty of increases in nuclear power facilities, it is vital that human exposure to radiation is strictly controlled and minimised. Without attempting alarmist propaganda, it seems likely that by 2000 A.D. we must be prepared to live in harmony with the expected minimum of 6×10^7 gallons of liquid reactor wastes (total β-activity in U.S.A. alone about 1.6×10^{11} Ci) and with approximately 10^{12} Ci of atmosphere Kr-85 and H-3 released to the environment by reactors [80]. There is obviously, then, a need for radiation monitoring! To this effect, maximum permissible doses have been established, levels of environmental exposure which are considered safe for the populace. For example, the suggested limit of permissible dose to the whole body from artificial radiation is 0.17 rad/year, although occupational exposure a hundred times this value is acceptable. The general population figure is about equal to the natural radiation dose. The controversy, then, at present surrounds the concept of an 'acceptable' or 'permissible' radiation level. For example, according to several specialists [81] if everyone receives this 'permissible. amount there will be 16,000 additional cases of cancer plus leukemia each year in U.S.A. and 4400 in U.K.. Of the specific nuclides present in fallout and reactor effluent, I-131 is known to concentrate in the thyroid and is associated with thyroid cancer, Sr-90 similarly is correlated with bone cancer, while its decay product Y-90 accumulates in the gonads and may cause genetic damage. C-14 is incorporated into all living matter, producing extensive long-lived, low-level exposure, Cs-137 concentrates in soft tissues, Kr-85 dissolves in body fluids and H-3 is intrinsically present in all aqueous systems. It is, however, impossible as yet to relate quantitatively exposure doses with such consequences as shortened life, cancer incidence, etc. These relationships will require extended surveys over many decades to gain statistical validity. It is therefore not known whether a fixed 'threshold' radiation dosage is necessary before biological damage is done. In other words, the unknown question is; is any dose, even the natural level, of radioactivity harmful to mankind? Although radiation control authorities appear to return a negative response to this question by their assumption that twice the natural dose is acceptable, from a consideration of the interactions at molecular and cellular level it does seem likely that even small dosages can cause health aberrations.

10.4 CONCLUSIONS

Radioisotope assay techniques can be applied with advantage to a wide spectrum of scientific problems. In general, the aim of such studies is to define more precisely the mechanisms which have been, are and will be, responsible for stabilisation of our planet. At the present time, when the recognised pressures of population, pollution and ecological damage threaten that stability, it seems imperative that terrestrial limitations are fully understood. It is to this end that many investigations involving radioisotopes are planned. The application of some such projects are not always obvious. Yet, for example, a study of river or ocean mixing rates can specify the capacity of that dynamic system to receive and disperse not only radioactive wastes but also common industrial effluents. Whatever happens in the future, radioactivity studies have a significant role to play. If we are to succeed in accommodating the nuclide accumulations from reactors then techniques of radioactivity assay must be emphasised. With understanding, stringent precautions and wise leadership a secure and safe future can yet be achieved.

REFERENCES

1. M. S. BAXTER, Ph.D. Thesis, Univ. of Glasgow, (1969).
2. S. GLASSTONE, The Effects of Nuclear Weapons, U.S. At. Energy Comm. (1962).
3. Fed. Rad. Counc., Rep. No. 4 (1963).
4. N. A. HALLDEN, I. M. FISENNE, L. D. Y. ONG and J. H. HARLEY, U.S. At. Energy Comm., Rep. HASL-117 (1967).
5. L. B. LOCKHART, Jr., and R. L. PATTERSON, Jr., The Natural Radiation Environment (J. A. S. Adams and W. M. Lowder, eds.), p. 279. Univ. Chicago Press, Chicago, (1964).
6. P. J. DREVINSKY and J. PECCI, Proc. 2nd Int. Conf. Radioactive Fallout from Nuclear Weapons Tests, Germantown, Maryland, p. 158. (1964).
7. L. B. LOCKHART, Jr., R. A. BAUS and I. H. BLIFFORD, Jr., *Tellus*, 11, 83 (1959).
8. J. R. LAI and E. C. FREILING, Radionuclides in the Environment. (R. F. Gould, ed.), p. 337. Am. Chem. Soc., Washington, (1970).
9. R. C. WOOD, *J. Appl. Meterol.*, 3, 194 (1964).
10. J. F. BLEICHRODT and E. R. van ABKOUDE, *J. Geophys. Res.*, 68, 5283 (1963).

11. D. P. KHARKAR, V. N. NIJAMPURKAR and D. LAL, *Geochim. et Cosmochim. Acta,* **30**, 621 (1966).
12. L. B. LOCKHART, Jr., The Natural Radiation environment. (J. A. S. Adams and W. M. Lowder, eds.), p. 331. Univ. Chicago Press, Chicago, (1964).
13. W. M. COX, R. L. BLANCHARD and B. KAHN, Radionuclides in the Environment. (R. F. Gould, ed.), p. 436. Am. Chem. Soc., Washington, (1970).
14. H. W. FEELY and H. SEITZ, *J. Geophys. Res.,* **75**, 2885 (1970).
15. D. D. HARKNESS, Ph.D. Thesis, Univ. of Glasgow, (1970).
16. H. De VRIES and G. W. BARENDSEN, *Physica,* **18**, 652 (1952).
17. R. A. SHARP and J. G. ELLIS, Proc. 6th Int. Conf. Radiocarbon and Tritium Dating, Pullman. p. 17. (1965).
18. M. A. GEYH, Proc. I.A.E.A. Symp. Radioactive Dating and Methods of Low-Level Counting, Vienna. p. 575. (1967).
19. B. N. AUDRIC and J. V. P. LONG, *J. Sci. Instr.,* **30**, 467 (1953).
20. J. F. CAMERON, Proc. I.A.E.A. Symp. Radioactive Dating and Methods of Low-Level Counting, Vienna. p. 543. (1967).
21. G. H. ÖSTLUND, M. O. RINKEL and C. ROOTH, *J. Geophys. Res.,* **74**, 4335 (1969).
22. H. H. LOOSLI and H. OESCHGER, *Earth Planet. Sci. Lett.,* **5**, 191 (1968).
23. H. H. LOOSLI, H. OESCHGER and M. WAHLEN, Proc. I.A.E.A. Symp. Radioactive Dating and Methods of Low-Level Counting, Vienna. p. 593. (1967).
24. L. MACHTA and H. F. LUCAS, Jr., *Science,* **135**, 296 (1962).
25. S. GOLD, H. W. BARKHAU, B. SHLEIEN and B. KAHN, The Natural Radiation Environment. (J. A. S. Adams and W. M. Lowder, eds.), p. 369. Univ. Chicago Press, Chicago, (1964).
26. D. D. HARKNESS and A. WALTON, *Nature,* **223**, 1216 (1969).
27. M. S. BAXTER, M. ERGIN and A. WALTON, *Radiocarbon,* **11**, 43 (1969).
28. W. F. LIBBY, Radiocarbon Dating. Univ. Chicago Press, Chicago, (1955).
29. M. EISENBUD, Environmental Radioactivity. p. 115. McGraw-Hill Book Co., New York, (1963).
30. P. L. PHELPS, K. O. HAMBY, B. SHORE and G. D. POTTER, Radionuclides in the Environment. (R. F. Gould, ed.), p. 202. Am. Chem. Soc., Washington, (1970).
31. P. K. WEYL, Oceanography: An Introduction to the Marine Environment. p. 103. John Wiley and Sons, Inc., New York, (1970).
32. R. GERARD and M. EWING, *Deep-Sea Res.,* **8**, 298 (1961).
33. A. W. YOUNG, R. W. BUDDENMEIER and A. W. FAIRHALL, *Limnol. Oceanog.,* **14**, 634 (1969).
34. W. S. BROECKER, E. R. BONEBAKKER and G. G. ROCCO, *J. Geophys. Res.,* **71**, 1999 (1966).
35. W. S. MOORE, *J. Geophys. Res.,* **74**, 694 (1969).
36. D. HEYE, *Geochim. Cosmochim. Acta,* **34**, 389 (1970).
37. S. G. BHAT, S. KRISHNASWAMY, D. LAL, RAMA and W. S. MOORE, *Earth and Planet. Sci. Lett.,* **5**, 483 (1969).
38. H. FAUL, Ages of Rocks, Planets and Stars. p. 17. McGraw-Hill Book Co., New York, (1966).

39. C. L. WILSON and D. W. WILSON, Comprehensive Analytical Chemistry. Elsevier Press, Amsterdam, (1959).
40. K. A. KRAUS and F. NELSON, Proc. 1st. Intern. Conf. Peaceful Uses At. Energy, 7, 118 (1956).
41. A. M. POSKANZER and B. M. FOREMAN, *J. Inorg. Nucl. Chem.*, 16, 323 (1961).
42. E. I. HAMILTON, *Nature*, 206, 251 (1965).
43. D. L. EICHER, Geologic Time, (A. L. McAlister, ed.). Prentice-Hall Inc., New Jersey, (1968).
44. O. A. SCHAEFFER and J. ZÄHRINGER, Potassium-Argon Dating. Springer-Verlag, Berlin, (1966).
45. W. HERR, R. WÖLFLE, P. EBERHARDTE and E. KOPP, Proc. I.A.E.A. Symp. Radioactive Dating and Methods of Low-Level Counting. p. 499. Vienna, (1967).
46. Proc. I.A.E.A. Symp. Nuclear Techniques and Mineral Resources, Vienna, (1968).
47. F. X. ROSER and T. L. CULLEN, The Natural Radiation Environment, (J. A. S. Adams and W. M. Lowder, eds.), p. 825. Univ. Chicago Press, Chicago, (1964).
48. O. A. SCHAEFFER and D. HEYMANN, *J. Geophys. Res.*, 70, 215 (1965).
49. R. W. STOENNER, W. J. LYMAN and R. DAVIS, Jr., *Science*, 167, 553 (1970).
50. D. M. HUNTER, *Science*, 145, 26 (1964).
51. R. J. MURGATROYD and F. SINGLETON, Quart. *J. Roy. Meteorol. Soc.*, 87, 125 (1961).
52. M. I. KALKSTEIN, *J. Geophys. Res.*, 68, 3835 (1963).
53. M. I. KALKSTEIN, *Science*, 137, 645 (1962).
54. A. WALTON, M. ERGIN and D. D. HARKNESS, *J. Geophys. Res.*, 75, 3089 (1970).
55. M. S. BAXTER and A. WALTON, *Proc. Roy. Soc. Lond.* A, 321, 105 (1971).
56. E. A. MARTELL, Radionuclides in the Environment. (R. F. Gould, ed.), p. 138. Am. Chem. Soc., Washington, (1970).
57. R. P. PARKER, *Nature*, 193, 967 (1962).
58. S. AEGERTER, N. BHANDARI, RAMA and A. S. TAMHANE, *Tellus*, 18, 212 (1966).
59. C. W. THOMAS, J. A. YOUNG, N. A. WOGMAN and R. W. PERKINS, Radionuclides in the Environment (R. F. Gould, ed.), p. 158. Am. Chem. Soc., Washington, (1970).
60. J. A. YOUNG, N. A. WOGMAN, C. W. THOMAS and R. W. PERKINS, Radionuclides in the Environment (R. F. Gould, ed.), p. 506. Am. Chem. Soc., Washington, (1970).
61. G. LINDBLOM, *Tellus*, 21, 127 (1969).
62. B. LINDELL and A. MAGI, *Arkiv för Fysik*, 29, 69 (1965).
63. United Nations, Second Comprehensive Report of the Scientific Committee on the Effects of Atomic Radiation, (1962).
64. A. W. FAIRHALL and J. A. YOUNG, Radionuclides in the Environment. (R. F. Gould, ed.), p. 401. Am. Chem. Soc., Washington, (1970).
65. R. BERGER and W. F. LIBBY, *Science*, 164, 1395 (1969).
66. H. CRAIG, *J. Geophys. Res.*, 74, 5491 (1969).

67. J. N. ROSHOLT, Proc. I.A.E.A. Symp. Radioactive Dating and Methods of Low-Level Counting. p. 299. Vienna, (1967).
68. M. S. BAXTER and A. WALTON, *Nature,* 225, 937 (1970).
69. E. L. FIREMAN, J. C. D'AMICO and J. C. De FELICE, *Science,* 167, 566 (1970).
70. R. W. PERKINS, L. A. RANCITELLI, J. A. COOPER, J. H. KAYE and N. A. WOGMAN, *Science,* 167, 577 (1970).
71. P. K. KURODA, Radionuclides in the Environment. (R. F. Gould, ed.), p. 83. Am. Chem. Soc., Washington, (1970).
72. D. HEYMANN and E. ANDERS, *Geochim. Cosmochim. Acta,* 31, 1793 (1967).
73. D. HEYMANN and O. A. SCHAEFFER, *Physica,* 28, 1318 (1962).
74. D. HEYMANN and O. A. SCHAEFFER, *J. Geophys. Res.,* 66, 2535 (1961).
75. J. ZAHRINGER, *Geochim. Cosmochim. Acta,* 32, 209 (1968).
76. F. BEGEMANN, J. GEIES and D. C. HESS, *Phys. Rev.,* 107, 540 (1957).
77. J. C. COBB, *Science,* 151, 1524 (1966).
78. H. VOSHAGE, Proc. I.A.E.A. Symp. Radioactive Dating and Methods of Low-Level Counting. p. 281. Vienna, (1967).
79. E. ANDERS, *Space Sci. Rev.,* 3, 583 (1964).
80. G. R. TAYLOR, The Doomsday Book. Thames and Hudson Ltd., London, (1970).
81. J. W. GOFMAN and A. R. TAMPLIN, Supplement to Testimony presented to Sub-Committee on Air and Water Pollution, Committee on Public Works, U.S. Senate, 18th Nov. (1969).

Chapter 11

Miscellaneous Methods

P. Martinelli

Commissariat a l'Energie Atomique, Gif-sur-Yvette, France

11.1   INTRODUCTION

The analytical methods already described in this book fall into two categories; basically their use involves either radioactivation of the elements of interest or radioactive tracing. In both cases the results are obtained by quantitative determinations of radioactivity.

Radioisotopic sealed sources, on the other hand, are used for most analytical methods which are based on interaction of radiation with matter. The richness of the phenomena involved is such that numerous methods have been proposed, and a variety of apparatus has been developed.

Industrial development of these methods and devices is justified whenever they offer particular advantages over other physical or physicochemical methods. This development is bound to that of electronic equipment, and in this field the scope, reliability and price are improving every year.

A large variety of radiation generators are available in the form of radioactive sealed sources, with a total volume generally less than one ml. The radiations are emitted with a stable spectral distribution, and their mean intensity in most cases depends only on the half-life of the generating radioisotope which is of known value. The main limitation in their use is their relatively low intensity which plays an important role in the elaboration of methods and in the design of apparatus.

Without going into details on the advantages and disadvantages of radioactive methods, we shall point out their potential interest in analysis for on-stream and on-line control, for the design of portable instruments, for non-contact and non-destructive control.

A complete description of every method and its applications is beyond the scope of this chapter and the reader will be referred to the literature for details of methods. Nevertheless, it is necessary to give a minimum number of theoretical formulae for the most important methods, so that the reader may judge whether they may be applied to his particular problem.

11.2   ABSORPTION METHODS

The term absorptiometry is usually applied to methods based on the transmission of radiation through the samples being analysed. They can be applied when mass absorption coefficients of the elements of interest differ sufficiently from those of other components of the matrix. They will be generally used to determine the content of a well defined element in matrixes of known mean composition. Selective methods using the Mössbauer effect are mentioned separately.

11.2.1   Gamma-Ray Absorptiometry

Fig. 11.1 is a schematic representation of the simplest device used in

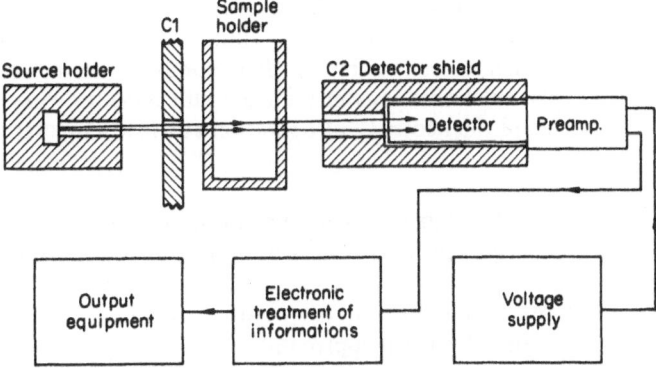

Fig. 11.1. Device for analysis by transmission of γ-rays.

gamma-ray absorptiometry. It consists of a source of radiation emitting a useful beam, collimators C_1 and C_2, a sample holder and a suitable detecting arrangement.

The radiation received by the detector is essentially made up of:

(i)   Direct photons from the source.

(ii) Photons that have been scattered in the source-holder, the collimators, the sample-holder, the sample and the materials surrounding the source and detector.

(iii) Characteristic X-rays of atoms that have been excited by direct or scattered gamma-rays.

The relative intensity of these radiations depends first on the quality of the collimation. With a well collimated device the absorption will not differ appreciably from that predicted theoretically for the absorption of direct photons. With very poor collimation, i.e. a beam of wide geometry, the detector will receive a larger proportion of Compton-scattered photons and, in some cases, characteristic X-ray lines of the excited elements.

Choice of Detector

(i)   For intensity measurements without energy discrimination Geiger-Müller counters are cheap, but the response time is limited by the permissible counting rate. Ionisation chambers allow a fast response time if intense sources are used. For photons of energy below 120 keV, argon or xenon-filled chambers with thin windows are sometimes used to give greater efficiency. For the same reason, high pressure ionisation chambers (up to 150 kg/cm²) are used for the detection of high energy γ-rays such as those from ^{60}Co or ^{137}Cs.

(ii) For intensity measurements of selected energy bands, where it is necessary to discriminate between pulses due to the detection of direct γ-rays and those due to scattered radiation, the scintillation counter is often preferred for its high efficiency, the relatively small volume of its detecting element, and because it can be used easily in collimated geometries. For low energy photons, the

argon-methane or xenon-methane proportional counter offers a
better resolution, a lower energy limit and is less sensitive to
unwanted high energy radiation. Semi-conductor detectors would
be the best but limitations in their use still remain owing to their
high price and the necessity of operation at a low temperature.

Theory

The intensity of the direct beam of γ-rays transmitted to the detector is
often used for the measurement of concentration of the wanted component
in the sample.

For a narrow beam of radiation, the number of photons, N, detected
during the counting time t is given approximately by

$$N = k \, A \, t \, T \, D_\gamma \, e^{-\mu\rho\ell} \tag{11.1}$$

or

$$N = k_\gamma t \, e^{-(\mu_j - \mu_M) x_j \rho\ell} \, e^{-\mu_M\rho\ell} \tag{11.2}$$

where k and k_γ are constants for source yield and geometry, k_γ being the
counting rate for the empty measurement cell or for zero sample thickness;
A is the activity of the source; T is the transmission coefficient of γ-rays
by all the materials other than the sample between the source and detec-
tor; D_γ is the detection efficiency for the direct γ-rays; ρ is the density
of the sample; ℓ is the distance travelled through the sample by the beam;
$\mu = \Sigma\mu_i \, x_i$ is the total mass attenuation coefficient of the sample; x_j is
the mass concentration of the element to be measured, and μ_j and μ_M
are the mass attenuation coefficients of the element and of the matrix,
i.e., all the other components of the sample.

In such collimated systems the intensity of the detected direct γ-rays is
dependent mainly on

(i)   the density of the sample for a given cell and a given path length.
(ii) the mass attenuation coefficient of the sample which depends on
      the energy of the γ-radiation used for the measurement
(iii) the chemical composition of the sample.

Sensitivity

The relative sensitivity, S_j, of the instrument to a small change in the
content of the wanted element or compound at concentration x_j, is, by
differentiating equation (11.2):

$$S_j = \frac{\partial \, N/N}{\partial \, x_j/x_j} = - (\mu_j - \mu_M) \, \rho\ell x_j \tag{11.3}$$

When considering the best conditions for measurement, the aim should
be to have maximum sensitivity although this alone is not sufficient.
Theoretical considerations on sensitivity have been published by
Dziunikowski [1] and by Lubecki [2].

Error in Evaluating x_j

The error made in the evaluation of x_j can be divided into two components: a random error from which the precision can be deduced, and a systematic error. A combination of the two gives the accuracy.

These errors, according to their nature, arise from the statistical fluctuations or long-term drift in counting-rate, matrix composition and density. If necessary the density can be measured separately and correction for counting-rate drift is possible. Drifts in counting rate are due to several causes which include electronic drift and variations of peak shape. This can also be minimised if necessary by temperature stabilisation of the detector, or by checking it periodically with a standard sample. Nul point methods can also be employed.

Drift in matrix composition can sometimes be overcome by comparison between the transmissions of two γ-ray energies but this is only possible in particular cases and the density has to be known. Drift due to matrix composition can be evaluated theoretically from equation (11.2) for the direct beam of γ-rays.

The Contribution of Compton-Scattered γ-Rays to the Counting Rate

Especially in the case of poorly collimated systems or with thick samples, γ-rays which have been scattered by single or successive Compton interactions throughout the sample accompany the direct beam. Their proportion increases with the thickness, volume and density of the sample and is also dependent on the ratio of Compton to photoelectric mass absorption coefficients of photons in the traversed materials. Equation (11.1) is no longer valid. Nevertheless theoretical prediction made by calculations on the direct beam can be considered valid if $\mu \rho \ell \leqslant 2$ or 3.

The influence of scattered γ-rays on the counting rate can be reduced by elimination of the low amplitude pulses when a proportional detector is used or by means of filters of appropriate thickness and generally of high Z materials placed at the entrance to the detector. γ-ray scattering by objects close to the apparatus can be minimised by shielding placed around the source and the detector.

The Contribution of Characteristic X-Ray Lines to the Counting Rate

X-ray fluorescence is excited in the elements of the sample and in the walls of the collimator preceding the detector. In most cases these X-ray lines do not contribute significantly to the counting rate but they can be eliminated by filters placed at the entrance to the detector and by appropriate coatings on the internal walls of the collimator.

The Effect of Heterogeneity and Grain Size

If the distribution of mass in the path of the γ-ray beam is variable, additional fluctuations occur in the counting rate due to the scattered γ-rays, e.g. the case of suspensions flowing through horizontal pipes.

In the case of powders or suspensions, even if the grain-size is constant and the grains are chemically identical, there will be an additional statistical fluctuation in the measured intensity. The effect is not an important one if the mean number of particles present in the path of the beam is large, but large variations in the measured intensity will be observed if the dimensions of the grains are variable and large compared with the total thickness of the sample. In such cases the integrated intensity at any moment will be the mean of exponential attenuations, it is difficult to calibrate the system, and the error is increased. Examples of this are liquids containing relatively large bubbles and muds or slurries when lumps of solid are carried with the stream.

The above considerations on the effect of grain size apply to chemical heterogeneity of the grains which will introduce an additional error.

All these effects result in modification of the intensity of scattered radiation, the effect being a function of the ratios of the size of the heterogeneous zones to the mean path of the radiation through the sample.

Choice of Source

For a source of a given activity and for a given source to detector distance it can be shown that the minimum statistical fluctuation on the transmitted intensity is obtained when:

$$\mu m = \mu \rho \ell = \left[(\mu_j - \mu_M) x_j + \mu_M \right] \rho \ell = 2 \qquad (11.4)$$

In practice the errors which are proportional to the transmitted radiation intensity, such as those due to gain or high voltage variations or to drift in the integrator, are minimum for $\mu \rho \ell = \mu m = 1$. Taking into account the fact that the curve of error versus μm has a rather broad minimum, and of the presence of scattered γ-rays in the measured beam, it is generally accepted that $0.3 < \mu m < 3$ which is by no means a rigorous condition.

The density is generally a fixed parameter and if the width of the sample is also imposed, the number of possible sources will be limited. If the thickness of the sample can be chosen, the choice of the source should be made so as to increase the contrast due to a given variation of the content x_j, or in such a manner that variations due to the matrix are minimised.

Applications of Gamma-Ray Absorptiometry

Reference to formula (11.2) shows that several typical cases occur:

(i) The content x_j can be measured where the attenuation coefficient, μ_j, of element j is significantly different from the attenuation coefficient μ_M of the matrix. In favourable cases a high sensitivity can be achieved. This is generally called 'preferential absorption'. An example is the measurement of the sulphur content in hydrocarbons [3]. The energy of the radiation is generally chosen so that the mass attenuation coefficient is the same for carbon and hydrogen, which occurs approximately at about 22 keV when $\mu_S = 5.5$ cm^2 g^{-1} and $\mu_C = \mu_H = 0.4$ cm^2 g^{-1}. The sources used are ^{109}Cd, which emits silver K X-rays, a silver target excited by an emitter such as ^{241}Am also giving silver K X-rays or a mixture of ^{147}Pm and aluminium which emits bremmstrahlung with a maximum energy about 25 keV. The length of the cell is often between 5 and 10 cm. As the density of the hydrocarbon may be variable the measurement is made with a constant mass of sample, apparatus being available to do this, or the density is measured and a correction applied. The latter method is used in on-stream sulphurimeters. The error is less than 0.03% for industrial apparatus used in the control of sulphur content between 0.3 and 6%. The same principle applies to the determination of tetra-ethyl lead in petrol where sulphur is absent. Other examples are the measurement of silicochloroform ($SiHCl_3$) in mixtures with hydrogen [4] or the sulphur dioxide content of natural methane at high pressure using ^{55}Fe radiation.

The use of low energy photons is often an efficient method in the analysis of binary compounds. Nozaki [5] has extended the range of usable sources with a special method for the elimination of hard γ-rays.

(ii)   Where the measurement has to be made through thick walls or a considerable thickness of sample, e.g. 10 to 50 cm., only high energy gamma-rays can be used and suitable sources are ^{60}Co, ^{137}Cs or sometimes ^{192}Ir. In this case the mass attenuation coefficients do not vary very much with Z, and the determination of the amount of the wanted element depends on density differences between the element and the matrix. Commercially available 'γ-ray densitometers' are used for this purpose.

(iii) In between these two limiting cases is that of the measurement of heavy elements in matrices with much lower density. A large variety of sources can be used, the choice depending on the particular problem to be solved. Preference is often given to low energy γ- or X-ray sources. Typical examples are the measurement of the uranium content of aqueous solutions by transmission of 59.6 keV γ-rays emitted by ^{241}Am and the measurement of lead in ore pulps flowing through pipes in flotation plants [6], see Fig. 11.2.

Special Techniques in γ-Ray Absorptiometry
The technique of adding an absorbing element to the sample has been applied by Norel [7] especially for the measurement of water in various porous materials.

The technique of $4\pi\gamma$-ray absorptiometry was developed by Whittaker *et al.* [8] in the assay of purified uranium and plutonium solutions in nuclear fuel reprocessing plants. Fig. 11.3 shows a schematic representation of one of the measuring heads.

11.2.2   Critical-Edge Gamma-Ray Absorptiometry
In this method the concentration of a given element is obtained by comparing the transmission of two gamma rays with energies selected on either side

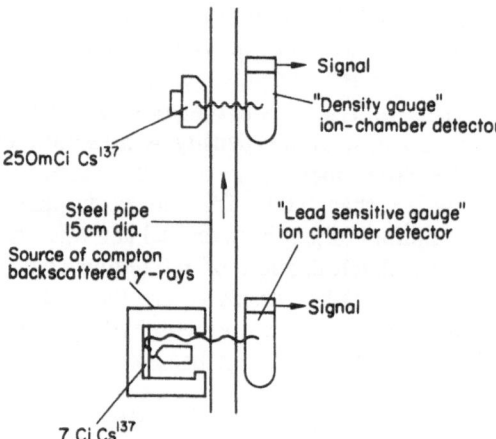

Fig. 11.2. Schematic representation of a device for on-line analysis of lead pulps (from J. S. Watt [6]).

of the K absorption edge of the wanted element, and very close to it. In the absence of the element the difference in transmission between the two photons will be very small. This applies also to LI, LII, LIII, MI, etc., discontinuities [9].

Using subscripts 1 and 2 for the photons energies and j and M for the element and matrix respectively we have:

$$R = \frac{N_1}{N_2} = \frac{k_1}{k_2} e^{(\mu_{2,j} - \mu_{1,j}) x_j \rho \ell} e^{(\mu_{2,M} - \mu_{1,M})(1 - x_j) \rho \ell} \tag{11.5}$$

where k_1 and k_2 are constants and N_1 and N_2 counting rates.

We see that for photons of nearly the same energy $(\mu_{2,M} - \mu_{1,M})$ is close to zero and $(\mu_{2,j} - \mu_{1,j})$ is fairly high, so that R will depend mainly on the total mass of the wanted element.

It can be shown that the optimum value for the thickness of the sample when using two sources of equal γ-ray intensity is given by $\Sigma_i \mu_i x_i \rho \ell \sim 2$, where the attenuation coefficients are taken for the highest γ-ray energy. Attenuation coefficients can be found in tables, or from general formulae such as those given by Leroux [10] and by Theisen [11], but in the neighbourhood of the absorption edges, corrections up to 15% must be applied.

Suitable sources consist of

(i) Pairs of γ-ray emitters although this is rarely feasible. It has been proved possible to measure uranium contents by comparison of transmitted intensities from ^{57}Co and ^{153}Gd sources, the γ-ray energies being on either side of the K absorption edge of uranium [12].

(ii) Secondary X-ray sources composed of a mixture of a β^- emitter with elements which emit secondary characteristic X-ray energies although unfortunately bremsstrahlung is also present.

(iii) Secondary X-rays produced by excitation of atomic energy levels

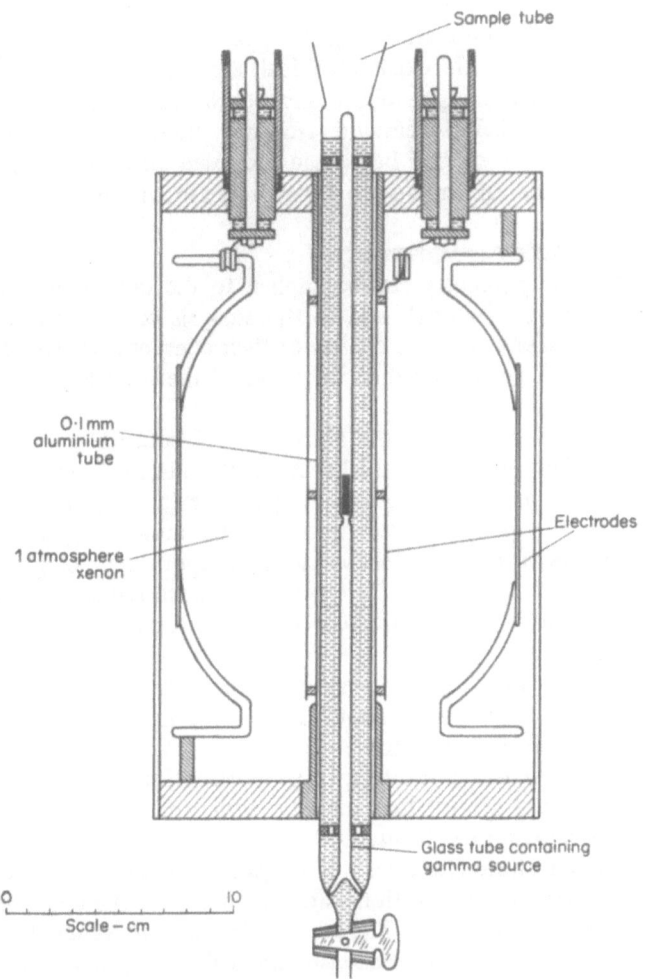

Fig. 11.3 Experimental 4π absorptiometer with cylindrical geometry (after Whittaker *et al.* [8]; reproduced with permission of I.A.E.A.).

in two appropriate targets by irradiation with X- or γ-rays. This technique has been used for the determination of zirconium in acid solutions in presence of hafnium [13].

Measurements may be made in the same way as in non-selective X-ray absorptiometry but improvement can be obtained by the measurement of each transmitted γ-ray intensity through pairs of balanced filters; three filters are sufficient for this purpose. For high atomic number elements the secondary emission of the filters is too high and it becomes necessary to have good collimation and to increase the distance between the filters and the detector.

Theoretically it is also possible to use a γ-ray source of variable energy making use of the Compton scattering of a γ-ray line at various angles. For a given primary energy there exists a critical angle, on either side of which scattered γ-rays have energies above and below the characteristic energy level of the element being measured. However, these systems have at the moment a very poor yield and have been proposed only for the analysis of rare earths for which high intensity γ-ray sources are available.

11.2.3 Beta Particle Absorptiometry

Beta particle absorptiometry can be applied to the determination of the hydrogen content of materials and to the analysis of binary compounds when there is a difference in the density of their components. The thickness of the sample must be smaller than the range of the particles and its value must be known.

Experience has shown that the absorption of beta particles in matter varies exponentially with the thickness of the absorber, except in the vicinity of their range. This absorption comes essentially from inelastic collisions with the peripheral electrons of the atoms. For low atomic number absorbers it has been shown to be proportional to the electron density Z/A. However, for high Z elements elastic scattering by the nucleus is not negligible and the absorption coefficient is proportional to Z/A times a function of Z.

Hydrogen has a value of Z/A very close to unity, while for the other elements this value varies from about 0.5 for the light elements down to about 0.4 for the heaviest elements.

The detector is generally a Geiger or proportional counter, or an ionisation chamber, but a scintillation counter with a plastic scintillator or a semi-conductor detector can be used.

Bremsstrahlung photons are also detected. The intensity of the bremsstrahlung spectrum increases with the atomic number of the target and with the energy of the particles. Though bremsstrahlung generally gives a relatively low background, greater sensitivity will be obtained if a detector has a low efficiency for photons and if the thickness of the samples is not too close to the range of the beta particles.

Measurement of the Hydrogen Content of Samples of Low Z

A well-known example is the measurement of the C/H ratio in liquid hydrocarbons. Two techniques have been used:

(i)   At constant sample thickness Smith and Otvos [14] and Jacobs *et al.* [15] have developed instruments using a nul point method. The sample and a standard are compared by means of two ionisation chambers. The temperature of the cells is kept constant. Density is also measured so that correction factors are provided by these instruments. The absolute error is about 0.1% by weight. According to the authors 1% oxygen or nitrogen are respectively equivalent to 0.07% and 0.02% of hydrogen. The instrument is calibrated by means of known hydrocarbons, ref. Fig. 11.4.

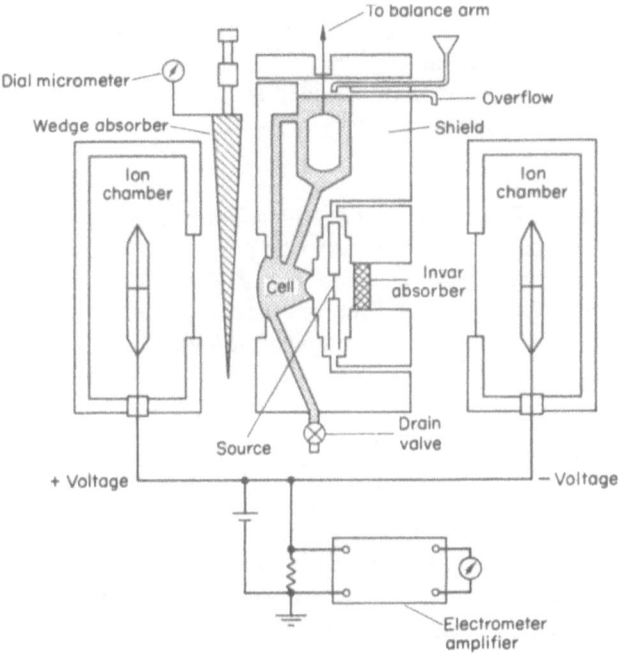

Fig. 11.4 Schematic diagram of a C/H gauge. (Anal. Chem., **28**(3), 325 (1956); copyrighted 1956 by ACS and reprinted with permission of copyright owner.)

(ii) At constant mass of sample per unit area along the beam of particles Berthold [16] has proposed the following formula:

$$\log \left(\frac{I_1}{I_0} \right) = k \, \rho d \, \frac{Z}{A} \tag{11.6}$$

where I_1 and I_0 are measured intensities with and without samples, and d is the thickness of the cell. ρ, A and Z are the density, atomic weight and atomic number of the sample. k is a constant depending on the beta particle energy, the geometry and on the method of measurement.

For a compound, Z/A is calculated according to

$$\frac{Z}{A} = \frac{\Sigma_i Z_i n_i}{\Sigma_i A_i n_i}$$

where the n_i values are the numbers of atoms of each element in the molecule of the compound.

In the apparatus of Berthold the source is mounted on a float immersed in the sample, so that the mass of sample between the source and the detector is constant. The detector is a Geiger counter.

It has been found that the logarithm of the transmitted intensity is a linear function of Z/A for various hydrocarbons and water. With a sample thickness of 0.7 g/cm^2 and a $^{90}Sr - ^{90}Y$ source the smallest detectable percentage of hydrogen is 0.16 provided an accuracy of radiation measurement of $\pm 1\%$ is guaranteed.

General Case

As the intensity of the transmitted particle beam is an exponential function of the mass of the sample traversed it is possible in practice to determine absorption coefficients of elements for a given beta spectrum and the formulae given in section 11.2.1 for γ-ray absorptiometry will be valid.

For aluminium absorbers the following formula has been given by Katz and Penfold [17] for primary beta particle energies between 0.1 and 3 MeV:

range R = 412 E^n mg/cm2

where n = 1,265−0.0954 $\log_e$ E and E = end-point energy (in MeV) of the beta particles.

The absorption coefficient in aluminium $\mu = 22\ E^{-1.33}\ cm^2/g$ or the half-thickness value $m_{1/2} = 0.032\ E^{1.33}\ g/cm^2$. In practice the values $m_{1/2} \sim 1/10$ R or 1/12 R are often used for rapid calculations.

For other elements the following formula given by several authors is convenient for a comparative study of various absorbers:

$$\mu = k\frac{Z}{A}\exp\left(\tfrac{1}{3} Z^{\tfrac{1}{3}}\right)$$

k being a constant for a given beta emitter counted with a given detector in a given geometry. If aluminium is taken as reference

$$\frac{\mu_Z}{\mu_{Al}} = \frac{Z}{A}\exp\left(\tfrac{1}{3} Z^{\tfrac{1}{3}}\right)0.95$$

However, using this procedure, the absorption coefficients will be underestimated for hydrogen and slightly overestimated for high atomic number elements such as lead.

Some Recommendations

In order to minimise the influence of bremsstrahlung on the measurement the use of high atomic number materials near the path of the particle beam must be avoided. The detector may be a Geiger counter, a proportional counter or an ionisation chamber but the last requires sources of higher activity. The method is generally applied to liquid samples. If powders are used the particle size must be sufficiently small compared with the sample thickness.

11.2.4 Alpha Particle Absorptiometry

This technique has limited application in analysis. It is well-known that alpha particles emitted by radioactive sources are distributed in one or more monoenergetic lines. For each energy the range of the particles is about the same. However, there is a dispersion or straggling of the range about a mean value. This straggling is about 3 to 4% of the range. Self-absorption in the source is also responsible for some spread in the ranges of the α-particles. The number of α-particles transmitted by an absorber having a thickness close to the range is very sensitive to variations in the mass or the chemical nature of the absorber. Tables giving α-particle ranges (R), in air are available; for a material having a mass number A they are given in mg cm^{-2} by $0.56 R A^{\frac{1}{2}}$, where R is expressed in cm. of air. These ranges are very small, the range of 5.3 MeV alpha particles in air for instance being 4.72 mg cm^{-2}.

Alpha particle absorptiometry can be applied to the measurement of heavy gas concentrations in air. The detector is generally either a scintillation counter with a zinc sulphide phosphor or a semiconductor.

One application of alpha particle absorptiometry has been described by Hallowes and Hodgson [18]. Their apparatus is designed for the control of the relative humidity of air independently of its SO_2 content. The system operates by comparing the transmission through a non-hygroscopic thin film, and through a sulphonated reticulated polystyrene film which is very hygroscopic but insoluble in water. A sensitivity of about 0.6% relative humidity has been obtained.

11.2.5 Neutron Absorptiometry

Thermal Neutrons
A particular feature of thermal neutron absorptiometry is that neighbouring elements in the Periodic Table can have very different transmission cross-sections.

TABLE 11.1. *Transmission Coefficients for Thermal Neutrons* $(v = 2200 \ m/s)$

Element	Microscopic cross section (barns/atom)		Total macroscopic cross sections Σ_t (cm^2 g^{-1})
	Absorption	Total	
H	0.33	36	21.67
Li	71	72.4	6.28
B	755	760	42.2
C	0.003	5	0.250
O	< 0.0002	4.2	0.160
Al	0.215	1.6	0.034
Si	0.13	2.4	0.052
Cl	31.6	48	0.814
Ca	0.43	9.4	0.143
Fe	2.43	13.4	0.145
Cd	3315	3322	17.8
Sm	8250	8255	33.1
Pb	0.17	11.2	0.033
U	7.6	15.9	0.040
H_2O		~ 107	3.51

Table 11.1 shows the total transmission coefficients of various elements and of water. The values for compounds can be calculated by the addition of elemental cross-sections. Compounds containing hydrogen are exceptional. Thus for water the total transmission coefficient given by the tables in 3.51 cm^2 g^{-1}, while the additivity formula gives 2.45 cm^2 g^{-1}.

Among the elements or compounds more commonly found in nature, boron, chlorine, cadmium and water have relatively high cross-sections while iron, lead and uranium are rather transparent to neutron beams.

The formulae which have been described for γ-ray absorptiometry can be used for the absorption of a collimated beam of neutrons.

A fraction of the neutrons removed from the beam by primary interactions is scattered in every direction. Successive scattering processes occur and a fraction of the multi-scattered neutrons is transmitted to the detector. This build-up is a function of the geometry, of the thickness of the sample and of the ratio of the diffusion cross section to the total cross section. In practice, for a given source-detector distance, the optimum thickness of sample giving the highest precision is $1 < \mu$ m < 3 where μ is the absorption coefficient and m is the mass per unit area in the path of the beam.

The apparatus usually employed has the following components (see Fig. 11.5):

(i) A neutron gun consisting of a block of paraffin, pure polythene or other plastic material, or water, in the centre of which is placed the radioactive source. A cylindrical or parallelepipedical hole between the vicinity of the source and the surface of the thermalising block serves to collimate the useful beam. Externally the block is covered with a sheet of neutron-absorbing material, e.g. 0.4 mm cadmium, in order to achieve biological protection and to collimate the useful neutron beam. The size of the thermalising block is calculated from data given for radiological protection. For a 50 mCi ^{226}Ra/Be

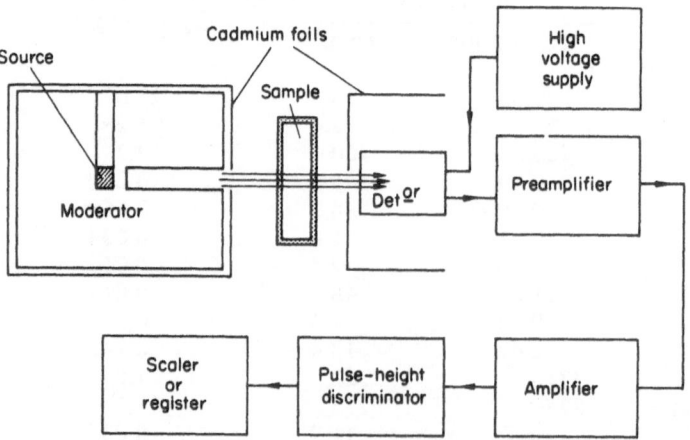

Fig. 11.5. Device for analysis by transmission of thermal neutrons.

source or a 300 mCi [241] Am/Be source, for example, a paraffin cube of 25 cm side is sufficient.

(ii)   A pair of collimating sheets is provided on either side of the sample, e.g. 0.8 mm sheets of cadmium.

(iii) The thermal neutron detector is protected against scattered neutrons by cadmium foil. The detector may be the well-known boron trifluoride proportional counter with a filling gas enriched in [10]B, a scintillation counter equipped with a scintillating glass containing lithium-6 or a helium filled proportional counter. Of these detectors, the first one is probably the least sensitive to epithermal neutrons.

(iv)   The method has been applied to the study of the variations in water content of soil samples [19, 20], the measurement of the boron content of rock samples and the measurement of the lithium content of Li/Al alloys. The application of this method has been limited, up to now, by the low intensity of available sources but this situation may change with the advent of intense radioisotopic neutron sources such as [252]Cf.

Epithermal and Fast Neutrons

This method has not been applied in collimated systems because the counting rate is too low. Nevertheless it has been used in combination with γ-ray absorptiometry for the measurement of the humidity of soils near the surface. Fig. 11.6 shows the apparatus used by Wack [20] for this purpose. The transmitted intensity of epithermal and fast neutrons depends on the water content of the soil. The measurement of the transmitted γ-ray intensity depends mainly on the density of the soil. Combination of the two measurements gives the value of the humidity. A careful calibration is necessary for every kind of soil.

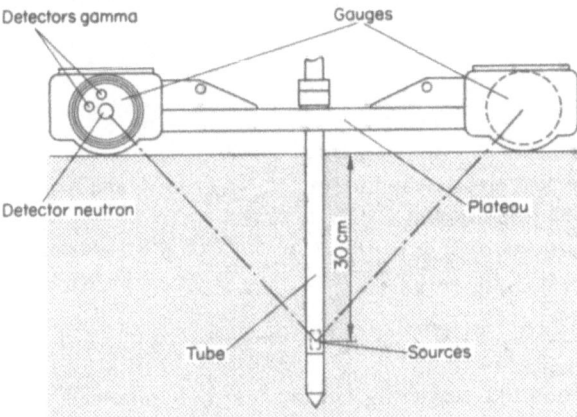

Fig. 11.6. Measurement head of a surface moisture probe based on epithermal neutrons and gamma ray absorptiometry (from B. Wack [20]).

11.3   ANALYTICAL APPLICATIONS OF
RADIATION BACKSCATTERING

Backscattering systems are those for which the radioactive source and the detector are on the same side of the sample. In many cases the sample is 'infinitely' thick, i.e. the intensity of the backscattered radiation is not changed significantly by increasing the thickness a further amount.

11.3.1   Gamma-Ray Backscattering

γ-ray scattering is essentially due to the Compton effect. The intensity of the backscattered photons will depend mainly on the scattering and absorption properties of the sample and on the geometry of the system.

Compton Scattering Coefficients

The differential coefficient for Compton scattering in a given solid angle is given by

$$\left(\frac{\partial \sigma_c}{\partial \Omega}\right)_\theta = \left(\frac{\partial \sigma_e}{\partial \Omega}\right)_\theta N_{Av} \sum_i \frac{Z_i}{A_i} x_i$$

where θ is the angle between the incident and scattered photons, N_{Av} is the Avogadro number (0.6023×10^{24}), and Z_i, A_i and x_i are the atomic number, atomic weight and concentration of the component i of the sample. $(\partial \sigma_e/\partial \Omega)_\theta$ is the differential electronic coefficient for Compton scattering; that is, the probability of scattering into unit solid angle at angle θ. It is given by the Klein and Nishina formula:

$$\left(\frac{\partial \sigma_e}{\partial \Omega}\right)_\theta = \left(\frac{h\nu'}{h\nu_0}\right)^2 \left(\frac{h\nu_0}{h\nu'} + \frac{h\nu'}{h\nu_0} - \sin^2 \theta\right) \frac{r_0^2}{2}$$

$$\text{cm}^2/\text{electron/steradian}$$

where

$$\frac{h\nu'}{h\nu_0} = \frac{1}{1 + \alpha(1 - \cos \theta)} \qquad \text{and} \qquad \alpha = \frac{h\nu_0}{m_0 c^2}$$

r_0 is the classical electron radius, $m_0 c^2 = 511$ keV, and $h\nu_0$ and $h\nu'$ are the energies of the primary and scattered γ-rays.

The total cross section per electron for Compton scattering is given for low values of α by the formula [21].

$$\sigma_e = 0.66527 \times 10^{-24} (1 - 2\alpha + 5\alpha^2 - 13.3 \alpha^3 \ldots)$$

The total Compton scattering coefficient σ_c is obtained from the above formula or from tables.

Table 11.2 shows the ratio of total Compton to total absorption coefficients for some elements and for three different γ-ray energies [22]. It

TABLE 11.2. *Ratio of Compton over Total Absorption Coefficients of Some Elements for γ-Rays*

Element	$\sigma_c\mu$ for $h\nu_o =$		
	10 keV	100 keV	1000 keV
H	0.996	0.997	~ 1
C	0.0839	0.984	~ 1
Al	0.0068	0.846	0.998
Fe	0.0010	0.371	0.990
Sn	0.0013	0.077	0.931
Pb	0.0018	0.021	0.717
U	0.0013	0.091	0.631

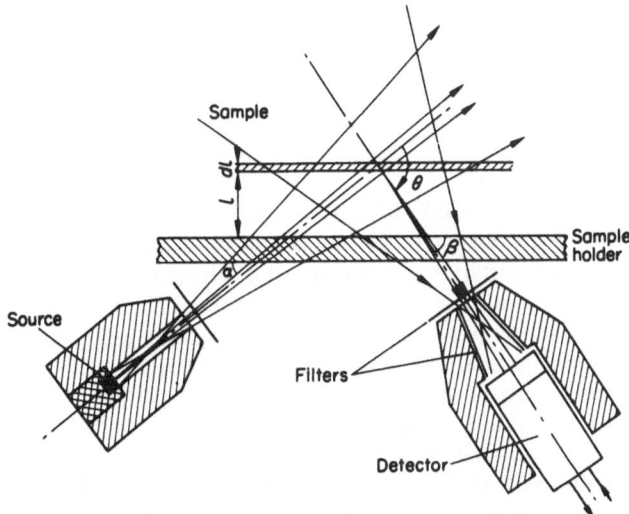

Fig. 11.7. Collimated geometry for analysis by γ-ray backscattering or by X-ray fluorescence.

is seen that σ_c/μ decreases with Z and approaches unity as the energy of the γ-rays increases.

Differential Expression of Compton-Scattered Intensity
Referring to Fig. 11.7 the number, N_c, of photons which are detected in a given direction after single (or primary) Compton interaction is given in differential form by the formula:

$$dN_c = kAtT_\gamma T_{\gamma'} D_{\gamma'} \frac{\Delta\Omega_1}{4\pi} \Delta\Omega_2 \left(\frac{\partial\sigma_c}{\partial\Omega}\right)_\theta \frac{1}{\sin\alpha} \times$$
$$e^{-\left(\frac{\mu_{\gamma e}}{\sin\alpha} + \frac{\mu_{\gamma'e}}{\sin\beta}\right)\rho\,d\ell} \rho\,d\ell \qquad (11.7)$$

where A is the activity of the source, t the counting time, T_γ and $T_{\gamma'}$ the transmission coefficients of the incident and of backscattered γ-rays through materials other than the sample, e.g. sample holder, $D_{\gamma'}$ is the detection efficiency, $\Delta\Omega_1$ and $\Delta\Omega_2$ are the solid angles of irradiation and detection, $\mu_{\gamma e}$ and $\mu_{\gamma' e}$ total mass absorption coefficients of the sample for incident and scattered γ-rays, ρ is the density of the sample, ℓ the distance between the surface of the sample and the scattering element and k is a constant which takes into account self-absorption and yield of source.

Systems with Source and Detector Both Collimated

When the sample volume determined by the two collimators is small relative to the mean free path of the radiation in the sample, equation (11.7) applies directly. When the size of the sample volume is large this formula has to be integrated. Infinite thickness is achieved in practice when

$$\ell = \frac{4}{\rho\left(\dfrac{\mu_{\gamma e}}{\sin\alpha} + \dfrac{\mu_{\gamma' e}}{\sin\beta}\right)}$$

The Compton and total absorption coefficients can be expressed as a function of the composition of the sample and the sensitivity, $\partial N_c/\partial x_j$ of the counting rate for a variation in the content of element x_j is easily derived.

The effects of density on the counting rate must be taken into account in calculating the error in the value of x_j. For systems using single or multi-hole collimators these theoretical formulae agree quite well. The shielding and collimators must be designed so as to minimise direct irradiation of the detector and the effect of the surrounding medium on the counting rate. Heavy metals are generally used in their construction.

The contribution of the counting rate due to γ-rays which have undergone several Compton interactions can be reduced by means of a filter placed at the entrance to the detector. Characteristic X-rays of elements in the sample can be eliminated in the same way. High efficiency detectors are often required by these systems because of their relatively poor geometry. An improvement is obtained by electronic selection of those pulses due to γ-rays which have only been scattered once in the sample.

Non-collimated Systems

Fig. 11.8a illustrates a device having a very compact geometry. This geometry is used preferably, but not only, with low energy photon sources. Lower activities are required but detection of multiple scattered γ-rays becomes important and equation (11.7) no longer applies. The spectrum of the backscattered γ-rays may show two peaks due partly to a shadow effect of the source and partly to a second scattering in the plane of the sample. Fig. 11.8b shows a device allowing the use of high energy γ-rays such as the 662 keV γ-ray of ^{137}Cs, with a good geometrical efficiency.

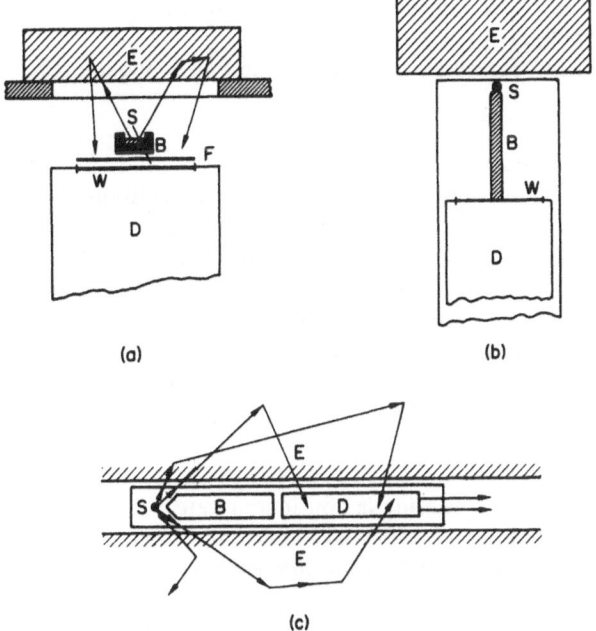

Fig. 11.8. Non-collimated geometries for measurements by γ-ray backscattering. (a) Compact geometry; (b) 'pencil' geometry; (c) probe. Symbols: S = sources; D = detector; F = filter; W = window; B = shielding; E = sample. The compact geometry is also the most commonly used one in X-ray fluorescence apparatus using radioisotopes.

Probes
Cylindrical probes (Fig. 11.8c) are used in mineral prospecting, but they can be used for continuous control. An optimum source to detector distance must be determined for each application, especially when high energy γ-ray sources are used. Here also the multiple scattering contribution to the counting rate is significant.

Applications
Among the few applications of γ-ray backscattering in analysis two will be mentioned. In the measurement of ash in coal several systems have been proposed using various photon sources. In one which has been developed by Rhodes [23] a compact geometry is used with a source consisting of tritium absorbed on zirconium and a scintillation detector. The source emits bremsstrahlung photons with a continuous energy distribution between 0 and 18 keV. The backscattered X-ray intensity decreases with ash content and with the iron content. To compensate for the latter a fraction of the K X-rays emitted by the iron, adjusted by an aluminium filter placed at the entrance to the detector, is allowed to enter the detector.

For measuring ash in coal γ-ray absorptiometry has also been used successfully [24].

Donhoffer has measured the zinc in boreholes by gamma backscattering [25].

11.3.2 Beta Particle Backscattering

The analytical applications of beta-particle backscattering were intensively studied some ten years ago by Müller [26] and by Danguy [27]. There is now a renewed interest in this technique [28–30]. Beta-ray backscattering can be studied and applied with low cost apparatus. If a Geiger-Müller detector is used the radioactive source may be of very low intensity while with high intensity sources used in conjunction with an ionisation chamber statistical fluctuations can be reduced to a very low level. Shielding presents no problem and the weight of the equipment is low.

Theory
Only saturation backscattering is considered here. At saturation, i.e. sample thickness > 0.2 times beta particle range, the intensity, I_s, of the back-scattered radiation is related to the intensity, I_0, of the incident radiation by the equation of Danguy [27]:

$$I_s = 1.3 \, I_0 \, \bar{R} \tag{11.8}$$

where

$$\bar{R} = \sqrt{\sum_i \left[\frac{Z_i \, (Z_i + 1)}{A_i} \, x_i \right]}$$

The coefficient 1.3 refers to 2π geometry. Under practical operating conditions a different coefficient applies while the figure is lower for beta particle energies below 0.6 MeV.

Equation (11.8) may be written

$$I = 1.3 \, I_0 \sqrt{x_j \, (R_j^2 - R_M^2) + R_M^2}$$

where the suffix M refers to the components of the matrix and j to the element of interest. The counting rate increases with x_j if $R_j > R_M$ and decreases if $R_j < R_M$. The relative sensitivity is given by

$$\frac{\partial I / \partial x_j}{I / x_j} = \frac{0.5 \, x_j}{x_j + \dfrac{R_M^2}{R_j^2 - R_M^2}}$$

The error in the determination will depend not only on counting rate precision and instrumental stability, but also on grain size and on the composition of the matrix. It may be noted that the crystalline form of the sample and the hardness of metallic samples have an influence on the intensity of the backscattered beta rays. Unwanted radiation is the same as for beta-ray transmission, i.e. bremsstrahlung and X-ray lines, from the source itself, the sample and the surrounding materials. Their contribution to the counting rate can become important if a collimated geometry is used. In this case the internal walls of the collimator holes should be coated with a low Z material.

Generally a non-collimated compact geometry (Fig. 11.8a) is used so that the influence of unwanted radiation on the counting rate is small. Filters placed on the detector window can be used to improve the sensitivity. Their thickness is usually chosen below 0.2 of the maximum range of the beta particles. The use of filters is equivalent to energy discrimination.

Applications
The method has been applied to the analysis of binary mixtures. Examples of applications which have been described are the measurement of the lead content of lead glass, tungsten in steel, the analysis of the alloys Ta-Nb, Cr-Nb, Ti-W, W-Nb, Cu-Ag, Al-Cu, etc., the examination of the lead content in antiquities and dies, the measurement of the ash content of coal, C/H determinations and applications in the control of solid or liquid pharmaceuticals. In the mineral industry the measurement of the iron + manganese content of iron ores, with an absolute error less than 0.5% may be mentioned.

Commercial equipment uses Geiger counters, ionisation chambers, or plastic scintillators. Because thin shielding is adequate and sources of small diameter are used, measurements on areas less than 1 mm diameter are possible if samples have negligible surface irregularities. Special cylindrical measuring heads with external diameter down to 16 mm have been developed for the examination of the internal coatings of pipes.

11.3.3 Alpha Particle Backscattering

Owing to the limited penetration of alpha particles, analysis by alpha backscattering only involves the superficial layers of the sample. The spectra of backscattered α-particles have characteristic shapes for each element. However, the probability for α-particle backscattering is so low that long counting times are needed and possible applications in industrial control appear to be rather restricted.

Extensive work has been done in this field by Patterson *et al.* [31] and by Semmler *et al.* [32]. A gauge has been developed for lunar soil analysis. Some of the conclusions of Patterson *et al.* will be described here using a ^{242}Cm emitter of 6.11 MeV α-particles.

For a given target element the spectral distribution (ref. Fig. 11.9) of backscattered α-particles is nearly constant in intensity up to a maximum end-point energy characteristic of the element. Above this there is a sharp decrease in intensity.

The value of this energy is given by the Rutherford formula:

$$f = \frac{E_{max}}{E_o} = \frac{\left[\dfrac{4\cos\theta}{A} + \left(1 - \dfrac{16}{A^2}\sin^2\theta\right)^{\frac{1}{2}} \right]^2}{(1 + 4/A)^2} \tag{11.9}$$

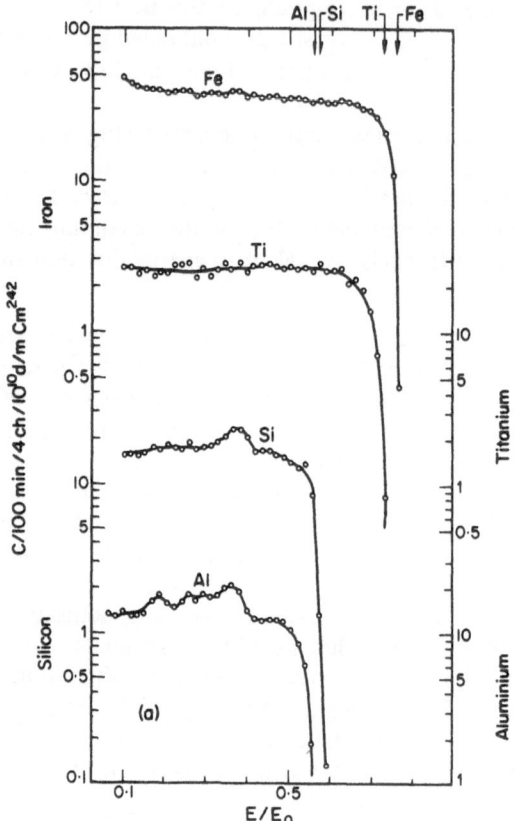

Fig. 11.9. Alpha-ray scattering spectra from thick targets of some pure elements. The
ordinates are intensities of the scattered particles in unit of counts per 100
minutes per 10^{10} d/m of ^{242}Cm in an α source per four channels of the
analyzer (approximately 0.02 E/E_{max}) (from Patterson et al. [31]).

in which E_0 is the energy of the α particles incident on the target, θ the
scattering angle and A the mass number. For $\theta \sim 180°$

$$\frac{E_{max}}{E_0} \doteq \left(\frac{A-4}{A+4} \right)^2$$

The intensity ΔI of the backscattered spectrum measured in a channel of
width ΔE near the energy cut-off is given by

$$\frac{\Delta I}{\Delta E} = k \, \frac{1}{E_0^{3/2}} \, \frac{Z^{3/2} \, f^{1/2}}{(1 - f^{3/2} \cos \theta)}$$

where k is a function of θ.

For $A > 20$, $\Delta I/\Delta E$ varies approximately as $Z^{3/2}$. Below $Z = 10$,

(11.9) does not apply because there is penetration of the coulomb barrier of the nucleus and the intensity is increased. Simultaneously low energy peaks appear in the spectrum due to nuclear resonance energies.

The intensity of spectra given by compounds or homogeneous mixtures can be obtained additively from those of their components. For the energy region of interest and for not too heavy elements, the following relation can be used to calculate the contribution of each element to the spectrum:

$$\frac{(\Delta I/\Delta E)}{^0(\Delta I/\Delta E)} = \frac{x_j \, Z_j^{\frac{1}{2}}}{<Z^{\frac{1}{2}}>}$$

where x_j is the fraction of atoms of type j in the compound, $<Z^{\frac{1}{2}}>$ means the average value of $Z^{\frac{1}{2}}$ for the compound and superscript ^{0} denotes the value for the pure element of type j.

Applications
The potential application to the analysis of rocks and the analysis of light elements ($Z < 12$) on thin layers has been studied. Fig. 11.10 shows details of the device used by Patterson *et al.* In their experiment, information from α-particle backscattering was completed by the measurement of protons from α, p reactions.

11.3.4 Thermalisation and Scattering of Neutrons
The thermalisation and scattering of neutrons, which is used mainly in measurements of moisture in soils, foundry sands and industrial products in bulk, can be combined with neutron absorption measurements to determine concentrations of highly absorbing elements in samples. This method allows a relatively rapid measurement of moisture in a large volume of sample, automatic survey of moisture variations with time and it can be operated under rather difficult conditions.

Most commonly used radioisotopic sources emit neutrons with a continuous distribution of energies up to ~ 11 MeV the maximum intensity often appearing between 4 and 6 MeV. By successive collisions with nuclei these neutrons are slowed down to thermal energies, i.e. about 1/40 eV at ambient temperature, where they are in thermal equilibrium with the scattering medium. The coefficient of mean energy loss of a neutron by collision with a nucleus of mass number A is equal to

$$\frac{A^2 + 1}{(A + 1)^2}$$

Thus collisions with hydrogen atoms are the most efficient for the slowing-down process. The density of the thermal neutron flux round the source will depend markedly on the hydrogen content of the scattering medium and also on the scattering cross sections of all the elements in the medium, the density of the medium and on the absorption cross sections of the same elements. Details on the theory will be found in the papers of Holmes [33], Gardner and Kirkham [34], Semmler [35] and Cameron [36].

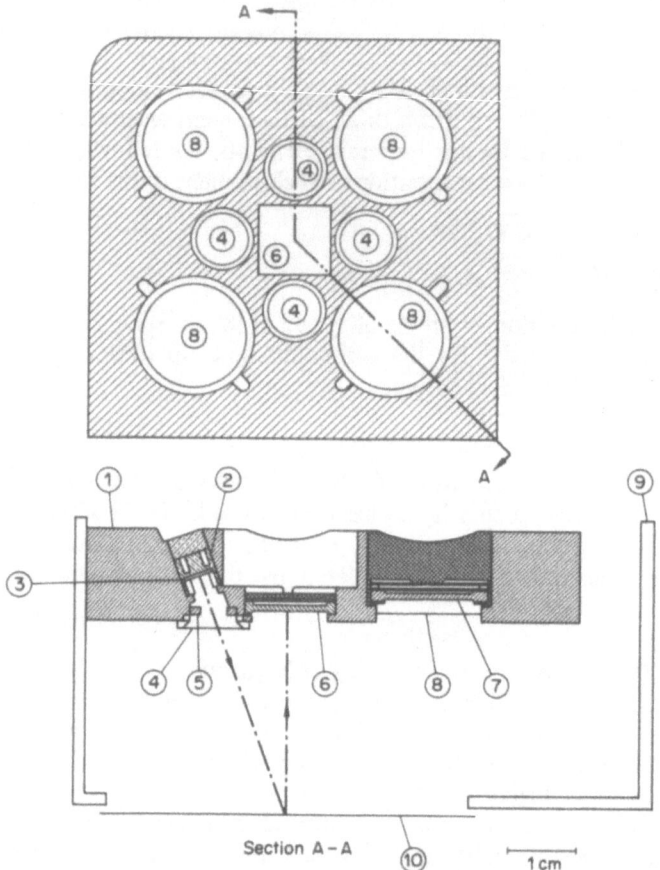

Fig. 11.10. Construction details of breadboard instrument for chemical analysis using
α-particle interactions. (1) head block; (2) source-holder assembly; (3)
source position; (4) VYNS film for containing recoil from source; (5)
source collimator; (6) α detector; (7) proton detector; (8) α absorber in
front of proton detector; (9) instrument case; and (10) sample position.
The breadboard instrument, including case and transistorized amplifiers
and mixing circuit, weighed about 0.3 kg. (from Patterson *et al.* [31];
reproduced with permission of the Journal of Geophysical Research, U.S.A.).

Subsurface Moisture Probes

These are usually equipped with a proportional counter filled with enriched
$^{10}BF_3$ gas, a preamplifier and a $^{241}Am/Be$ or $^{226}Ra/Be$ source which is placed
at mid-level of the counter sensitive volume. The probe is connected by
cable to a high voltage supply, a pulse amplifier and a scaler or recorder. The
envelope of the probe is made of a thin steel or aluminium cylinder (Fig.
11.11).

The sensitive volume corresponding to 95% of the maximum possible

Fig. 11.11. Subsurface moisture gauge (courtesy of Saphymo-Srat, France).

counting rate has a radius of about 20 cm for high moisture soils but can increase up to 40 cm for low moisture soils. In order to obtain reproducible results, drift of the response must be compensated from time to time. This compensation is made after placing the probe inside a paraffin or polythene block which also serves as a biological shield.

Neutron absorbing elements such as boron, cadmium, chlorine and calcium, and even iron and manganese, influence the measured neutron flux, making it necessary, for precise work, to calibrate the gauge for the type of soil being examined.

In practice it is difficult to obtain an accuracy better than ± 0.01 g/cm^3 H$_2$O. Such gauges are used in agriculture for irrigation studies, in hydrology and in civil engineering. Further details will be found in a survey by Cameron [37].

Surface Gauges
Here the basic principles are the same as above and subsurface probes can frequently be converted to surface gauges. Such gauges are used in the measurement of soil moisture before construction of buildings, roads, airfields, etc.

Measurement of Relative Moisture of Soil
For this determination, results obtained with the neutron probe must be combined with density measurements made at the same location with a γ-ray density gauge.

Moisture Measurements of Bulk Products
Moisture gauges have been used for the control of water content, e.g. in foundry sands, wheat grains, wood chips, etc. A simultaneous density measurement is necessary. Problems arise in the circulation of material round the gauges in continuous control.

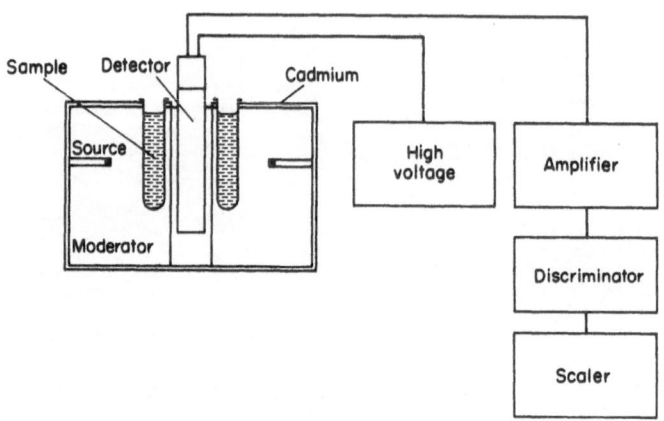

Fig. 11.12. Typical arrangement for analysis by absorption of thermal neutrons.

Concentration Measurements by Absorption of Scattered Neutrons

Fig. 11.12 shows a typical arrangement used for this purpose. For aqueous solutions the moderating block is not necessary. The counting rate of thermal neutrons decreases as the content of the absorbing element increases. If the element to be determined is present in a solid matrix, a known weight of the sample is placed between the probe and the moderating medium, e.g. in the control of boron in samples of crushed glass.

11.4 X-RAY FLUORESCENCE ANALYSIS

First reported in 1946 by J. E. Edwards, the excitation of characteristic X-rays by radiation from a radioisotope source was first applied practically by Reiffel [38] in 1955. Among methods using radioactive sources X-ray fluorescence analysis is the most promising and will certainly be extended in the near future. Theoretically, as is the case for activation analysis, it allows selective analysis of each particular element in a sample based on the intensity of its characteristic X-radiations. In the present state of the art, it can be easily applied to many industrial problems of analysis. Every year the limits of its selectivity and field of application are improved, aided by the evolution of electronic technology. The advent of new inexpensive processes of information treatment will increase the possibilities.

The limitations of the technique should not be ignored. Its main drawback, at the moment, is its lack of energy selection, but the advent of high resolution detectors will eliminate this inconvenience.

11.4.1 General Theory

This is developed here only for the case of 'backscattering' geometry, i.e. when the source and detector are placed on the same side of the sample. In this case the thickness of the sample can easily be made 'quasi-infinite' and its mass need not be measured. The method involves the three steps:

(i)   Ejection of electrons from the internal shells of atoms.
(ii)   Measurement of characteristic X-ray line intensities of the elements of the sample, emitted as a consequence of atomic readjustments.
(iii) Calculation of the amount of the elements of interest in the sample from the X-ray line intensities corrected for interference effects, i.e. matrix absorption, matrix enhancement, particle size effect, density effect and for the selectivity, stability and reproducibility of the detector.

Excitation and Emission

Excitation of the sample may be made by γ-rays, X-rays, alpha particles, beta particles, protons, etc. The binding energies of electrons in atoms are given in tables [39, 40]. The deepest electronic energy levels within atoms are not affected by the chemical or lattice environments, except for elements of low atomic number such as carbon.

The number of electrons ejected from a given energy level belonging to a particular element is proportional to the ratio of the photoelectric absorption cross-section for that level to the total mass absorption coefficient of the sample for the exciting radiation concerned.

The de-excitation of atomic energy levels takes place by transition of electrons coming from lower energy levels followed either by emission of Auger electrons or of photons with discrete energies. The probability of de-excitation followed by emission of photons is called 'fluorescence yield'. Various formulae have been proposed to obtain the fluorescence yield as a function of the atomic number. We shall mention one of them due to Arends [41] for the K level:

$$\omega_K = \frac{0.957\ Z^4}{0.984 \times 10^6 + Z^4}$$

The energy of an X-ray photon emitted during de-excitation is equal to the difference in energy of the two levels between which the electronic transition occurs. In the case of the K level, the various possible photon energies are designated by: K_{α_1}, K_{α_2}, K_{β_1}, K_{β_2} .... etc. Obviously photons corresponding to the L, M, ... series are also emitted as a result of excitation of the corresponding energy levels. The energies of the most abundant of these characteristic X-ray lines are given in tables [38, 39]. For the K X-ray lines of common metals the emission probability is about 50 for K_{α_2} and 17 for $K_{\beta_{1,2}}$ per 100 K_{α_1}; but these proportions vary with the atomic number of the element considered.

The emission of characteristic X-ray lines by an atom is isotropic. The intensity of X-ray lines emerging at the surface of the sample depends on: the mass attenuation coefficient of the exciting radiation and of the emitted radiation in the sample, the excitation efficiency of the atomic levels considered and on their fluorescence yield, the amount of the element concerned in the sample, the matrix enhancement and on the grain size effect. The mean thickness of a sample giving X-ray lines emerging at its surface depends on the absorption coefficients just mentioned; it may range

from a fraction of a micron to several nm according to the atomic numbers
of the element and of the matrix.

X-Ray Line Spectrometry

Because of the relatively low intensities of radioactive sources used for
excitation, non-dispersive methods of spectrometry must be employed and
we shall deal with the use of filters and/or electronic pulse height selection.

Simple filtration is sometimes sufficient to attenuate an interfering X-ray
line. Two cases are illustrated in Fig. 11.13. In the first one the K_α X-ray
line of iron at 6.4 keV is transmitted by an iron filter to a much greater
extent than the K_α X-ray of nickel at 7.47 keV. In the second case,
illustrated by the transmission curve of an aluminium filter, the nickel K_α
X-ray is best transmitted. In practice, the emission by the filter of its own
X-ray lines results in poorer filtration.

The principle of the use of balanced filters, a method due to Ross [42],
is illustrated in Fig. 11.14. In this example, where the intensity of nickel K
X-rays is measured in the presence of iron lines, let us compare the total

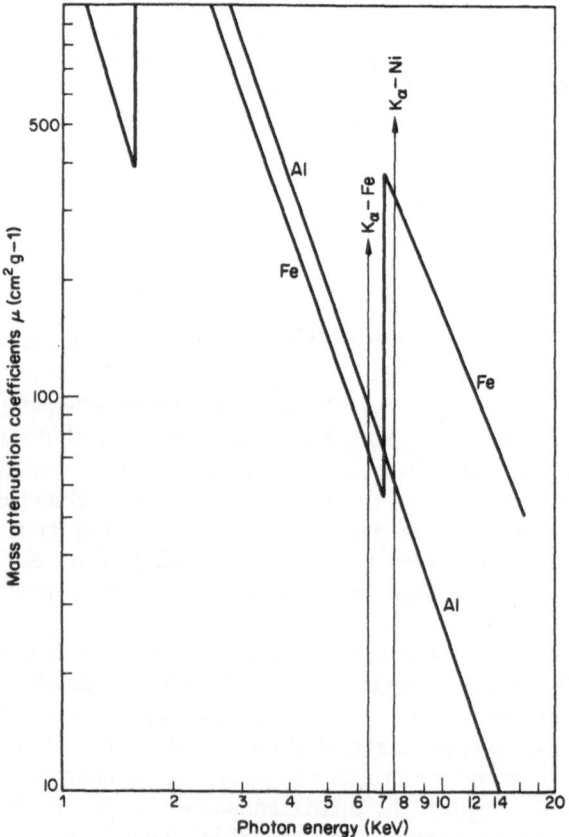

Fig. 11.13. Attenuation coefficients of aluminium and of iron for low energy photons.

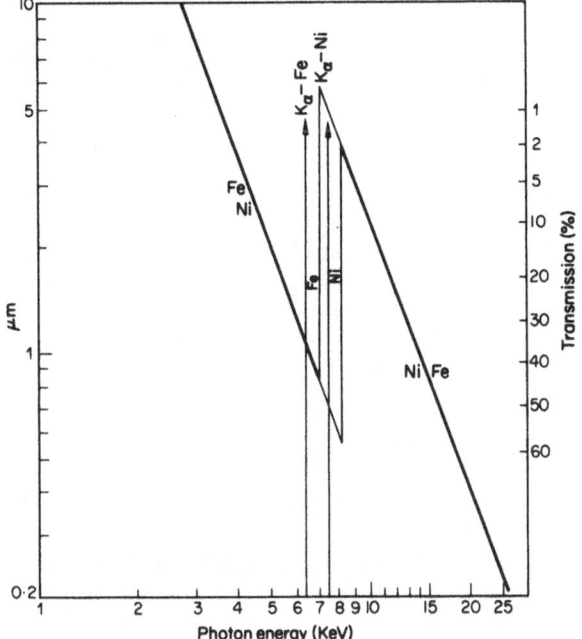

Fig. 11.14. Transmission of photons by a couple of balanced filters iron-nickel (m_{Fe} = 14.5 mg cm^{-2}; m_{Ni}= 11.7 mg cm^{-2}).

intensities transmitted by a nickel and by an iron filter. If the mass per unit area of these filters is chosen such that, for an energy below 7.111 keV or above 8.331 keV, we have the relation

$$\frac{m_{Ni}}{m_{Fe}} = \frac{\mu_{Fe}}{\mu_{Ni}}$$

the transmission is about the same for photon energies on either side of this gap but differs between 7.111 and 8.331 keV. The difference in the two counting rates is due only to the nickel K_α X-ray lines. In this case a Fe-Cu pair of filters can also be used. Such pairs of balanced filters are commercially available and are frequently used in apparatus sold for radioactive X-ray fluorescence analysis.

An interesting method combining the use of filters and radiators has been proposed by Brinkerhoff and Forsyth [67] for element identification. After passage through a filter which cuts out energies above that of the wanted element, the X-rays emitted by the sample are allowed to excite a radiator which cannot be excited by X-rays of energy lower than those of the wanted element. The X-rays of the radiator are detected with good selectivity.

Statistical fluctuations, background and imperfections of filters limit the sensitivity of devices based only on the comparison of total counting rates measured with pairs of balanced filters. An improvement is obtained when pulse-height selection is used, with or without filters, according to the problem. Simultaneous measurement of several X-ray peaks and comparison with backscattered γ-ray intensities is then possible. Further improvement will follow the use of small computers for spectral analysis.

The detectors used most frequently are scintillation counters, gas-filled proportional counters and semi-conductor detectors. The first are used in portable apparatus where rugged and cheap instruments are needed, and almost exclusively for photons above 20 keV. The lower limit of energy detection is about 2 keV. The gas-filled proportional counter has a better energy resolution ($\sim$ 18% for 5.9 keV), and a lower sensitivity for high energy photons. It can be used between $\sim$ 0.1 keV and $\sim$ 20 keV. The gas-filling of sealed counters is often a 90%—10% argon-methane mixture, sometimes carbon dioxide or mixtures of xenon or krypton with methane. In flow counters methane, propane, butane or helium + methane are also used for the detection of low energy photons.

Semiconductor detectors have the best resolving power ($\sim$ 3% for 5.9 keV) and are rapidly developing. Commercial analysers using these detectors are available and allow greater sensitivity, precision and accuracy in analysis. Surface barrier lithium-drifted silicon detectors are used for the detection of photons of energy between 1.5 keV and about 20 keV. Above this energy the detection efficiency is poor and lithium-drifted germanium diodes are preferred. The main limitations in the use of the latter are their relatively high price and the need to keep them at a very low temperature.

For very low energy photons, e.g. below 2 keV, their absorption by any materials interposed between the sample and the detecting element is often prohibitive and windowless flow proportional counters must be used. Figs. 11.15 and 11.16 give examples of spectra obtained with a proportional counter and with a semi-conductor device.

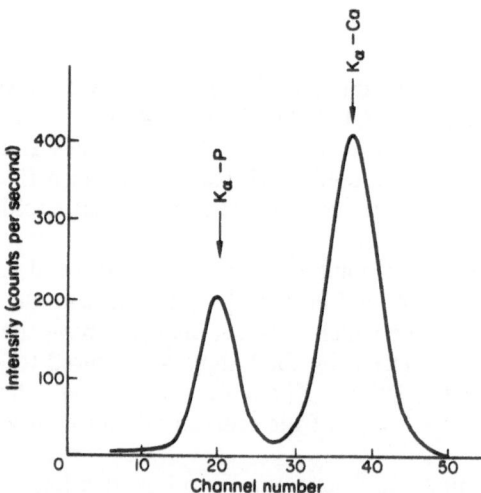

Fig. 11.15. Spectrum of a PO₄HCa sample obtained by excitation with an iron-55 source and detection with a windowless proportional counter.

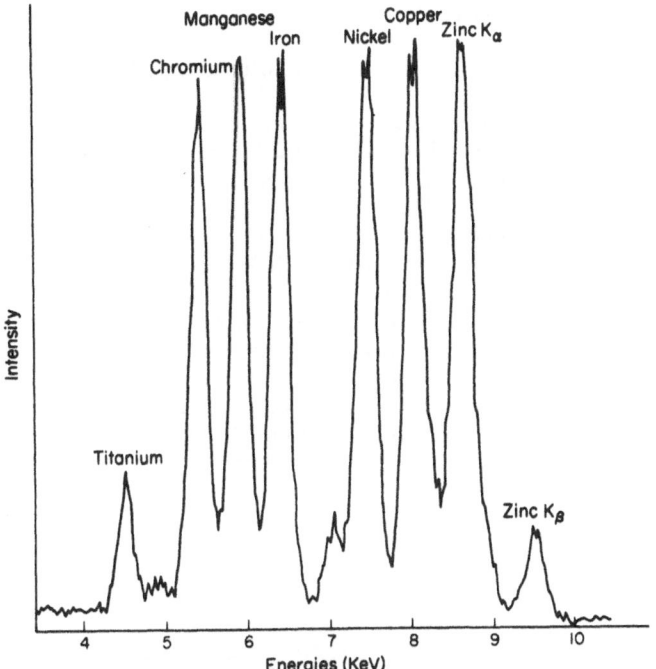

Fig. 11.16. Spectrum obtained with a lithium-drifted silicon detector (courtesy of Kevex, U.S.A.).

Interference Effects

Density effects appear when the mean free path of the X-ray line in the sample is relatively large compared with the source-sample distance, the sample-detector distance or the thickness of sample under examination. In most cases it need not be considered. Matrix absorption effects occur, due to variations in the chemical composition of the sample, independent of variations in the elements being measured. Matrix enhancement is an increase in an X-ray line intensity due to secondary excitation by X-ray lines of other elements in the sample. Variations in the content of these interfering elements modify the intensity of the 'enhanced' X-ray line. Inhomogeneity effects include the particle size effect encountered in the case of powders or slurries, when the absorption of a fluorescent X-ray line in a single particle is greater than a few per cent. The effect disappears for fine grains and for relatively large grains but there is an intermediate grain size for which very large variations in X-ray line intensities can be expected if the grain size of the sample is not constant. Variation in particle composition also has an effect on the X-ray line intensities emitted by the sample and, like the effect of particle size, it can be reduced by grinding or fusion of the sample.

11.4.2 Excitation by Means of Photons

The sources used most frequently are listed in Table 11.3. Because of the relatively low energy of the photons emitted by these sources, a heavy metal

TABLE 11.3. *Photon Sources Commonly Used in X-Ray Fluorescence Analysis*

Source	Half-life		Photon energy (keV)	
^{55}Fe	2.7	years	5.9	(Mn K X-rays)
^{238}Pu	86.4	years	12 to 17	(U L X-rays)
^{109}Cd	1.27	years	22.1	(Ag K X-rays)
			87.5	γ
^{210}Pb	22	years	47	γ
			11 to 13	(Bi L X-rays)
^{241}Am	458	years	59.6	γ
			14 to 21	(Np L X-rays)
^{3}H/Ti	12.3	years	0 to 18	(bremsstrahlung)
			4.5	(Ti K X-rays)
^{3}H/Zr	12.3	years	0 to 18	(bremsstrahlung; max at $\sim$ 5.5 keV)
^{147}Pm	2.6	years	0 to 220	(bremsstrahlung; max at $\sim$ 25 keV)

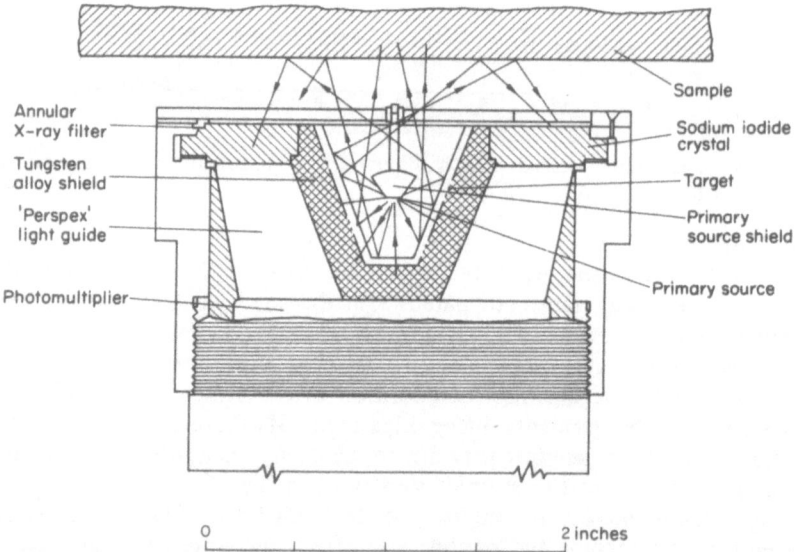

Annular
X-ray filter

Tungsten
alloy shield

'Perspex'
light guide

Photomultiplier

Sample

Sodium iodide
crystal

Target

Primary
source shield

Primary source

0   2 inches

Fig.11.17. Source-target combination mounted on a scintillation counter (reprinted
with permission from I.A.E.A. Technical Report No. 115).

shielding, 1 or 2 mm thick, is often sufficient to protect the detector
against direct radiation, so that very compact geometries can be used.

Secondary low energy X-ray sources allow the choice of a more appropriate energy for excitation. As illustrated in Fig. 11.17 they consist of a
radioactive photon source and a target which emits its characteristic X-ray
lines.

With photon excitation the sample surface emits not only the characteristic X-ray lines of its elements but also photons that have been scattered by

single or multiple Compton effects. Elastic scattering by the Rayleigh effect also contributes to the counting rate. The best conditions for the excitation of K X-rays of high atomic number elements have been investigated by Enomoto [43].

Calculation of Intensity of X-ray Lines

In the case of compact geometry the complete and precise calculation requires a medium size computer. Here we shall consider only the case of a collimated beam of exciting radiation and a collimated detector looking at the same area of sample. The results can be extended to the case of a compact geometry if a proper choice is made of the mean direction of the radiation.

The differential expression of the counting rate, N_{K_j}, due to the detection of the K group of characteristic X-ray lines of element j is given by equation (11.7) where data relative to secondary radiations are changed to the corresponding data for X-ray production and absorption. It is noted that the product

$$\Delta\Omega_2 \left(\frac{\partial \sigma_c}{\partial \Omega} \right)_\theta \quad \text{becomes} \quad \frac{\Delta\Omega_2}{4\pi} \, \tau_{\gamma,K_j} \, \omega_{fK_j,x_j}$$

where τ_{γ,K_j} is the photoelectric mass absorption coefficient of the exciting γ-rays on the K levels of atoms of element j, ω_{fK_j} is the fluorescence yield for the K levels of element j, and x_j is the mass content of element in the sample.

The partial photoelectric absorption coefficients τ_{K_j} for a given element are obtained: a) graphically from the difference between the total mass attenuation coefficient and the value which is read on the extrapolated part of the lower energy branch of the absorption curve, or b) from the value of the K jump of this particular element.

For accurate calculations, account has to be taken of the probability for the emission of every particular line of the K series. The same considerations apply to the L and M series.

In practice samples of quasi-infinite thickness are used so that:

$$N_{K_j} \doteq Ct\omega_{fK_j} D_{K_j} T_{K_j} T_\gamma \frac{\tau_{\gamma,K_j}/\sin\alpha}{\dfrac{\mu_{\gamma e}}{\sin\alpha} + \dfrac{\mu_{K_j e}}{\sin\beta}} x_j \tag{11.10}$$

where D_{K_j} is the efficiency of the detector for K X-rays lines of element j and

$$C = kA \frac{\Delta\Omega_1}{4\pi} \cdot \frac{\Delta\Omega_2}{4\pi}$$

the first subscripts relate to radiations and the second to absorbers.

Now let us represent by the subscript M data concerning the matrix, i.e. all components of the sample other than the element j.

Equation (11.10) becomes

$$N_{K_j} \doteq K_{\gamma,j} t \; \frac{x_j}{(a-b)\,x_j + b} \tag{11.11}$$

where

$$a = \mu_{\gamma,j} + \mu_{K_j,j} \; \frac{\sin \alpha}{\sin \beta}$$

$$b = \mu_{\gamma,M} + \mu_{K_j,M} \; \frac{\sin \alpha}{\sin \beta}$$

and

$$K_{\gamma,j} = C \omega_{f K_j} D_{K_j} T_{K_j} T_\gamma \, \tau_{\gamma,K_j}$$

Equation (11.11) can be written

$$\frac{N_{K_j}}{N^\circ_{K_j}} = a \; \frac{x_j}{(a-b)x_j + b}$$

where $N^\circ_{K_j}$ is the counting rate for a target of pure element j.

The calibration curve is linear if $a = b$, or if x_j is of very low value. If $a \neq b$ linearisation may be obtained by dilution of the sample, e.g. by fusing with a lower atomic number diluent such as lithium tetraborate.

The same relations apply in the case of excitation of L, M.... levels.

Sensitivity and Precision

In the case of a sample of quasi-infinite thickness the relative sensitivity to change in the content x_j is given by

$$\frac{\partial N_{K_j}}{N_{K_j}} \bigg/ \frac{\partial x_j}{x_j} = \frac{b}{(a-b)\,x_j + b}$$

The error made on the determination of x_j is made up of a random error and a systematic error.

As an example, the precision due to a 1% error in mean counting rate of the X-ray line is equal to

$$\left(\frac{\Delta x_j}{x_j} \right)_\% = \left(\frac{a}{b} - 1 \right) x_j + 1$$

This precision can often be obtained in practice. For a high accuracy, a calibration should be carried out with standard samples of known composition.

Commercially Available Systems

A large number of radioactive X-ray fluorescence analysers are now commercially available. The major interest was originally for portable, light and cheap instruments. More elaborate instruments have been designed for laboratory use in the routine control of major components in industrial products.

Portable instruments are composed of a light measuring head that can be handled easily and an electronic unit usually weighing less than 12 kg. The detector is generally a photomultiplier tube and a NaI(Tl) crystal. The measurement is made by comparison of the counting rates measured through pairs of balanced filters corresponding to the elements of interest. The auxiliary electronic equipment consists of a high voltage power supply, pulse amplifier, pulse-height selector and either a scaler or ratemeter (Fig. 11.18). They have been suggested for ore analysis in the field, for sorting metals and for a number of industrial analytical applications.

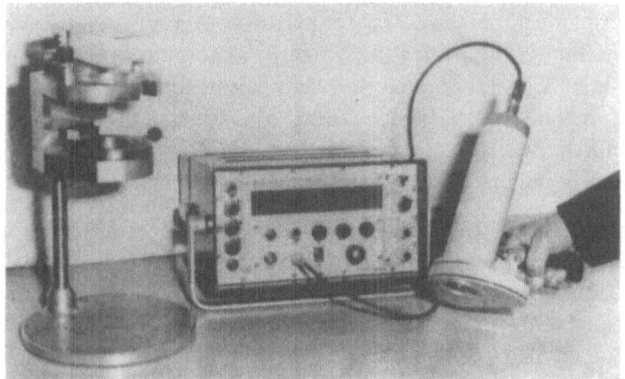

Fig. 11.18. Portable X-ray fluorescence analyzer (courtesy of Nuclear Enterprises, U.K.).

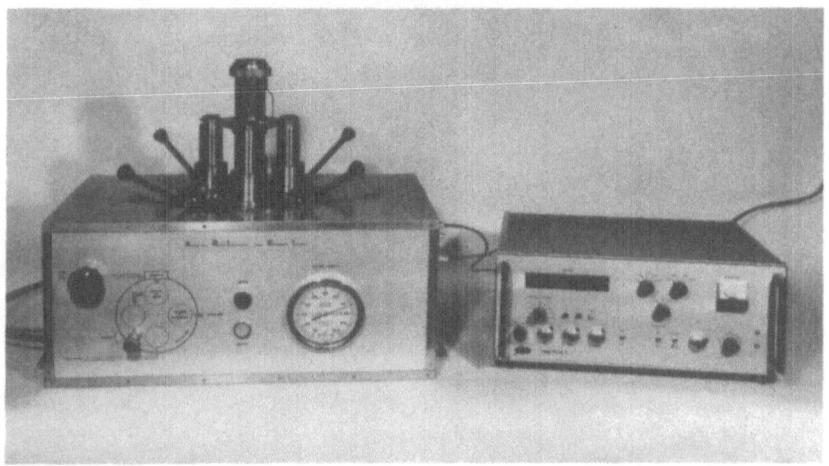

Fig. 11.19. Spectrometer for X-ray fluorescence analysis in the low energy X-ray region (courtesy of C.E.A., France).

While portable instruments can be used in laboratories and workshops, more elaborate instruments are available having a thermoregulated measuring head and capable of more precise measurements using pulse-height selection. These instruments offer greater possibilities for the output, control and treatment of results. We shall consider three types:

(i)   Those with sealed proportional counters or scintillation counters.
(ii) Those with a gas-flow proportional counter specially designed for the detection of low energy X-rays. An instrument of this type, designed at Saclay, may be operated without a window between the sample and the detecting gas. By the choice of an optimum gas pressure, high detection efficiencies are obtained (Fig. 11.19).
(iii) High resolution instruments referred to previously, having a lithium-drifted silicon detector or, more rarely, a lithium-drifted germanium detector. These instruments are of relatively high price but their resolution is now nearly as good as that given by conventional X-ray apparatus and they allow simultaneous measurement of several X-ray peaks. These are already used in industry but improvements can be expected as larger crystals of improved quality become available (Fig. 11.20).

Applications have been studied for continuous or sequential analytical control of industrial processes. Sample presenters, automatic sample changers, circulation apparatus for liquids and slurries, and instruments with which complementary measurements, such as density, can be made have been developed for this purpose [44]. It is possible to have several measuring heads connected to the same central electronic equipment. Apparatus has been proposed for bore-hole logging [45].

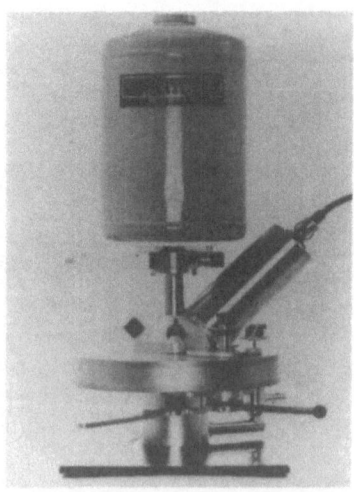

Fig. 11.20. High resolution instrument for radioactive X-ray fluorescence analysis (courtesy of Nuclear Diodes, U.S.A.).

Applications

A very large number of applications have been described, not always of a routine nature, and it is only possible to mention a few in this short outline. Details will be found in the Conferences of the I.A.E.A. and in its Technical Reports Series No 115. These include 38 examples of analysis of elements in metalliferous mineral exploration and development, including the measurement of tin with a precision of ± 0.05% in powdered ores using Pd/Ag filters. For on-stream analysis of slurries, crushed minerals and solutions, applications and feasibility studies have been made for 17 industrial problems. In alloy analysis 26 applications have been made mainly for the identification of the elements V, Cr, Ni, Cu, Nb, Mo, Sn, W, Pb, U. For example, tin in Cu/Sn alloys has been determined with a precision of 0.1%.

Other applications have been made since this compilation including the measurement of titanium in paper and the measurement of silver on photographic plates.

The duration of a measurement generally ranges from 0.5 to 5 minutes. The highest sensitivities are 50 ppm or 25 ppm or better when using a proportional counter filled at the optimum pressure. An example of what is called 'Multielement Analysis' is given by Uchida *et al.* [46] who developed a gauge for the control of Al, Si, Ca, and Fe in raw cement mix.

11.4.3 Excitation by Means of Beta Particles

This means of excitation has only been explored in the laboratory. Compared with $\gamma \to X$ excitation $\beta \to X$ excitation has several advantages: a greater range of elements can be excited, smaller areas of sample can be examined and matrix absorption effects reduced. The sensitivity is somewhat limited, however, by background due to bremsstrahlung generated in the sample. The operator must be carefully protected against direct irradiation because of the high ionising power of the beta emitting sources. Backscattered beta rays must not enter the detector window and they are eliminated by screens or by magnetic deflection.'Sources that have been used for these experiments are ^{147}Pm, ^{90}Sr + ^{90}Y, ^{204}Tl and ^{85}Kr. The theory and possibilities of this method have been developed by Filosofo [47] and Leontiadis [48] and systematic investigations of beta excited X-ray spectra have been made by Preuss [49]. Although the method has been largely abandoned renewed interest may result from developments in detection techniques.

11.4.4 Excitation by Means of Alpha Particles

For targets of atomic number less than 29, alpha ray excitation gives very pure X-ray lines, as practically no other kind of secondary radiation is emitted. Another interest in such sources lies in the excitation of low atomic number elements, especially below $Z = 12$, as there are at present no satisfactory photon sources for these elements.

The excitation efficiency of characteristic X-rays is approximately proportional to the fourth or fifth power of the energy of the alpha particles and decreases rapidly as the atomic number of the target element increases. Such sources are particularly necessary for X-ray fluorescence analysis of the very important C, N and O.

The available pure alpha emitters are scarce. Polonium-210 has been used successfully but, because of its high toxicity and relatively short half-life, the use of ^{242}Cm and ^{244}Cm sources is now considered. Papers on α-ray excitation have appeared by Sellers [50] and by Robert [51].

Apparatus and Applications
Because of the high absorption coefficients of very low energy X-rays and the short range of alpha particles the measuring system must be in vacuo or in the atmosphere of a light gas such as hydrogen or helium. Alternatively a windowless proportional counter operated at optimum pressure can be used as shown in Fig. 11.19. In the case of carbon, background due to the direct excitation of the characteristic K X-rays of carbon in the detecting gas must be avoided.

11.4.5 Excitation by Means of Protons
Methods using accelerating tubes are well established in conventional X-ray fluorescence, direct emission and microanalysers. It is worth noting that many promising results have been obtained by proton excitation using Cockcroft-Walton accelerators. Very high signal-to-noise ratios are obtained using 50 to 300 keV protons. Developments can be expected in the X-ray energy range, 0.05–1 keV, in quantitative analysis and in studies of the physics and chemistry of surfaces.

11.5 ANALYTICAL APPLICATIONS OF THE MÖSSBAUER EFFECT
The Mössbauer effect offers a powerful means of studying molecular structure, the width and life of excited nuclear levels, magnetic nuclear moments and magnetic and gravitational fields. It can also be applied to the qualitative and quantitative study of structure and chemical bonding in crystals. It makes possible a quantitative chemical analysis of compounds included in a crystal, i.e. molecular analysis. For details of the theory of the Mössbauer effect and on its applications in chemistry the reader is referred to a paper by Goldansky [52].

11.5.1 Theory
The first attempts to obtain resonance γ-fluorescence of nuclear energy levels were made in 1929 with very little success and only under very special conditions.

The natural width, Γ, of an excited nuclear energy level, E_r, is related to its mean life, τ, by the uncertainty principle, i.e. $\Gamma\tau = \hbar = 1.05 \times 10^{-27}$ erg/second.

For resonant excitation of these levels the energy of the photon must be equal, within a tolerance $\pm\,\Gamma$, to that required to bring the target nucleus from the ground state to one of its excited states.

When such an excitation occurs de-excitation of the nucleus takes place by the emission of a γ-ray or an internal conversion electron.

Because of the thermal motion of the target nucleus, the relative energy of the incident photon of energy $h\nu$ is equal to

$$h\nu - h\nu\left(\frac{v}{c}\cos\theta\right)$$

where v is the velocity of the nucleus and θ the angle between its direction of motion and that of the photon. There is thus a 'Doppler broadening' of the resonance energy E_r and the resonance width at half maximum becomes

$$D = \frac{1}{c}\sqrt{\frac{2\,kT}{m}} \cdot E_r$$

when $D \gg \Gamma$, where k is the Boltzmann constant and m the mass of the nucleus.

Furthermore there is a shift in the mean energy E_r required for resonance, because of the recoil of the target nucleus in the gamma absorption process. This shift is equal to

$$R = E^2/2mc^2$$

Doppler broadening and energy shift occur in the same manner for the emission of the gamma ray and the same formulae apply. The observed cross section for resonance is determined by the overlap of the two lines (see. Fig. 11.21). Even in the case of complete overlap, however, this cross section would be proportional to the value of Γ/D which can be as low as 10^{-5}.

In 1957 Mössbauer applied to gamma interactions the former theories of Debye on crystal vibration states and of Lamb for neutron capture by the

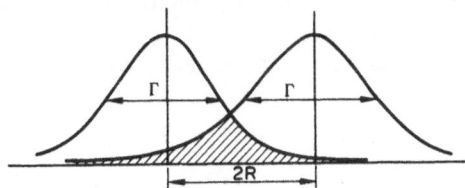

Fig. 11.21. Overlap of two resonance lines separated by recoil during emission and absorption.

atoms of crystal lattices. If the recoil energy of an emitting nucleus bound in a crystal lattice is not sufficient to break the chemical bonds, it becomes the property of the whole crystal. This energy absorption by the crystal, as well as its de-excitation, occurs only in the form of quanta of collective motion, i.e. phonons. In this case Doppler broadening and recoil energy becomes negligible and the gamma line is reduced practically to its natural width. The same considerations apply to the absorption of a gamma ray line by a nucleus bound in an atom.

Thus the emission and absorption spectra of the atoms in a crystal lattice show two components; a rather broad distribution due to the thermal motion of the lattice atoms and to changes in the vibrational states of the lattice, and a narrow far more intense resonance line of natural width.

11.5.2   Transmission Systems for Study of the Mössbauer Effect

Mössbauer used the Doppler shift due to source or absorber movement to obtain small controlled differences of energy about the resonance value between the incident gamma rays and the energy levels to be excited.

Fig. 11.22 shows one of the devices used In this example, the absorber is moved at constant velocity towards the source (positive velocity) then towards the detector (negative velocity). If the source and absorber are identical Mössbauer absorption is a maximum for $v = 0$. For other velocities the apparent energy of the photon is displaced by a magnitude

$$\Delta E = E_r \; \frac{v}{c}$$

and the absorption decreases (see. Fig. 11.23). Thus we have a negative peak in the curve relating measured intensity versus velocity of absorber. In another system the velocity of the absorber is varied continuously, e.g. by vibration induced by the coils of a loudspeaker, and the Mössbauer spectrum is obtained directly on a multichannel pulse height analyser. In other similar systems the absorber is in a fixed position and the source is displaced or vibrated.

Convenient sources are scarce. In practice the emitter nucleus must be the same as the absorbing nucleus, the electronic conversion coefficient must not be too high and the ratio Γ/E_r must be neither too large nor too small. According to Goldansky the most promising range of Γ/E_r values is between 10^{-10} and 10^{-14}, though the Mössbauer effect has been used mainly with cobalt-57 sources for the study of iron-57 and with tin-119 for the study of tin [53]. The velocities used have varied from a fraction of a μm per second up to several centimetres per second.

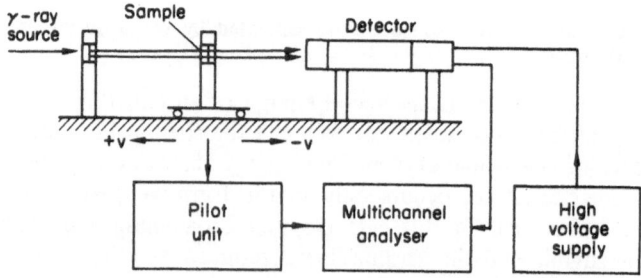

Fig. 11.22. Schematic representation of a device for Mössbauer experiments.

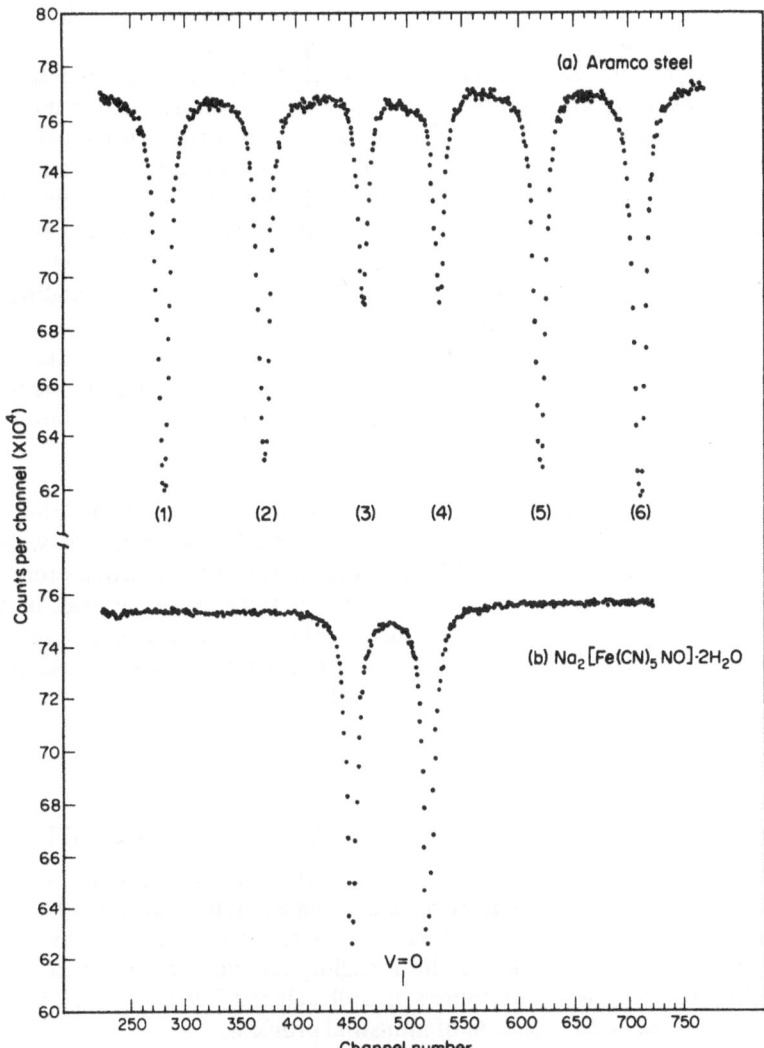

Fig. 11.23 Mössbauer spectra (courtesy of Elsint-France).

Chemical Applications

Applications use the measurement of line shifts due to chemical (or isomeric) effects and of line splittings due to interaction between an inhomogeneous axial electric field given by the crystal itself and the quadrupolar moment of the nucleus. Splitting of the lines can also be obtained by the Zeeman effect due to the magnetic field of the crystal.

The chemical shift of the resonance line is essentially dependent on the density of the electrons round the nucleus, especially the s-electrons. Unless there is a perfect chemical and crystalline identity between the emitting and the absorbing nucleus, there is no resonant absorption for v = 0 but there will be a velocity v ≠ 0 for which one will appear. Shifts are usually given in millimetres/second. A slight increase of the chemical shift is due to a second order Doppler effect resulting from the relativistic shortening of time for moving bodies. Chemical line shifts are slightly dependent on temperature.

Two lines are often observed instead of one because of interaction between the electric field gradient round the nucleus and the quadruple electric moment of the excited nucleus This quadruple splitting is even more sensitive than the chemical shift on the nature of the chemical bonding. Splitting due to the Zeeman effect brings some additional complexity in the interpretation of Mössbauer spectra.

In favourable cases crystalline compounds of iron and tin can be analysed quantitatively, but quantitative analysis is rather difficult on a routine basis and this procedure is likely to be used mainly for the study of structure.

11.5.3 The Backscattering System

Several attempts have been made to measure the intensity of the photons re-emitted by the excited nuclei. The backscattering geometry allows the examination of samples of quasi-infinite thickness and of industrial products with limited preparation. A study made in 1970 under the auspices of the United States Atomic Energy Commission showed that austenite could be determined in steel samples with a precision at least as good as those made by X-ray diffraction techniques. This is the way in which future practical developments of the Mössbauer effect in chemical analysis might be expected to develop.

11.6 ANALYTICAL APPLICATIONS OF AUTORADIOGRAPHY

In chemical analysis autoradiography has been used as a qualitative method for the visualisation of elements or compounds, even at trace level, at the surface of suitably prepared samples. Its success has been considerable in scientific research especially in the metallurgical and biological fields. For quantitative analysis its usefulness is still debated but some applications have been made to the solution of industrial problems

The method consists in taking a radiographic image by contacting a film with the surface of the sample containing a radioactive isotope of the element of interest. This image is developed, enlarged and interpreted quantitatively. Labelling of the sample is generally done by irradiation in a neutron flux or by incorporation of the radioactive tracer in the sample during preparation. The first method can be used for elements having a large cross-section for neutron absorption, or for isotopes of short half-life and it can be used for samples already prepared. When it can be applied, the second method is more convenient for accurate research work because the activity and amount of the added isotope can be exactly known. Other methods are of little interest in analysis.

11.6.1   The Conditions Required for a Good Autoradiograph

Contact between the sample and the emulsion must be close and the energy of the ionising particles, generally beta particles, should be low. Both the sample and the emulsion (including its protective coating) should be as thin as possible. The resolving power of the emulsion should be large, i.e. low speed emulsions are preferable.

The types of photographic emulsion commercially available are:

(i)   Stripping films on glass plates with very fine grains. After separating from the base they are transferred to a water surface where they are allowed to swell before placing them on the surface to be autoradiographed.

(ii) Non-stripping films on glass plates ready for use.

(iii) Radiographic films which are of practical use but their sensitivity to low energy particles is lower. Low speed films must be used if good definition is required.

(iv) Liquid emulsions which are painted on the sample surface or in which the sample can be immersed.

11.6.2   Particular Problems of Quantitative Autoradiography

(i)   Exposure. According to Matteoli *et al.* [54] the density of blackening, i.e. $D = \log I_o/I$, should be about 0.5 but densities up to 2.0 can be used with some emulsions. Depending on the emulsion, the integrated number of particles required to obtain an optical density of unity, range from 10^6 particles/cm^2 for very fast radiographic films to 10^9 for the highest resolution nuclear emulsions.

(ii) Fading. For long exposure times, especially when using emulsions having a high density of silver halide, the recombination rate of products from photochemical reactions must be taken into consideration.

(iii) For quantitative determinations correction factors have to be applied to allow for self-absorption and backscattering in the sample.

(iv) Emulsion thickness. Commercially available stripping films have a thickness between 1 and 5 μm with possible variations of ± 10%. The density of silver halide is very high. Dipping films cannot be used because their thickness is not known with sufficient accuracy.

(v)   Grain size. The diameter of the grains must be as regular as possible. For nuclear emulsions this diameter ranges from 0.02 to 0.5 μm according to the sensitivity while for X-ray films the diameters of the grains are between 0.2 and 3.0 μm.

(vi) Specific activity of the tracer. An additional error is that due to the uncertainty of the specific activity of the tracer element so that calibration is necessary.

11.6.3   Quantitative Interpretation of Autoradiographs

For grain counting or track counting the activity of the sample must be low and the 'grain yield' of known value. With these methods relative values of concentration will be measured rather than absolute values. For track counting relatively thick emulsions have to be used.

For densitometry the blackening of the developed film is measured by transmission of light with a microdensitometer. Linear variation of blackening density versus irradiation time can be obtained up to densities of 2.0 but $D = 0.5$ is sometimes given as the most favourable value. For quantitative chemical analysis Matteoli and Logi (see. [54], p. 56) used known radioisotopic sources for comparison.

A combination of grain counting and densitometry can be used when the grain yield is known.

11.6.4   Applications

Examples of quantitative determinations which have been made include a study of phosphorus distribution in a carbon steel, the analysis of W, S, Sn, Ce, P, etc., impurities in nickel alloys and the measurement of the uranium content of alloys.

11.7   ANALYTICAL APPLICATIONS OF IONISATION METHODS

These methods belong to the category of analytical procedures which are based on modifications in the characteristics of radiation detectors induced by the substance being analysed. They are essentially applied to the analysis of gases and vapours. When the composition of the gas filling in a gas detector is modified, differences in the output current appear due to variations in the absorption of radiation or in the mobility and recombination rates of the ionisation products..

For the physical basis of these methods the reader is referred to papers by Lovelock [55] and for their application to those of Clayton [56]. The most important applications are discussed here.

Ionisation chamber smoke detectors were suggested by Greinacher in 1922 and the method was developed by Meili [57] in 1952. In an air-filled ionisation chamber containing an α-ray source the current decreases when smoke particles are mixed with the air. For 100 ppm by weight of smoke the variation of current intensity is 10%. Fig. 11.24 shows a block diagram of this apparatus which is now widely used in industry. The radioactive source generally used is Americium-241, the activity being less than 100 μCi.

Early work on gas analysis using ionisation methods was carried out in 1938 by Malsallez et al. [58] and apparatus for this purpose has been described [55, 59–61]. When filling gas in an ionisation chamber is permanently irradiated by means of an alpha or beta source the intensity of the saturation current depends, among other parameters, on the mass per

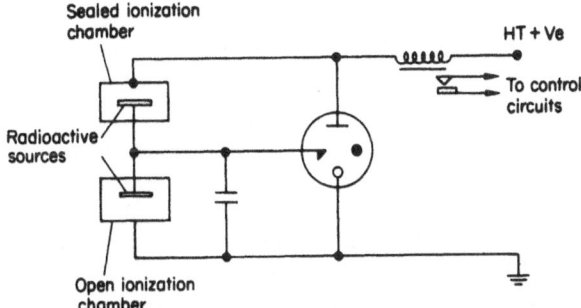

Fig. 11.24. Circuit diagram of ionization-chamber smoke detector (from C. G. Clayton [56]; reproduced with permission of I.A.E.A.).

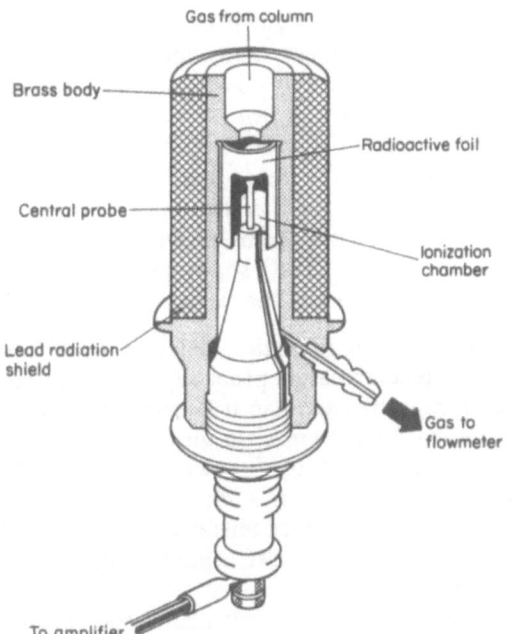

Fig. 11.25 Argon ionization detector (from C. G. Clayton [56]; reproduced with permission of I.A.E.A.).

unit volume and on the ionisation cross-section of the filling gas. Smith *et al.* [61] have developed a system using two ionisation chambers, one being filled with a standard reference gas. By this means they measured the argon content of argon-ammonia mixtures with an accuracy better than 0.1%

11.7.1 Ionisation Detectors in Gas Chromatography

The use of radioactive ionisation detectors has expanded considerably, concurrently with flame ionisation detectors. Following Lovelock they may be classified:

(i)   The cross-section ionisation detector which operates on the same principle as the detector described previously for gas analysis. It allows gas concentration measurements up to 100% but the minimum detectable limit is about 100 ppm of gas by volume in the carrier. The carrier gas is hydrogen or preferably helium.

(ii) The argon ionisation detector is the most commonly used ionisation detector. In the argon carrier gas the following reactions take place:

a) Excitation of argon atoms to their metastable states.

b) Exchange reactions between these metastable argon atoms and molecules having lower ionisation potentials, i.e. below 11.7 eV, when these are ionised.

The current in the ionisation chamber increases very rapidly with the concentration of the vapour. Space charge limits the value of this current. The dimensions of the ionisation chamber, as well as the magnitude of the applied voltage must be carefully chosen.

The ionisation efficiency of this detector is 1000 times greater than for the cross-section detector but the upper limit of measurable concentrations is between 10^{-3} and 10^{-5} by volume. The presence of water vapour can introduce errors and not all gases can be analysed by this means. Fig. 11.25 shows such a detector. Usinga 20 mCi strontium-90 source a gas concentration of 0.01 ppm can be detected in the carrier.

(iii) The electron-capture ionisation detector takes advantage of the fact that the recombination probability is 10^5 to 10^8 times greater for ions of opposite sign than it is between electrons and positive ions. There will be a decrease in the current in presence of a gas or a vapour having an affinity for electrons. In this detector, due to Lovelock, the gas flows in a direction opposite to that of the negative ions giving increased sensitivity. He, H_2 and N_2 are used as carrier gases. Argon is also suitable though its use is not recommended.

This detector is used particularly for the measurement of halogen compounds, ozone and oxygen containing gases. With some systems samples of 1 μg are sufficient for analysis.

(iv) The electron mobility detector is used mainly for the analysis of permanent gases. This detector depends on the variations in electron mobilities in noble gases when other gases are mixed with them. Radiations having a small range in matter (given by an alpha emitter or a beta emitter such as tritium) irradiate the gas near the cathode. At the anode electrical pulses are applied, having a duration such that, for pure noble gas, electrons have insufficient time to reach the anode and the current is very low. If, however, another gas is present the mean agitation velocity of the electrons is reduced and their drift bulk velocity towards the anode and the current are increased. By suitable adjustment of the pulse duration the detector can be made insensitive towards various types of gases.

This detector is insensitive to nitrogen and to highly electropositive gases. Examples of applications are the measurement of H_2, CO, CO_2, methane and other gases in mines, the detection of traces of pesticides in agriculture and the control of the acetylene content of liquid oxygen in the storage tanks of production plants for which the limit of detection is 0.05 ppm.

11.8 COMBINATIONS OF METHODS

The solution of industrial problems sometimes requires the combination of several radioactivity methods in order to eliminate the influence of an uncontrolled variable on the measurement or simply to extend the field of application of a particular method. Some examples are given below.

11.8.1 X-Ray Fluorescence and Gamma-Ray Backscattering

In many cases variations in the intensity of characteristic X-ray lines, induced by geometry or density changes in the sample, can be compensated by comparison with the intensity of backscattered gamma-rays, particularly in collimated systems. According to the problem the two measurements can be made either with a single source and a single detector with pulse height selection, or by using two separate sources and detectors. The compensation is very good when the absorption coefficients of the two measured energies in the sample are about the same.

11.8.2 X-Ray Fluorescence and Gamma-Ray Transmission

The transmission measurement is generally made for density corrections but, in this case, the mean density of the sample is involved instead of 'local' density as in the preceding method.

11.8.3 Gamma-Ray Transmission and Gamma-Ray Scattering

Measurements of concentration made by gamma-ray transmission may be usefully complemented by a measurement of the scattered intensity made, for instance, at an angle of $90°$ to the direction of the main beam.

11.8.4 Beta-Particle Backscattering and Transmission

The combination of these two procedures may be used in the analysis of ternary mixtures. The method has been applied in the study of liquid mixtures in organic chemistry [62].

11.9 OTHER METHODS

Although they have been less systematically considered other methods using radioactivity have interesting applications to specific problems or have shown considerable possibilities of future development.

11.9.1   Nuclear Reactions

Various nuclear reactions have been used in the analysis of elements of
atomic number less than about ten.

Gamma Emission Following an n,α Reaction

For the measurement of low concentrations of boron in aqueous solutions
the following reaction has been proposed:

$$^{10}B + n \rightarrow {}^{7}Li + \alpha + \gamma \text{ (477 keV)}$$

This has been applied to the on-line determination of perborate at concentrations about 12% in detergents with a relative accuracy of about ± 15%.

α,n Reactions

The α,n reaction is used for the detection of beryllium oxide in air. The
sample is collected on a filter which is then irradiated by the α-rays from
^{210}Po. γ-rays emitted by the reaction ^{9}Be $(\alpha,n\gamma)^{12}C$ are counted in the
range 2.5 to 5.0 MeV. In this way 2 μg beryllium per cubic metre of air can
be detected in about 30 minutes.

The same type of reaction has been used for measuring Be, B and F at
the surface of samples, counting the neutrons emitted in the α,n reaction.

γ,n Reaction

The following γ,n reaction has been used in prospecting for beryllium ores:

$$^{9}_{4}Be + \gamma \rightarrow {}^{1}_{0}n + {}^{8}_{4}Be$$

The energy threshold of this reaction is 1.67 MeV. The source used for the
irradiation is antimony-124. Neutrons emitted by the samples are
thermalised and counted by means of a boron trifluoride proportional
counter. The thickness of rock involved in the measurement is about 10 cm.

Prompt γ-ray Emission

The prompt gamma rays emitted in n,γ reactions and by the inelastic
scattering of neutrons, i.e. n,n'γ reaction, have energies and intensities
which are characteristic of the target nuclei. Their energies can be rather
high, and values up to 12 MeV have been observed.

Several systematic investigations have been carried out on the γ-ray
spectra given by these reactions for most natural elements [63]. It is known
that the spectrometry of prompt or capture γ-rays can be used to supplement the results of activation analysis using a neutron beam from a nuclear
reactor or particle accelerator. Potential applications to on-line analysis of
raw materials have been investigated [64]. Isotopic neutron sources have
been used to test the possibility of continuous measurement of iron in iron
ores [65] and in the determination of the oil-brine interface in oil wells.

The range of applications of this method could be increased with the
advent of less expensive and more intense neutron sources such as ^{252}Cf and
by improvements in semi-conductor detectors.

11.9.2 Rayleigh Effect

Brinkerhoff *et al.* [66] have shown that coherent scattering of photons could be applied to the analysis of high Z components in low Z matrices.

Mass absorption coefficients for elastic scattering of photons have values proportional to Z^2. Because the energies of such scattered photons are practically equal to those of incident photons, it is relatively easy to separate the signals due to these photons from those due to Compton scattered photons.

This method has been used in the measurement of uranium in uranium aluminium alloys. The radioactive source was a mixture of U_3O_8 and krypton-85 hydroquinone clathrate with an activity of 1.25 Ci. The uranium K X-rays emitted by the source were used. The method has been applied to continuous non-destructive control of nuclear fuel plates.

The use of semi-conductor detectors will open up new possibilities in this field also, especially for photon energies below 120 keV.

11.9.3 Radioluminescence Induced by Alpha or Beta Particles

It is worth mentioning that light emission produced by the irradiation of samples with α-rays from a 70 mCi polonium-210 source, has been investigated for the analysis of rock and ore samples although, up to now, the method does not appear to have been used in practice.

11.10 CONCLUSION

Radioactive methods of analysis using sealed sources have been described only briefly in this chapter and not all their possibilities have been reviewed. As mentioned already, their development depends partly on the evolution of electronic equipment and partly on the development of theoretical predictions. 'Gaps' left by non-radioactive methods are often uneconomic but there are specific advantages and, sometimes, industrial problems cannot be solved satisfactorily by other means.

The engineer will be sufficiently aware of the state of the art in this field to consider their possibilities for a particular analytical problem. Details will be found in the proceedings of specialised symposia, such as those of the International Atomic Energy Agency, in publications of the Division of Isotope Development of the United States Atomic Energy Commission or of Euratom, etc.

A particularly striking example of the interest in these methods is the increasing use of radioactive X-ray fluorescence analysers coming after years of laboratory work and of theoretical and technological development.

REFERENCES

1.   B. DZIUNIKOWSKI, *Nukleonika*, **10**, 107 (1965).
2.   A. J. LUBECKI, *J. Radioan. Chem.*, **1**, 211 (1968).

3. H. K. HUGHES and J. W. WILCZEWSKI, *Anal. Chem.*, 26, 1889 (1954).
4. A. ROSSI and P. G. CALDERA, Euratom Report (to be published).
5. T. NOZAKI, *Bull. Chem. Soc., Japan*, 34, 1769 (1961).
6. J. S. WATT, Proc. 2nd Symposium on low energy X- and Gamma Sources and Applications, U.S.A.E.C. Report ORNL- 11C-10, p. 663. Austin Texas, (1967).
7. G. NOREL, Thesis, Paris, (1965).
8. A. WHITTAKER, G. GREEN and J. E. GARNETT, Symposium on Radioisotope Instruments in Industry and Geophysics, p. 271. I.A.E.A. Vienna, (1966).
9. G. D. McPHERSON, *Trans. Amer. Nucl. Soc.*, 4, 246 (1961).
10. J. LEROUX, Advances in X-ray Analysis, Vol. 5. Plenum Press, New York, (1962).
11. R. THEISEN and D. VOLLATH, Tables of X-ray Mass Attenuation Coefficients. Verlag Stahleisen, Dusseldorf, (1967).
12. M. LAVAUD, Rapport CEA-R-3708 (1969).
13. U.K.A.E.A., British Patent No. 23704/1962.
14. V. N. SMITH and J. W. OTVOS, *Anal Chem.*, 26, 359 (1954).
15. R. A. JACOBS, L. G. LEWIS and F. J. PIEHL, *Anal. Chem.*, 28, 324 (1956).
16. R. BERTHOLD, Proc. 2nd Intern. Conf. Peaceful Uses At. Energy, 19, 288 (1958)
17. L. KATZ and A. S. PENFOLD, *Revs. Modern Phys.*, 24, 28 (1952).
18. K. H. HALLOWES and A. E. HODGSON, Symposium on Radioisotope Instruments in Industry and Geophysics, Vol. 1, p. 317. I.A.E.A. Vienna, (1966).
19. P. LÉVÊQUE, R. HOURS, P. MARTINELLI, S. MAY, J. SANDIER and J. BRILLANT, Proc. 2nd. Intern. Conf. Peaceful Uses At. Energy, 19, 34 Vienna, (1958).
20. B. WACK, Bull, de la dir. des etudes et recherches de l'EDF 'La Houille Blanche' Grenoble Serie A No. 2 (1969).
21. R. D. EVANS, The Atomic Nucleus. McGraw Hill, New York, (1955).
22. E. STORM and H. I. ISRAEL, Nuclear Data, (US), Sec. A, 7, 565 (1970).
23. J. R. RHODES, J. G. DAGLISH and C. G. CLAYTON, Symposium on Radioisotope Instruments in Industry and Geophysics, Vol. 1. 447 I.A.E.A. Vienna, (1966).
24. A. TROST, Symposium on Radioisotope Instruments in Industry and Geophysics, Vol. 1. 435 I.A.E.A. Vienna, (1966).
25. D. K. DONHOFFER, Symposium on Nuclear Techniques and Mineral Resources, Vol. II. 23 I.A.E.A. Vienna, (1969).
26. R. H. MÜLLER, Radioisotopes in the Physical Sciences and Industry, Vol. II. 65 I.A.E.A. Vienna, (1962).
27. L. DANGUY, Monographie No. 10, Institut Interuniversitaire des Sciences Nucléaires, Bruxelles, (1962).
28. L. M. BOYARSHINOV, *Dokl. Akad. Nauk.* USSR, 178, 573 (1968).
29. R. JIRKOVSKY, *Z. anal. Chem.*, 184, 35 (1961).
30. G. C. SNEEMAN and C. G. CLAYTON, *Int. J. Appl. Radiat. Isotopes*, 14, 183 (1963).
31. J. H. PATTERSON, A. L. TURKEVITCH and E. FRANZGROTE, *J. Geophys. Res.*, 70, 1311 (1965).

32. R. A. SEMMLER, J. F. TRIBBY and J. E. BRUGGER, U.S. At. Energy Comm. Report COO-712-89 (1964).
33. J. W. HOLMES, *Austr. J. Appl. Science,* 7, 1 (1956).
34. W. GARDNER and D. KIRKHAM, *Soil Science,* 73, 391 (1951).
35. R. A. SEMMLER, U.S. At. Energy Comm. Report COO-712-73 (1963).
36. J. F. CAMERON, Symposium on Nuclear Techniques and Mineral Resources, p. 81. I.A.E.A. Vienna, (1969).
37. J. F. CAMERON, Symposium on Isotope and Radiation Techniques in Soil Physics and Irrigation Studies, I.A.E.A. Vienna, (1967).
38. L. REIFFEL, *Nucleonics,* 13, 22 (1955).
39. J. A. BEARDEN, U.S. Report No. NSRDS-NBS 14 (1967).
40. Encyclopaedia of X-ray and Gamma Rays. (G. L. Clark, ed.) Reinhold, London, (1963).
41. E. ARENDS, *Ann. Physik,* 22, 281 (1935).
42. P. A. ROSS, *J. Opt. Soc. Am.,* 16, 433 (1928).
43. S. ENOMOTO, Rapport CEA-R-3369 (1968).
44. K. G. CARR-BRION and J. R. RHODES, *Instruments Practice,* 19, 1087 (1965).
45. J. R. RHODES, T. FURUTA and P. F. BERRY, Symposium on Nuclear Techniques and Mineral Resources, p. 353. I.A.E.A. Vienna, (1969).
46. K. UCHIDA, H. TOMINAGA, H. IMAMURA and H. MIWA, Symposium on Radioisotope Instruments in Industry and Geophysics, Vol. I, p. 113. I.A.E.A. Vienna, (1966).
47. I. FILOSOFO, U.S. At. Energy Comm. Report ARF-1122-27 (1961).
48. I. LEONTIADIS, Rapport CEA-R-2970 (1966).
49. L. E. PREUSS, U.S. At. Energy Comm. Report TID-22361 (1966), TID-22361 (Part 2), (1967).
50. B. SELLERS and C. A. ZIEGLER, U.S. At. Energy Comm. Report ORNL-II-C-5, p. 353 (1964).
51. A. ROBERT, Rapport CEA-R-2539, p. 98 (1964).
52. V. I. GOLDANSKI, *Atomic Energy Review,* 1, 3 (1963).
53. E. FLUCK, W. KERLER and W. NEUWIRTH, *Angewangte Chemie,* 75, 461 (1963).
54. L. MATTEOLI and P. LOGI, Euratom Sci. and Tech. Report EUR-3173 e, (1966).
55. J. E. LOVELOCK, *Anal. Chem.,* 33, 162 (1961).
56. C. G. CLAYTON, Industrial Radioisotope Economics, I.A.E.A. Vienna, (1965).
57. E. MEILI, *Bull.* SEV., 43, 3 (1952).
58. P. MALSALLEZ and L. BREITMANN, *Rev. Gén. Élect.,* 43, 279 (1938).
59. D. J. POMPEO and J. W. OTVOS, U.S. Patent 2,641,710 (1953).
60. J. W. OTVOS and D. P. STEVENSON, *J. Am. Chem. Soc.,* 78, 546 (1956).
61. V. N. SMITH, J. W. OTVOS and D. J. POMPEO, 11th ISA Conference, Paper No. 56-11-1 (1956).
62. P. R. GRAY, D. H. CLAREY and W. H. BEAMER, *Anal. Chem.,* 32, 6 (1960).
63. R. C. GREENWOOD, U.S. At. Energy Comm. Report IITRI-1193-53 Vols 1 and 2, (1965).

64.   T. C. MARTIN, J. D. HALL and I. L. MORGAN, Symposium on Radioisotope Instruments in Industry and Geophysics, Vol. II, p. 411. I.A.E.A. Vienna, (1966).

65.   R. CHRISTELL and K. LJUNGGREN, Symposium on Radiochemical Methods of Analysis, Vol. I, p. 263. I.A.E.A. Vienna, (1965).

66.   J. M. BRINKERHOFF, U.S. At. Energy Comm. Report TID-20643 (1963).

67.   J. M. BRINKERHOFF and R. FORSYTH, U.S. At. Energy Comm. Report NYO 3160-1 (1965).

Index